Prosimian Anatomy, Biochemistry & Evolution

Prosimian Anatomy, Biochemistry & Evolution

edited by

R.D. MARTIN, G.A. DOYLE and
A.C. WALKER

Duckworth

Publisher's Note

This book, *Prosimian Anatomy, Biochemistry and Evolution*, is the second half of *Prosimian Biology*, first published in 1974. The first half, *Prosimian Behaviour*, and the complete work *Prosimian Biology* are also available as separate publications.

For convenience of reference the preliminary pages (Contents, Preface, and Foreword) and the Index are retained from the complete work in full. This volume begins on page 413. All references to pages before that number apply only to the first volume.

This volume first published 1977
Published by Gerald Duckworth & Co. Ltd.
The Old Paino Factory,
43 Gloucester Crescent, London N.W.1

ISBN 0 7156 1179 7 (hardcover)
ISBN 0 7156 0982 3 (paperback)

Printed in Great Britain by
Redwood Burn Limited
Trowbridge & Esher

CONTENTS

Section B: *Laboratory Studies of Behaviour*

Section C: *Olfactory Communication in Prosimians*

Section D: *Physiology of Behaviour in Captivity*

Part I

This Vol. begins here

PART II: PROSIMIAN ANATOMY, BIOCHEMISTRY AND EVOLUTION

Section A: *General Studies of Prosimian Evolution*

Section B: *Anatomy and Function of Skull and Teeth*

Section C: *Morphology of the Brain*

Section D: *Prosimian Locomotion*

Section E: *Chromosomes, Proteins and Evolution*

Preface

On 14, 15, 16 and 17 April 1972, an international meeting was held at the Institute of Archaeology, University of London, to discuss the general field of prosimian biology. Following a suggestion made by Dr P.J. Ucko, coupled with much valuable advice and encouragement, the meeting was held within the framework of the series of Research Seminars in Archaeology and Related Subjects, and the organisers of the present meeting were accordingly able to benefit from the accumulated expertise of what has so far proved to be a highly successful pattern of seminars. One happy result was that the Research Seminar on Prosimian Biology profited greatly from the hospitality and ready assistance of the Institute of Archaeology, for which thanks are especially due to Professor W.F. Grimes. In particular, the success of the meeting depended heavily on the invaluable services of Miss S.E. Johnson, who undertook almost all of the secretarial work for the meeting, provided seasoned advice on the running of the Research Seminar, and assisted in the preparation of this book. Thanks are also due to Miss Lorna Mullings and Mrs Penny Wyatt for their willing assistance in the Research Seminar secretariat, which dealt with all the practical problems of running a meeting with over fifty active participants.

The meeting – like its three predecessors in the Research Seminar series – was designed to enable research workers concerned with different specialised aspects of the subject to meet in an attempt to bridge gaps and achieve a synthetic approach through personal contact and lively discussion. For this reason, emphasis was placed on informality of presentation and debate. Following the practice established in previous Research Seminars, all papers were circulated prior to the meeting in order to reduce the stifling influence of formal reading of papers during periods which could be more profitably used for informal discussion. Several sessions were arranged in which particular topics were thrown open to general discussion, and participants made full use of the opportunities provided. Judging from the various comments which were made after the meeting, the air of informality was generally greatly appreciated, and many participants felt that they had gained far more than would have been possible with more formal conference organisation. It must be admitted, however, that the general friendly tone of the meeting was set from the beginning by the warm official welcome provided by Professor Grimes in the form of a reception to open the Research Seminar.

The important task of chairing discussion sessions was undertaken by Professor R. Andrew, Professor N.A. Barnicot, Professor F. Bourlière, Dr W. Bishop, Professor M. Day, Dr K. Hiiemae, Dr J.R. Napier, Professor W.C. Osman Hill, Dr L. Rosenson, Dr E.L. Simons and Dr L. Wolin. All these people deserve our deep gratitude for encouraging constructive debate and ensuring that the participants did not wander too far from the subject of each session. In addition, the editors would like to express their thanks to all of the more than fifty participants who came from various parts of the world to contribute papers and to join in the discussion of prosimian biology. They have helped to produce a text which represents one of the first really broad-based attempts to consider the general biology of the prosimians. The success of this attempt is reflected in the fact that many of the participants extensively expanded and modified their contributions as a result of discussion held during the meeting, and that a number of new lines of research seem to have developed following the meeting.

The general organisation of the Research Seminar was dependent upon the assistance of Mr J.F. Dahl, who operated the tape-recorder, and of Mr D.G. Lewis, who was in charge of the projection facilities. Additional secretarial assistance was provided by Mrs P.M. Blair. The Institute of Archaeology provided many different services, which were invaluable to the smooth running of the meeting, and thanks are due to the many un-named people who provided assistance on behalf of the Institute.

Finally, a special note of thanks must be made to the initiating publishers, Duckworth & Co. Ltd., without whose help the Research Seminar could not have been held. Among the several unique features of this meeting, it was notable that our publisher attended the greater part of the proceedings himself, and in the preparation of the book even insisted on the inclusion of *more* material in order to make the text as comprehensive as it is.

R.D. Martin
G.A. Doyle
A.C. Walker

W.C.OSMAN HILL

Foreword

The foundation of our knowledge of the prosimians was laid during the latter half of the nineteenth century, and it is there that we first meet with monographers of the stature of Burmeister (1846) and Owen (1863) through their respective studies of the tarsier and of the aye-aye. There have also been numerous and outstanding contributions in the form of more comprehensive works, and here I should especially mention the labours of Murie and Mivart and the great explorers of Madagascar, such as Schlegel and Pollen (1867-77) and particularly Grandidier and Edwards (1876-1900), who produced a magnificent set of volumes on the natural history of the island.

While speaking of the giants of our discipline, it is opportune to include one of the more recent exponents – the late Sir Wilfred Le Gros Clark. Though he also made major contributions to many aspects of the anatomy, neuroanatomy and palaeontology of the Prosimii, he will perhaps best be remembered for his advocacy of the inclusion among them of the tree-shrews. These animals clearly still have a place in our deliberations, however controversial it may be, as is evidenced by several contributions to the present symposium.

It is now almost a quarter of a century since the first overall assessment of the status of the prosimian primates was made. At that time, I had the privilege of partaking in a discussion which was opened by my revered colleague F. Wood Jones, on the occasion of the meeting of the section on Anatomy and Anthropology at the annual meeting of the British Medical Association in Cambridge, 1948.

Since then, much has happened that is of vital importance in the field of prosimian biology, which is now not only wider, but has inevitably expanded to incorporate applications of disciplines almost unheard of at that time. It is, therefore, a matter of no little difficulty for any single author to write with authority on the whole gamut of prosimian biology. Thus the four-day research seminar on the subject held in April 1972 was particularly welcome, in that it enabled a wide range of prosimian biologists to take stock of their own advances and to profit from the researches of their colleagues from many different disciplines.

Needless to say, the problems posed by the position of *Tarsius* and its fossil allies proved to be as intriguing as ever, and they would have still delighted Wood Jones and engaged his powers of debate in no small measure, just as they did at the memorable meeting held at the London Zoo in 1918. But the problems set by the tarsiers are by no means the only ones that have engaged the interests of workers in recent decades. Our attention is automatically drawn towards 'la grande île' of Madagascar, where the activities of conservationists have emphasised time and again the pressing urgency for field-work on the behaviour of all of the lemurs. In spite of the laudable endeavours of the Malagasy Republic to protect their unique fauna, it may soon be too late to study at least some of the lemur species in the natural state. Fortunately, heed has been taken of the warnings, and there are now a number of studies that have been published or are in progress, as demonstrated by several of the contributions to the Research Seminar.

Many lemur species still demand special attention, however, and there is an even more pressing need for detailed study of their ecology so that regional schemes may be examined, which may permit the establishment of local breeding units. Madagascar is ideally situated as a natural workshop for the student of evolution, and it behoves us to adopt early measures for the protection of the existing fauna – an urgent necessity already recognised by the enlightened Malagasy government – and also to provide the means of enhancing the existing facilities for internal laboratories where local and visiting students may avail themselves of the unique opportunities to study the rich fauna.

That definite advances in prosimian biology have accrued in many different directions is amply shown by the contents of the present symposium. The wide range of problems which have now been considered makes this abundantly clear, and I have myself gained valuable knowledge both of a wealth of new facts and of reassessment of earlier problems. It would be invidious for me to single out any one paper for special praise, since all are remarkable in their various approaches and treatments; I will, therefore, deal with the different papers in an order that appears to link the subject matter in a logical fashion, as a guide to the mass of information brought together by the Research Seminar.

The Seminar was broadly divided into two sections: studies of prosimian behaviour, and studies of prosimian anatomy and evolution, including reference to cytogenetics and biochemical studies. Behaviour in the field justly received a major share of attention from the participants, and papers were given on representatives from all the major prosimian groups. Contributions by Frances D'Souza and

Michael Fogden brought in the two most controversial prosimians – namely, *Tupaia* and *Tarsius* – and thus provided welcome additions to the field of prosimian research.

In Madagascar, field studies on *Lemur catta* have been undertaken both by Alain Schilling and by Robert Sussman, the latter providing a very elegant study of the ecological distinctions between sympatric populations of *Lemur catta* and *Lemur fulvus*. Alison Richard elaborated on the patterns of mating in *Propithecus* and provided us with one of the first detailed accounts of the natural mating behaviour of any lemur species. Marcel Hladik and Pierre Charles-Dominique have significantly expanded our knowledge of lemur feeding ecology and social dynamics with their intensive study of *Lepilemur*, while Jean-Jaques Petter and A. Peyriéras have added to our understanding of *Indri*, specifically with respect to population density and home range. This latter paper is again one of particular interest, in that it deals with a little-known genus.

Turning to mainland Africa, the bushbabies and pottos were very comprehensively covered. All the prosimians outside Madagascar are nocturnal, and special techniques have been developed for studying the behaviour of these night-active forms over the last few years. These techniques are illustrated in the papers on *Tarsius* (Fogden) and *Lepilemur* (Hladik and Charles-Dominique); but in the papers on the African prosimians, the full possibilities of these techniques are displayed. Bearder and Doyle have given a comprehensive account of the South African species *Galago senegalensis* and *Galago crassicaudatus*, whilst Charles-Dominique has achieved the magnificent feat of considering the ecology and feeding behaviour of all 5 prosimian species in Gabon (*Galago demidovii, Galago alleni, Euoticus elegantulus, Arctocebus calabarensis* and *Perodicticus potto*). Finally, a link with laboratory studies is provided by Georges Pariente's able contribution .dealing with two little-studied nocturnal Malagasy lemurs, *Phaner furcifer* and *Lepilemur mustelinus*, and illustrating the relationship of their activity patterns to ambient light intensities.

Laboratory studies of prosimian behaviour were equally well represented. The paper presented by Jan Bergeron on the Duke University Primate Facility, set up by John Buettner-Janusch, which I had the privilege of visiting a few years ago, summarises a very interesting large-scale project on laboratory observation of prosimians. We may expect a stream of far-reaching research projects on prosimian biology from this well-equipped facility. The lesser galago has been further studied in South Africa by Gerald Doyle in a very comprehensive programme aimed at obtaining quantifiable data over long periods, and his paper summarises our knowledge about the behavioural repertoire of this species. Judy Epps and Jocelyn Tandy, respectively, have given detailed attention to social behaviour of captive *Perodicticus potto* and captive *Galago crassicaudatus*, while

Peter Klopfer has looked specifically at aspects of lemur mother-infant relationships. Ursula Cowgill reported on some unusual observations indicating a possibility that pottos may exhibit a form of cooperative behaviour in the laboratory. On a more experimental slant, Howard Cooper investigated the possibility that *Lemur macaco* might exhibit learning sets in captivity, and reported on evidence that this capacity does in fact exist in at least some prosimians.

The functions of the glandular apparatus associated with the external genitalia of the two slow-moving African lorisoids (*Perodicticus* and *Arctocebus*) provided an interesting topic for Gilbert Manley, who exploited the subject to the full, drawing conclusions from behaviour, morphology and biochemistry. The perennial subject of urine-washing has been studied on a comparative basis by Richard Andrew and Robert Klopman, who have underlined the practical difficulties of providing a conclusive interpretation of a specific aspect of behaviour of this kind by presenting a wide-ranging examination of the various possibilities. The subject of olfactory communication, which had been considered in Alain Schilling's paper on *Lemur catta*, was also tackled in Jonathan Harrington's paper on *Lemur fulvus*.

A more physiological approach to the behaviour of captive prosimians was provided by two papers relating to a laboratory colony of Mouse Lemurs (*Microcebus murinus*). Arlette Petter-Rousseaux submitted a throughly competent study of the relationships between photoperiodicity, sexual activity and body-weight in this small nocturnal lemur, whose behaviour (like that of all other Malagasy lemurs) has a strict seasonal basis. The endocrine glands, especially the hypophysis, have been explored in detail by Martine Perret, who relates various changes to aspects of behaviour and the seasonal reproductive cycle. To round off the laboratory studies of behaviour, Dietrich von Holst provides a salutory warning for those who maintain animals of any kind in captivity. In his paper on social stress in tree-shrews (*Tupaia belangeri*), he provides ample evidence of the wide scale of pathological changes that may be brought about by interaction between conspecifics in captivity. He also raises the interesting possibility that such interaction between conspecifics – probably in a far milder form – may play a part in the natural situation.

In the second section, dealing with anatomy, biochemistry and evolutionary aspects, special praise is due to a number of papers which have a functional anatomical emphasis. This is evident in Tattersall's investigation of the mechanics of the temporo-mandibular joint and accessory structures of the subfossil *Archaeolemur*. Matt Cartmill casts a wide net in his behavioural and ecological consideration of anatomical features of such biological aberrants as *Daubentonia, Dactylopsila* and various woodpeckers, which are functionally linked by virtue of their adaptation to a wood-boring insectivorous regime, as a result of which they all exhibit some degree of klinorhynchy. P.D. Gingerich, in 'Dental function of the Palaeocene primate *Plesiadapis*', has produced cogent arguments for the retention of this genus within the Order Primates, basing his reasoning on imputed feeding behaviour, dental morphology and molar faceting. Dental occlusion mechanisms were likewise the subject of the joint contribution by Seligsohn and Szalay on Lemurinae, with its bearing on the systematics of *Lemur* and *Varecia*. They confirm the separation of *Varecia* from *Lemur* at the generic level and the relegation of the subfossil species *insignis* and *jullyi* to the former genus. Other significant odontological studies with a functional basis are the contribution on *Galago* presented by R.F. Kay and Karen Hiiemae, illustrating their cinefluorographic technique of analysis, and Ron Every's magnificently illustrated paper on thegosis. Jeffrey Schwartz, in a theoretical reinvestigation of premolar loss among primates, resorts to the 'field theory' to help resolve the problems posed by diminishing premolar representation in primate evolution.

Three contributions are devoted to cerebral anatomy in prosimians. One, by Leonard Radinsky, deals with functional and phylogenetic implications of prosimian brain morphology, and two others are concerned with the neuroanatomy of lorisids. These are Duane Haines' study of the cerebellum in various lorisids and the joint contribution by Haines and his colleagues from the Anatomy Department of the Medical College of Virginia, dealing with the external cerebral morphology of galagos, lorises and pottos.

There are also three papers dealing in various ways with the topic of prosimian locomotion. Clifford Jolly and Ann Gorton examined the proportions of the intrinsic foot muscles in some lorisid prosimians, while Mme F. Jouffroy used cineradiographic techniques to examine the biomechanics of jumping in *Galago alleni*. Combined in one paper, E.C.B. Hall-Craggs has collected both physiological and histochemical parameters of locomotor behaviour in *Galago* and the slow-moving lorisids, including data on the differing activities of red and white muscular tissue and the mechanics of saltation.

Passing now to evolutionary studies, there are a number of papers dealing with general palaeontological and comparative anatomical treatments of the prosimians. Elwyn Simons has provided notes on Early Tertiary prosimian fossils, thus bringing our knowledge of these forms up to date. Alan Walker tackles the other major category of fossil prosimians, presenting a review of the Miocene lorisids of East Africa, including reference to the postcranial skeleton. The new associated fossil lorisid material has confirmed the value of size analysis in allocating some of the remains. A particularly thought-provoking paper linking both behaviour and evolutionary mechanisms is that of Lee Wolin, with the intriguing title of 'What can the eye tell us about behaviour and evolution? or: The aye-ayes have it, but what is it?'. His only complaint is that *Tupaia* has, so far, not received enough attention. The aye-aye is also made the subject, among other notable matters, of Colin Groves' evocative paper on the taxonomy and phylogeny of prosimians. Finally, Patrick Luckett draws upon evidence from the morphogenesis of the foetal membranes and the placenta in order to establish the phylogenetic relationships of the prosimians, whilst Caroline Berry has utilised non-metrical features of the prosimian skull for evolutionary interpretation in the manner of Wood Jones' important review (1930-1).

A particularly valuable addition to the discussion of evolutionary relationships was provided by a series of papers on chromosomal and biochemical studies of prosimians. Chromosomal evolution in the prosimians is the subject of Josei Egozcue's paper, which provides further indication of the distinctiveness of *Tarsius*, in that it has the highest diploid number so far found in any primate. The results of Rumpler's study of lemur chromosomes with a view to revising their classification indicate that it would be justifiable to raise the cheirogaleines to family rank, with the family Cheirogaleidae containing two subfamilies – Cheirogaleinae and Phanerinae. Rumpler also confirms the distinctiveness of *Daubentonia*, suggesting that this genus should be separated from all other lemurs at least at the family level. A general survey of chromosomal features in prosimians, provided by Brunetto Chiarelli, rounds off this triplet of cytogenetic studies.

Nigel Barnicot and David Hewett-Emmet have contributed a paper on electrophoretic studies of red cell and serum proteins in prosimians, which covers a wide range of species. It was of great interest to learn that the haemoglobin, adenylate kinase and phosphatase patterns of *Tarsius* are so like those of man. The properties of the Cheirogaleinae studied are also of systematic value, emphasising the distinctiveness of *Microcebus* and *Cheirogaleus*. Electrophoretic properties of haemoglobins were also the subject of a paper by Francis Bush and colleagues. Other phylogenetically directed protein studies concern the amino acid sequences of individual proteins and the use of serological tests to determine affinity. Christopher Cook and David Hewett-Emmett provide a general paper examining the uses and limitations of tree-building techniques based on amino acid sequence data, while Russell Doolittle provided an account of recent work on the short-chain fibrinopeptides, which are of great significance because of their evident lability in evolution. Klausdieter Bauer reports on his recent work based on identification of the actual number of determinant sites on proteins which are involved in serological cross-reaction, and a paper by Morris Goodman and his colleagues rounds off the subject with a treatment of immunodiffusion systematics in *Tarsius*, Lorisidae and Tupaiidae, in which he introduces the concept of antigenic distance.

This collection of papers can thus be seen to cover an extremely broad spectrum, and the Research Seminar has accordingly provided a general basis for future discussion of the biology of prosimians and of their evolution. In conclusion, I can only commend to all the overall value of such a broad-based synthesis in dealing with a particular topic, such as that of prosimian biology.

W.C. Osman Hill
Folkestone, September 1973

PART II
PROSIMIAN ANATOMY, BIOCHEMISTRY AND EVOLUTION

SECTION A
General Studies of Prosimian Evolution

E. L. SIMONS

Notes on Early Tertiary prosimians

Introduction

Several of my prior papers constitute reviews of Early Tertiary primate groups and more recent studies are incorporated in my book *Primate Evolution.*[1] What follows will be a series of notes and observations supplementary to the conclusions of that book. I will comment also on the most recent publications dealing with Early Tertiary prosimians. Because the scope of early primate history is so broad, and since more recent conclusions as to the relationships of evolving prosimians are stated at length in my book, I will take up only a limited series of topics in this contribution. Initially, I should say that discussions of prosimian evolution have, increasingly, been punctuated in recent years by sterile debates about higher category arrangements. Before discussing sequentially the Early Tertiary prosimians, some remarks on this subject seem warranted.

Use and misuse of higher categories

There has been some impetus given recently to revival of the allocation of Tarsiiformes to Anthropoidea, not Prosimii. This ranking was first seriously suggested by Hubrecht,[2] but actually even he was only making the point that *Tarsius* was more like Anthropoidea than are the lemurs. Such ranking was later espoused by Wood Jones, whose principal motivation for so doing seems to have been that he considered tarsiers more aesthetically pleasing as animals closely allied to man's ancestry than the monkeys and apes. Pocock proposed a suborder Haplorhini to include Mivart's Anthropoidea plus *Tarsius* and, by implication, its fossil allies.[3,4] Pocock, however, substantiated the transfer almost entirely on the basis of a few soft tissue characters. These were the dry rhinarium, the structure of the upper lip and nose and the nature of placentation. The latter had been urged by Gadow as resembling that of Anthropoidea.[5] Pocock gave only the post-orbital partition (or closure) of *Tarsius* as an osteological character allying it with monkeys, apes and man. Nevertheless, Simons and Russell[6] pointed out some time ago that the manner of bone growth accomplishing the partial postorbital closure of *Tarsius* is different from that of Anthropoidea. Thus partial postorbital closure and metopic fusion in it are probably consequences of the extraordinary orbital enlargement of that genus, and not of relatedness to higher primates. Starck reviewed the significance of the manner of placentation in *Tarsius* and concluded that the resemblance of its placenta in late stages of development to those of monkeys is based on convergence, since earlier in development and in implantation the two systems are quite different.[7] He concluded that the manner of placentation can only be used to bolster taxonomic systems that are founded on more secure data than that provided by the study of placentae, and should never be made the basis of classification on its own. Pocock's use of Haplorhini was more recently supported by Hill, but he founded his definition entirely on the nasal characters.[8] In any case, most authors have preferred to sustain the definition of Anthropoidea and Prosimii accepted by Simpson.[9]

A new transfer of Tarsiiformes into Anthropoidea is unwarranted for a whole series of cogent reasons. Some of these will perhaps not be taken as seriously by the neontologist as by students of primate history. Simply stated, this transfer of tarsioids to Anthropoidea would cancel out most of the familiar characteristics of higher primates that are to be found in their osteological and dental anatomy. If assignment to either suborder is to be considered possible for any extinct form, the bony parts (being the only portions of the animal concerned that are available) must include the defining characters of the two suborders. With tarsioids ranked among Anthropoidea, subordinal assignment of many extinct primate species would immediately be uncertain. It would be necessary to judge in each case whether the species concerned was significantly affined to *Tarsius* or not. Placing of *Tarsius* in Anthropoidea would thus raise the question with every single genus and species of Early Tertiary primates: whether it should or should not be carried along with tarsiers into Higher Primates. The futility of such procedures is well demonstrated by Hill, who assigned 31 extinct genera to Haplorhini.[8] My investigations show that with 23 of these genera there is no reason whatever to believe that there was (within Prosimii) a special relationship to *Tarsius*. Our quite inadequate knowledge of many Early Tertiary primate species and even genera would leave a very large number of them in an *incertae sedis* position between the two suborders. Retention of Tarsiiformes within Prosimii eliminates this problem. There is not even one species of past primates (with the possible exception of *Amphipithecus mogaungensis*) which would be in doubt as to its subordinal position if Simpson's usage is maintained.

That *Tarsius* possesses several interesting approximations toward Anthropoidea, not seen in other living prosimians, has been known at least since Burmeister published in 1846.[10] Nevertheless, these 'advanced features' are nearly all to be found only in the embryology and soft anatomy of *Tarsius* and are thus not applicable to fossils. It is to be expected that whenever an order of animals is divided up into two or more subsets (in this case suborders) each containing many members, that one or more members of one such subset will approximate more closely to the other subset (suborder) than does any other. Such consequences are inevitable whenever a broad spectrum of animals is ranked.

Transfer of Tarsiiformes to Anthropoidea would render non-diagnostic most characteristics that we consider today essential to the definition of Higher Primates. Characteristics that would have to be dropped include such features as enlarged, convoluted brains. The brain of *Tarsius*, although large in proportion to body size, is simple in construction and lacks well-defined convolutions. Postorbital closure and symphyseal fusion in the mandible would have to be discarded as distinctive of Higher Primates, since these advances are not fully expressed in *Tarsius*. The dental construction otherwise characteristic of Anthropoidea, such as presence of hypocones, would have to be dropped since *Tarsius* possesses teeth that are extra-ordinarily simple and primitively constructed and could hardly be less like that of living higher primates and still belong to a primate. The frontal fusion of *Tarsius*, resembling Anthropoidea, is most probably a parallelism or coincidence since it develops relatively late in juvenile life, not early as in Anthropoidea (defined excluding tarsiers). Moreover, as already stated,[6] the manner by which the parietal postorbital plates grow out in *Tarsius* differs from such growth in monkeys, apes and men. Frontal fusion in *Tarsius* occurs together with the marked flanging of the orbital margins and the full growth of the comparatively enormous orbits. The fusion is as easily interpreted as a result of the unique extreme of visual development seen in this animal as it is interpreted as a character of common heritage linking it with Anthropoidea. As will be discussed later, the character of possession of an ossified tubular auditory meatus in tarsiers and catarrhines can also be attributed to parallelism. Earliest catarrhines (such as the Yale skull of *Aegyptopithecus*) do not possess such a meatus. Plesiadapids, which on dental grounds alone cannot be related to catarrhines, do have an ossified tubular auditory meatus. Quite apart from the foregoing osteological consideration is the powerful evidence that derives from all recent biochemical investigation.[11] This shows that *Tarsius* is about as distinct from any Anthropoidea as is any other prosimian. Thus there is no longer any value in retaining the term Haplorhini in usage.

Such considerations as the above lead me to the plea that scholars in the future try to devote more time to the actualities of palaeontological discovery: the attempt to unravel, describe and explain the course and processes of the history of life. Such investigations are far more useful than the continued pedantic rearrangements of higher categories, the delimitation of which is ultimately a matter of taste. An example of the puerile nature of such efforts is the recent attempt by two or three authors to introduce the taxonomic rank of tribes into the literature on primates, although these have been little used before. Their widespread introduction into taxonomic discussions of the order now would serve no useful purpose. There is another reason, however, for *now* coining tribal group names among primates: it is the only rank where a present-day author can originate a higher category name and then refer frequently to himself as the author of such a new higher rank of primates. It is unlikely that arguments now in support of *new* families, superfamilies, infraorders or suborders for primates would gain wide acceptance. The practice of coining and discussing higher category names is rendered even more futile by Article 36 of the Code of Zoological Nomenclature, which gives authorship of all categories from subfamily through superfamily to whoever was the first proposer of any of the three. With primates, this means that authorship of a new higher category by a living author is now almost impossible to achieve, since so many prior alternative names are already available.

Szalay has recently argued that the primitive condition of the tympanic in primates is 'external to the bulla' and that therefore fossil forms like *Pronycticebus*, or the living lorisids, in which the ectotympanic is at or just inside the auditory opening are not derived from forms with a free ring inside the bulla like the living lemurs, but from forms like *Plesiadapis* which had a tube located largely outside the petrosal bulla.[12,13] Whatever the primitive position of the ectotympanic (which, developmentally, arises lateral to the bulla) was among primates, we know that *Pronycticebus* is ranked by all as close to species of the genera *Adapis, Notharctus* and *Smilodectes.* All three of these genera share with the Malagasy lemurs the condition of having a free ring within a petrosal bulla.

Discussions as to whether the ectotympanic was primitively located 'inside' the bulla as in lepticitid insectivores, tree shrews and lemurs, or 'outside' as postulated for other early insectivores, are unresolved, and are plagued by semantic problems, lack of direct observation and of precise definition. The ear drum and its supportive osseous ring functionally close the bulla. In a sense, the ring is no more inside nor outside the bulla than a window or door is inside or outside a house. Known fossils prove that, among primates, ectotympanic tubes have arisen independently, as among plesiadapids and hominoids; that in many forms the tympanic ring lies in the aperture of the bulla (which may thus be the primitive condition); and that extinct and living species that there is reason to consider primitive, such as adapids and lemurs, have a free *annulus tympanicus* located inside the bulla. We know that developmentally in animals as different as man and tarsier (which share an osseous, tubular ectotympanic) the tympanic begins development as a ring. I doubt that there are enough fossils documenting the ear structure, and enough understanding of the meaning of structure in the ear region, on which to base sound or far-reaching phylogenetic interpretations at present. Thus, I doubt the effectiveness of all such considerations as those of Szalay on the derivation of the tympanic in primates.

Palaeocene primates

There are four families of Palaeocene primates which have been dealt with in the recent literature: Plesiadapidae, Carpolestidae, Paromomyidae, and Picrodontidae. The first of these is the only group in which we know the postcranial anatomy well. *Contra* Szalay[13] no one, certainly not Napier and Walker,[14] has ever thought that this group, Plesiadapidae, represented vertical clinging and leaping primates. Rather, they have always been represented as heavily built, perhaps terrestrial, clawed, and rodent-like. What is important to remember is that we know that almost all of the members of the above listed families died out without descendants. Thus, when Napier and Walker discussed vertical clinging and leaping locomotion as being primitive for primates they, of course, had reference to the second, Eocene, radiation of primates of modern aspect. It is this second radiation and not the Palaeocene first radiation which is basal or 'primitive' to all the living primates.

The criticism of Charles-Dominique and Martin[15] that the position of Plesiadapidae as primates needs review has been dealt with adequately by Gingerich (this volume) and will not be taken up further here. Plesiadapids are of course primates. The striking similarities between molars of *Elphidotarsius* (Carpolestidae), *Pronothodectes* (Plesiadapidae), and *Pelycodus* (Adapidae, s.l.) justifies allocation of all to one order. The cranial and postcranial anatomy of the Adapidae in turn undoubtedly rank them as primates – a consideration that is quite apart from any similarities to modern forms in their dental structures.

Contra the discussion of the dental occlusion of *Plesiadapis* by Szalay,[12] the figure provided by that author of the lower incisor occlusion (when the molars are fully contacted) is not new and the figure not significantly improved over that illustrated by Simons.[16] Moreover, my 1960 drawing was made directly from the original materials in Paris, and Russell and I spent some time judging the issues, newly raised as novel by Szalay. Gingerich has fully demonstrated the facile nature of Szalay's argument that he had something new to report about the skull and dentition of this animal.[17] I am mystified that Szalay, in a more recent review,[13] entirely fails to take into account, or even mention, Gingerich's discussion. Instead, he lets stand his whole set of wrong assertions that prior workers never showed proper occlusion in *Plesiadapis.*

Justification of the picrodontids as primates recently attempted by Szalay and by others, nevertheless remains uncertain.[13,18] This is particularly true if groups with far more primate-like dental patterns, such as Microsyopidae, are excluded from the order by Szalay. Picrodontids are very small Palaeocene mammals of the genera *Picrodus* (middle) and *Zanycteris* (late) Palaeocene. Matthew originally considered *Zanycteris* a bat related to modern *Phyllonycteris* and the stenodermine and sturnirine bats.[19] Szalay carried out the type of one-sided transfer of these animals often now made, but which is outmoded. That is, he presented arguments purporting to show resemblance to and derivation from archaic primates, but he presented no complementary arguments to show why these genera are not bats. Instead, he deferred any discussion of why picrodontids are not bats to the prior comment of McGrew and Patterson.[20] Nevertheless, he neglected the cautions of these authors and of Simpson that there is very little to recommend transfer to the primates.[21,22] The derivation of the dental pattern of the teeth of *Picrodus* from primitive primate molars discussed by Szalay is topologically possible, but the lack of fossils documenting such a transition renders the derivation highly conjectural. What if the ear region of picrondonts, when discovered, proves to be even less like that of primates than the ear region of *Microsyops* (a non-primate for Szalay) is supposed to be? The final allocation of picrodonts depends on discovery of their now unknown basicranial structure.

Analyses currently being made at Yale by Brown, Gingerich, and myself on the Middle Palaeocene primate *Plesiolestes* suggest that *Plesiolestes* and *Palaechthon* are closely related (see Simons, 1963),[23] and that both are related to the ancestry of the family Microsyopidae (including *Microsyops, Niptomomys*, and *Uintasorex*). To the extent that this view is correct, all the above mammals may be considered primates.

Studies of the Carpolestidae, carried out recently under my direction by Rose, will lead to the publication of a definitive work on these poorly known animals. His investigations indicate that Jepson (personal communication) was correct in proposing that the three genera of these animals are all valid, *contra* Szalay,[13] and that they represent a time-successive series documenting exceptionally well a small and intriguing phylogenetic branch of Primates in extreme detail. Interestingly, there is high individual variability in the quarry samples of these primates. The genera concerned in this family are successively: *Elphidotarsius, Carpodaptes,* and *Carpolestes.* In Rose's judgement these appear to contain six valid, named species; two of *Elphidotarsius*, three of *Carpodaptes* and one of *Carpolestes.* The total number of specimens representing individuals of this primate family is approximately 225, including isolated teeth – of these only about a dozen specimens have been reported on in previous publications. All will be included in Rose's monograph. Because they document microevolution in action and ancient population variability, these animals, which left no descendants, are nevertheless of interest. The main trend in their evolution was toward extreme specialisation of the P_4 and P^{3-4}. This development placed the lineage as the principal primates to acquire the plagiaulacoid type of dentition, working on the assumption that its slight development in the German Palaeocene *Saxonella* is convergent. Interestingly, the molars, save for M_1, do not specialise away from the primitive primate condition, these teeth in all three genera being particularly close to those of plesiadapids such as *Pronothodectes.* The P^{3-4} of *Elphidotarsius*, not previously mentioned in print, are also close to those of *Pronothodectes* – the earliest plesiadapid. The resemblances confirm the view that carpolestids are primates, not *incertae sedis* as they were considered by Matthew and Granger,[24] and Saban.[25] They could have had a North American Puercan (early Palaeocene) common ancestry with plesiadapids. As mentioned, these animals make an outstanding example of evolution in action because of the documentation they provide of successive stages in the development of their basic dental adaptation. Trends are: (1) Reduction in size and number of the anterior premolars: P_2 is small in *Elphidotarsius* and is lost in the latter two genera; (2) P_3 decreases in size as P_4 enlarges; (3) The multiplication of cuspules on P_4 from about four in *Elphidotarsius* to eight or nine in *Carpolestes*; (4) The heel of P_4 is distinctly expressed in *Elphidotarsius*, but by the stage represented by *Carpolestes* entirely merges with the blade; (5) Trigonid of M_1 opens out and becomes blade-like in successive stages, thus extending the blade of P_4 posteriorly: by the time of *Carpolestes* this blade is doubled in length; (6) In P^{3-4} there is enlargement and relative complication of structure: the external row of cusps rises from two cuspules in *Elphidotarsius* to four or five in the later species.

There is little doubt that these fossils document the actual details of the successive evolutionary elaboration of the carpolestid dental mechanism. What this mechanism evolved to do is less certain. Direct attribution of the feeding habits of extant plagiaulacoid mammals (*Bettongia, Hypsiprymnodon, Burramys* or *Epiprymnis*) to carpolestids may not be relevant, since they share resemblances only in the lower blade and the teeth which oppose this above in these modern marsupials are nothing like those of carpolestids. Carpolestids were quite small, mouse-sized mammals. The smallest mandibles are little more than a centimetre and a half long – the largest, of *Carpolestes*, ranging perhaps up to 2.5 cm. long. Because of small size, their fossilisation and recovery could be considered less likely than with larger mammals. Nevertheless, in those few Palaeocene quarries where careful collection of small mammals has been conducted they occur in considerable numbers.

Finally, the feeding adaptations and diet of Palaeocene Primates have recently been speculated about by Szalay.[13] I doubt his conclusion (p. 27) that: 'It appears (but of course it cannot be proven) that the various Palaeocene families of primates fed predominantly on plant materials.' The Palaeocene primates exhibit a range of incisor, premolar, and molar configurations – and these configurations apparently represent a wide range of feeding adaptations. No living primates have procumbent incisors morphologically or functionally similar to those seen in specialised Palaeocene primates, though the Australian marsupial radiation provides close morphological parallels. Although some of the specialised Palaeocene primates, particularly certain Plesiadapidae, such as *Chiromyoides*, have often been analogised with the living Madagascan aye-aye, *Daubentonia*, their incisor function is clearly not the same. Daubentonia, unlike Palaeocene primates, has rodent-like, ever-growing incisors with chisel-like, sharpened tips. These front teeth function very differently from the occlusion plesiadapid incisors achieved. Kangaroos (Macropodinae) have a cropping incisor mechanism in which the lower incisors shear within a U-shaped arc formed by the upper incisors. They have no prominent projecting premolars and are herbivorous, feeding mainly by grazing and browsing. The Palaeocene primate *Plesiadapis* has a functionally similar incisor cropping mechanism, has no projecting premolars, and (based on analogy with macropodines) is apparently adapted to a predominantly herbivorous grazing feeding mode (see Gingerich, this volume).

The remaining Palaeocene primates, however, almost certainly did not feed predominantly on plant materials. The dentitions of Palaeocene and earliest Eocene primates of the genus *Phenacolemur*, for example, closely resemble the marsupial Sugar Glider *Petaurus* (Gingerich, in preparation), which feeds on insects, insect larvae, and small birds, as well as buds, blossoms, nectar, sap, and juices. The arrangement of cusps and crests of upper and lower molars of many of the Palaeocene primates (*Palenochtha, Elphidotarsius, Paromomys, Navajovius*) are closer to those of the living *Tarsius* than to those of any modern lemurs and lorises (some of which are mainly herbivorous). *Tarsius*, of course, is a predaceous insectivore with practically no plant component in its diet (see Fogden, this volume). Moreover, it is clear from all palaeo-ecological evidence that the regions from which these primitive primates come were warmer and wetter in the Palaeocene than they were again subsequent to the middle Eocene. This would ensure a year-round insect supply. The same tarsier-like molar conformation also characterises earliest anaptomorphids, adapids and microchoerines. Some of them, such as *Tetonius, Pseudoloris*, and *Nannopithex* are so tarsier-like as to strongly suggest that they had closely similar feeding habits; that is, they were mainly insectivorous. It is only with the evolution of larger body size, and much more broadened, flattened, and cuspidate cheek teeth in such genera as *Microchoerus* and *Notharctus* or *Smilodectes* that there is a suggestion of a shift toward plant-feeding in some Eocene primates. *Contra* Szalay,[13] these Palaeocene primates (and many of the early Eocene primates as well) *do*, by their clear-cut cheek tooth resemblance to *Tarsius*, substantiate the postulate made by Crook and Gartlan[26] that primitive primates were primarily insectivorous and carnivorous, with diets probably resembling those of *Microcebus, Galago demidovii* and *Tarsius.* Szalay was also incorrect in implying that long insect-stabbing upper front incisors, such as *Tarsius* has, need be a necessary correlate of insect-feeding preferences. Such evidence as is preserved in broken and fragile specimens shows that in many early Tertiary primates the upper front teeth must have been adequate for piercing or grasping insects. The long, pointed upper central incisors of *Tarsius* are unique among primates and may be a relatively recent acquisition. Upper incisor elongation is certainly not present in the smaller galagos nor in *Microcebus*, yet these animals are largely insectivorous (see Charles-Dominique, this volume).

Eocene prosimians

It has long been recognised that the various ancient families of primates have been split up too much into subgroups. This came about because of the wide adaptive range of their dental mechanisms, and because so little is known of their skulls and postcrania whereby group resemblances might be more easily demonstrated. Without such cranial and postcranial knowledge, any non-dental similarities between early groups of primates (apparently distinctive as superfamilies, families, or subfamilies) of course cannot be discussed. More postcranial and cranial remains are described from time to time, but stronger associations between groups of primates, particularly Eocene groups, have been made possible by recent improvements in understanding the topology of teeth among these early primates and by the reinterpretation of antemolar dental formulae in several different groups. Just as the new studies of Rose show that carpolestids and plesiadapids are much more closely related than had been thought, many of the smaller primates of Eocene times can now be drawn closer together. They share tarsioid dental features.

It is clear that there has always been a preferable case for ranking the mainly North American notharctine primates along with the European adapine primates in one family: Adapidae. Gazin chose to separate these two groups into two families Notharctidae and Adapidae and at one time I followed this, but do so no longer.[23, 27] Gazin essentially took the position adopted by Stehlin.[28,29] However, Gregory had already given cogent reasons for grouping the two as subfamilies within a single family.[30] Apart from the anatomical basis for this grouping, these subfamilies are now known to overlap geographically (in part) since *Pelycodus*, a notharctine, has been discovered in Europe. Moreover, the latest thinking on early Tertiary zoo-geographic relationships strongly weighs in favour of a direct North Atlantic connection between Europe and North America when *Pelycodus* existed (early Eocene).

Another slight change in grouping will be discussed later: that of the anaptomorphid (or tarsioid) prosimians and their relationship to two associations of genera which have often been treated as distinct families: the Omomyidae and the Microchoeridae (= Necrolemuridae). My present view is that, in the case of the above-mentioned five families (which have sometimes been recognised for various Eocene primates) a more probable ranking would recognise only three families. Two of these are extinct, while the third, Microchoerinae, is a group which has already been assigned to the family Tarsiidae.[31] *Contra* the statement in Szalay that their relationship to tarsiids has been 'alleged',[13] it is quite easy to demonstrate such an allocation for this particular Eocene subfamily. The assignment is founded on two principal sets of facts: 1. All details of dental and palatal morphology of the early microchoerine *Pseudoloris* are extraordinarily like those of *Tarsius*, save for its smaller size (see Simons, Fig. 66).[1] The upper teeth of another microchoerine, *Nannopithex*, are also extraordinarily like those of *Tarsius*, as are also those of at least one species of *Necrolemur*. 2. Craniological similarities shared by microchoerines and tarsiines are far too extensive to have their taxonomic assignment termed an allegation (see Simons, Table 1).[31]

In view of the astonishingly close dental similarities between *Pseudoloris* and *Tarsius*, any attempt to show that the groups are not clearly interrelated would merely be an attempt to create the illusion that there is something new to say about the relationships of these animals, and would certainly waste the time of students and scholars alike. No significant new microchoerine fossil material has been discovered during the past forty-five years, and in fact almost all the points and issues which could be raised in regard to this taxonomic matter are already dealt with in Stehlin,[28] Simpson,[32] Hürzeler,[33] and Simons.[31] In addition to this, a casual reading of the foregoing papers might fail to disclose a curious piece of motivation lying behind the strongly contrasting assessment of the microchoerines (or

Necrolemuridae as they termed them) held by Stehlin and Hürzeler. Stehlin had presented evidence for a relationship between *Necrolemur* and *Tarsius.* Later on, Hürzeler, who had studied with him, differed catagorically with the elder scholar, and argued that *Necrolemur* and its allies were not like tarsiers but resembled lemurs. All his positions for so arguing were taken in opposition to those of Stehlin, but when I investigated them, Stehlin always proved correct, Hürzeler wrong. As I have pointed out before, the science of palaeoprimatology is likely to operate at a level below the standards of competing sciences as long as hasty analyses, half-truths, and quotes out of context are the stock-in-trade of some of the workers in this field. This problem extends far beyond the analysis of *Necrolemur* and its allies. There have always been cogent reasons for sustaining close taxonomic ranking between tarsiines and microchoerines, and although representations to the contrary have been made, they can be sustained only by sophistry, not by facts. To some extent it may be a matter of taste whether one ranks the microchoerines in modern Tarsiidae or reserves a separate early Tertiary family rank for them, but their resemblances to *Tarsius* are real, not alleged.

It has been common practice since 1958, when Gazin raised the omomyines to a family rank separate from Anaptomorphidae, to deal with these families as distinct entities.[23] More recently Russell, Louis and Savage[34] pointed out that four of the five subfamilies included by Simpson[32] in the Anaptomorphidae were subsequently raised to family rank. Countering this trend, they again placed anaptomorphines and omomyines together in Anaptomorphidae, but gave little reason for this other than that 'they overlap in many features'. In Simons, the view is also adopted that both should be considered subfamilies of one family, Anaptomorphidae.[1] Reasons for this are that the type and referred species of *Anaptomorphus* do not even exhibit some of the few characters, such as distinctly enlarged P_4, commonly accepted as distinguishing anaptomorphines. In addition, the lower dental formula for *Tetonius* of 0.1.2.3. proposed by Matthew and Granger was wrong.[24] This was revised by Gazin to an interpretation of $\frac{?}{1 \text{ or } 2}, \frac{1}{1}, \frac{2 \text{ or } 3}{2}, \frac{3}{3}$ for at least some *Tetonius.*[27] Recent studies by Szalay and at Yale (Simons; Bown and Gingerich)[1,35] show that in at least some *Tetonius* the position is $\frac{2}{2}, \frac{1}{1}, \frac{3}{3}, \frac{3}{3}$ and is therefore exactly the same as in the omomyines. There is thus clearly no wide divergence in dental formula, at least between the best-known members of the two subfamilies. Moreover, anaptomorphines are more closely similar to microchoerines in their dental formula than has been thought. Specimens of *Absarokius* represent a time sequence of three successive species. The earliest of these, *Absarokius abbotti*, is known from about 70 mandibles and maxillae recently found by my expeditions in the Bighorn Basin, Wyoming and about 60 more in other collections. This species first appears in the Lysite faunal zone, early Late Wasatchian stage, North America. It was apparently derived from a species of *Tetonius* of the earlier Wasatchian. Members of the latter are represented by about 75 new jaws in the Yale collection from the Bighorn Basin. These include half a dozen or so specimens preserving anterior teeth or teeth sockets. From all these new materials it is clear that the lower dental formula of *Absarokius noctivagus nocerai* (2.1.2.3.) arose from that of *Absarokius abbotti* (2.1.3.3.). This in turn is the same as is found in relevant Yale *Tetonius* specimens. Presumably the dental formula of the earliest anaptomorphids, s.l., was 2.1.4.3, as in *Teilhardina belgica.*[1] The dental formula derivation of *Absarokius noctivagus nocerai* shows the most plausible interpretation of the lower dental formula of the European microchoerine genera is the same as in advanced *Absarokius.* That is, *Pseudoloris, Nannopithex, Necrolemur,* and *Microchoerus* may well all have had the lower dental formula of 2.1.2.3., *contra* that proposed by Stehlin and supported by Simons.[29,31] This change in dental formula interpretation, however, does not cast doubt on a relationship of these animals to *Tarsius.* During the time represented by the late Wasatchian and Bridgerian North American provincial ages, the ancestors of microchoerines are not known. Nevertheless such forms, if derived from something like *Teilhardina*, would require an intermediate ancestor with a dental formula of 2.1.3.3. From such a stage the lower dental formula of *Tarsius* could be derived through loss of the lateral incisor. Clearly, the known microchoerines need not be in the direct lineage leading to *Tarsius* in order to be placed in the same family. In fact, *Pseudoloris* may yet prove to be a direct ancestor of *Tarsius*,[1] but this cannot be fully demonstrated without better knowledge of its anterior dentition.

The term Anaptomorphidae was coined by Cope, but without diagnosing it.[36] Therefore the first work which gave the group considerable discussion was that of Wortman, based on the Yale fossil prosimian collections.[37] He treated these animals as a family also including the omomyines and *Necrolemur.* Simpson indicated a similar concept unifying the three groups represented by *Anaptomorphus, Omomys*, and *Necrolemur* as subfamilies under Anaptomorphidae.[32] He also placed there the mainly Palaeocene paromomyines, together with another subfamily coined by him: Pseudolorisinae. In the latter he placed only *Pseudoloris.* Simpson was the only author to rank *Pseudoloris* separately in this manner. Simons demonstrated that *Pseudoloris* belongs in Microchoerinae (= Necrolemurinae).[31] Paromomyidae was raised in rank by Simons and has subsequently been considered as including Phenacolemuridae[38] by Simons.[1] As mentioned already, Gazin removed the omomyines to a separate family.[27] With these various removals the

Anaptomorphidae became a very much restricted taxon containing only the genera *Anaptomorphus, Tetonius, Absarokius*, and *Uintanius*, along with a few less-well-known, more doubtfully referred, or possibly synonymous genera, such as *Trogolemur, Uintalacus, Anemorhysis*, and *Tetonoides.* Simons, following Szalay in part, has suggested that *Tetonoides, Anemorhysis*, and *Chlororhysis* are probably all synonyms of *Tetonius.*[1] Sustaining the separation of anaptomorphines from omomyines are the following characteristics, most of which occur in species of the short list of anaptomorphine genera, *sensu stricto,* given above: Anaptomorphines have, relative to omomyines, cheek tooth cusps that are rounder or more inflated at the base, with talonids on the lower molars that are relatively less broad. Trigonids are typically shorter and narrower than in omomyines. As a result of these differences in anaptomorphines, molar cusps are somewhat more centrally shifted. In anaptomorphines, *s.s.*, such as *Tetonius*, the shape of the trigonid changes from front to back. The paraconid is progressively de-emphasised posteriorly, but in a manner different from that most often seen in omomyines. In *Tetonius* the M_1 trigonid cusps are distinctly formed and the angle formed by the crests running from paraconid and metaconid to the protoconid flares widely; in M_2 the paraconid and metaconid are brought closer together but are in line with each other; in M_3 they are even closer or may be coalescent. Omomyines differ from this in that the posterior de-emphasis of the paraconid in the successive molars comes about because this cusp is usually shifted more towards the midline, and decreases in size in the series from M_1-M_3. The P_4 in most anaptomorphines is typically much larger than are adjacent teeth, but *Anaptomorphus* (the type of the group) forms an awkward exception to this. The anaptomorphine genera *Tetonius* and *Absarokius* exhibit the *Nannopithex*-fold, which is also retained in microchoerines and is seen as well in the adapid *Pelycodus*, but not normally in omomyines. This unusual feature seems to bind all these groups together. Knowledge of anaptomorphine incisors is based mainly on the lower dentition, but where this is known the calibre of the central pair is usually great relative to the lateral pair of incisors. The same holds true of omomyines, but the disparity between central and lateral incisors is less great. Since anaptomorphines such as *Tetonius* are clearly related to modern *Tarsius* it is interesting that their lateral lower incisors are small compared to those of omomyines. This could be relevant to their complete loss in the modern genus. Interestingly, in the earliest Anthropoidea where lower incisor structure can be documented (*Apidium, Parapithecus, Aeolopithecus*), the lateral incisor root clearly has a far larger calibre than does the central root pair. This, then, is inconsistent with derivation of higher Primates from known Eocene prosimians, unless lower incisor size relationships somehow reversed in one or more lineages.

That some member of the 'Omomyidae' might have given rise to at least the platyrrhines was suggested by Gazin,[27] but he in turn was following or echoing the presentation of *Omomys* in this role by Wortman.[37] Now that assignment of the omomyines to Anaptomorphidae has been better justified, the question of whether or not omomyines were at least as tarsier-like as anaptomorphines is raised again. The attempt, on Simpson's part, to show that the postcranial structure failed to prove that the animal had ties with *Tarsius*, was probably well taken.[32] Nevertheless, his effort was slightly misleading inasmuch as close parallels to the unique adaptive features of the tarsier hind-limb may never be found in any early Tertiary form. This would certainly be the case if tarsier ancestors went through some of their final refinements in postcranial adaptation (such as development of tibiofibula) in the late Tertiary. Probably more relevant to this question is the unique skull of *Rooneyia* from the earliest Oligocene of West Texas. This animal is generally classified as an omomyine. Although the bullae are large and there is an ossified meatus, the pterygoid alae are long and the foramen magnum is so backwardly directed as to be quite unlike its orientation in *Tarsius*. In sum, there seems to be little in the craniology of *Rooneyia* that could be urged as being more tarsier-like than like galagines.

Recently Szalay (personal communication) has suggested that there are no Eurasian omomyines and that *Teilhardina* from the early Eocene of Belgium and France, long considered as a very similar, if not congeneric, relative of *Omomys*, is instead an anaptomorphine close to *Tetonius*. My studies of *Teilhardina* indicate that the situation is not so simple as that.[1] Although Teilhard originally referred this species to *Omomys*,[39] Simpson recognised its generic distinctiveness and coined the name *Teilhardina* for it.[32] However, he confused a mandible, edentulous save for P_4, of an insectivore with the hypodigm and so determined the wrong lower dental formula, 3.1.3 or 4.3. for *Teilhardina*. My study of the extensive new material, discovered since Simpson wrote, showed that the lower dental formula is 2?.1.4.3. The presence of four premolars, the relatively unenlarged P^4_4, and other unspecialised features, makes this species the most generalised anaptomorphid, *s. l.* known. The assessment of *Teilhardina* has been varied. Hürzeler[33] suggested that it was conceivably in or near the basal ancestry of the European Eocene microchoerines (Necrolemurinae). Later Quinet[40] proposed that *Teilhardina belgica* was at the base of the whole radiation of Anthropoidea. Although it seems to me too generalised for definite placement among anaptomorphines, *Teilhardina* may well be near the base of the Anaptomorphidae, *s. l.*

There are several other Eurasian Eocene prosimians, as well as a possible member of Anthropoidea, *Pondaungia,* from the late Eocene of Burma, which are currently being investigated at Yale; but there is not sufficient space to go into a discussion of all of these here. Valid

Eocene genera of adapids, anaptomorphids and microchoerines are reviewed in Simons.[1] It should be pointed out, however, that the pongid-like features of M^{1-2} of *Pondaungia cotteri*, Pilgrim,[41] tend to reinforce the original suggestion of Colbert that *Amphipithecus*, from the same general region and time in Burma, can be considered a higher primate.[42] One cannot explain away one of these as an anthropoid without discounting the other.

Oligocene primates

At the close of the Eocene, prosimian primates decline rapidly in variety and number in the northern hemisphere. The occurrence of some individuals of *Microchoerus ornatus* from Europe appears to extend the range of this genus into the early Oligocene. In North America only two genera and species, each known only from a single individual specimen, have been shown to exist in Oligocene times. These are a cranium of *Rooneyia viejaensis* from earliest Oligocene deposits of Presidio county in west Texas, and a single mandible of *Macrotarsius montanus* from a locality at Pipestone Springs, Jefferson county, Montana, again of earliest Oligocene age. Long after these had disappeared, in the Miocene, one additional primate *Ekgmowechashala* occurs in North America. Apart from these very few fossils from the northern hemisphere, all that we know about primates between about 35 and about 15 million years ago comes from the southern continents of South America and Africa.

At the outset of the record only higher primates, not prosimians, occur in either of these continents. In consequence, my concluding remarks will depart from the subject of this volume. Nevertheless, it seems advisable to stress to students of prosimians that the early Oligocene primates of the southern continents all show that they have clearly reached the grade we call Anthropoidea. Moreover, they exhibit far fewer features that could be taken as linking the two suborders than one might theoretically expect in view of their intermediate temporal position. In addition, as far as the African Oligocene primates are concerned, ample material from our recent expeditions to the Egyptian Oligocene deposits of the Fayum, Egypt now proves that the literature that has grown up around *Apidum* and *Parapithecus* is in many points wrong. Major errors about them have been published as recently as the *lapsus* of Hürzeler[43] suggesting that species of both of these genera were not even primates. As has been briefly mentioned elsewhere,[1,44–47] new *Parapithecus* material from the Fayum of Egypt shows that its initial describer, Schlosser, wrongly concluded that the type and only specimen was a juvenile with an unfused symphysis and a lower dental formula of 1.1.3.3.[48] As he then noted, the latter two features also typify modern tarsiers. The new Yale specimens of *Parapithecus* and *Apidium* show that the two are most closely related to each other, although differing distinctly in molar structure. Specimens of both show symphyseal fusion at an early age and have sockets indicating that large lateral incisors were present. Therefore, the dental formula in both is actually $\frac{2?}{2}, \frac{1}{1}, \frac{3}{3}, \frac{3}{3}.$ These facts make it clear that, as far as the type of *Parapithecus* is concerned, breakage before or during collection shattered the symphysis and considerable bone from the incisor region was lost. When the horizontal rami were subsequently glued together, the remaining crack between the horizontal rami (resembling an unfused symphysis) and the high angle of posterior divergence of the tooth rows (exaggerated by the loss of symphyseal bone) mistakenly give it the dental formula, the high posterior mandibular divergence, and the unfused symphysis seen in *Tarsius.* These false evidences were the foundation for the conclusion of Schlosser, and later of Kälin,[49] that *Parapithecus* was an animal somehow intermediate between tarsioids and anthropoids. Actually, *Parapithecus* (at least in such parts as can be compared) is as different from any prosimians in its dental and mandibular structures as are many South American monkeys. In a longer discussion than can be detailed here (see Simons)[47] I have given the evidence that *Parapithecus* is neither a prosimian nor a hominoid, but is a monkey. It could be in or near the ancestry of later Old World monkeys. This conclusion is supplemented by a second account[50] which shows that the dental eruption sequence of *Apidium* and apparently also of *Parapithecus* is the same as in monkeys, not prosimians. In addition to this, Kay has been able to demonstrate that there are extensions of certain wear facets of both upper and lower molars in *Parapithecus* which are to be found elsewhere only among Old World monkeys.[51]

In view of the importance of the construction of the mandibular symphysis (whether fused or unfused) as one of the features used in segregating the catarrhine and platyrrhine primates, on one side, from the Prosimii on the other, it is surprising to read in Szalay[13] that:

> It is a well-known fact that living primates have solidly fused mandibular symphyses, yet the earliest members of the order, like all primitive mammals – with the possible exception of *Chiromyoides* – had a mobile joint at the symphysis. In all later lineages, however, including genera of the Madagascan radiation, the symphysis fuses early in ontogeny. It appears that among the primates, symphyseal fusion is an adaptive response to withstand great, horizontally directed stresses during chewing of plant materials, as opposed to the primitive placental conditions where the stresses of orthal shearing occlusion were primarily vertically directed.[13]

The foregoing observation conveys a misunderstanding of one of the fundamental generalisations of primatology: that all Anthropoidea, living and fossil, have fused symphyses, but that no living prosimians and few fossil ones show such fusion. Contrary to Szalay's remark that *Chiromyoides* is the only one of the earliest primates which might show fusion – that is, might lack the mobile, unfused symphysis – even casual inspection would show that *Chiromyoides* lacks symphyseal fusion. The actual early exceptions, that is, real cases of ancient prosimians with fused symphyses, are not mentioned. There *are* certain Eocene adapid prosimians where fusion occurs. That this is a convergent development, only parallel to the case in Higher Primates, is confirmed because earliest adapids, *Pelycodus*, and *Protoadapis*, lack symphyseal fusion. Again some few, certainly not all, of the Malagasy extinct prosimians show fused symphyses, but none of the living forms does. These large, extinct Malagasy prosimians are convergent in their dental mechanisms either to certain higher primates or towards other types of herbivores. Thus divergent, insular prosimians, like certain adapids, constitute an understandable (and equally independent) exception to the general rule that prosimians typically have unfused symphyses. To put this situation in numerical terms, there are approximately 190 species of Anthropoidea, living and fossil, and none show unfused symphyses. Enumerating prosimians on the other hand (and setting aside the less than twenty species of adapids and extinct Malagasy lemurs which do show fusion) there are no known species of living prosimians with such fusion. Moreover, there is a combined total of the remainder, some 140 living and fossil species, which have the primitive, unfused condition.

NOTES

1 Simons, E.L. (1972), *Primate Evolution: An Introduction to Man's Place in Nature*, New York.

2 Hubrecht, A.A.W. (1971), *The Descent of the Primates*, New York.

3 Pocock, R.I. (1918), 'On the external characters of the lemurs and of *Tarsius*', *Proc. Zool. Soc. Lond.* 55, 19-53.

4 Mivart, St. G. (1964), 'Notes on the crania and dentition of the Lemuridae', *Proc. Zool. Soc. Lond.* 1, 611-48.

5 Gadow, H. (1898), *A Classification of Vertebrata, Recent and Extinct*, London.

6 Simons, E.L., and Russell, D.E. (1960), 'Notes on the cranial anatomy of *Necrolemur*', *Breviora* 127, 1-14.

7 Starck, D. (1956), 'Primitiventwicklung und Plazentation der Primaten', *Primatologia* 1, 723-886.

8 Hill, W.C.O. (1955), *Primates, Comparative Anatomy and Taxonomy: 2. Haplorhini: Tarsioidea.* Edinburgh.

9 Simpson, G.G. (1945), 'The principles of classification and a classification of mammals', *Bull. Amer. Mus. Nat. Hist.* 85, 1-350.

10 Burmeister, H. (1846), *Beiträge zur Kenntnis der Gattung Tarsius*, Berlin.

11 Sarich, V. (1970), 'Primate systematics with special reference to Old World monkeys: A protein perspective', in Napier, J.R., and Napier, P.H. (eds.), *Old World Monkeys*, New York.

12 Szalay, F.S. (1971), 'Cranium of the Late Palaeocene *Plesiadapis tricuspidens*', *Nature* 230, 324-5.

13 Szalay, F.S. (1972), 'Paleobiology of the earliest primates', in Tuttle, R. (ed.), *The Functional and Evolutionary Biology of Primates*, Chicago and New York.

14 Napier, J.R. and Walker, A.C. (1967), 'Vertical clinging and leaping: A newly recognized category of locomotor behaviour of primates', *Folia primat.* 6, 204-19.

15 Charles-Dominique, P. and Martin, R.D. (1970), 'Evolution of Lorises and Lemurs', *Nature* 227, 257-60.

16 Simons, E.L. (1960), 'New fossil primates: a review of the past decade', *Amer. Sci.* 48, 179-92.

17 Gingerich, P.D. (1971), 'Cranium of *Plesiadapis*', *Nature* 232, 566.

18 Szalay, F.S. (1968), 'The Picrodontidae, a family of early primates', *Amer. Mus. Novitates* 2329, 1-55.

19 Matthew, W.D. (1917), 'A Paleocene bat', *Bull. Amer. Mus. Nat. Hist.* 37, 569-71.

20 McGrew, P.O. and Patterson, B. (1962), 'A picrodontid insectivore(?) from the Palaeocene of Wyoming', *Breviora* 175, 1-9.

21 Simpson, G.G. (1935), 'The Tiffany fauna, upper Palaeocene: I. Multituberculata, Marsupialia, Insectivora and ?Chiroptera', *Amer. Mus. Novitates* 795, 1-19.

22 Simpson, G.G. (1937), 'The Fort Union of the Crazy Mountain Field, Montana, and its mammalian faunas', *U.S. Nat. Mus. Bull.* 169, 1-287.

23 Simons, E.L. (1963), 'A critical reappraisal of Tertiary primates', in Buettner-Janusch, J. (ed.), *Genetic and Evolutionary Biology of the Primates*, New York.

24 Matthew, W.D. and W. Granger (1921), 'New genera of Palaeocene mammals', *Amer. Mus. Novit.* 13, 1-7.

25 Saban, R. (1961), 'Carpolestidae Simpson 1935', in Piveteau, J. (ed.), *Traité de Paleontologie*, Paris.

26 Crook, J.H. and Gartlan, J.S. (1966), 'Evolution of primate societies', *Nature* 210, 1200-3.

27 Gazin, C.L. (1958), 'A review of the middle and upper Eocene primates of North America', *Smithson. Misc. Coll.* 136, 1-112.

28 Stehlin, H.G. (1912), 'Die Säugetiere des schweizerischen Eocäns', *Verhdl. schweiz. paläont. Gesell.* 38, 1165-298.

29 Stehlin, H.G. (1916), 'Die Säugetiere des schweizerischen Eocäns', *Verhdl. schweiz. paläont. Gesell.* 41, 1299-552.

30 Gregory, W.K. (1920), 'On the structure and relations of *Notharctus*, an American Eocene primate', *Mem. Amer. Mus. Nat. Hist.* 3, 53-243.

31 Simons, E.L. (1961), 'Notes on Eocene tarsioids and a revision of some Necrolemurinae', *Bull. Brit. Mus. Nat. Hist.* (Geol. Ser.) 5, 45-69.

32 Simpson, G.G. (1940), 'Studies on the earliest primates', *Bull. Amer. Mus. Nat. Hist.* 77, 185-212.

33 Hürzeler, J. (1948), 'Zur Stammesgeschichte der Necrolemuriden', *Schweizer. paläont. Abhand.* 66, 1-46.

34 Russell, D.E., Louis, R. and Savage, D.E. (1967), 'Primates of the French Early Eocene', *Univ. Calif. Publ. Geol. Sci.* 73, 1-46.

35 Bown, T.M. and Gingerich, P.D. (1972), 'The dentition of the Early Eocene primates *Niptomomys* and *Absarokius*', in press.

36 Cope, E.D. (1883), 'On the mutual relations of the bunotherian Mammalia', *Proc. Acad. Nat. Sci. Phila.* 35, 77-83.

37 Wortman, J.L. (1903, 1904), 'Studies of Eocene mammalia in the Marsh collection, Peabody Museum', *Amer. J. Sci.* (1903), 15, 163-76, 399-414, 419-36; 16, 345-68; (1904), 17, 23-33, 133-40, 203-14.

38 Simpson, G.G. (1955), 'The Phenacolemuridae, new family of early primates', *Bull. Amer. Mus. Nat. Hist.* 105, 417-41.

39 Teilhard de Chardin, P. (1927), 'Les mammifères de l'Eocene inférieur de la Belgique', *Mém. Mus. roy. Hist. nat. Belgique* 36, 1-33.

40 Quinet, G.E. (1966), '*Teilhardina belgica,* ancêtre des Anthropoidea de l'ancien monde', *Bull. Inst. roy. Sci. nat. Belgique* 42, 1-14.

41 Pilgrim, G.E. (1927), 'A *Sivapithecus* palate and other primate fossils from India', *Mem. Geol. Surv. Ind.* 14, 2-26.

42 Colbert, E.H. (1937), 'A new primate from the Upper Eocene Pondaung Formation of Burma', *Amer. Mus. Nov.* 951, 1-6.

43 Hürzeler, J. (1969), 'Questions et reflexions sur l'histoire des Anthropomorphes, *Ann. Paléont.* 54, 13-41.

44 Simons, E.L. (1967), 'The significance of primate paleontology for anthropological studies', *Am. J. Phys. Anthrop.* 27, 307-31.

45 Simons, E.L. (1969), 'The origin and radiation of the primates', *Ann. New York. Acad. Sci.* 167, 319-31.

46 Simons, E.L. (1971), 'A current review of the interrelationships of Oligocene and Miocene Catarrhini', in Dahlberg, A.A. (ed.), *Dental Morphology and Evolution,* Chicago.

47 Simons, E.L. (1974), 'A new species of *Parapithecus* from the Oligocene of Egypt, with notes on the initial differentiation of Cercopithecoidea', in press.

48 Schlosser, M. (1911), 'Beiträge zur Kenntnis der oligozänen Landsäugetiere aus dem Fayum (Ägypten)', *Beitr. Paläont. Österreich-Ungarns* 24, 51-167.

49 Kälin, J. (1961), 'Sur les primates de l'Oligocène inférieur d'Egypte', *Ann. Paléont.* 47, 1-48.

50 Conroy, G.C. (1974), 'Dental eruption patterns in earliest Cercopithecoidea', in press.

51 Kay, R. (1974), 'Dental wear patterns in *Parapithecus*', in press.

A.C. WALKER

A review of the Miocene Lorisidae of East Africa

Introduction

The Miocene lorisids of East Africa are the only known African fossil prosimians sampled from populations older than the Pleistocene. As such, they offer the only palaeontological clues to prosimian evolution on the African mainland and may help in understanding the separate evolution of the lemurs of Madagascar. The African Pleistocene galagines seem so close, morphologically, to living species of *Galago* that they only serve to demonstrate that a species very close to, or the direct ancestral species of, *Galago senegalensis* was established 2 million years ago in northern Tanzania.[1] The subfossil Madagascan lemuroids are so recently extinct[2] that they must be considered as biologically contemporaneous with modern species and their importance as evolutionary indicators is at the same level as modern lemuroid skeletal remains. Although Miocene lorisids are our only knowledge of earlier African prosimians, it must be stressed that they are sampled from very local populations from a very few sites peripheral to early central volcanos and that each site has only sampled from populations living over a fairly short period of time. Strictly speaking, we are dealing with parts of isolated prosimian populations in a very limited area of East Africa, which were living at separate times over a fairly limited period. We are fortunate in that the fossil sites have steadily yielded new specimens and the number of isolated parts has now increased considerably since the first specimen was described by MacInnes. The deductions that can be made from the fossils have increased with the specimens. The evolutionary story has not, by the same token, become simpler.

Historical

The first, and for some time only, specimen was described by MacInnes as the holotype of *Progalago dorae.*[3] More material was added to the hypodigm of this species by Le Gros Clark and Thomas,[4] who created at the same time two new species of the genus, *P. robustus* and *P. minor*. Le Gros Clark later described what is still the most complete cranium, but refrained from designating it specifically.[5] Leakey made *Mioeutocus bishopi* as a new taxon to accept a facial skeleton from Napak.[6] Simpson, in his revision, divided the material available to him into two species of *Progalago, P. dorae* and *P. songhorensis*, and created a new genus, *Komba*, to receive *P. robustus* and *minor.*[7] At the same time, he proposed a new genus and species *Propotto leakeyi*, but the specimens concerned were later removed from the order.[8] The first postcranial remains were described in 1970.[9] Since Simpson's revision, more specimens have been discovered and full descriptions of these will be published shortly, only brief notes on the most important ones being given here.

The problems of interpretation

One of the basic problems of interpretation involves the dating of the different fossiliferous deposits. Simpson, while discussing what was then known about the localities, pooled all his available specimens.[7] The differences he could see in his samples could be accounted for, he suggested, by differences in facies or chances of discovery. In fact, now that larger samples are known, there do seem to be meaningful differences between the Songhor, Napak and Rusinga assemblages, and some of these differences may be accounted for on the suggested difference in age. The radioisotopic evidence has been summarised by Bishop,[10] and Bishop et al.,[11] and it appears likely that the Napak and Rusinga assemblages come from deposits roughly 18-19 million years old, while the available evidence points to the Songhor deposits being perhaps as much as a million years older. Differences in sampling in two contemporaneous deposits can produce the effect of mild assemblage differences that need not reflect the living assemblages. Similarly, even minor palaeoecological differences may reflect themselves in slightly different mammalian assemblages. The unknown factors are so many that, without detailed knowledge of the sedimentological and taphonomical features of the individual fossil sites, practically any statement that purports to deal with the 'palaeoecology' is meaningless.

Another basic problem is the fragmentary nature of the material.

There are the remains of only two individuals that probably have parts of the associated upper and lower dentitions. Equally, apart from these two cases, there is only one other in which dental and postcranial remains are associated. In some cases, therefore, the upper and lower dentitions of a species have to be matched by morphology and 'fit' between different individuals and this has in the past led to difficulties. The postcranial remains are assigned by size using modern species as a guide, and as chance would have it not one of the commoner postcranial elements is matched by any of the associated material.

On the comparative side, when dealing with the modern Lorisidae, there is a high proportion of monotypic genera (four out of five). These are distributed over enormous geographical ranges and hence representative samples across the ranges show fairly high variability, a complicating factor when comparing them with fossil samples from successive populations in small areas.

Taxonomy

The taxonomy of the Miocene lorisids is based mostly on dental and cranial remains. The scheme used here differs slightly from that of Simpson, since use is made of associated partial upper and lower dentitions that were not available to him.[7] Simpson felt that without postcranial remains he could not assign any species to either subfamily. I have attempted to do so, with the postcranial evidence now available, although the reasons for placement in a subfamily do not rest entirely on postcranial evidence.

Family Lorisidae Gregory 1915
- Subfamily Galaginae Mivart 1864
  - *Progalago* MacInnes 1943
    - *Progalago dorae* MacInnes 1943
    - *Progalago songhorensis* Simpson 1967
  - *Komba* Simpson 1967
    - *Komba robustus* (Le Gros Clark and Thomas 1952)
    - *Komba minor* (Le Gros Clark and Thomas 1952)
- Subfamily Lorisinae Flower and Lydekker 1891
  - *Mioeuoticus* Leakey 1962
    - *Mioeuoticus bishopi* Leakey 1962
    - *Mioeuoticus* sp. nov.

The size range of the fossil species, as far as can be judged from the known parts, is as great as seen today in the genus *Galago*. *K. minor* is the size of *G. demidovii*, *K. robustus* the size of *G. alleni*, while *P. songhorensis* is a little smaller and *P. dorae* a little larger than *G. crassicaudatus. M. bishopi* is roughly the size of *Arctocebus calabarensis*, and *Mioeuoticus* sp. nov. is a somewhat larger species, about the size of *Perodicticus potto.*

Because *Mioeuoticus bishopi* is the only species with a type based on upper dentition, difficulties of comparison arise since it can be compared only with referred upper dentitions of *Progalago dorae*, the only species similar to it in size. Simpson believed that the cranium described by Le Gros Clark belonged to *Progalago dorae*,[5,7] but I cannot accept this in view of new material from Songhor. I do accept the close similarity of the type of *M. bishopi* with the cranium and think that the minor differences are at a specific level. For this reason, whereas Simpson almost (but not quite) removed the genus *Mioeuoticus* in favour of *Progalago*, I have kept the genus and include the Rusinga cranium as a new species of it. This species will be named and described elsewhere.

An isolated tooth (R.649'49) was placed as a P_2 in *P. dorae* by Le Gros Clark and Thomas,[4] and later referred by Simpson to lorisid indet., *not P. dorae*, on the grounds of size.[7] Reading Simpson's note on this specimen, it is implied that there is present, represented by one tooth, another species of lorisid larger than *P. dorae.* This tooth is, in fact, a hominoid deciduous canine, probably a left lower deciduous canine of *Limnopithecus macinnesi.*[12]

Cranial and dental material

Twenty-seven specimens are known that include part or all of the mandibular body. As far as is known, no pair of these come from one individual. Of this number, roughly half preserve some part of the symphyseal region. Clark and Thomas believed that the degree of procumbency of the tooth-comb was less than in modern lorisids.[4] Simpson, using the same specimens, thought that the degree of procumbency was probably the same as in modern species.[7] Walker put forward new evidence from Uganda to support Simpson's view that the tooth-comb was fully developed in the Miocene forms.[13] Several new specimens are now known that give a picture of symphyseal morphology and the roots of the six anterior teeth for *Progalago dorae* and *songhorensis* and *Komba robustus* and *minor.*

All specimens that include the symphysis show that the two bodies were not fused. The shape of the symphyseal surface is nearly elliptical in all specimens. An inferior symphyseal tubercle is developed in all specimens, the strength of the tubercle being proportionate to the size of the mandible – *P. dorae* having the greatest tubercle and *K. minor* the smallest. This tubercle is for the insertion of the anterior belly of *m. digastricus* and is also developed in modern lorisids to a greater or lesser extent depending on size. The symphyseal surface of the larger species is also more rugose than the

Table 1 Depth of mandible at various levels in fossil lorisids

(Ranges of measurements in millimetres, numbers of observations in brackets)

	at P_3	at P_4	at M_1	at M_2	at M_3
K. minor	–	3.8-4.0(2)	2.8-4.0(5)	2.6-3.0(4)	3.0-3.5(2)
K. robustus (Songhor)	4.3-4.8(2)	3.8-4.5(3)	3.8-4.7(5)	4.0-4.3(3)	4.0-4.5(2)
K. robustus (Rusinga, Napak and Mwafanganu)	5.6(1)	5.3(1)	5.0-5.1(3)	4.4-5.3(6)	4.4-5.5(6)
P. songhorensis (Songhor)	–	6.0(1)	6.4(1)	6.2-6.5(2)	6.7(1)
P. songhorensis (Rusinga)	–	6.0(1)	6.1-6.6(2)	7.3(1)	7.5(1)
P. dorae	–	6.7-7.2(2)	7.1(1)	7.9(1)	8.7(1)
Mioeuoticus sp. nov.	–	8.5 (1)	8.4 (1)	9.0 (1)	9.0 (1)

smaller ones. This region does not differ in any meaningful way between *Progalago* and *Komba*. The mental foramina of *Progalago* and *Komba minor* are single and relatively large, whereas those of *K. robustus* are double and each is relatively small.

The form and dimensions of the mandibular bodies of *Progalago* and *Komba* species are different. Le Gros Clark and Thomas did not consider those features to be of generic significance, but Simpson did.[4,7] The mandibles of *Progalago* species deepen posteriorly to gain their maximum depth posterior to M_3, but the thickness of the mandibles is not increased proportionately so that the posterior part of the body is thin and plate-like. In species of *Komba*, the posterior deepening is not seen and roughly the same thickness of body is maintained along the length. Measurements are given in Table 1. A mandible that I have referred to *Mioeuoticus* sp., that was previously included in *P. dorae*, is extremely deep and of roughly constant depth along the length. It has a double mental foramen with each foramen of fair size.

Alveoli, or alveoli with broken roots, of the anterior six teeth are seen in ten specimens. In all, the canine and incisor roots were laterally compressed and packed tightly together. The angle of inclination of the alveoli or roots is within the range of modern species. Although no complete grooming comb tooth is known, the evidence strongly suggests that the comb in galagines was as well developed in the Miocene as at the present time.

The rest of the mandibular dentition is now known for the species in the following parts: *P. dorae* P_3-M_2 ; *P. songhorensis* P_3-M_3 ; *K. robustus* P_3-M_3 ; *K. minor* P_4-M_3 ; cf. *Mioeuoticus* sp. M_2-M_3. Scale drawings of reconstructed mandibles and dentition are given in Fig. 1. The general aspect of the lower dentition is very much like that found in modern galagines, except that the P_4 is not molariform in the fossils. The differences between the two *Progalago* and two *Komba* species lie mostly in the more bunodont character of the former against the relatively more acrodont character of the latter. There is, however, the same difference (albeit slightly less marked) between the small and large members of the modern genus *Galago* and the possibility that the differences are largely due to size cannot be excluded. Specimens of *K. robustus* from Rusinga and Mwafanganu that are larger than the Songhor ones have wider trigonids and talonids as well as less acrodont cusps. The relative size of the trigonid also decreases with increasing size of tooth in *K. robustus*, and in the Rusinga and Mwafanganu specimens, which are increased in size to nearly as large as the Songhor *P. songhorensis*, the

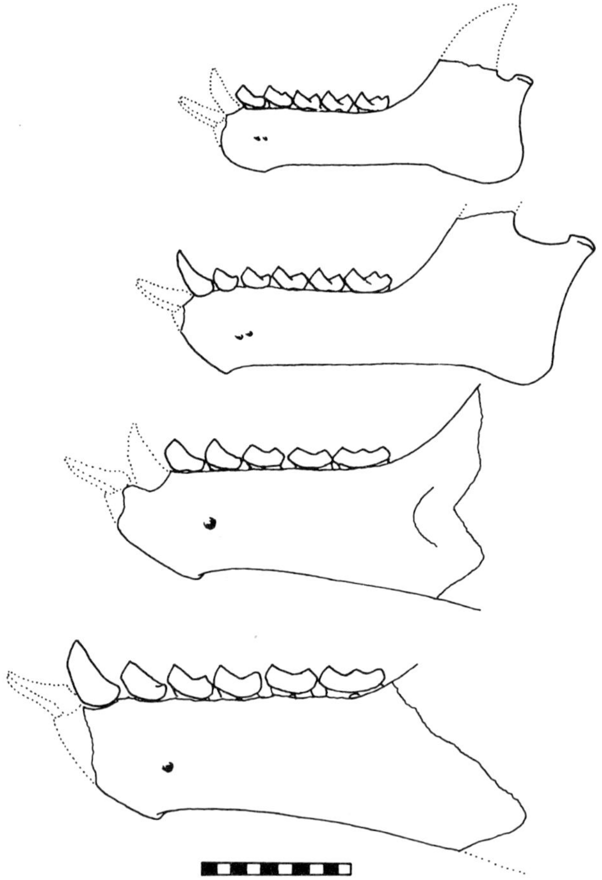

Figure 1. Lateral views of mandibles and dentition of Miocene lorisids. Species illustrated are, from above downwards, *Komba minor, Komba robustus, Progalago songhorensis* and *Progalago dorae.* The scale is in millimetres.

proportions of the talonids to trigonids are almost identical to those of that species. This may be taken as evidence that the dental differences between *Komba* and *Progalago* are not striking enough to warrant generic separation of the forms.

The lower dentition referred to *Mioeuoticus* has teeth larger than *P. dorae* with very expanded trigonids and talonids that indicate a quadrate form of upper molar.

The number of specimens of upper dentition has been increased since Simpson's revision and the finding of associated upper and lower teeth of *P. songhorensis* and *K. robustus* from Kathwanga, Rusinga Island leads to clearer associations of the known upper and lower dentitions. The upper dentitions are known from parts that include: *P. dorae* P^3-M^3; *P. songhorensis* P^2, P^4-M^3; *K. robustus* P^4-M^3; *K. minor* – nil; *Mioeuoticus bishopi* P^2, P^4-M^3; *M.* sp. nov. P^4-M^3. Information can be gathered about the anterior dentition of *P. dorae* and species of *Mioeuoticus*, and one deciduous premolar of *P. dorae* is known. The two most complete upper dentitions and facial skeletons are both of *Mioeuoticus*, a fact that is of some importance. Knowledge of the palate and facial skeleton of other species is limited to fragments of bone still in place around the tooth roots. *Progalago* and *Komba* species had, evidently, lightly built facial skeletons that have fragmented quickly prior to fossilisation or when weathering out of the sediments.

The upper dentition of *Progalago* and *Komba* shows the general situation as found in modern galagines, with the following exceptions: the P^4 is not as molariform, and the development of the molar hypocones is not as extreme. The degree of reduction of the third molar is less in the smaller species and in all forms is a little less than in modern galagines; a feature that is correlated, of course, with the fact that the M_3 in the fossils are relatively larger than their modern counterparts. Of the upper premolars, only P^2 is two-rooted in *P. dorae* and *K. robustus*, the remainder are three-rooted. In *Mioeuoticus bishopi* the same is true, but in *Mioeuoticus* sp. nov. both P^3 and P^4 are two-rooted as Le Gros Clark noted, not as Simpson recorded.[5,7] Simpson noted that the P^2 and P^3 preserved in one specimen assigned to *P. dorae* were 'clearly similar to *G. crassicaudatus* and markedly unlike the shorter, more transverse teeth of all Recent lorisines'. *Mioeuoticus bishopi, M.* sp. nov. and *Progalago dorae* have a depression in the palate just posterior to the canine that would have accepted the caniniform P_2. P^4 of *P. dorae* and *K. robustus* are basically bicuspid teeth and do not show the molarisation seen in Recent galagines. The P^4 of *Mioeuoticus* is also bicuspid but differs from the others in being lower crowned and more rectangular in outline due to the strong development of a distolingual cingulum. This cingulum is seen in *P. dorae* and *K. robustus*, but is very weakly developed.

The upper molars of *P. dorae* and *P. songhorensis* as they are known now are quite like those of *Galago* spp., except that the hypocone, though distinct, is more closely applied to the protocone and that the teeth are less wide transversely. The few known upper teeth of *P. songhorensis* are a little more transversely widened than in *P. dorae.* Upper molars of *K. robustus* are more acrodont with a more isolated hypocone than in *Progalago* spp. These teeth are also wide transversely. The M^3 is less reduced than in *P. dorae.* Upper molars of *Mioeuoticus* are strikingly different from those of *Progalago* and *Komba.* M^1 and M^2 are large and almost square in outline, with well developed buccal and distolingual cingula. M^3 is more triangular in outline due to the absence of the hypocone, but the distobuccal cingulum is still present. All the molars are low-crowned, bunodont teeth. *Mioeuoticus* molars show a striking resemblance to those of *Arctocebus*, but the latter are more acrodont. The resemblance to *Perodicticus* that Le Gros Clark pointed out is, I think, overstressed.[5]

Facial and palatal parts of *Progalago* and *Komba* are limited to small fragments supported by tooth roots. A fragment of maxilla shows that in *P. dorae* the canine root made a moderately strong jugum on the lateral surface, fairly similar to that of *G. crassicaudatus.* As far as can be judged, the palate of *P. dorae* was flat and not strongly arched. The orbits of *Progalago* and *Komba* were seemingly as large as in modern species of *Galago*, with the exception of *G. elegantulus,* but the orbital floor approaches the molars more in *P. dorae* than in the other species. The facial skeleton of *Mioeuoticus* is strongly built, with the canine root making a strong maxillary jugum. The orbits were large and their floors very thin. The frontal bones have a depressed area either side of a gentle sagittal swelling at the level of the zygomatic process. The ethmoid contributes to the medial wall of the orbit.

The rest of the cranium is known from two specimens only, a partial cranium with natural endocast and a cranium of *Mioeuoticus*, both from Rusinga. These two are strikingly different and are certainly from different genera. Now that the nearly complete cranium is seen to be of *Mioeuoticus*, it seems that the partial cranium is most probably either *K. robustus* or *P. songhorensis.* The two species are of roughly the same size and it is impossible to say, at present, to which species this belongs. Simpson thought that this specimen was probably of *K. robustus*, but he was using his assignation of the nearly complete cranium as *P. dorae* at the time. This no longer applies and the cranium could be of either *K. robustus* or *P. songhorensis.* The implications of recognising that the nearly complete cranium is of *Mioeuoticus* are considerable, for features of that cranium have an unmistakeable lorisine stamp, while the other cranium has, as Le Gros Clark and Thomas suggested, a definitely galagine aspect. Among the lorisine characters of the cranium of *Mioeuoticus* are the following:

1. Cranium strongly constructed.
2. Temporal ridges raised and very distinct.
3. Orbits more upwardly directed.
4. Bulla and mastoid weakly inflated.
5. Internal nares wide.
6. Basicranial flexion slight.

Among the galagine characters of the endocast specimen are the following:

1. Cranium lightly constructed.
2. Temporal lines present but not developed as strong ridges.
3. Bulla strongly inflated.
4. Basicranial flexion moderate (as in Recent *Galago* of same size).

The ear regions of both specimens have been cleaned on one side to expose the periotic. In all features except for the degree of bulla inflation they resemble each other as well as modern lorisids. Several features of the ear region show that these two specimens have unequivocal lorisid affinities. The details of the round and oval windows are exactly the same as in modern lorisids. The promontorial artery lay in a groove that runs across the surface of the periotic and was not encased in bone as in lemuroids. There is a clearly developed bony division of the tympanic cavity into hypotympanic and tympanic ones, unlike the lemurid condition where only one large tympanic cavity is found. On the cranial base there is in both specimens a well developed *foramen lacerum medium*, as in lorisids, but unlike the lemurid condition (with the exception of the cheirogaleines). The ectotympanic ring in the larger specimen lies in a position very similar to that seen in *G. crassicaudatus*, but the ring itself is thicker and less closely moulded to the bulla wall. The smaller specimen seems to have a thinner, more closely applied, ectotympanic.

Postcranial remains

Since my note on the first limb bone specimens, others have been found to give a more complete picture.[9] As far as can be told, no specimens are known of the postcranial remains of *Mioeuoticus*. Two sets of fragments are associated with dentition of *P. songhorensis* and *K. robustus* and one pelvic fragment with teeth of *P. dorae*. The following parts are now known, using the associations available and the size analysis that I applied to the original fragments.[9]

P. dorae	Left acetabulum and ischium.

cf. *P. dorae*	Proximal part humerus, proximal and distal parts femur, distal hallucal phalanx, calcaneum.
P. songhorensis	Distal end tibia, capitate, a terminal phalanx, talus.
cf. *P. songhorensis*	Proximal and distal parts humeri, proximal part tib
K. robustus	Two vertebrae.
cf. *K. robustus*	Distal part humerus, proximal parts femora, calcaneum.
cf. *K. minor*	Distal part humerus, proximal and distal parts femora, calcaneum.

The new associated material confirms the usefulness of the size analysis that I used earlier. However, I made the size analysis using Simpson's dental references and the reassigning of dental material and the finding of new specimens has caused some minor changes in my placement of the limb bones. I no longer believe that we have any limb bone material of *Mioeuoticus* (I had, following Simpson's hints, dropped that genus in favour of *Progalago*) and the three fragments assigned to *M. bishopi* probably belong to other species of either *Progalago* or *Komba*, both of which genera are represented at Napak.

The major features of the limb bones of *Progalago* and *Komba* are galagine and even in some minute details the bones are remarkably like those of Recent species. In only one major respect do they differ and that is in the lesser elongation of the calcaneum. It may be, as I suggested before, that all tarsal elements were elongated and that the total foot length might have been approximately the same as in living species. However, the test of that hypothesis, the finding of an elongated cuboid, has not yet come about.

The major limb bone epiphyses in galagines do not fuse until after the eruption of the third molars. The pelvic specimen associated with teeth of *P. dorae* shows that the same was true in that species. I have been unable to confirm whether the same is true of the lorisines.

At all events the basic galagine locomotion was established by the early Miocene in the East African species and, as far as I can judge, no fossil material exists that can give us any indication as to whether or not the lorisines had by then developed their locomotor propensities, although the chances of their having done so are quite high.

Summary

The new specimens of fossil lorisid from Kenya help to show that both the Lorisinae and Galaginae were established by lower Miocene times. The lorisines are represented by two species of one genus separated in time and space, and although the sample is extremely small the specimens are the best we have of all the lorisids. The galagines are represented by four species and their fossil remains are more common but more fragmentary. Charles-Dominique reports mean densities of 10 lorisines of two species and 80 galagines of three species per square km. in the forests of Gabon.[14] On the basis of minimum numbers of individuals, and using all fossils, the ratio of lorisines to galagines in Miocene collections is close to 1:9. The close agreement of these two ratios may be due to chance or differences in preservability between the two subfamilies, but the small number of lorisine specimens remains a fact and it could possibly be that this represents a density situation such as is present today. A size range is seen among the galagines as great as is seen among living East African species.

There is a strong possibility that the ancestors of at least some modern *Galago* species are present in the Miocene assemblage. *K. minor*, as Le Gros Clark and Thomas noted, is especially close dentally to *G. demidovii* and *P. dorae* seems very close to *G. crassicaudatus.* Postcranially, all four galagine species appear to be very similar, despite the substantial size differences, and all closely resemble modern species except for the shortness of the calcaneum. It is very likely that the locomotion of the extinct forms was extremely similar to that of modern *Galago* species. Because the two smallest species are placed in *Komba* and the two largest in *Progalago*, and because of the possibility that some distinguishing features are merely those of size, the wisest step might prove to be to remove Simpson's genus *Komba* and leave all these Miocene galagines in one genus. The taxonomy as it stands implies that species of one or the other of the Miocene genera cannot be ancestral to *Galago* species, or, conversely, that the genus *Galago* is not a natural genus and that to reflect phylogeny it should be divided. Removing the taxon *Komba* would solve this difficulty.

Simpson's criteria for creating *Komba* as a distinct genus have a basis in his believing *Euoticus* and *Galago* to be distinct genera, and were further complicated by his acceptance of the Rusinga cranium as being *P. dorae.* It is very likely, given the apparently stable condition of the equatorial forest belt of Africa since before the Miocene, that the ancestors of East African species are to be found in East Africa and, similarly, of West African species in West Africa. Adaptation to a more arid environment, such as is seen in *Galago senegalensis,* could have been caused by a forest species colonising the peripheral areas or, alternatively, by adaptation of one species to the changing conditions of the peripheral areas as the forests contracted. *Galago inustus*, in this regard, can be viewed as representing either something like the *G. senegalensis* forest ancestor or as a forest species developed from *G. senegalensis. G. inustus* is not, as far as I can determine from the morphology, closely related to *G. elegantulus*, and those who favour a subgeneric rank for *Euoticus* should not place *G. inustus* therein.

Mioeuoticus might be considered as closest to *Arctocebus* among the lorisines, but equally plausibly it could be part of the ancestral African lorisine populations, or for that matter the ancestral lorisine one. Why lorisines and not galagines crossed into Asia is a vexing question. I can think of no reasonable explanation except that the galagine types of niches were occupied by tarsiids of some sort. In the absence of any fossil evidence the question remains unanswered.

Table 2 Sites from which Miocene lorisids have been recorded

	Songhor	*Rusinga*	*Napak*	*Moroto I*	*Mwafanganu*	*Koru*
K. minor	+	+	+	+	—	—
K. robustus	+	+	+	—	+	+
P. songhorensis	+	+	—	—	—	—
P. dorae	+	—	+	—	—	+
M. bishopi	—	—	+	—	—	—
Mioeuoticus sp. nov.	—	+	—	—	—	—

Acknowledgments

I thank the Director and Trustees of the National Museums of Kenya for permission to examine specimens in the National Museum, Nairobi. The late Dr L.S.B. Leakey and P.J. Andrews kindly gave me permission to study specimens collected by them.

NOTES

1 Simpson, G.G. (1965), 'Family Galagidae', in Leakey, L.S.B. (ed.), *Olduvai Gorge 1951-1961*, London.

2 Walker, A. (1967), 'Patterns of extinction among the subfossil Madagascan lemuroids', in Martin, P.S. and Wright, H.E. (eds.), *Pleistocene Extinctions*, New Haven, Conn.

3 MacInnes, D.G. (1943), 'Notes on the East African Miocene primates', *J. East. Afr. Uganda Nat. Hist. Soc.* 17, 141-81.

4 Le Gros Clark, W.E. and Thomas, D.P. (1952), 'The Miocene lemuroids of East Africa', *Fossil Mammals of Africa* 5, London.

5 Le Gros Clark, W.E. (1956), 'A Miocene lemuroid skull from East Africa', *Fossil Mammals of Africa* 9, London.

6 Leakey, L.S.B. (1962), in Bishop, W.W. (ed.), 'The mammalian fauna and geomorphological relations of the Napak volcanics, Karamoja, *Rec. Geol. Surv. Uganda (Entebbe)*, 1-18.

7 Simpson, G.G. (1967), 'The Tertiary lorisiform primates of Africa', *Bull. Mus. Comp. Zool.* 136, 39-61.

8 Walker, A. (1969), 'True affinities of *Propotto leakeyi* Simpson 1967', *Nature* 223, 647-8.

9 Walker, A. (1970), 'Post-cranial remains of the Miocene Lorisidae of East

Africa', *Am. J. Phys. Anthrop* 33, 249-61.
10 Bishop, W.W. (1971), 'The late Cenozoic history of East Africa in relation to hominoid evolution', in Turekian, K.K. (ed.), *The Late Cenozoic Glacial Ages*, New Haven, Conn.
11 Bishop, W.W., Miller, J.A. and Fitch, F.J. (1969), 'New potassium-argon age determinations relevant to the Miocene mammal sequence in East Africa', *Amer. J. Sci.* 267, 669-99.
12 Andrews, P.J., personal communication.
13 Walker, A. (1969), 'New evidence from Uganda regarding the dentition of Miocene Lorisidae', *Uganda Journal* 33, 90-1.
14 Charles-Dominique, P. (1971), 'Eco-éthologie des prosimiens du Gabon', Biol. Gabon. 7, 121-228.

C. P. GROVES

Taxonomy and phylogeny of prosimians

Phylogenetic systematics

The aim of the present paper is to present gaps in our knowledge as much as facts. The unsatisfactory state of primate taxonomy today is due, at least in part, to the explosion of research on primates (especially on prosimians) in recent years; the old certainties are gone and a host of new information is available, still largely undigested and still incomplete, but sufficient to point the way to radically new assessments. New areas of research are indicated in their turn, and it is the role of taxonomy to provide the synthesis and indicate possible avenues of useful future investigation.

It is necessary at the outset to ask oneself the question: what, precisely, is the function of taxonomy? The phenetic/phyletic and grade/clade controversies show no signs of abating; the theoretical approaches of the different schools are as wide apart as ever, but strange to say the effect of this in practical terms in minimal. Taking the mammals as an example, it is probably fair to say that the internal arrangement of most orders is widely accepted, only the Rodentia, Insectivora and Primates being sources of major disagreement. In the first case the phylogeny itself, and the roles of convergence and parallelism, are at stake; in the other two cases, there are radical divergences of opinion over how the agreed phylogenetic data should be employed to yield the most useful taxonomic scheme.

It is widely agreed that phylogenetically the lemurs (*sensu lato*) are more divergent from the tarsiers, monkeys, apes and man than these latter are from each other, and this theme recurred throughout the Research Seminar. In spite of this general agreement, taxonomic schemes vary widely; three in common use are presented in Table 1. Simpson's is a frankly grade-type of classification: for all its phyletic association with Anthropoidea, the tarsier is 'primitive', so it is ranked alongside the other 'primitive' primates in Prosimii.[1] Hill's scheme is strictly clade (phylogenetic):[2] the tarsier is related to the Anthropoidea (Pithecoidea) more closely than to the lemurs, so it must be placed nearer them in a classification. Romer's is essentially a compromise between these two opposing viewpoints.[3]

In the last analysis, a grade classification like Simpson's is essentially anthropocentric. When one asks *how* lemurs and tarsiers are primitive, the only possible answer is that they have small brains: they are less advanced up the ladder to man. The other 'primate trends' [4,5] – enhancement of digital mobility, orbital frontality, olfactory reduction and so on – which are also said to lead up the ladder to man, are consequently as anthropocentric as brain size; but all of them are subject to exceptions (frontal orbits of *Loris*, pseudo-opposable thumbs of many lemurs, hypocones present in many lemurs but absent in marmosets), so that we are left with brain size alone as the basis for subordinal division. Such a situation is of course an occupational hazard of grade taxonomy. If one attempts to characterise living species as 'primitive' or 'advanced' overall and then to make taxonomic groupings on this basis, there is always a strong risk that the members of one's resultant taxa may in fact be 'primitive' or 'advanced' in different ways, leaving the taxa undefinable and very probably polyphyletic.

It is here insisted that a phylogenetic ('clade') type of classification is more meaningful in all respects than a grade one. It avoids subjective assessments of overall evolutionary 'advancement'; its degree of objectivity varies only with the state of knowledge of the group's evolution; it excludes polyphyletic taxa; and perhaps most important, it maximises the amount of information content in a classification, enabling it to be used as a kind of one-dimensional phylogeny. The drawback to such a scheme is that its validity is totally dependent on the validity of the evolutionary scheme on which it is founded. It might at times be useful to retain an old classification in a case where phylogenetic views were in a state of flux: within the primates, for example, it is probably preferable for the moment to retain a grouping – Platyrrhini, Ceboidea or whatever – for the New World primates, although their homogeneity is by no means certain; but such an arrangement should only be an interim measure, and as far as suborders are concerned the relevant phylogenetic data are perfectly clear, and the advantages of a strict clade taxonomy should not be lost.

The study of fossil primates is progressing rapidly, but there are still so many gaps, uncertainties and disputes – not to say polemics – that comparative anatomy must still be the prime basis for phylogenetic reconstruction. For example, the detailed similarity of the foetal membranes of *Tarsius* and monkeys is so remarkable,[6] and is supported by such a wealth of other resemblances,[7,8] that to complain that the fossil record is inadequate to demonstrate the

Table 1 Three currently used classifications of Primates

Simpson, 1945[3] (grade)			Hill, 1953[4] (clade or phylogenetic)		Romer, 1967[5] (compromise scheme)	
SUBORDER PROSIMII	Infraorder Lemuriformes	Superfamily Tupaiodea	(not Primates)		(not Primates)	
		Superfamily Daubentonioidea	GRADE STREPSIRHINI	Suborder Lemuroidea	SUBORDER LEMUROIDEA	Family Daubentoniidae
		Superfamily Lemuroidea				
	Infraorder Lorisiformes			Suborder Lorisoidea		Family Lorisidae
	Infraorder Tarsiiformes		GRADE HAPLORHINI	Suborder Tarsioidea	SUBORDER TARSIOIDEA	Family Tarsiidae
SUBORDER ANTHROPOIDEA	Superfamily Ceboidea			Suborder Pithecoidea	SUBORDER PLATYRRHINI	Families Callithricidae, Cebidae
	Superfamily Cercopithecoidea				SUBORDER CATARRHINI	Family Cercopithecidae
	Superfamily Hominoidea					Families Pongidae, Hominidae

origin of monkeys and apes from a tarsier-like group is irrelevant;[9] nor is it really feasible to envisage separate derivation for Catarrhines and Platyrrhines from paromomyiform ancestors.[10]

There remains the problem of deciding the validity of shared characters. Detailed homology, as in the placentation example, can generally be relied upon to exclude convergence especially if backed up by other similarities. The same can probably be said of parallelism. The question of symplesiomorphy v. synapomorphy ('characters of common inheritance' and 'characters of independent acquisition') demands some knowledge of animals outside the group one is considering, as has become all too clear in the case of the tree-shrews, and their inclusion in, or exclusion from, the Primates.[11,12]

The species concept

The critical question for deciding the specific status of two extant forms is whether or not they form hybrid populations in a state of nature (if they share portions of their range). This is the meaning of reproductive isolation. Obviously, if the two are allopatric and never meet, then other means must be found to decide their status; but if they are sympatric and show no signs of forming a hybrid population between them, i.e. they do not form a morphological continuum when museum specimens or living examples are examined, then reproductive isolation is indicated. It is worth stressing this point, since many reports in the literature on primates give the impression that the two parent types are shown to be conspecific by the mere fact that they can hybridise under captive conditions.[13] Nor is the production of occasional wild hybrids indicative. The following cases can be used to indicate respectively the hybridisation of forms in captivity which are sympatric but reproductively isolated in the wild, and the production of hybrids in the wild without destroying reproductive isolation:

1. A recent checklist records the production, in captivity, of hybrids between taxa whose coexistence in the wild indicates their status as good, separate species;[14] among the Primates, these include:

   Hylobates lar x *H. pileatus* (two gibbons sympatric in Thailand)
   Cercocebus torquatus x *Papio sphinx* (redcap mangabey and mandrill, sympatric in Río Muni and Gabon)
   Cercopithecus cephus x *C. mona* (two guenons, sympatric in Gabon)
   Cercopithecus hamlyni x *C. lhoesti* (two guenons, sympatric in Kivu)

Cercopithecus mona x *C. nictitans* (two guenons, sympatric in Cameroun)
Macaca arctoides x *M. assamensis* (two macaques, sympatric in N. Burma)
Macaca nemestrina x *M. fascicularis* (two macaques widely sympatric in South-East Asia)
Papio hamadryas x *Theropithecus gelada* (sacred baboon and gelada, sympatric in Ethiopia)
Cebus albifrons x *C. apella* (two capuchin species, sympatric in Colombia and N.W. Brazil)

Especially noteworthy is the fact that several of these hybrids were capable of producing viable offspring when backcrossed with one or other of the parental species: in other words, the reason for their failure to form hybrid populations in the wild is purely ethological, not because of karyotypic incompatibility or for any other morphological cause. Instructive examples from outside the Primates can be found among the Cervidae: hybrids between *Cervus elaphus* (Red deer) and *Elaphurus davidianus* (Père David's deer), two species formerly sympatric in China and belonging to distinctly different genera by any standards, are not only obtained with great ease in captivity, but are 'endlessly fertile'!

2. There are, on the other hand, several cases of the production of occasional hybrids in the wild without the formation of hybrid populations, i.e. without actual gene-flow between the two parental species. Three such cases are known from primates:[14] *Cercopithecus mona* x *C. pogonias* in Cameroun, *C. mitis* x *C. ascanius* in Uganda, and *Macaca nemestrina* x *M. fascicularis* in Malaya. In at least two of these cases, habitat interference by man was suspected; it is likely that, in the macaque instance, all the males in a *nemestrina* troop had been shot, and the females joined a *fascicularis* troop. There is also at least one case, from the Bovidae this time, of regular hybridisation between two sympatric species which nonetheless manage to retain their separateness: it is claimed that the two ibex species *Capra caucasica* and *C. cylindricornis* frequently hybridise where their ranges overlap in the central Caucasus;[15] but the hybridisation goes no further than F_1, so that there is no gene-flow between the two species.

The major divisions of the Strepsirhini

There has been a tacit assumption that, because Madagascar is an island, and because its fauna is by and large very unlike that of the

African mainland, the Malagasy lemurs must be a homogeneous group: that there can have been only one invasion of Madagascar per mammalian order, and that this must have been in the far distant past. Hence, the aye-aye forms with the Lemuridae and Indriidae a single infraorder, Lemuriformes, while the lorises and galagos are set apart as Lorisiformes. Starting from this assumption, primatologists have searched for differences between the two groups and claimed to discover them; on this insecure basis, zöogeographers have felt at liberty to use the lemurs as illustrative of the distinctness of the Malagasy fauna, thereby completing the circularity of the argument![16]

Lorisiform-Lemuriform differences

The differences of the Lorisiform lemurs from the Lemuriform ones were first emphasised by Osman Hill,[17] who listed 39 characters said to distinguish them. Many of these however are at once qualified by such statements as 'except *Galago*', while many (perhaps most!) of the others certainly should have been. The most important of these characters, in that they have been given prominence in later works such as Le Gros Clark's great textbook,[4] would seem to be:

1. Tympanic ring external to bulla and fused with it.
2. Os planum (ethmoid plate) on orbital wall.
3. Large stapedial, small promontory artery, branching external to bulla and only stapedial entering bulla.
4. Microstructure of tooth enamel.
5. First ethmoturbinal enlarged, overlapping maxilloturbinal.
6. Vibrissae reduced.

Of these features, nos. 4 and 5 are of somewhat uncertain value since they have been investigated only in relatively few species. The first three are the 'key' features, quoted time and again in this regard. Table 2 shows however that, even if they are taken at their face value, only the tympanic region distinguishes the Malagasy and non-Malagasy groups as a whole. The characters are all more complicated than the list would indicate, however, and should be examined in detail.

Table 2: 'Lorisiform' characters in the Strepsirhini

	Lemuridae, Indriidae, Daubentoniidae	*Cheirogaleidae (Phanerinae)*	*Cheirogaleidae (Cheirogaleinae)*	*Lorisidae*
Tympanic ring	–	–	–	x
Os planum	–	–	x	x
Carotid arteries	–	x	x	x

1. The position of the tympanic ring, and its relation to the bulla, has been widely misunderstood or misinterpreted. Alan Walker has discussed the matter with the author, and permits the preliminary results of his researches to be mentioned here:

   (*a*) In the Malagasy lemurs the bulla is enlarged and inflated so that it overgrows the tympanic ring. The ring is therefore inside the bulla, and its inferior edge is separated from the bulla by a recess (except in *Microcebus*, according to Saban;[18] but Walker suggests that this latter condition may be an artifact of postmortem shrinkage of the extra-tympanic membrane).
   (*b*) In the Lorisidae the bulla is small and fails to overgrow the tympanic ring, coming into contact with it; and the latter grows laterally, forming the inferior margin of the mouth of the bulla.
   (*c*) In *'Galago' crassicaudatus* (probably forming genus *Otolemur*, see below), there is an intermediate condition: the bulla does not extend appreciably lateral to the tympanic ring, so there is little tympanic component in the bulla. This condition seems to occur in the *Mioeuoticus* cranium too.[19]

   This character therefore, though given great weight in most taxonomic schemes, is seen to have a very simple basis, being largely dependant on the size of the bulla itself. There is, however, a rather more fundamental difference, which Walker has demonstrated to me: in the Lorisidae, alone among the Strepsirhini, there is a partition within the bulla separating the hypotympanic cavity from the tympanic cavity proper; this occurs in the African Miocene crania as well.

2. The presence or absence of an ethmoid component (*os planum*) in the orbital wall is largely a matter of orbital frontality.[20] The Lemuridae and Indriidae lack it; the Lorisidae possess it; while the Cheirogaleidae are divided, *Phaner* lacking it and the other three genera possessing it. (Its evident existence in *Allocebus* has been ascertained from the holotype in the British Museum by Cartmill and myself.)

3. The most complex of the three characters, the disposition of carotid branches in the ear region, divides not the Lemuriformes from the Lorisiformes but the Cheirogaleidae and Lorisidae from the Lemuridae and Indriidae.[18] In the latter two families the common carotid artery divides into two major branches, the external and internal carotids (with the occipital artery arising from the same junction); the latter then, after giving off a 'rameau tubaire', enters the bulla via the carotid foramen, and on the cochlea it divides into a large stapedial and a small promontory branch. (The stapedial feeds the middle meningeal

arteries, while the promontory joins the *circulus arteriosus.*)

In the Cheirogaleidae and Lorisidae, on the other hand, the first junction gives off an additional artery, called the anterior carotid by Saban,[18] which runs forwards outside the bulla and enters the cranial cavity via a large foramen in the position of the human *foramen lacerum*, but undoubtedly not homologous with it (it is called the promontory foramen by Szalay).[21] The internal carotid enters the bulla and, as in the Lemuridae and Indriidae, divides on the cochlea into stapedial and promontory branches; but the stapedial is this time a very small artery – in fact it is absent in *Phaner* – while the promontory is large. The promontory does not enter the braincase right away, but leaves the bulla and goes through the promontory foramen along with the anterior carotid. The Lorisid-Cheirogaleid pattern therefore involves not an earlier stapedial-promontory branching as Le Gros Clark and others have stated (for this occurs in the usual place, *inside* the bulla), but the existence of an entirely new artery! It is tempting to identify this with the medial entocarotid – a primitive mammalian artery not otherwise recorded in primates – as do Van Valen and Szalay;[21, 22] but since its existence (as well as the changed pathway of the promontory) is closely bound to that of a promontory foramen, which is not present in fossil primates in which this region is discernable (*Plesiadapis, Phenacolemur, Adapis, Notharctus* etc.), this seems unlikely. It is best interpreted as a specialisation, probably derived from a modification of the 'rameau tubaire', which in the Lemurid type is a similar extrabullar branch of the internal carotid but in the Lorisid type arises from the anterior carotid.

It should finally be mentioned that the existence of the promontory foramen is associated also with a specialisation of the venous system, serving as the exit for veins draining the cavernous sinus into the external jugular vein.[18]

4. The final differentiating feature, vibrissal reduction, also falls down on closer examination.[8] The full complement of vibrissae – supraorbital, genal, mystacial, interramal and carpal – occurs in the Cheirogaleidae and in the lemurid genus *Varecia. Lemur* and *Hapalemur* lack only the interramal group. *Lepilemur*, the Indriidae and the Galaginae lack interramals and carpals. Finally the Lorisinae lack all except the mystacial group. There is therefore no clear-cut division that can be made.

The upshot of all this is that there is only one feature – bullar construction – in which a clear-cut Lemuriform/Lorisiform division exists; and no common specialisation uniting all the Malagasy lemurs can be found. However the Afro-Asian Lorisidae and the Malagasy

Cheirogaleidae do share a specialisation in the carotid system. Szalay underplays this resemblance, and argues that for mammals as a whole a tympanic external to the bulla is primitive;[21] but for lemurs, if they are truly derived from the Adapidae, it would have to be the other way around.

Zöo-geographic considerations

If, then, one can detect probable specialisations in common (synapomorphy) between two groups separated by the Mozambique channel, it is worth asking whether multiple invasions of Madagascar, or perhaps re-invasions of Africa from Madagascar, would have been possible. The mammalian fauna of Madagascar belongs to six orders: one, the Primates, is the order in question: another, the Chiroptera, is represented on the island only by worldwide genera. The other four are:

1. *Carnivora.* All Malagasy carnivores belong to the family Viverridae; they are assigned to seven genera, belonging to three subfamilies,[23] of which two (Cryptoproctinae and Galidictinae) are indigenous, the third (Hemigalinae) having members outside Madagascar – not indeed in Africa, but in South-East Asia! There is a suggestion of a close relationship between the Malagasy *Eupleres* and the Indochinese *Chrotogale* (Owston's civet). Later studies have extended this picture somewhat. Karyotype data[24] suggest that the Malagasy subfamilies are in fact closely interrelated and belong in a group with another Oriental subfamily, Paradoxurinae, contrasting with the true civets, Viverrinae, on the one hand, and the mongooses, Herpestinae, on the other. The most recent research finds that the Malagasy viverrids do in fact all form a homogeneous group, even with regard to the Oriental Hemigalines, and that one of the supposed hemigalines, *Fossa* (the Fanalouc) is the most primitive living form.[25]

2. *Rodentia.* After reviewing monophyletic and polyphyletic theories of the origin of Malagasy rodents, F. Petter concludes that they, like the carnivores, do derive from a close-knit common stock (probably among the Cricetidae), but have reached Madagascar via multiple invasions.[26]

3. *Lipotyphla.* Except for the ubiquitous *Suncus*, clearly of recent origin on Madagascar and probably introduced by man, all Malagasy insectivores belong to a single family, Tenrecidae, and to a single indigenous subfamily, Tenrecinae. The other subfamily, Potamogalinae, occurs on the mainland of Africa; a

morphological intermediate, *Protenrec*, is known from the early Miocene of East Africa.[27]

4. *Artiodactyla.* The African bush-pig is represented on Madagascar by a form currently classified as a distinct species, *Potamochoerus larvatus*, but probably conspecific with the East African *P. koiropotamus.*[28] It has evidently been there long enough to have formed subspecies, there being a large form (*edwardsi*) in the mountains, and two small ones (*larvatus* and *hova*) on the west and east coasts respectively. A hippopotamus, *H. lemerlei*, decidedly distinct from the African species, formerly lived on the island but is now extinct.

It would seem from the above survey that multiple invasions of Madagascar would have been possible, but at the same time all groups seem to have come from rather limited stocks. This would indicate that the island was populated from the nearest mainland – probably Africa – at a time when the latter was itself rather isolated, and had a restricted fauna. The absence of catarrhines, bovids, felids and so on may simply indicate that none were present in eastern Africa at the time; or else some factor such as size may have placed a limit on the fauna that was available for dispersal. The pre-Miocene fauna of Africa is poorly known, with the exception of the Oligocene Jebel-el-Qatrani formation of the Fayum, Egypt, whence remains are available of both Primates (primitive catarrhines, not strepsirhines) and rodents (family Phiomyidae, not closely related to the Cricetidae), but no viverrids; we have no means of knowing, of course, whether this was a pan-African fauna and the lemurs, cricetids and viverrids had not yet arrived, or whether it was in some way atypical of the more southern parts of Africa. Currently, however, the entry of viverrids and cricetids into Africa is provisionally dated to the end of the Oligocene.[29] A detailed consideration of the separation of Madagascar has recently appeared,[30] and an early date for separation is favoured, although a possible avenue of exchange may have remained open until the Oligocene.

Since the early Miocene *Mioeuoticus* had a lorisid bulla construction (see above), it would appear that whichever of the two types of bulla is derived from the other (probably the extra-bullar tympanic from the intra-bullar, see above) had already begun to evolve by then. This would put the cheirogaleid-lorisid divergence prior to 20 million years before present. The cheirogaleid-lorisid carotid type, and therefore the joint divergence of these two families from the lemurid-indriid stock, must therefore have been established even earlier.

Daubentonia

There is still one Malagasy 'lemur' which fits uneasily with the others, and indeed with the lemurs as a whole. This is *Daubentonia*. Many of its unique features can be explained as a consequence of its extraordinary incisor specialisation – high facial skeleton, large premaxillae, lack of interorbital depression[31] – and probably its brain form can be similarly explained;[32] but there are other features which certainly cannot be explained in this way:

1. In the orbit, there is a big maxillary plate which makes contact with the frontal wing, and no fronto-palatine suture or ethmoid exposure. This condition is similar to most of the Lipotyphla as well as to the Plesiadapidae and Adapidae, where this region has been preserved. It seems therefore to be a symplesiomorph primate feature. All living primates other than the aye-aye have either a long fronto-palatine suture, or else an ethmoid plate.

2. In the nasal cavity, *Daubentonia* has five endoturbinals and at least three ectoturbinals, as well as the single naso- and maxilloturbinals. The other lemurs all have four endo- and two ectoturbinals; while the Haplorhini have only two endo-, one ectoturbinal. In this morphological series, the aye-aye stands at the base of the primate stem and resembles the Lipotyphla.

3. In the claws (which occur on all digits except the hallux) the deep stratum comprises 46% of the thickness near the tip, and the terminal matrix giving rise to its comprises 40% of the total germinal matrix.[33] These proportions approach those of non-primates (marsupials, tree-shrews) at 75-79% and 73-81% respectively; lemurs (which have nails on all digits except the second toe) have neither deep stratum nor terminal matrix, not even on the 'toilet claw', the whole of the claw or nail being composed of superficial stratum, and the whole of the germinal matrix being basal matrix. The aye-aye claws cannot therefore be derived from the nails or claws of lemurs. (It is interesting that the lemurs resemble catarrhines in this, while tarsiers and platyrrhines retain at least a remnant of deep stratum and germinal matrix).[33,34]

4. In the muscular system there are numerous anomalies;[35] some are connected with the uniqueness of the third manual digit, but others cannot be so explained:

   (*a*) *M. trapezius* inserts on the occiput, and overlies the 'Acromio-trachelian' (the *M. levator claviculae* of apes, an anomaly in man), as it does in higher primates, but unlike any other lemur.

(*b*) *M. coraco-brachialis medius* originates from the coracoid itself as in higher primates, not from the tendon of of the short head of the *M. biceps brachii* as in other lemurs.
(*c*) *M. flexor digitorum sublimis* has a double origin like higher primates, with an ulnar head as well as the epitrochlear one found in lemurs.
(*d*) *M. obturator externus* is completely divided into two (except at the tendon of insertion), the two portions being separated by the obturator nerve. The overlying portion, essentially a separate muscle *(M. obturator intermedius),* occurs in rodents and insectivores but is quite absent from all other primates.
(*e*) In at least some specimens of the aye-aye, there are two *Mm. peroneo-tibiales*, an anomaly not seen in any other primates.

5. The carotid pattern is simpler than that of other lemurs, with an initial bifurcation into two branches, internal and external; the occipital artery then arises from the latter. This pattern is the one seen in tree-shrews, also in the orang utan and man; in other 'higher primates' the external carotid gives rise first of all to the posterior auricular, which in turn gives rise to the occipital.[18]
6. The mammae, two in number, are inguinal in position: unique among primates.
7. There is persistent oögenesis in the adult, as in the Lorisidae, but unlike any other Malagasy form.[36]

There are in this list a number of features in which the aye-aye resembles higher primates, and others in which it resembles a pre-primate mammalian condition. Sometimes, indeed, these features are the same, implying the existence of characters in which the 'Anthropoidea' have remained primitive while the lemurs as a whole (i.e. except *Daubentonia*) have acquired a specialised condition – a salutory antidote to the view of lemurs as 'lower primates'.

It is in fact none too easy to find specialisations that the aye-aye shares with other lemurs. The common possession of an epitheliochorial placenta may be a specialisation or it may not;[6,12] the M-loop of the colon ('ansa coli') is probably a specialisation;[4] and there appear to be common specialisations in the foot skeleton.[10] On the other hand there are features in common between the 'higher primates' and the remaining lemurs, which the aye-aye does not share: reduced deep stratum in nails and claws; olfactory reduction; pectoral mammae; simple form of *M. obturator externus*; and so on. The evidence is perfectly clear, in any case, that the aye-aye is the earliest lineage to split off from the Strepsirhine stem; it could even conceivably have been the first line of split off the primate stem as a whole, although it is still most probable that the initial division is the Haplorhine-Strepsirhine one.

The infraorders of Strepsirhini should therefore not be Lemuriformes and Lorisiformes, but Lemuriformes (including lorises) and Daubentoniiformes (Vallois, 1955 = Chiromyiformes Anthony and Coupin, 1931 = Chiromyoidea Pocock, 1918). Both are strongly specialised groups, the former with dental comb and toilet claw, the latter with its woodpecker-like functional complex.[31] Within the former group the superfamilies should be Lemuroidea (to include Lemuridae and Indriidae) and Lorisoidea (including Lorisidae and Cheirogaleidae), with the abandonment of any grouping that attempts to separate the Lorisidae from the Malagasy lemurs as a whole.

The phylogenetic scheme on which this classification is based is figured here in Fig. 1. It will be seen that Madagascar is shown to have been invaded three times by strepsirhine primates. An alternative would be to suppose that Madagascar was the original home of the Strepsirhini – their presence there and that of the rodents and viverrids dating perhaps from its original connection with India[37] – and that the presence of Lorisidae in Africa and Asia is due to one single dispersal event from Madagascar to Africa.

Taxonomic problems within families

1. Lorisidae

Traditionally this family has been divided into Galaginae (for the galagos) and Lorisinae (for the slow climbers). Not only, however, are there large differences between the African and Asiatic sections of the Lorisinae,[38] but there are even some indications of specialisations shared by the pottos and galagos but lacking in the lorises.[39] It is possible that the two lorisine groups have converged in their locomotor anatomy.

Within the Galaginae, studies in progress show that the genus *Euoticus* is highly distinctive and should be retained: although not unique in possessing 'needle-claws' (*Galago inustus* also has them), its dentition, skull form, body build, colour pattern and mammae (1 pair, pectoral, as opposed to 3 pairs, the norm in all other galagos and in pottos) set it well apart. Though restricted to a relatively small part of the Central African forest bloc, there are two very well-marked geographical forms (north-west and south-east of Mt. Cameroun), which are distinguishable by dental characters as well as by size and sharply distinct colour-patterns.

The next most distinctive form is *Galago crassicaudatus*, which can be separated as the genus or subgenus *Otolemur* (see Table 3). It appears to share features with the pottos (karyotype, serological reactions, skin structure) which the other galagos lack, and may perhaps be related to the stock which gave rise to them;[40] in this context its 'intermediate locomotion patterns' and the possible existence of a glandular area on the scrotum, like pottos, are particularly interesting.[40,41] Among its own unique features its peculiar nails, with their concave free edges, are especially interesting

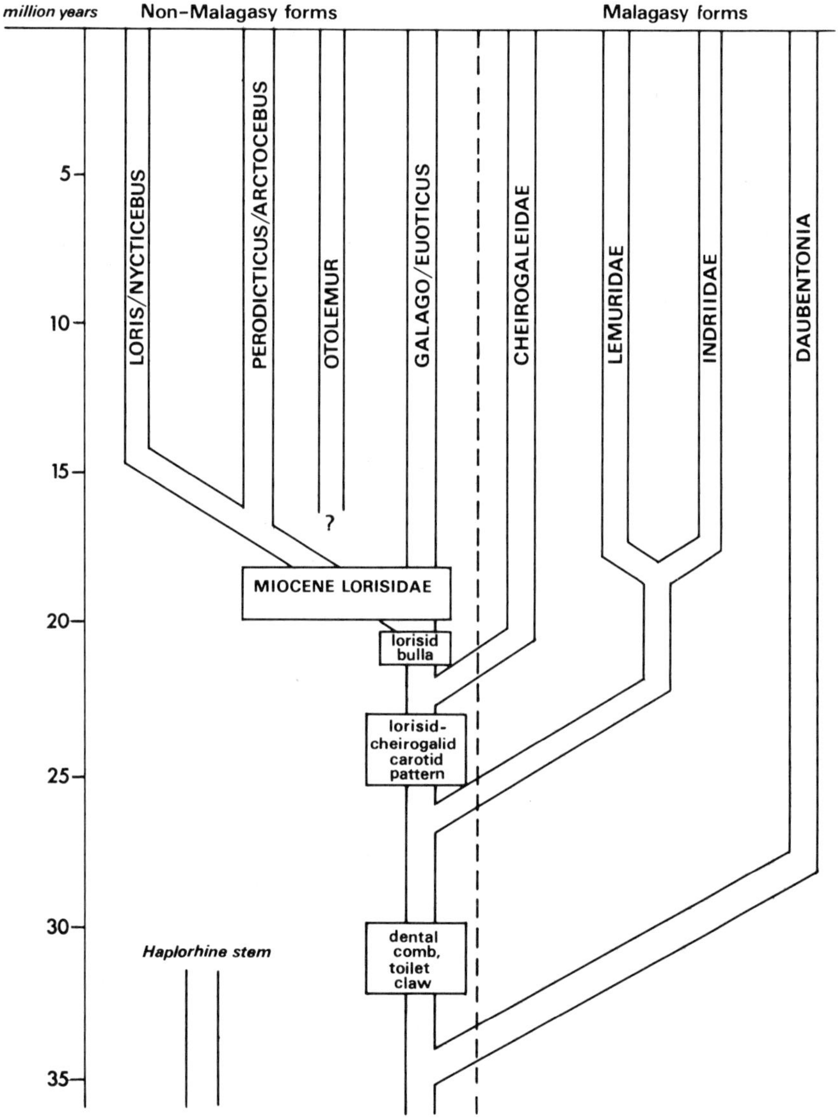

Figure 1. Suggested interrelationships and times of divergence of major strepsirhine groups. Taxa and lineages to the left of the dotted line are Afro-Asian; those to the right of it, Madagascan.

and are as yet unexplained. It is possible that there are more than one species in the *Otolemur* group: the small East African coastal races (*panganiensis, lasiotis* etc.) are very different animals from the much bigger south-eastern *crassicaudatus* and its relatives with their very long ears, while the beautiful silvery *monteiri* (= *argentatus*), also very large, from Angola and Zambia and the Western Rift lakes region as far north as Lake Victoria, tending to melanism in some parts of its range, is different yet again.

The other galagos form a close-knit group of at least 5 species:

(*a*) *G. demidovii* occurs, like the potto, throughout the forest bloc of Africa from Senegal to Uganda. From Senegal east to about Mt. Cameroun occur small races of about 115-130 mm. head and body length (N.B. these are approximately 95% limits, not observed ranges), in two colour phases, greyish and red. From Cameroun to south of the Congo River is found a larger type, about 130-145 mm., more of a grey-brown colour but with occasional foxy-red individuals. North and east of the Congo and in Uganda is found the distinctive light brownish form *thomasi*, which goes up to 160 mm. in length.

(*b*) *G. senegalensis* replaces *demidovii* in the bush-savannah belt of Africa. It is larger on average, greyish in colour, with lighter (yellow to white) limbs and underparts. The biggest subspecies (head and body 160-190 mm.) are the northern series, found from Senegal to Ethiopia, Kenya and northern Tanzania. South of about Dodoma a very different type is found, perhaps representing a different species from which the prior name is *moholi*: much smaller (130-160 mm., no bigger than *G. demidovii thomasi*), grey with pinkish hues, the face white with striking dark eye-rings. There are also slight differences in the skull and teeth.

(*c*) *G. zanzibaricus.* The apparent coexistence in the same small forest in the Ulugurus (Tanzania) of the most Senegal-like race of Demidoff's bushbaby (*G.d. orinus*) with the most Demidoff-like race of the Senegal bushbaby (*G.s. zanzibaricus*)[42] has been recently called into question by Kingdon,[43] who suggests that they are actually the young and the adult respectively of a distinct species, *G. zanzibaricus*, whose range interdigitates with that of *senegalensis* which it replaces in forested coastal areas. It is about the same size as *G.s. moholi* or *G.d. thomasi* – 130-160 mm. – with cranial and dental characters intermediate between the two, but nearer to Demidoff's; it is yellow-brown to olive in colour with lighter coloured limbs and a brown tail; the hindlimbs are not as strongly elongated as those of *G. senegalensis*, the intermembral index being about 65 like *demidovii* and *alleni*, not 50-55 like *senegalensis.* In Fig. 2 the apparent distribution is mapped, on the basis of specimens in the British Museum and of Kingdon's records; the Mozambique *granti* seems likely to be a southerly representative of *zanzibaricus*, and has been so mapped. It will be seen that the distribution is more or less continuous down the coast, with inland outliers: Mt. Marsabit, Ugogo, and Ukinga. (Lawrence and Washburn record three specimens of *G.d. thomasi* from Ukinga,[42] while a dot in approximately this position is marked on Kingdon's map for *zanzibaricus*; whether Kingdon is thereby implying that the three '*thomasi*', like the type of *orinus*, are young *zanzibaricus*, is uncertain. Provisionally, both species are mapped here in Fig. 2).

(*d*) *G. inustus* is not, as claimed by some authors[8,42] an eastern representative of *Euoticus*, but despite its needle-claws is a forest version of *senegalensis.* From the latter it differs (apart from the shape of its nails) in its very dark colour, shorter hindlimbs and larger size (head and body 190-200 mm.).

(*e*) *G. alleni* is another, but rather more distinctive, forest version of *senegalensis.* Like *Euoticus* and *Arctocebus* it is restricted to the forest between the Niger and the Congo; unlike them it appears to be a more recent immigrant and does not show such striking geographic variation north-west and south-east of Mt. Cameroun, varying only in size (165-192 mm. in Nigeria, 205-230 mm. in Gabon).

2. *Cheirogaleidae*

That the dwarf lemurs have nothing in common with the Lemuridae has long been recognised,[8,44] and it is pleasant to find that this situation has at last been formally recognised by the recognition of a full family for them,[45] quite independently of the present contribution.

A new genus, *Allocebus*, has recently been added to the family.[46] Within the genus *Microcebus*, the species *M. murinus* apparently contains at least two species (a larger grey one weighing 75 g., and a smaller red one weighing 55 g.) which are sympatric in some areas;[47] for the vastly bigger *M. coquereli* (weighing 385 g.) the name *Mirza* Gray, 1870, is available (Table 3) if subgeneric division is required, although according to Petter (personal communication) it is really quite close to the small species.

The genus *Phaner* is somewhat isolated from the rest of the family: it lacks an os planum in the orbit,[20] has no stapedial artery,[18] and differs greatly in its foot structure and in its behaviour.[47] Subfamilial separation seems perfectly justified.[45]

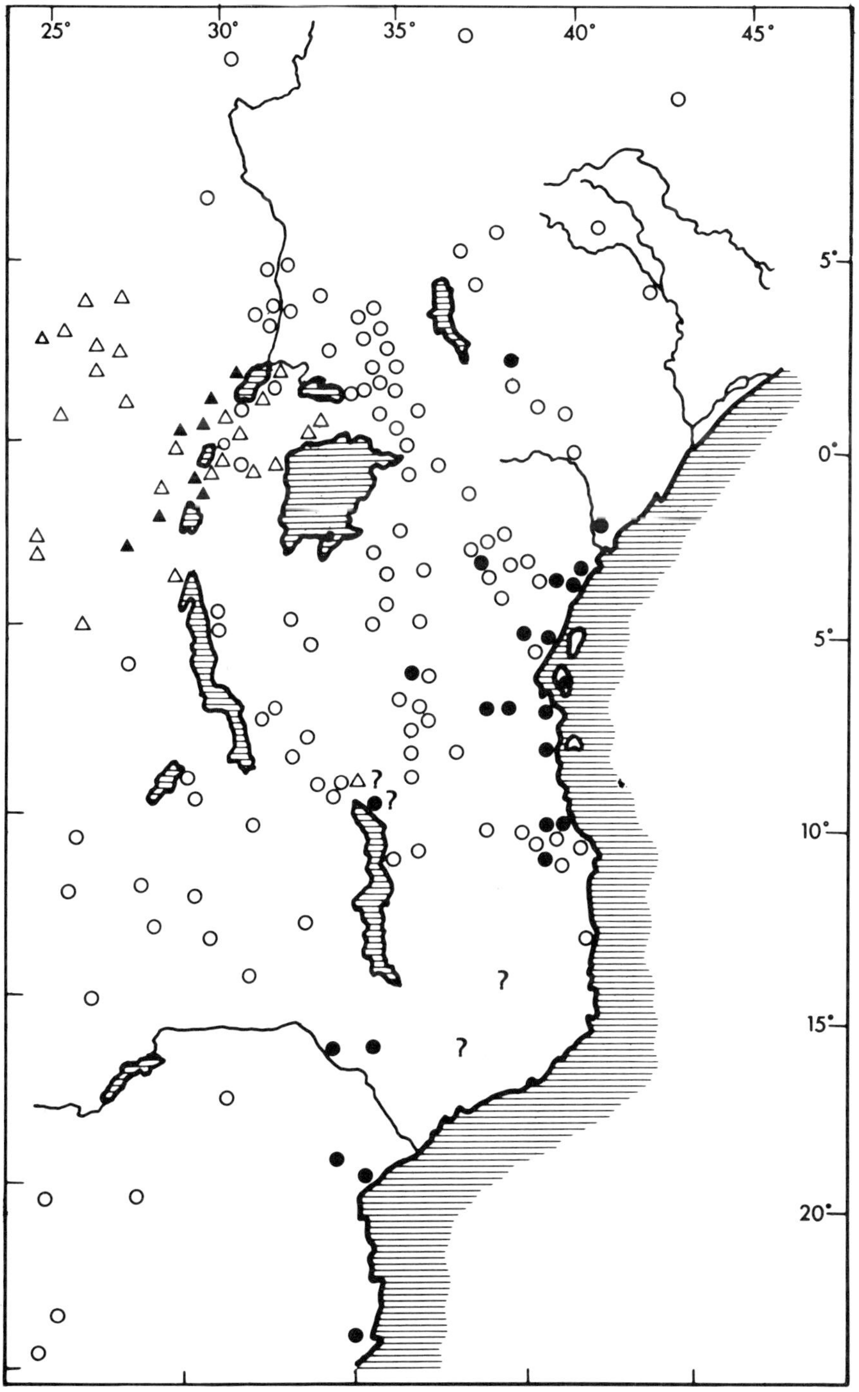

O *G. senegalensis* ▲ *G. inustus* ● *G. zanzibaricus* △ *G. demidovii*

Figure 2. Distribution of members of the subgenus *Galago* in East Africa. Note that the identification of certain specimens from Ukinga (northern shore of Lake Malawi) is equivocal (see text).

Table 3: Generic names available in polytypic genera (as defined in Napier and Napier),[5] in the Strepsirhini

GALAGO	
G. alleni	*Sciurocheirus* Gray, 1872
G. senegalensis	*Galago* E. Geoffroy, 1796; *Otolicnus* Illiger, 1811; etc.
G. demidovii	*Galagoides* A. Smith, 1833; *Hemigalago* Dahlbom, 1857
G. crassicaudatus	*Otolemur* Coquerel, 1859; *Callotus* and *Otogale* Gray, 1863
G. elegantulus	*Euoticus* Gray, 1863
CHEIROGALEUS	
C. major	*Cheirogaleus* E. Geoffroy, 1812; *Cebugale* Lesson, 1840; etc.
C. medius	*Opolemur* Gray, 1872; *Altililemur* Elliot, 1912
C. trichotis	*Allocebus* Petter-Rousseaux and Petter, 1967
MICROCEBUS	
M.m. rufus	*Microcebus* E. Geoffroy, 1828; *Azema* Gray, 1870
M.m. murinus	*Scartes* Swainson, 1835; *Myscebus* Lesson, 1840
M. coquereli	*Mirza* Gray, 1870
HAPALEMUR	
H. griseus	*Hapalemur* I. Geoffroy, 1851
H. simus	*Prolemur* Gray, 1871
LEMUR	
L. catta	*Lemur* Linnaeus, 1758; *Procebus* Storr, 1780; etc.; *Odorlemur* Bolwig, 1958
L. mongoz	*Prosimia* Boddaert, 1785 (here designated as genotype)
L. variegatus	*Varecia* Gray, 1863

3. Lemuridae

As in the case of full family status for the Cheirogaleidae, the apartness of *Lepilemur* has long been hinted at,[44] but quite how extraordinary it is has only recently become apparent.[48] It is aptly removed to a subfamily of its own.[45] The remaining members of the subfamily are currently classed in the two genera *Lemur* and *Hapalemur*; to these at least one more, *Varecia*, should be added.[44] Additionally, there is the question of the relationships of *Lemur catta* to be considered: it shares characters of both skin and karyotype with *Hapalemur griseus*,[49,50] differing from other species of *Lemur*. Formerly it was thought that the little-known *Hapalemur simus* differed strongly from *H. griseus*, especially in lacking brachial and antebrachial glands; but it has recently been discovered[51] that it does in fact possess the glands, though very small and much nearer the elbow. (A diligent search on skins in the British Museum did successfully reveal the brachial gland; so inconspicuous is it that until my attention was directed to it (Petter, personal communication) I had not suspected its existence although perfectly familiar with the skins!)

Future studies may reveal whether these glands, and the other features that are in common, are primitive lemurine features that the *macaco*-group have lost, or whether they are common specialisations of *L. catta* and *Hapalemur*. If the latter, then the animals in question should be associated taxonomically at a closer level than to the *macaco*-group.

Within *Hapalemur*, the Lake Alaotra form is distinct and awaits formal diagnosis;[52] within *Varecia*, the problem of the red form, whether colour phase, subspecies or species, is still unsolved.[44]

The name *Prosimia* is available for the *mongoz-macaco-rubriventer* group of the genus *Lemur*, dating from Boddaert, 1785 (*Elenchus Animalium*, 43 and 65). It is first mentioned on p.43; further on, on p.65, the following species are listed within it: *mongoz, macaco, catta, volans* (i.e. *Galeopithecus*!), *spectrum* (i.e. *Tarsius*) and *minima* (i.e. *Microcebus*). The action of Gray[53] in excluding *L. catta* from this genus precludes the selection of the latter as genotype, and leaves the name available in the sense in which it is used by Osman Hill[8] and in the present survey, with *mongoz* as genotype. Provisionally, it will be ranked as a subgenus. The number of species that may be recognised is arguable. Study of the skins in the British Museum (Natural History), American Museum of Natural History, and Smithsonian Institution yields the following results:

(*a*) *Lemur (Prosimia) mongoz* group. These are distinguished primarily by the white facial 'mask', but are not easy to define by comparison with the next group. The recognisable forms, probably subspecies, are:

1. *mongoz*. Found north-east and south-west from Maroantsetra. Male reddish on head and cheeks; foreparts reddish, hindparts grey. Female light grey foreparts, yellow-red hind parts; cheeks not red. Both sexes have yellow belly, white chest.
2. *albimanus*. Comoro Is. Cheeks maroon-red; body maroon-grey, underparts yellower.
3. *coronatus*. Found north from Vohémar. Male grey, often grizzled or mottled, with a V-shaped red mark on crown, which is black behind; the red extends to the cheeks. Female is like *mongoz*, but with less contrast between grey foreparts and red hindparts, and has male's red chevron but no black on crown. Underside is all yellowish white.

(*b*) *Lemur (Prosimia) macaco* group. These generally have a black facial mask. The map (Fig. 3) is compiled from records of specimens on museum labels, and from the literature, especially Jolly and Schwarz;[54,55] it shows some apparent distributional overlaps, suggesting, as Hill opines,[8] that more than one species may be involved – a conclusion already hinted at by chromosome studies.[56] It should be noted, however, that if one follows Hill in separating *macaco* from the rest, then the name *fulvus* cannot be used for the composite species, as it is by no means the earliest available name. The valid names (as far as I can ascertain) in this group are, in order of priority: *macaco* Linnaeus, 1766; *albifrons* E. Geoffroy, 1796; *rufus* Audebert, 1799; *fulvus* and *collaris* E. Geoffroy, 1812; *mayottensis* Schlegel, 1866; *cinereiceps* Grandider and Milne-Edwards, 1890; *sanfordi* Archbold, 1932. It will be seen from the map that the forms *macaco-sanfordi-albifrons-rufus-cinereiceps-collaris* replace each other around the island starting at Majunga on the north-west coast and going north-east, then down the east coast. But *fulvus* appears to overlap with *rufus* in the west (Majunga district) and with *macaco* in the east (Tamatave district), with no sign of intermediacy. Moreover, *rufus* appears to approach *macaco* at Tamatave, and there may well be a ring-species effect here. On the other hand, specimens in the American Museum of Natural History appear to represent intergrades in certain cases: between *albifrons* and *rufus* at Maroansetra, and between *rufus* and *cinereiceps* at Ankazoabo; so some of the forms, at least, are conspecific.

1. *macaco.* Male black with maroon tints; female reddish-yellow becoming greyer with age, facial mask red. The typical form with long ear-tufts is found over most of the range marked on the map, while the form described as *flavifrons* (female) or *nigerrimus* (male), lacking ear-tufts, 'inserts' into this range at Maromandia.
2. *sanfordi.* Greyish, browner on back; head darker with a light grey zone on forehead. Sexes alike, except that male has ear-tufts. This form is unlike either of its geographical neighbours, most closely resembling *fulvus* in colour.
3. *albifrons.* Wood-brown, the tail becoming black towards the tip; hair round ears and on cheeks slightly elongated. Male is darker grey-brown, nearly black in midline of back; whole head, except facial mask, whitish; grey-white below; crown of head darker than body. Female is lighter, chocolate-grey-brown; crown darker, nearly black; supra-orbital region dark grey; light brown below, or orange; interramal space white.
4. *rufus.* Like the last, sexually dichromatic – but each sex has two varieties which do not seem to be geographic. In male, Variety A is grey, yellow below; Variety B is lighter, browner, and off-white below. Both have big white supraorbital spots, and the crown may be reddish or

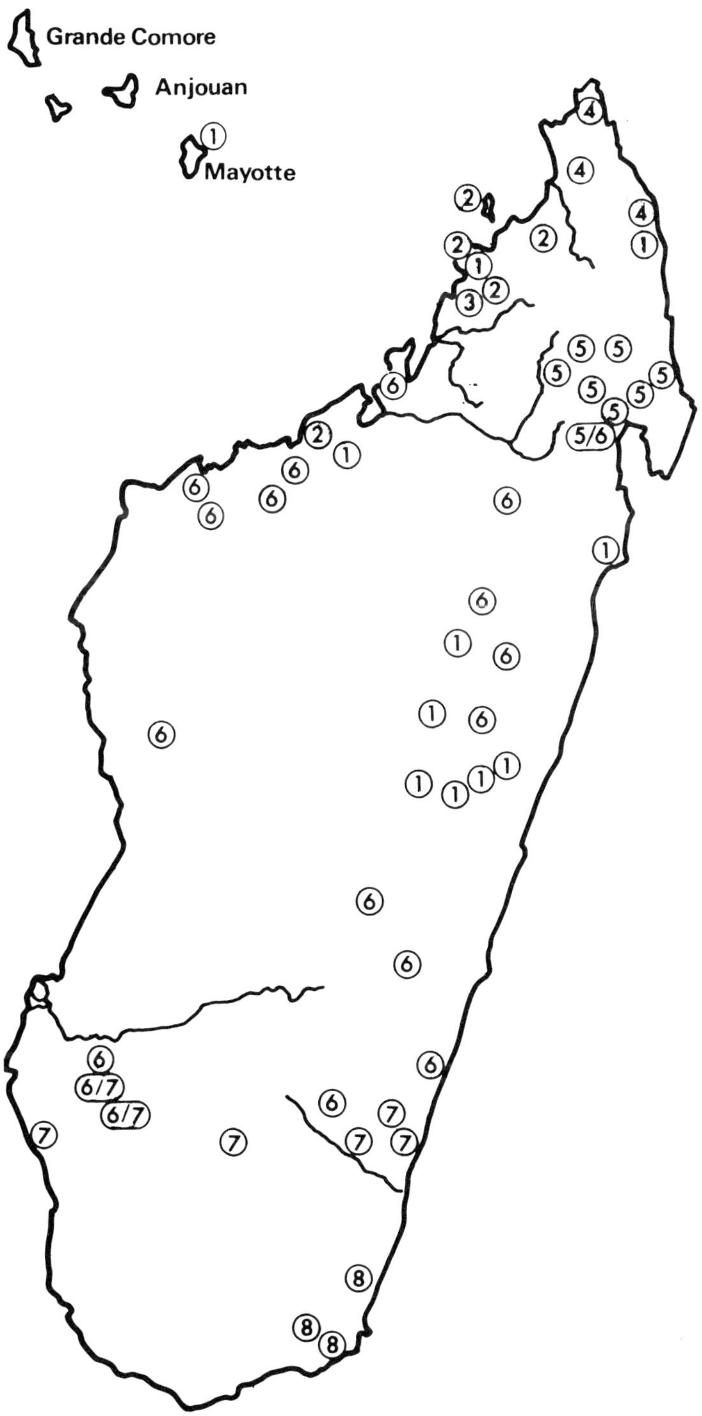

Figure 3. Distribution of members of the *Lemur macaco* group in Madagascar.

1. *fulvus* (? = *mayottensis*)
2. *macaco*
3. *flavifrons* (a localised variant of *macaco*)
4. *sanfordi*
5. *albifrons*

5/6. intergrade between *albifrons* and *rufus*

6. *rufus*

6/7. intermediate between *rufus* and *cinereiceps*

7. *cinereiceps* (the white-cheeked variant of *collaris*)
8. *collaris*

blackish. In female, Variety A is grey-red, becoming very red on underside and rump, with some red on cheeks; Variety B is foxy-red, yellower below, greyish crown.

5. *cinereiceps.* This has the supraorbital spots in the male like *rufus.* The male is yellow-grey or yellow-brown, with hindparts redder; the crown blackish; underside light red-brown; cheeks white. Female is greyish, becoming suffused with yellow on the foreparts and redder towards the hindparts; hindlimbs are red-yellow; underparts yellow; crown dark grey; face light grey.

6. *collaris.* Like the last, but with cheeks orange-red instead of white. The two have not usually been separated, but museum skins show the geographic separation. According to Petter (personal communication) the relatively slight differences between this race and the last may well reflect the situation elsewhere, where neighbouring populations thus far included under one head in fact can be distinguished by small colour attributes.

7. *fulvus.* Grey-brown, slightly redder in midback region. Ochery underparts; tail-root black; a black cap on crown; grey supraorbital spots. Not sexually dichromatic. Jolly says there are two isolated populations of this type, an eastern one (round Perinet) which are darker, with lighter white 'beards', and a western (around Ankarafantsika) which are grizzled, lighter grey, with tan or creamy beards.[54] All specimens in museums seen by me are of the eastern type, and agree approximately with Jolly's description. She comments that the existence of these two isolated populations may account for the reported chromosome polymorphism and also for the so-called *Lemur* sp. nov. of Chu and Swomley;[56,57] but Egozcue (personal communication) has informed me that this latter animal was in fact *collaris.*

8. *mayottensis.* This appears, though on the basis of only two specimens, to be but feebly differentiated from the last. One of the two falls in the range of variation of *fulvus*, the other is more intensely brown in colour.

(*c*) *Lemur (Prosimia) rubriventer.* The most distinctive of the three species-groups; it has a wide distribution, but no evident geographical variation.

The above notes may perhaps prove of use in working towards a total reassessment of the interrelationships of all lemurs. It will be seen that, while special studies on individual species and genera are still urgently needed, wide-ranging studies across the whole strepsirhine suborder are vital to a proper understanding of relationships.

Acknowledgments

I would like to acknowledge some very valuable discussions held during the course of the conference. Particular thanks for discussion and information are due to J.-J. Petter; additionally, to F.-K. Jouffroy, J. Egozcue, F.S. Szalay and W.W. Bishop. I would also like to acknowledge fruitful discussions on this and many other occasions with Robert Martin and Alan Walker.

NOTES

1 Simpson, G.G. (1945), 'The principles of classification and a classification of mammals', *Bull. Amer. Mus. Nat. Hist.* 85, 1-350.

2 Hill, W.C.O. (1953), *Primates: Comparative Anatomy and Taxonomy*, 1. *Strepsirhini*, Edinburgh.

3 Romer, A.S. (1967), *Vertebrate Paleontology*, 3rd ed., Chicago.

4 Le Gros Clark, W.E. (1959), *Antecedents of Man*, London.

5 Napier, J.R. and Napier, P.H. (1967), *Handbook of Living Primates*, London.

6 Luckett, W.P., this volume.

7 Pocock, R.I. (1918), 'On the external characters of lemurs and of *Tarsius*', *Proc. Zool. Soc. Lond.* 19-53.

8 Hill, W.C.O. (1955), *Primates: Comparative Anatomy and Taxonomy*, 2. *Tarsioidea*, Edinburgh.

9 Simons, E.L., this volume.

10 Szalay, F.S., in discussion.

11 Martin, R.D. (1968), 'Towards a new definition of Primates', *Man*, n.s., 3, 377-401.

12 Martin, R.D. (1968), 'Reproduction and ontogeny in tree-shrews (*Tupaia belangeri*) with reference to their general behaviour and taxonomic relationships', *Z.f. Tierpsychol.* 25, 409-95.

13 Albignac, R., Rumpler, Y. and Petter, J.J. (1971), 'L'hybridation des lémuriens de Madagascar', *Mammalia* 35, 358-68.

14 Gray, A.P. (1972), *Mammalian Hybrids: a Checklist with Bibliography*, Slough.

15 Heptner, V.G., Nasimowitch, A.A. and Bannikov, A.G. (1966), *Die Säugetiere der Sowiet-Union*, 1: *Artiodactyla and Perissodactyla*, Jena.

16 Darlington, P.J. (1957), *Zoogeography*, New York.

17 Hill, W.C.O. (1936), 'The affinities of the Lorisoids', *Ceylon J. Sci.*, B, 19, 287-314. See also Charles-Dominique, P. and Martin, R.D. (1970), 'Evolution of Lorises and Lemurs', *Nature* 227, 257-60.

18 Saban, R. (1963), 'Contribution à l'étude de l'os temporal des Primates', *Mem. Mus. Nat. Hist. Nat. Paris*, ser. A, 29, 1-378.

19 Walker, A., this volume.

20 Cartmill, M. (1971), 'Ethmoid component in the orbit of primates', *Nature* 232, 566-7.

21 Szalay, F.S. (1972), 'Cranial morphology of the Early Tertiary *Phenacolemur*, and its bearing on Primate phylogeny', *Amer. J. Phys. Anthrop.* 36, 59-75.

22 Van Valen, L. (1965), 'Treeshrews, primates and fossils'. *Evolution*, 19, 137-151.

23 Petter, G. (1961), 'Le peuplement en carnivores de Madagascar'. In 'Problèmes actuels de paléontologie (évolution des vertbrés)', *Coll. Int.. CNRS* 104, 331-42.

24 Wurster, D.H., and Benirschke, K. (1968), 'Comparative cytogenetic studies in the order Carnivora', *Chromosoma (Berlin)* 24, 336-82.

25 Albignac, R. (1972), 'Ethoécologie des Carnivores malgaches', in press.

26 Petter, F. (1961), 'Monophylétisme ou polyphylétisme des Rongeurs malgaches: problémes actuels de paléontologie', *Coll. Int. CNRS* 104, 301-10.

27 Butler, P.M. (1969), 'Insectivores and bats from the Miocene of East Africa, in Leakey, L.S.B. (ed.), *Fossil Vertebrates of Africa*, 1, London.

28 Mohr, E. (1960), *Wilde Schweine*, Wittenberg-Lutherstadt, 83-4.

29 Cooke, H.B.S. (1968), 'Evolution on Mammals on Southern Continents: 2. The fossil mammal fauna of Africa', *Quart. Rev. Biol.* 43, 234-64.

30 Walker, A. (1972), 'The dissemination and segregation of early primates in relation to continental configuration', in Bishop, W.W. and Miller, J.A. (eds.), *The Calibration of Hominoid Evolution*, Edinburgh.

31 Cartmill, M., this volume.

32 Radinsky, L., this volume.

33 Le Gros Clark, (1936), 'The problem of the claw in primates', *Proc. Zool. Soc. Lond.* 2, 1-24.

34 Thorndike, E.E. (1968), 'A microscopic study of the marmoset claw and nail', *Amer. J. Phys. Anthrop*, 28, 247-53.

35 Jouffroy, F.K. (1962), 'La musculature des membres chez les lémuriens de Madagascar', *Mammalia*, A, 3860.

36 Petter-Rousseaux, A. and Bourlière, F. (1965), 'Persistence du phénomène d'ovogénèse chez l'adulte de *Daubentonia madagascariensis*', *Folia primat.* 3, 241-4.

37 McElhinny, M.W. (1970), 'Formation of the Indian Ocean', *Nature* 228, 977-9.

38 Groves, C.P. (1971), 'Systematics of the genus *Nycticebus*', *Proc. 3rd. Int. Congr. Primat. Zürich 1970*, 1, 44-53.

39 Series: 'The skin of primates', see in particular Montagna, W. and Ellis, R.A., 'The skin of the potto', *Amer. J. Phys. Anthrop.* 17, 137-62; Montagna, W., Yasuda, K. and Ellis, R.A., 'The skin of the slow loris', *ibid.* 19, 1-22; Montagna, W. and Yun, J.S., 'The skin of the great bushbaby, *ibid.* 20, 149-66.

40 Bearder, S.K. and Doyle, G.A., this volume.

41 Tandy, J., this volume.

42 Lawrence, B. and Washburn, S.L. (1936), 'A new eastern race of *Galago demidovii*', *Occas. Papers Boston Soc. N.H.* 8, 255-66.

43 Kingdon, J. (1971), *Atlas of East African Mammals*, 1, London.

44 Petter, J.J., (1962), 'Recherches sur l'écologie et l'éthologie des lémuriens malgaches', *Mém. Mus. nat. Hist. nat. Paris,* sér. A, 27, 1-146.

45 Rumpler, Y., this volume.

46 Petter-Rousseaux, A. and Petter, J.J. (1967), 'Contribution à la systématique des Cheirogaleinae (Lémuriens malgaches): *Allocebus*, gen. nov., pour *Cheirogaleus trichotis* Günther 1875', *Mammalia* 31, 574-82.

47 Petter, J.J., Schilling, A. and Pariente, G. (1971), 'Observations éco-éthologiques sur deux Lémuriens malgaches nocturnes: *Phaner furcifer* et *Microcebus coquereli*', *Terre et Vie* 3, 287-327.

48 Charles-Dominique, P. and Hladik, C.M. (1971), 'Le *Lepilemur* du Sud de Madagascar: écologie, alimentation et vie sociale', *Terre et Vie* 3, 3-66.

49 Montagna, W. and Yun, J.S. (1962), 'The skin of primates: 10. The skin of the ring-tailed lemur', *Amer. J. Phys. Anthrop.* 20, 95-118.

50 Egozcue, J. (1967), 'Chromosome variability in the Lemuridae', *Amer. J. Phys. Anthrop.* 26, 341-8.

51 Petter, J.J., in press.

52 Petter, J.J. and Peyriéras, A. (1970), 'Observations éco-éthologique sur les lemuriens malgaches du genre *Hapalemur*', *Terre et Vie* 3, 356-82.

53 Gray, J.E. (1870), *Catalogue of Monkeys, Lemurs and Fruit-eating Bats in*

the Collection of the British Museum, London.
54 Jolly, A. (1967), *Lemur behaviour*, Chicago.
55 Schwarz, E. (1931), 'A revision of the genera and species of Madagascan Lemuridae', *Proc. Zool. Soc. London.* 399-428.
56 Chiarelli, B., this volume.
57 Chu, E.H.Y. and Swomley, B.A. (1961), 'Chromosomes of lemurine lemurs', *Science* 133, 1925-6.

W. P. LUCKETT

The phylogenetic relationships of the prosimian primates: evidence from the morphogenesis of the placenta and foetal membranes

Introduction

Comparative studies have demonstrated that the developmental pattern of the mammalian foetal membranes and placenta can provide considerable insight into the phylogenetic relationships among higher taxonomic categories.[1] Certain foetal membrane characteristics have undergone relatively little change during the evolutionary history of mammals, as suggested by their great similarity within the higher categories. Such conservative characters are the best kind for revealing phylogenetic relationships, because they are easier to trace across the wide gaps in the evolutionary record.[2] The conservative nature of the foetal membrane characteristics may be due in part to their relative isolation from the selective effects of the external environment,[1] but it is also likely that intense centripetal selective forces are acting upon them to maintain a relatively constant pattern.[3] There is no clear evidence that body size, locomotor activity or feeding habits have had any pronounced effect on the type of placentation that develops within any order of eutherian mammals.

The fossil record alone provides the essential time dimension for directly observing evolutionary changes, despite its limited nature. However, broad comparative analyses of characteristics in living animals provide additional insight into possible phylogenetic relations, particularly when an attempt is made to determine the primitive and specialised (derived) features of these characters. Convergence is less likely to occur in those characters or character complexes that are the product of the interaction of an extensive amount of genetic information.[4] The mammalian foetal membranes form such an interacting complex; they are derived from all of the embryonic germ layers and their interrelationships can be followed throughout their entire life history.[5] The terms 'primitive' and 'specialised' are relative; characteristics of common ancestry are primitive within a particular taxon, whereas others are more or less specialised in proportion to their departure from the ancestral condition.[6] Speculations concerning the primitive and specialised features of the developmental pattern of the foetal membranes in eutherian mammals are based upon a consideration of their nature in reptiles, birds, prototherians, marsupials, and all the orders of eutherian mammals.

It should be emphasised that mammalian classifications ought to be based upon a consideration of all of the available evidence, and the foetal membrane characters discussed here comprise only a portion of this evidence. The comparative development of the foetal membranes and placenta in primates has recently been discussed in detail,[3] and the present report will be limited to a summary of the main features of the foetal membranes in prosimian primates, and to a discussion of the possible relationships among the three major living taxa that have been included by many authors within the Prosimii: the Tupaiidae, Strepsirhini (=Lemuroidea and Lorisoidea), and the Tarsioidea.

Implantation

The early uterine blastocyst is fundamentally similar in all eutherian mammals prior to implantation; it consists of a peripheral epithelial layer of *trophoblast* surrounding a fluid-filled cavity, and a small cluster of *inner cell mass* attached to the inner surface of the trophoblast at one pole of the blastocyst (Fig. 1). A flattened epithelial layer of primitive endodermal cells becomes segregated from the internal surface of the inner cell mass, and this layer subsequently spreads peripherally to line the internal surface of the blastocyst in pre-implantation stages of tupaiids, strepsirhines, and tarsiids. Among the strepsirhines, the initial attachment of the bilaminar blastocyst is non-invasive in the lorisines *Nycticebus* and *Loris* and in the lemurid *Microcebus*, whereas there is a limited and variable amount of invasive activity by the trophoblast during the early stages of attachment in the two galagine species that have been examined (Figs. 2, 3, 4). Implantation in *Galago senegalensis* is central, and the invasive giant cell trophoblast of the attachment site is localised and transitory.[7] The invasive giant cell trophoblast degenerates by the time of the early primitive streak stage, and there is a complete replacement of the uterine epithelium at the previously denuded attachment site. The further development of the foetal membranes and placenta in this species is essentially identical to that of the lorisines.

The localised invasive attachment of the blastocyst in *Galagoides*

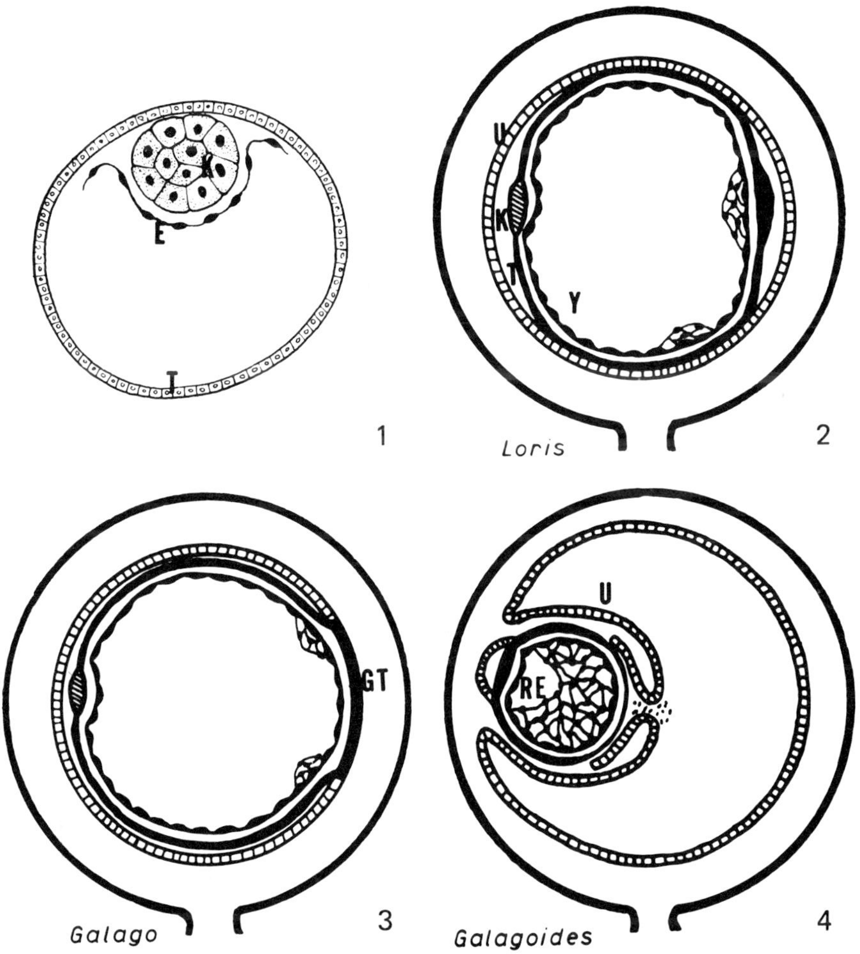

Figure 1. A schematic figure of a generalised eutherian blastocyst, prior to implantation, with peripheral trophoblast (T), inner cell mass (K), and endoderm (E).

Figure 2. Early implanted blastocyst of *Loris.* In this and all remaining figures the mesometrium of the uterus is oriented toward the bottom. Uterine epithelium (U), embryonic knot (K), trophoblast (T), yolk sac (Y).

Figure 3. Early implanted blastocyst of *Galago senegalensis*, with invasive giant cell trophoblast (GT).

Figure 4. Early implanted blastocyst of *Galagoides demidovii*, with reticulated endoderm (RE) in the yolk sac.

demidovii occurs within an implantation chamber that forms as the result of the closure of annular folds of the uterine mucosa about the blastocyst (Fig. 4). However, this condition is not strictly homologous to the interstitial implantation characteristic of the Hominoidea. The invasive activity of the para-embryonic trophoblast is more persistent than in *Galago senegalensis*, and it apparently contributes to the layer of giant cell trophoblast in the intimate zone of the definitive placenta,[8] although the critical intermediate stages are lacking. It seems likely that the implantation specialisations in *Galago senegalensis* and *Galagoides demidovii* have evolved independently from a non-invasive and centrally implanted blastocyst similar to that of *Nycticebus*, *Loris*, and *Microcebus*. However, further discussion of the significance of these specialisations is unwarranted without a knowledge of the mechanism of implantation in other species of the subfamily Galaginae.

Initial attachment of the blastocyst in *Tarsius* elicits a peculiar reaction in the necks of the underlying uterine glands, and this results in their proliferation and the formation of a unique nodular mass of modified epithelium (Fig. 5).[3] Subsequently, there is a proliferation of the trophoblast at the localised attachment site to form a trophoblastic preplacenta whose interstices soon become filled with maternal blood. This developmental relationship foreshadows the differentiation of the definitive haemochorial placenta in later stages, and this contrasts sharply with the condition in strepsirhines. Strepsirhines and *Tarsius* share a number of presumed primitive primate features, including the exposure of the embryonic disc to the contents of the uterine lumen by the rupture of the overlying layer of trophoblast, the participation of the para-embryonic trophoblast in the initial attachment, and the lateral orientation of the embryonic disc to the endometrium at the time of implantation (Figs. 2, 5). The first two of these features also appear to be primitive eutherian characteristics.

The tupaiid blastocyst implants superficially and bilaterally on specialised, gland-free endometrial pads, and the exposed embryonic disc is oriented antimesometrially (Fig. 6). Excluding the specialised nature of the endometrial pads, the relationships of the early implanted tupaiid blastocyst are quite similar to those of the insectivore families Soricidae and Talpidae. It is possible that these resemblances are due to primitive retentions in all three families.[9]

Development of the primitive streak

The initial differentiation of the primitive streak in strepsirhine primates and tupaiids follows the pattern characteristic for most other mammalian taxa. The earliest mesoderm derived from the primitive streak is axial (embryonic), and only later do cells of

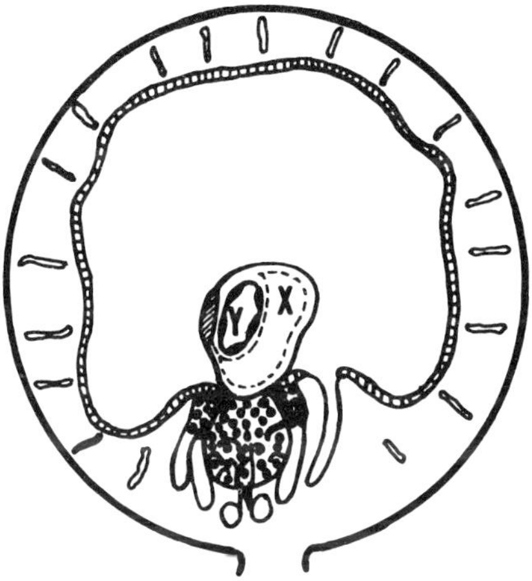

Tarsius 5

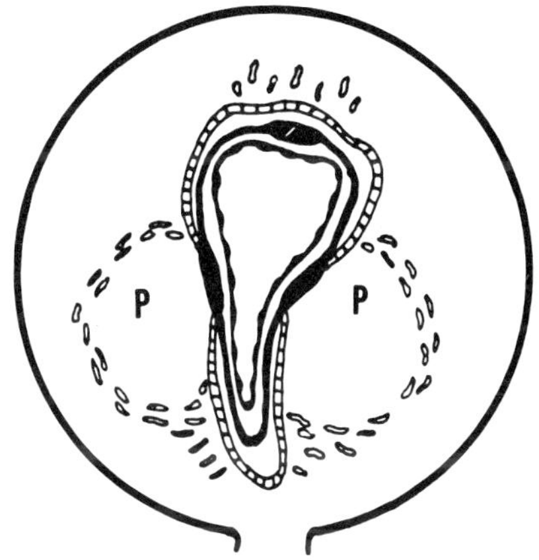

Tupaia 6

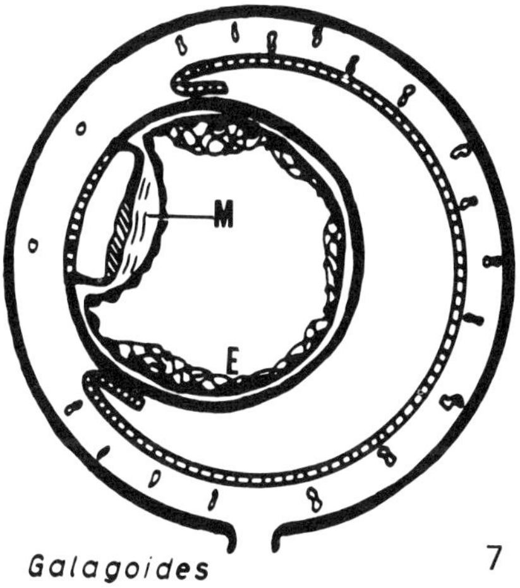

Galagoides 7

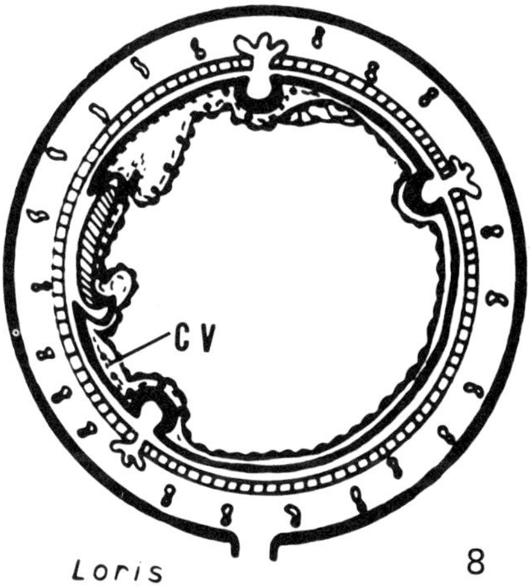

Loris 8

Figure 5. Early implanted blastocyst of *Tarsius*, with exocoelom (X), and nodular epithelial mass (N).

Figure 6. Early implanted blastocyst of *Tupaia*, with bilateral endometrial pads (P).

Figure 7. Early primitive streak stage of *Galagoides demidovii*, with embryonic mesoderm (M).

Figure 8. 12-14 somite stage of *Loris*, with amniotic folds and the development of a chorio-vitelline placenta (CV).

primitive streak origin spread beyond the margins of the embryonic disc to form the extra-embryonic mesoderm (Fig. 7). By contrast, there is a precocious differentiation of the caudal margin of the primitive streak in *Tarsius,*[10] so that extra-embryonic mesoderm is formed before the appearance of embryonic mesoderm (Fig. 5). This precociously differentiated extra-embryonic mesoderm gives origin to the mesodermal lining of the exocoelom and to the primordium of the body stalk, prior to the differentiation of more cranial portions of the streak. This specialised pattern of primitive streak formation is strikingly similar in *Tarsius* and the Anthropoidea (Platyrrhini and Catarrhini),[3] and it is correlated with the development of the mesodermal body stalk characteristic of all these taxa.

Development of the yolk sac and chorio - vitelline placenta

The migration of extra-embryonic mesoderm between the trophoblastic and endodermal layers of the strepsirhine and tupaiid blastocyst and the subsequent differentiation of blood vessels in this mesoderm result in the development of a chorio-vitelline placenta during the 12-14 somite stage (Figs. 8, 9). The chorio-vitelline placenta apparently functions as a site of foetal-maternal exchange until it is displaced topographically by the spread of the exocoelom and the allantoic mesoderm; this occurs at variable times after the 30 somite stage. The development of a vascular chorio-vitelline placenta at the equatorial pole of the blastocyst appears to be a primitive characteristic of therian mammals. It forms the definitive placenta for the great majority of marsupials that have been examined, as opposed to its transitory nature in eutherians.

The interior of the early implanted blastocyst of *Galagoides demidovii* is partly filled with a cellular meshwork of reticulated endoderm,[8] and the interstices of the meshwork are in continuity with a more distinct, eccentric primary yolk sac cavity (Fig. 4). The reticulated meshwork of endoderm becomes reduced as the expanding blastocyst increases in size. Most of the abembryonic endoderm is reduced to a simple squamous layer by the time of the early primitive streak stage, with only small isolated patches of persisting reticulated endoderm (Fig. 7). The continued expansion of the *Galagoides* blastocyst results in the rupture of the 'capsular' endometrium overlying its abembryonic pole. Thus, the abembryonic wall of the blastocyst secondarily acquires a relationship with the endometrium, similar to that seen in comparable stages of other strepsirhines. The subsequent development of the chorio-vitelline placenta in *Galagoides demidovii* appears to be identical to the general strepsirhine conditon.

The precocious development of extra-embryonic mesoderm and

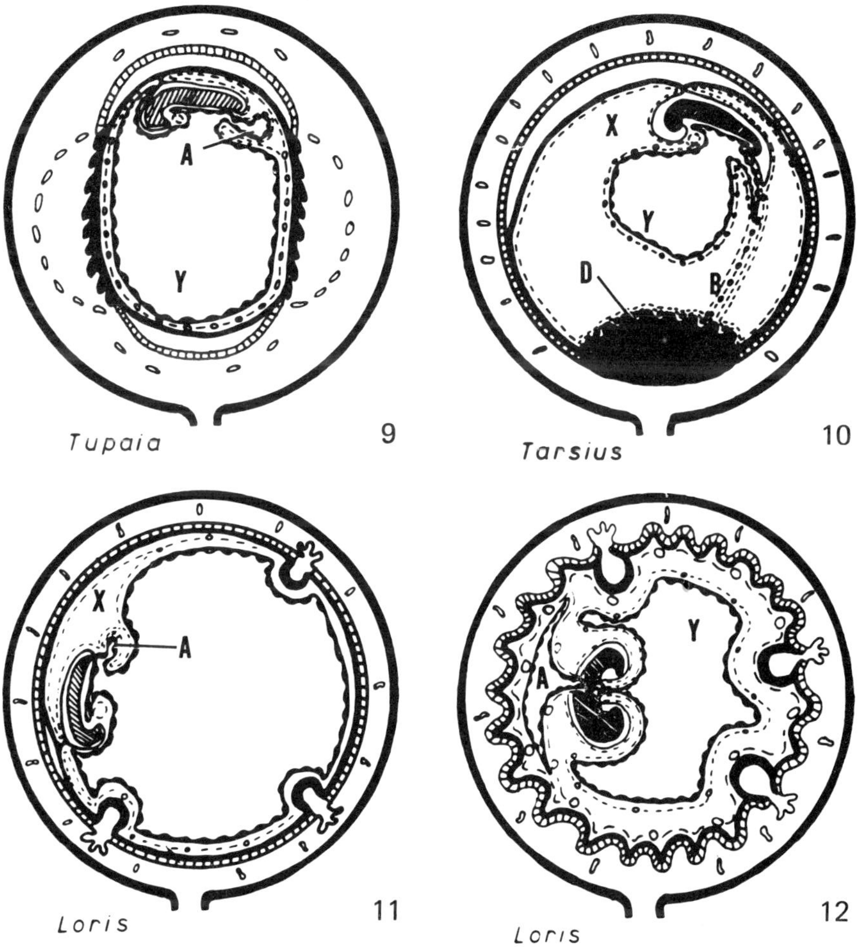

Figure 9. 23 somite stage of *Tupaia*, with chorio-vitelline placenta, recently fused amniotic folds, and allantoic diverticulum (A).

Figure 10. 17 somite stage of *Tarsius*, with free vascular yolk sac (Y), tubular allantoic diverticulum, recently fused amniotic folds, mesodermal body stalk (B), and early stage in the development of the definitive placental disc (D).

Figure 11. 22 somite stage of *Loris*, with chorio-vitelline placenta, recently fused amniotic folds, and allantoic diverticulum.

Figure 12. 39 somite stage of *Loris*, with free vascular yolk sac, allantoic vesicle, and the complete vascularisation of the chorion by allantoic mesoderm.

the exocoelom in *Tarsius* results in the separation of the endodermal yolk sac from its contact with the adjacent chorion prior to the development of blood islands in the splanchnic mesoderm of the yolk sac (Fig. 5). Therefore, a vascular chorio-vitelline placenta (that is, the fusion of the vascular yolk sac with the chorion) does not occur in *Tarsius*. When vitelline vessels subsequently develop in later stages, the vascular splanchnopleure of the yolk sac lies in contact with the large fluid-filled exocoelom (Fig. 10). It seems likely that the free vascular yolk sac of *Tarsius* plays an important role in foetal nutrition by its absorption of exocoelomic fluid, prior to the establishment of the chorio-allantoic placenta. The relationships of the yolk sac and exocoelom in *Tarsius* thus differ sharply from the condition in strepsirhine Primates, whereas the tarsiid condition is essentially identical with the pattern in platyrrhines and catarrhines.[3]

Allantoic diverticulum

The allantoic primordium arises as an endodermal diverticulum from the caudal end of the developing hindgut in strepsirhine and tupaiid embryos that bear about 13-20 pair of somites (Figs. 9, 11). The allantoic diverticulum, surrounded by a globular mass of highly vascularised mesoderm, projects freely into the exocoelom at this stage. The continued expansion of the allantoic vesicle and the exocoelom results in the displacement of the vascular splanchnopleure of the yolk sac from its contact with the chorion. The fusion of the vascular allantois with the chorion results in the replacement of the chorio-vitelline placenta, both topographically and functionally, by the developing chorio-allantoic placenta. Following the fusion of the allantois with the chorion in strepsirhines, the allantoic mesoderm rapidly spreads beyond the margins of the allantoic vesicle to vascularise the entire inner surface of the chorion (Fig. 12). The further expansion of the allantoic vesicle and its subsequent lobulation appear to be secondary phenomena.

The precocious spread of allantoic mesoderm has been interpreted as the initial step in the evolution of the mesodermal body stalk in haplorhine Primates (Tarsioidea, Platyrrhini, and Catarrhini).[11] However, it is equally plausible that this pattern of chorionic vascularisation is a secondary specialisation, because the chorio-allantoic placenta of strepsirhines is not established at a relatively earlier developmental stage than it is in tupaiids.[3] The expansion of the allantoic vesicle in tupaiids is directly involved with the vascularisation of the chorio-allantoic placenta; this appears to represent a primitive retention of the ancestral eutherian condition that is shared with carnivores, ungulates, and some insectivores.

The precocious development of a mesodermal body stalk between the caudal margin of the embryonic shield and the chorion in early primitive streak embryos of *Tarsius* occurs prior to the differentiation of the allantoic diverticulum. Nevertheless, the precociously developed extraembryonic mesoderm of the body stalk is clearly homologous to the allantoic mesoderm of strepsirhines and most other mammals.[11] Both are derived from the most caudal portion of the primitive streak and are ultimately associated with the allantoic diverticulum, and both regions serve as the pathway for the distribution of allantoic vessels to the chorio-allantoic placenta. The allantoic diverticulum of *Tarsius* invaginates the proximal portion of the body stalk in presomite stages; it retains a thin tubular appearance and never reaches the surface of the developing placental disc (Fig. 10).

The specialised pattern of body stalk and allantoic development in *Tarsius* is strikingly similar to the condition in the Anthropoidea, and differences in the length of the body stalk in the two groups are believed to be due to basic differences in the method of implantation.[3] The evolution of the haplorhine body stalk from the vesicular allantoic condition is considered to be the result of developmental acceleration in order to facilitate the vascularisation of the chorion at the earliest possible moment.[11] This concept is supported by a comparison of the initial vascularisation of the chorion in strepsirhines and *Tarsius*. The chorion is vascularised by the yolk sac mesoderm during the 12-14 somite stage in strepsirhines (as in tupaiids), prior to the differentiation of the allantois, whereas the initial vascularisation of the chorion in *Tarsius* is effected by umbilical vessels in the body stalk (allantoic) mesoderm during the 9-10 somite stage.[3] Thus, the process of chorionic vascularisation in *Tarsius* is clearly accelerated, as it is in other haplorhines, and a chorio-vitelline stage is bypassed.

Chorio-allantoic placenta

The fusion of the allantoic vesicle with the chorion and the vascularisation of the chorion by allantoic mesoderm initiate the differentiation of a non-invasive, diffuse, epitheliochorial placenta in all families of strepsirhine Primates. Much of the surface of the chorion differentiates into simple and branched villi that interdigitate with endometrial septa. There are two primary mechanisms involved in foetal-maternal exchange in strepsirhines: (1) exchange between the foetal and maternal capillaries across the intervening layers of the epitheliochorial placenta; and (2) absorption of uterine gland secretions by specialised chorionic vesicles or by less specialised 'bare areas' of the chorion. Histochemical evidence indicates that mucopolysaccharides and iron are absorbed by the trophoblastic lining of the chorionic vesicles in *Galago senegalensis*,[12] and this is doubtless true for strepsirhines in general.

In addition to the diffuse, epitheliochorial portion of the placenta in *Galagoides demidovii*, there is a small specialised zone overlying the site of the initial fusion of the allantoic vesicle with the chorion. Foetal and maternal tissues adhere more intimately in this zone, and a layer of columnar, pale-staining giant cells lies between the cuboidal layer of chorionic cytotrophoblast and the maternal stroma. Although the critical intermediate stages in the formation of this intimate zone are lacking, it seems most probable that the layer of giant cells is of trophoblastic origin.[3,8] If this is true, then the specialised intimate zone in *Galagoides* exhibits an endotheliochorial relationship, because maternal capillaries are in intimate apposition with the giant cell layer. Other authors have suggested that the layer of giant cells may represent hypertrophied uterine epithelium, and this region of the placenta would then be only a specialised epitheliochorial zone.[12,13]

A localised, discoidal, haemochorial placenta develops in *Tarsius*, as it does in all haplorhine primates, in contrast to the diffuse epitheliochorial condition in strepsirhines. Mesoderm derived from the body stalk invaginates the foetal surface of the trophoblastic preplacenta, and the appearance of blood cells within the allantoic vessels of these mesodermal villi during the 9-10 somite stage initiates the functional stage of chorio-allantoic placentation in *Tarsius*. Further growth and rearrangement result in the development of a continuous system of trophoblastic-covered foetal trabeculae which surround a lacunar system of circulating maternal blood. The definitive placenta of *Tarsius* is therefore labyrinthine and haemochorial.

The definitive chorio-allantoic placenta of tupaiids develops bilaterally and utilises the specialised gland-free sites of the initial attachment and chorio-vitelline placentation. The uterine epithelium and some of the underlying maternal stroma is destroyed by the trophoblast, but the endothelium of maternal vessels remains intact where it comes into contact with the advancing trophoblast. Recent re-examinations demonstrated that the maternal endothelium of tupaiids remains intact during the differentiation of the definitive chorio-allantoic placenta, and the latter is classified as bidiscoidal, labyrinthine, and endotheliochorial.[13,14]

Amniogenesis

In all prosimian primates and tupaiids the amnion is formed by the elevation of somatopleuric folds at the margins of the embryo proper and by the subsequent fusion of these folds over the dorsal aspect of the embryo (Figs. 8, 9, 10). Following this fusion, the amnion and chorion became separated by the exocoelom, although the site of fusion may persist for some time as a chorio-amniotic raphe.

Amniogenesis by folding appears to be the primitive condition in mammals and in other amniotes. The more specialised pattern of amniogenesis by cavitation, characteristic of Anthropoidea, is probably related to the attachment of the blastocyst to the endometrium by the trophoblast that persists over its embryonic pole.[3]

Phylogenetic considerations

The evidence for including the Tupaiidae within the order Primates has been based primarily upon certain resemblances in their brain, visual system, auditory bulla, and placentation.[15] However, resemblances may be due to homology, parallelism or convergence, and recent re-examinations of all of these tupaiid features suggest that their resemblances to Primates are the result of convergence or the retention of primitive eutherian characters.[16] The only foetal membrane characteristics shared by tupaiids and strepsirhines are those that are retentions of the primitive eutherian condition; these features are also shared with carnivores, some insectivores, and ungulates.[9] Immunological comparisons of serum proteins also support the removal of the Tupaiidae from any special relationship with the Strepsirhini.[17]

Many of the similarities between strepsirhine and tarsioid Primates may be due to the retention of primitive primate or eutherian features, and it is possible that these resemblances reflect a general prosimian *grade* of organisation when compared with a more advanced simian grade. These grade levels of organisation in Primates are generally considered to also represent a basic cladistic division into the suborders Prosimii and Anthropoidea.[6] However, the characteristics that have been used to evaluate primate subordinal relationships do not necessarily reflect the most conservative features, and an examination of other characteristics within the prosimian grade reveals that the Tarsioidea show closer affinities with the Anthropoidea than with the Strepsirhini.[3]

Table 1 summarises the major features in the development of the foetal membranes and placenta among the higher taxonomic categories of Primates. The developmental pattern of the Tarsioidea is most similar to that of the Anthropoidea, particularly in the specialised interrelationship between the yolk sac and allantois (including body stalk). The vascularisation of the chorion by either the yolk sac or the allantois is essential for the functional development of the mammalian placenta, and there is probably a fundamental interrelationship in their developmental patterns. Intense centripetal selective forces doubtlessly act to maintain a delicate balance of functional interrelationship between the yolk sac and the allantois during ontogeny.[3] The nature of the yolk sac and the allantois are two of the most conservative features during the evolution of the mammalian foetal membranes,[1] and the specialised nature of their developmental pattern in the Tarsioidea and Anthropoidea provides good evidence of affinities between these taxa. On the other hand, the similarities between *Tarsius* and the strepsirhines with respect to the initial attachment pole of the blastocyst and the mechanism of amniogenesis appear to be primitive primate and eutherian features; in this respect they have no bearing on the taxonomic relationships of these two groups.

The evidence from the developmental pattern of the foetal membranes supports the subordinal division of the Primates into Strepsirhini and Haplorhini. The foetal membrane evidence also supports the concept,[18] based primarily on dental similarities, that certain Eocene tarsiiforms may have been ancestral to the Anthropoidea. Immunodiffusion studies utilising an antiserum prepared against the serum proteins of *Tarsius* also suggest that *Tarsius* shared a more recent common ancestry with the Anthropoidea than with either the Lemuroidea or Lorisoidea.[19]

Table 1 Major features of the foetal membranes and placenta of Primates

	Strepsirhini	*Tarsioidea*	*Anthropoidea*
Implantation:			
Lateral Orientation of the Disc	x	x	x
Attachment Trophoblast:			
Para-embryonic Pole	x	x*	
Embryonic Pole			x
Amniogenesis:			
Folding	x	x*	
Cavitation			x
Yolk Sac:			
Large, Free, Reduced Later	x		
Small, Free		x	x
Chorio-vitelline Placenta:			
Present	x		
Absent		x	x
Allantoic Vesicle:			
Large, Permanent	x		
Small, Rudimentary		x	x
Chorio-Allantoic Placenta:			
Diffuse, Epitheliochorial	x		
Discoidal, Haemochorial		x	x

* It has been suggested that the differences in initial attachment and amniogenesis between Tarsioidea and Anthropoidea are the result of the development of a simplex uterus in Anthropoidea.[3]

Meaningful evolutionary classifications cannot be based upon the nature of any single feature or character complex, and it is essential that features of the dentition, nervous system, skeleton and other systems be examined on a broad comparative basis in order to determine their primitive and specialised features. Such determinations would contribute considerably to our understanding of the interrelationships among Primates at the higher taxonomic levels.

NOTES

1 Mossman, H.W. (1937), 'Comparative morphogenesis of the fetal membranes and accessory uterine structures', *Contrib. Embryol.* 26, 129-246; Mossman, H.W. (1953), 'The genital system and the fetal membrances as criteria for mammalian phylogeny and taxonomy', *J. Mammal.* 34, 289-98.

2 Farris, J.S. (1966), 'Estimation of conservatism of characters by constancy within biological populations', *Evolution* 20, 587-91.

3 Luckett, W.P. (in press), 'Comparative development and evolution of the placenta in Primates', in Luckett, W.P. (ed.), *Reproductive Biology of the Primates: Contributions to Primatology*, vol. 3, Basel.

4 Simpson, G.G. (1964), 'Organisms and molecules in evolution', *Science* 146, 1535-8.

5 Mossman, H.W. (1967), 'Comparative biology of the placenta and fetal membranes', *in* Wynn, R.M. (ed.), *Fetal Homeostasis*, vol. 2, New York.

6 Simpson, G.G. (1961), *Principles of Animal Taxonomy*, New York.

7 Butler, H. (1964), 'The reproductive biology of a strepsirhine (*Galago senegalensis senegalensis*)', *Int. Rev. Gen. and Exp. Zool.* 1, 241-96.

8 Gérard, P. (1932), 'Etudes sur l'ovogénèse et l'ontogénèse chez les lémuriens du genre *Galago*', *Archs. Biol. Paris* 43, 93-151.

9 Luckett, W.P. (1969), 'Evidence for the phylogenetic relationships of tree shrews (family Tupaiidae) based on the placenta and foetal membranes', *J. Reprod. Fert. Suppl.* 6, 419-33.

10 Hubrecht, A.A.W. (1902), 'Furchung und Keimblattbildung bei *Tarsius spectrum*', *Verh. K. Akad. Wet.* 2, (8), 1-113; Hill, J.P. and Florian, J. (1963), 'The development of the primitive streak, head-process and annular zone in *Tarsius*, with comparative notes on *Loris*', *Bibl. Primat.* 2, 1-90.

11 Hill, J.P. (1932), 'The developmental history of the primates', *Phil. Trans. Roy. Soc.* 221, 45-178.

12 Butler, H. and Adam, K.R. (1964), 'The structure of the allantoic placenta of the Senegal bush baby (*Galago senegalensis senegalensis*)', *Folia Primat.* 2, 22-49.

13 Hill, J.P. (1965), 'On the placentation of *Tupaia*', *J. Zool.* 146, 278-304.

14 Luckett, W.P. (1968), 'Morphogenesis of the placenta and fetal membranes of the tree shrews (family Tupaiidae)', *Am. J. Anat.* 123, 385-428.

15 Le Gros Clark, W.E. (1962), *The Antecedents of Man*, 2nd ed., Edinburgh.

16 van Valen, L. (1965), 'Treeshrews, primates, and fossils', *Evolution* 19, 137-51; Campbell, C.B.G. (1966), 'The relationships of the tree shrews: The evidence of the nervous system', *Evolution*, 20, 276-81; Noback, C.R. and Shriver, J.E. (1966), 'Phylogenetic and ontogenetic aspects of the lemniscal systems and the pyramidal system', in Hassler, R. and Stephan, H. (eds.), *Evolution of the Forebrain*, Stuttgart; Campbell, C.B.G. (1969), 'The visual system of insectivores and primates', *Ann. N.Y. Acad. Sci.* 167, 388-403.

17 Goodman, M. (1967), 'Deciphering primate phylogeny from macromolecular specificities', *Am. J. Phys. Anthrop.* 26, 255-76.

18 Simons, E.L. (1963), 'A critical reappraisal of Tertiary Primates', in

Buettner-Janusch, J. (ed.), *Evolutionary and Genetic Biology of Primates*, vol. 1, New York; Simons, E.L. (1969), 'The origin and radiation of the Primates', *Ann. N.Y. Acad. Sci.* 167, 319-31.

19 Goodman, M. et al., this volume.

L. R. WOLIN

What can the eye tell us about behaviour and evolution? or: *The aye-ayes have it, but what is it?*

Introduction

Primates are predominantly visual animals in that they appear to rely more heavily on vision than on the other sense modalities in guidance of behaviour. While the primate visual system may not be superior in all aspects to that of other vertebrate orders, the overall functional capacity of this system, in combination with the advanced development of the primate brain, results in a highly effective system.

In spite of the primacy of vision in primate behaviour, only relatively few species have been studied in any detail, and extensive study has been limited to only a few selected species of higher primates. With regard to the evolutionary aspect of vision in primates, virtually nothing is known, and the several aberrant genera which have aroused particular interest are more sources of contention than of enlightenment. The sub-order Prosimii is even less understood. To my knowledge, only *Tupaia* has been subject to anatomical, physiological and behavioural study with regard to vision.

The purpose of this paper will be to summarise what is known about certain characteristics of the prosimian eye and to attempt to relate this information to what is known about certain aspects of the behaviour of prosimians. Some comment will also be made with respect to our knowledge of the eye and behavioural patterns, as they may bear on our understanding of prosimian evolution. Specifically, with respect to the eye, I will deal with the following: pupil, tapetum, pigment layer, neuro-retinal structure and fundus patterns. In so far as behaviour is concerned, light sensitivity (nocturnal vs. diurnal habits), acuity and colour vision will be considered. Finally, some comments will be made regarding our need for information in each of these areas, so that those who attempt to repeat this task in the future may meet with somewhat less frustration than I have.

The pupil

Prince dealt with the relationship of pupillary patterns and visual function.[1] He notes that 'there are a surprising number of variations' of pupil shapes 'the reasons for which are not always immediately obvious'.

Among the prosimians, we find two major divisions with respect to pupil shape; circular and elongated. Within the second group there are variations which may be described as oval and elliptical, with variations in horizontal vs. vertical orientation as well. The following observations are based on my own examination of the eyes of a number of prosimian species and a survey of photographs of many species published in a variety of texts on vision, primatology and mammalian taxonomy.[2-8]

It has sometimes been assumed that a circular pupil is associated with a diurnal life style, whereas elongated or slit pupils are associated with nocturnal habits. There appears to be little relationship, however, between pupil shape and activity patterns. Among the prosimians, *Tupaia*, which is diurnal in habit, has an elongated (oval) pupil which is essentially horizontal in orientation. This is probably functional in terms of the lateral setting of the eyes and the temporal location of the central retinal area. Most of the Lorisiformes have elongated pupils, and most are vertically oriented. *Nycticebus coucang*, however, has a round pupil which is capable of contracting almost to pinpoint size. All of the Lorisiformes are reported as nocturnal. Most of the Lemuriformes about which information is available appear to have round pupils. Prince describes the crown lemur (= *L.mongoz coronatus*) as having an oval pupil with vertical orientation. *L. mongoz* is diurnal and probably also crepuscular in habit. However, *Indri*, which is diurnal in habit, has a round pupil. *Tarsius syrichta* and *Tarsius spectrum*, both of which are nocturnal, have horizontally oriented oval pupils, while *Microcebus*, which is also nocturnal, has a round pupil.

While it is obvious that information regarding pupil shapes among prosimians is scanty, it nevertheless appears reasonable to conclude that (as with other vertebrate orders) pupil shape shows no systematic relationship to observed living habits or behaviour. I am also inclined to suggest that pupil shape will not be found to be of any major significance in attempting to describe prosimian evolution.

Tapetum and pigment epithelium

The tapetum and pigment layers are both associated with light sensitivity. The tapetum is regarded as increasing light sensitivity by reflecting light back through the receptors, while pigments located just behind the neuroretina presumably absorb light. Many nocturnal species and some species categorised as crepuscular in habit have a tapetum.[2,9,10] Most diurnal primates do not have a tapetum. The diurnal Tupaiiformes appear not to have any tapetum, while the Lorisiformes, all of which are reported as being nocturnal, appear to have a tapetum. There are, however, some interesting exceptions. The tarsiers appear to have no tapetum,[10] although they are nocturnal. Among the Lemuriformes, *Propithecus* and *Lemur catta*, both reportedly diurnal in habit, have tapeta.[10] I should add that the tapetum of *L. catta* is *at least* as highly reflective as that of any Lorisiform which I have examined.

Information regarding retinal pigmentations is perhaps slightly less confusing, but data is also somewhat sparse. In general, it may be said that the pigment deposit tends to be somewhat greater in the diurnal than in the nocturnal species and that there is usually relatively less pigmentation in the centre than in the periphery of the retina.[9,10] Where a tapetum is found, no pigment or only very slight pigmentation occurs over the tapetum, but there may be relatively dense pigmentation at the periphery.

It thus appears that ocular characteristics such as the existence of a tapetum and the occurrence of retinal pigment cannot tell us, with any certainty, about behaviour. It may well be that the apparent discrepancies between anatomy and behaviour noted above may have some evolutionary significance in terms of changing patterns of adaptation; but in the absence of any information regarding the ancestors of living species, interpretation of these data may be extremely difficult.

Neuroretina

The neuroretina, containing the light-sensitive receptors and various transmitting and integrating elements, should show some more direct relationship to visual capacity and to visual guided behaviour. Definitive statement of such relationships will, however, be dependent on more detailed study of anatomy, physiology and behaviour, than are presently available. Certain areas of investigation, such as the detailed study of synaptic relationships within the reticular layers of the retina have only recently begun.

1. Receptors

The duplicity theory of vision based on the distinction between rods as low light level, achromatic receptors and cones as higher intensity chromatic receptors, appears to be well established. Walls, however, states that 'as a guide to the interpretation of habits in terms of 'structure', the Duplicity Theory, faced with transmutation, has come to the end of its usefulness'.[11] This statement is based upon the conclusion that many species have become nocturnal secondarily and that accordingly cones have in effect been 'transmuted' into functional rods.

Among the prosimians, rods are predominant. Only *Tupaia* and *Urogale* are reported as having essentially a pure cone retina.[10,12] Both genera are diurnal,[2,6] and *Tupaia glis* is reported to have colour vision.[13] No reports on the retina of *Ptilocerus* appear to be available, though Polyak states that this genus is nocturnal.[6] The tarsiers appear to have pure rod retinas,[10,14] and are nocturnal in habit. No information was uncovered regarding acuity or colour vision of this group. The uniformly nocturnal Lorisformes have predominantly rod retinas, but Kolmer reports some cones in *Nycticebus coucang,*[9] and Rohen reported some 'cone-like structures' in the retina of *Loris tardigradus.*[15] The Lemuriformes include species which are nocturnal, species which are diurnal and some that are crepuscular.[2,4,5,8,16] The eyes of several species have been studied and are reported to be primarily rod retinas, but with some cones.[9,10,17] The ratio of rods to cones varies from 1000:1 in *Cheirogaleus* to 5:1 in *Lemur catta.*[9] Appropriately enough, *Lemur catta* (and *L. mongoz*) have been reported to have colour vision. While not unusual, it is interesting to note that the nocturnal *Cheirogaleus* (species unspecified) apparently has some cones or cone-like receptors.[9] Where cones are found (except in Tupaiaformes), they appear to be localised in the central retinal area, but there is no report of a 'rod-free area' in any prosimian.

Receptor type and/or structure also has some relationship to visual acuity. While the overall structure of the retina also determines minimum acuity, absolute size of the receptors and their density per unit area, or centre-to-centre spacing, is the first limiting factor in determining acuity. Most prosimian species studied show some structural variation in receptors between the central and peripheral retina. There is frequently an elongation of the receptors in the central retina,[9,10,17] and almost always the density of receptors in the central retina is much greater than in the periphery.[10] As more behavioural studies of prosimian species become available, it will be interesting to see to what extent these anatomical reports correlate with behavioural results.

2. Cellular layers of the retina

A great deal of information processing apparently occurs within the retina. The relative numbers of receptors, bipolar cells and ganglion cells in various species has been regarded as intimately related to the limits of visual acuity and light sensitivity. While this is undoubtedly true, it is only a part of the story, as more recent studies of fine structure of the reticular layers of the retina have indicated.[18-21]

I shall not attempt to restate here all of the detailed information about each species which has been studied. Most of this work has been done by Detweiler,[3] Kolmer,[9] Rohen,[10,15,22] and Woollard.[17,23] The most recent and detailed work has been presented by Rohen and Castenholz,[10] while I have previously attempted a general survey of anatomic studies of the retina of primates.[14] On the basis of their excellent studies, Rohen and Castenholz have proposed four basic types of retinal organisation.[10] The classification is based on the ratio of receptor to bipolar to ganglion cells and the variation in these relationships between different portions of the retina. Group I, which includes *Tupaia glis* and *Urogale everetti*, would appear to be structured for high visual acuity. There is some differentiation between the central area and peripheral retina, with the central area better designed for high acuity. The second group is exemplified by *Microcebus murinus, Cheirogaleus medius* and *Lemur fulvus.* No pattern is illustrated for a central retinal area, and the general structure is one apparently designed for enhancement of light sensitivity. The third type of retina, exemplified by *Galago senegalensis* and *Indri indri*, reveals a central area which should manifest higher acuity and lesser light sensitivity than the more peripheral portions. The fourth type includes all the higher primates (except possibly *Aotes*), and is divided into three subtypes, the first of which is represented by *Tarsius bancanus.* This type of retina shows a distinct three-part organisation with structural organisation for acuity decreasing in a central to peripheral direction, while light sensitivity would be expected to be increased toward the periphery. The significance of these anatomical patterns will be more apparent, if and when single cell studies of the retinas of representative species are undertaken. Behavioural studies of visual acuity, light sensitivity and colour discrimination of a sample of species representative of different patterns of retinal organisation would of course be invaluable.

3. The reticular layers

These layers of the retina, which contain the synaptic structure of the retina, have only recently been subject to detailed study through the use of electron microscopy.[18-21] Although earlier work suggested that the horizontal and amacrine cells of the inner nuclear layer should have some relationship to the processing of visual information,[6] only in the last several years have such relationships in fact been demonstrated.[19] These synaptic relationships, and the cellular constituents of the inner nuclear layer from which they stem, indicate that the interpretation of retinal function based upon ratios of the elements of the cellular layers may be greatly modified by more detailed understanding of synaptic patterns. The horizontal cell synapses in the outer reticular layer and the amacrine cell synapses of the inner reticular layer have been implicated in various integrative processes including colour and brightness contrast, contour enhancement and motion sensitivity (including the directionally selective patterns revealed by unit studies). Dowling has also described three different types of amacrine cells with different areal distributions, the functional significance of which is as yet speculative.[19]

I must note at this point that I have not attempted any comprehensive survey of this particular topic. I will defer to those with expertise in this field for more detailed analysis and critical evaluation of this work. In spite of (or perhaps because of) my lack of expertise in these fields, I cannot but feel that analysis of the fine structure of the reticular layers of the retinas of prosimians will contribute substantially to our knowledge, and may make less obscure some of the relationships between ocular anatomy, behaviour and, perhaps, even evolution.

4. *Fundus patterns*

I will comment on this topic only briefly, since most of what I have to say has been said already.[14,24] The fundus pattern is a very gross index of structure and as such contains only very limited information. Even what appears obvious may not, in fact, be what it appears.

The eye of *Lemur catta* would appear from funduscopic examination to be that of a truly nocturnal animal, since it is but sparsely pigmented and has a highly reflective tapetum; but this is contradicted by published observations. *Tarsius syrichta*, on the other hand, has a moderately pigmented retina with a well developed macula and fovea and thus should be a diurnal animal with excellent acuity; but we know that the first point is untrue, while the second (to my knowledge) remains unknown.

I have never yet examined the eye of any primate which did not show an area which appeared to be a central visual area, this being defined by either the vascular pattern, the pigmentation, or both. Nevertheless, histological studies apparently fail to reveal a specialised central retinal area in some prosimian species.[10] Perhaps more detailed anatomical study or studies of receptive fields by single cell methods will change this picture. In any event, the correlations between fundus patterns and life style of prosimians are far from impressive, and with regard to any interpretation of the evolution of prosimians, I sorrowfully conclude that my limited studies of the ocular fundus have nothing to offer.

Some general comments

The subtitle of this paper may have aroused some curiosity on the part of the reader, and I feel at this point that an explanation is in order. I, like millions of other people, have never faced *Daubentonia* eyeball to eyeball, so to speak, and may never have the opportunity to do so. Information on this prosimian is scant. Nevertheless, I often get the feeling that these rare and exotic and little-known animals can provide a wealth of knowledge for us.

The survey of literature and the interpretations which have been made are not very positive. For each generalisation about possible relationships between anatomy and behaviour, there appears to be *at least* one significant contradiction. In attempting to find why this should be so, it occurred to me that I was dealing with categories, both anatomical and behavioural, which are not so absolute as they appear.

Reference has already been made to Walls' statement that rods and cones should perhaps be functional classifications, rather than purely structural ones.[11] The same approach is obviously applicable to behavioural categories. Animals may be nocturnal or diurnal or crepuscular as a preferred mode of life. Most of them would not be totally incapacitated if forced to operate under other than the preferred conditions. Some (e.g. *Lemur catta*) might be found, in fact, to be quite versatile. Colour vision is another area where greater flexibility of classification is in order. In the case of humans, we no longer think in terms of two or three types of colour blindness, but rather a very complex system of degrees and mechanisms of colour deficiency. I think it will be the exceptional species which shows no adaptability with respect to living habits, just as it will be the rare species with reveals total achromatopsia. Field studies as well as laboratory studies of animal behaviour should be designed to provide us with greater understanding of both the flexibility and the limitations of the subjects under study.

I have indicated in various places throughout this paper kinds of information which were lacking. Anatomical data (except for electron microscopic studies) on some species is fairly good, but the eyes of many species have been studied not at all. Behavioural data on many species is sketchy and experimental studies of visual function in prosimians is the exception rather than the rule. Most lacking (perhaps for valid reasons) are physiological studies of vision of prosimians. Within the limitations imposed by the decreasing populations of many prosimian genera, I hope that some of the newer techniques in each of these areas of study may be employed. If this can be done, I think that some of the questions regarding the relationships between the structure of the eye and behaviour may be answered. With regard to the other question of what the study of the eye can tell us about evolution, I again quote Walls, who said: 'The structural variations of the eyes are largely expressions of habit differences rather than stages in a majestic phylogenetic progress of increasing complexity and perfection, such as one sees in the heart, the brain, the ear or indeed in almost any other organ.'[25]

NOTES

1 Prince, J.H. (1956), *Comparative Anatomy of the Eye*, Springfield, Illinois.

2 Buettner-Janusch, J. (1966), *Origins of Man*, New York.

3 Detwiler, S.R. (1939), 'Comparative studies upon the eyes of nocturnal lemuroids, monkeys and man', *Anat. Rec.* 74, 129-45.

4 Morris, D. (1965), *The Mammals*, New York.

5 Napier, J.R. and Napier, P.H. (1967), *A Handbook of Living Primates*, New York.

6 Polyak, S. (1957), *The Vertebrate Visual System*, Chicago.

7 See *Primate News* (1971), Lyon, C.E., Ito, J., and West, E.S. (eds.), 9, 3-9, Oregon Regional Primate Research Center, Beaverton.

8 Walker, E.P. (1964), *Mammals of the World*, 1. Baltimore.

9 Kolmer, W. (1930), 'Zur Kenntnis des Auges der Primaten', *Z. Anat. Entwicklungsgesch.* 93, 679-722.

10 Rohen, J.W. and Castenholtz, A. (1967). 'Über die Zentralisation der Retina bei Primaten', *Folia primat.* 5, 92-147.

11 Walls, G. (1934), 'The reptilian retina', *Am. J. Opthal.* 17, 892-915.

12 Samorajski, T., Ordy, J.M., and Keefe, J.R. (1966), 'Structural organization of the retina in the tree shrew (*Tupaia glis*)', *J. Cell Biol.* 28, 489-504.

13 Shriver, J.W., and Noback, C.R. (1967), 'Color vision in the tree shrew (*Tupaia glis*)', *Folia primat.* 6, 161-9.

14 Wolin, L.R. and Massopust, L.C. (1970), 'Morphology of the primate retina', in Noback, C.R., and Montagna, W., (eds.) *Advances in Primatology: The Primate Brain*, New York.

15 Rohen, J.W., (1962), 'Sehorgan', *Primatologia* 2 (6), 1-210.

16 Petter, J.J., Schilling, A. and Pariente, G. (1971), 'Observations éco-éthologiques sur deux lemuriens malgaches nocturnes: *Phaner furcifer* et *Microcebus coquereli*', *Terre et Vie* 3, 287-327.

17 Woollard, H.H. (1926), 'Notes on the retina and lateral geniculate body in *Tupaia, Tarsius, Nycticebus* and *Hapale*', *Brain* 49, 77-104.

18 Allen, R.A. (1969), 'The retinal bipolar cells and their synapses in the inner plexiform layer', in Straatsma, B.R., Hall, M.O., Allen, R.A. and Crescitelli, F. (eds.), *The Retina*, Los Angeles.

19 Dowling, J.E. (1970), 'Organization of vertebrate retinas', *Invest. Ophthal.*' 9, 655-80.

20 Dowling, J.E. and Boycott, B.B. (1969), 'Retinal ganglion cells: a correlation of anatomical and physiological approaches', in Straatsma, B.R., Hall, M.O., Allen, R.A. and Crescitelli, F. (eds.), *The Retina*, Los Angeles.

21 Sjöstrand, F.S. (1969), 'The outer plexiform layer and the neural organization of the retina', in Straatsma, B.R., Hall, M.O., Allen, R.A., and Crescitelli, F. (eds.), *The Retina*, Los Angeles.

22 Rohen, J.W. (1966), 'Zur Histologie des *Tarsiusauges*', *Graefe Arch. Klin.*

Exp. Ophthal. 169, 299-317.

23 Woollard, H.H. (1927), 'The differentiation of the retina in Primates, *Proc. Zool. Soc. Lond.* 1, 1-17.

24 Wolin, L.R. and Massopust, L.C. (1967), 'Characteristics of the ocular fundus in primates', *J. Anat.* 101, 693-9.

25 Walls, G. (1939), 'Origin of the vertebrate eye', *Arch. Ophthal.* 22, 452-84.

PART II SECTION B

Anatomy and Function of Skull and Teeth

R. F. KAY and K. M. HIIEMAE

Mastication in Galago crassicaudatus*: a cinefluorographic and occlusal study*

Introduction

The extant genus *Galago* is a lorisiform prosimian of somewhat uncertain lineage. It does, however, have molars with sufficient affinity to some Eocene prosimians for it to serve as a reasonable functional analogue for the masticatory apparatus of those forms.

The ancestry of *Galago* is by no means clear. The oldest unequivocally lorisiform fossils known are the two genera *Komba* and *Progalago* from the lower Miocene of East Africa. *Propotto*, described by Simpson,[1] is now considered to be a fruit bat.[2] A single isolated second lower molar, the type specimen of *Indraloris* found in Siwalik deposits of Nagri age,[3] is poorly documented and is possibly an adapid.[4] On present evidence, it appears that the lorises stem from a family which had become clearly differentiated by the beginning of the Miocene.[5]

There is a considerable literature on the behaviour of the modern Galaginae but, as far as the authors are aware, no detailed investigation of their occlusion or masticatory behaviour has been undertaken. Mills has briefly described the wear facets produced on *Galago* molars and attributed their distribution to the presence of a 'balanced occlusion'.[6] The normal diet of *G. crassicaudatus*, as for *G. senegalensis*, is 'especially fruit and insects, but (the animal) is more or less omnivorous'.[7] The use of the hands in feeding has been discussed by Bishop.[8]

Description and discussion of feeding behaviour and the movements of the skull, lower jaw, tongue and food involved, have been complicated by the multiplicity of terms used to describe various aspects of the process. To clarify the report of the findings obtained in this investigation, the terms most commonly used, and their functional and morphological implications, must be discussed.

There are three stages in feeding: *ingestion* in which food is transferred from an extra-oral position; *mastication* in which 'the bite' is triturated and *deglutition* in which the reduced food is transferred from the mouth across the pharynx and into the oesophagus. Movements of the lower jaw and tongue are always rhythmic and cyclical in mastication and may be so in certain types of ingestion such as licking and lapping.[9–14]

Each masticatory cycle consists of three types of movement or *strokes* which follow each other in a smooth sequence:[14]

(*a*) A *preparatory stroke* or 'closing stroke',[9] in which the mouth is closed, either by upwards movement of the lower jaw alone or in combination with cranial movement at the atlanto-occipital joint (usually depression, but extension can occur). The effect is to bring the lower molars of the *active side* below the uppers and into a vertical relationship such that their buccal surfaces (external or lateral) are aligned, or such that those of the lower molars are lateral to the uppers. (It follows that in animals such as *Didelphis* and *Tupaia*, where the post-canine tooth row is linear, the anterior teeth may be well lateral to the corresponding upper teeth). The preparatory stroke is completed when tooth-food-tooth contact is reached; this point is not always possible to determine accurately from cineradiographs.

(*b*) A *power stroke* in which the food is triturated. The methods by which the food is reduced can be divided into two broad categories: food may be *crushed* between the teeth by forces acting more or less at right angles to their occlusal surfaces. In this case, the stroke begins at the point of initial tooth-food-tooth contact and is completed when that contact ceases. Both points may be determinable cinefluorographically and can occur within a wide range of gapes. The net effect of this type of power stroke on the teeth is the blunting of the tips of cusps. Alternatively, food may be *sheared, cut* or *ground* by the movement of the lower tooth across the upper which is produced by forces acting more or less parallel with the direction of movement. This type of power stroke normally begins with food between the opposing tooth surfaces, but throughout the contact between them is sufficiently close for its direction to be governed by their morphology. Shearing mastication produces wear facets on the slopes of cusps or ridges and sometimes in the basins between them, which are striated in a direction parallel to that of the movement producing them. (These movements are the 'contact' or 'glide' movements forming the upper limit of the envelope of motion.)[15] The stroke commences at the point of initial occlusal or tooth-food-tooth contact and is completed when the teeth come out of occlusion. The former point may be determinable cinefluorographically, the latter is not normally visible. However, dorso-ventral or antero-posterior projection cineradiographs may allow horizontal or transverse movements in this stroke to be analysed. The power stroke may or may not involve occlusal contact on both active and balancing sides of the jaw.

(*c*) A *recovery stroke* ('opening stroke') in which the mouth is opened to the maximum gape for the cycle. As in the preparatory stroke, this may result from predominantly vertical movement of the lower jaw only, or from such movement combined with cranial extension (usually) or depression.[12] The downward movement of the mandible may involve some lateral, protrusive or retrusive excursion. The stroke begins when occlusal or tooth-food-tooth contact is observed and completed when the upwards movement of the next preparatory stroke begins. The latter point is readily determinable from cineradiographs, the former not necessarily so.

A *stroke* may have one or more *phases* distinguished by at least one feature such as a significant change of direction of jaw movement. This situation normally occurs only in the power stroke, and is most clearly seen in a chewing stroke where the first upward and medial movement of the lower teeth on the active side brings them into centric occlusion (protocone in talonid basin). This is followed by a second downward and medial movement taking the lower molars out of centric occlusion and further towards the midline. The first of these movements may be regarded as 'phase I' and the second as 'phase II'. It does not follow that such phases always occur, since one or other may be suppressed depending on the architecture of the teeth and jaws.

At the completion of any masticatory activity, the lower jaw normally returns to the *rest position.*[6,12] This involves the curtailment of the recovery stroke of the terminal cycle. Similarly, when the animal begins to feed, the jaw is depressed from the rest position to maximum gape for the ensuing ingestive cycle.

The terminology used to describe the power stroke and derived from wear facet analysis has introduced two terms, the 'buccal phase' and the 'lingual phase', into the literature.[6,16,17] The former describes the movement of the jaw from the buccal position to centric occlusion on the active side and from the lingual position to centric occlusion on the balancing side. Mills then describes a 'lingual phase' in which the lower molars on the active side move from centric occlusion towards the midline whilst those on the balancing side are moving buccally. In other words a 'buccal phase' on one side of the jaw may occur simultaneously with a 'lingual phase' on the other.[16] This is a 'balanced occlusion'. Since the conclusions on which the terminology is based have been disputed,[13,14] and given that the terms may be justifiable only if applied to the active side, with no implications for balancing side occlusion, they are not used in this study. Instead two phases in the power stroke on the active side are recognised.

It follows from the above that any investigation into mastication must utilise two techniques; the movements of the cranium, lower jaw and to some extent the tongue and food can be recorded cinefluorographically; the occlusal relationships of the teeth during a chewing stroke have to be determined from the morphology of the molar crowns and their pattern of wear. This study has used both these methods to establish the pattern of feeding and mastication in *Galago*.

Materials and methods

1. *Cinefluorography.* The specimen of *Galago crassicaudatus* (a male) had radio-opaque markers inserted in upper and lower jaws (see Fig. 1) under general anaesthesia. Cinefluorographic recordings were made in the lateral, dorso-ventral and antero-posterior projections at 60 frames per sec. on 16 mm. Eastman Kodak Plus X negative film no. 7231, at 55-70 kilovolts and 14-18 milli-amperes, depending on the projection. The animal was fed on apple, 'Purina' monkey biscuits,[18] either plain or previously soaked in water and dipped in honey to give a softened but still crunchy consistency. The recordings were analysed by the methods described by Crompton and Hiiemae,[14] and Hiiemae and Crompton,[12] and as indicated in the text.

2. *Tooth morphology and occlusion.* Specimens of *Galago crassicaudatus* from the osteological collection at the Museum of Comparative Zoology, Harvard University, have been examined to determine the morphology of the teeth, the extent and definition of the wear facets and the geometrics of occlusal relations in the power stroke. This last method is based on a geological technique described by Lahee.[19] The terms used are defined in the following text. The method has been adapted for the determination of the direction of movement between occluding upper and lower teeth during tooth-tooth contact. Three points were chosen to define the occlusal plane, selected from the tips of the hypoconids of the left and right M_1, and of the hypoconids of the left and right M_3. (In the text, the letters I, C, PM or M refer to Incisors, Canines, Premolars or Molars, respectively. The superscript numbers refer to upper teeth, the subscript numbers to lower teeth.) A separate mesiodistal axis was chosen for the body of the dentary: a line drawn between the hypoconids of M_1 and M_3. The dentary was oriented normal to the occlusal plane and examined under a stereomicroscope with a camera lucida. The strike of each wear plane examined was estimated as a line drawn between two points on each wear plane parallel to the occlusal plane. The dentary was then oriented laterally so that the wear plane was parallel to the line of vision and the angle of dip could be measured. These procedures were repeated ten times for each wear plane studied on a sample of four specimens of *Galago*

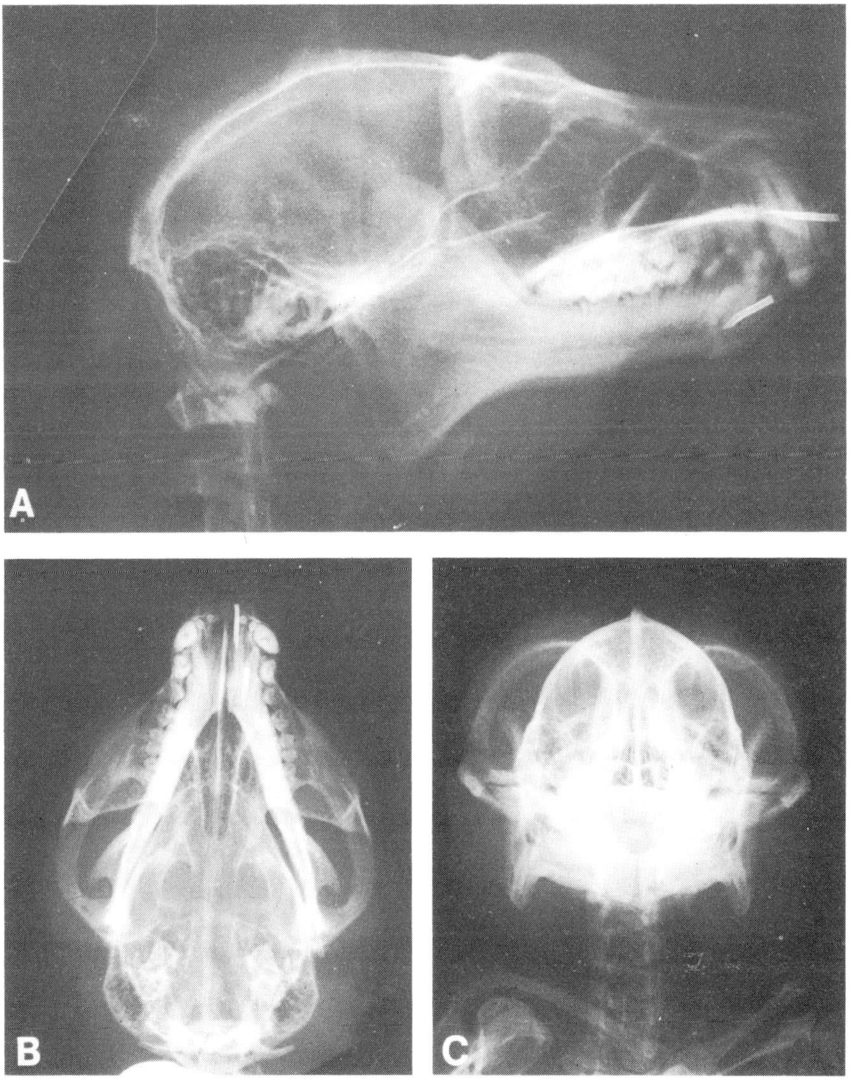

Figure 1. Laterial (*A*), dorso-ventral (*B*), and antero-posterior (frontal) (*C*) radiographs showing the appearance of the head of *Galago* in these projections and the positions of the markers. The radiographs are at the following approximate magnifications: *A* x 2.5, *B* and *C* x 2.

crassicaudatus. Given this raw data, the orientation and plunge for each facet were estimated by the geometrical method described by Lahee (Tables 1 and 2: see fig. 8).

Upper and lower molar dentitions of *Palenochtha* from the Middle Palaeocene, Gidley Quarry, Montana were used for comparative purposes.[20]

Table 1 **The dip and strike of facets 1 and 3. The figures represent the average of the mean of ten readings for each facet on each of four specimens**

	Wear Facet 1		*Wear Facet 3*	
	Dip°	*Strike°*	*Dip°*	*Strike°*
M_1	60	61	72	43
M_2	59	65	76	32
M_3	57	67	73	25

Table 2 **Movement during phase I of the power stroke in *Galago*, based on averaged data from four specimens**

	Tooth Position		
	M_1	M_2	M_3
Orientation°	66	59	50
Plunge°	54	52	55

Results

The experimental and analytical results obtained are described, using the terminology defined in the introduction, under three main headings: *Feeding Behaviour; The Masticatory Cycle*; and *Occlusion.*

The jaw apparatus in *Galago* is distinguished by the presence of a dental comb with a concomitant caninisation of the P_2 (P_1 is not present). The upper incisors are small and separated by a median diastema. The upper canine and the P_2 are large and protrude beyond the occlusal plane. P_2 and P^3_3 are somewhat sectorial, P^4_4 are semi-molariform. The post-canine teeth from $P^4 - M^2$ increase in their transverse width so that the outer profile of the tooth row bulges laterally (Fig. 1*B*). There is a less marked lateral bulge in the lower jaw. The mandible in *Galago* consists of two parts (dentaries). The mandibular symphysis is not ossified and fibrous material can be seen between the two dentaries in osteological material, suggesting that intra-mandibular movement is possible. The body of each dentary is stout, although shallow, whilst the ramus is much more slender with well developed and somewhat hooked coronoid and angular processes. The mandibular condyles have their longest axis at right angles to the midline of the skull and are sharply convex antero-posteriorly. There is a well-developed post-glenoid flange on the articular fossa. The radiographic appearance of the hard tissues is shown in each projection in Fig. 1.

Feeding behaviour

Relatively few animals have been studied using cinefluorography to record mastication: *Tupaia*,[21] *Didelphis*,[12] *Saimiri*,[13] *Ateles*,[22] and *Rattus*.[11] By comparison with the first two, *Galago* is a slow feeder, since a single chewing cycle takes longer, but it is more 'efficient' in that fewer cycles are needed to reduce the food sufficiently for deglutition.

1. *Ingestion*

Ingestion is normally preceded by inspection: the food is sniffed and/or licked. The actual method by which the bite is transferred to the mouth depends on the consistency of the food: liquids or semi-liquid 'mush' are ingested by licking or lapping; hard foods are ingested through the immediate post-canine region rather than through the front of the mouth. The incisors have never been seen to function as a biting organ during ingestion. Small bite-sized pieces of hard food are often directly picked up from the ground by rotation and tilting of the head to bring the canine and premolars of one side into position to pick up the bite; the latter is then 'tossed' distally onto the molars by a rapid movement of the head. Alternatively, one or other hand is used to manipulate the food into position for ingestion. A bite of hard food is separated from the remainder by a series of 'puncture-crushing' chewing cycles: this is 'ingestion by mastication'.[12] Throughout this procedure the bulk of the food is held in the ipsilateral hand. Tentative 'efficiency counts' indicate that three puncture-crushing cycles were required to separate a bite from the biscuit supplied in this study. At no time during these cycles do the teeth come into contact. Once the bite has been separated, the hand is dropped and mastication follows.

2. *Mastication*

As in *Didelphis* and *Tupaia*, mastication takes place in two stages for hard food – an initial group of puncture-crush cycles followed by normal chewing. Soft mush is only chewed. Frequently, towards the end of any chewing sequence, the tongue is protruded and the tip flipped up over the snout. At the completion of each cycle sequence, a bolus is passed. For hard food there appears to be a 1:1 relationship between bite and deglutition; i.e. one bite is ingested, completely reduced and the bolus passed before ingestion recommences. This is not necessarily true for soft food.

Preparatory Stroke

Power Stroke

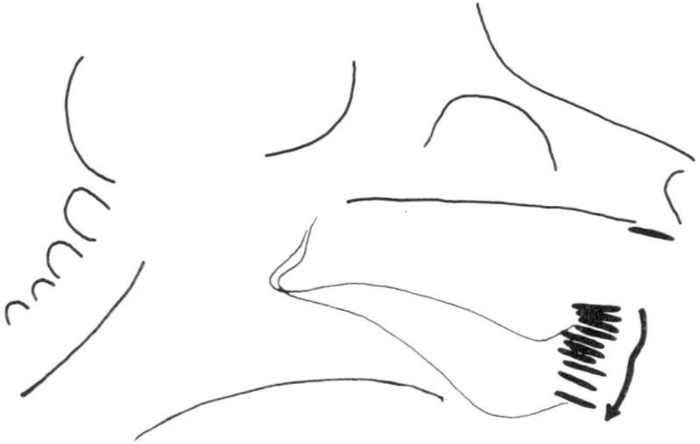

Recovery Stroke

Puncture-crushing

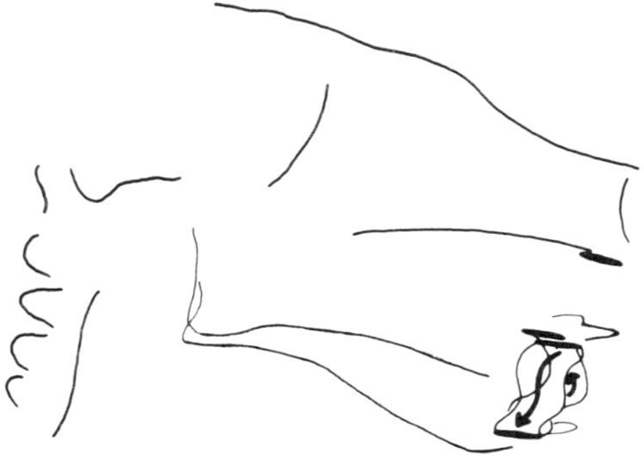

Chewing

The masticatory cycle

The 'efficiency' with which *Galago* feeds means that the number of consecutive chewing cycles required to reduce a bite of soft food is small, and even for the hard biscuit used in this study rarely exceeds ten. As might be expected, fresh apple constitutes a greater problem, the number of cycles required rising towards twenty (see Figs. 4 and 5). Chewing proceeds rhythmically until mastication is completed and is followed by deglutition.

The amplitude of movement of the lower jaw and the degree to which the mouth is opened at the completion of the recovery stroke depends on the type of mastication in progress (see Figs. 4 and 5). The widest gapes are recorded during the ingestion of hard food, followed by those for puncture-crushing. Small gapes are associated with chewing and lapping.

The mandibular symphysis of *Galago* is very clearly mobile. It mediates rotation of either or both dentaries along their long axes so that they can evert or invert relative to each other. In addition, the symphysis may allow either dentary to slip further forward than the other, so that although the paths of movement of the dentaries on the active and balancing sides tend to parallel each other, this is not always the case. In fact there is some cinefluorographic evidence that independent movement on active and balancing sides does occur during the power stroke.

1. *The preparatory stroke*

During this stroke, the lower jaw moves upwards until tooth-food-tooth contact is reached. This movement, particularly from large gapes, normally involves a combination of cranial depression and mandibular elevation. As can be seen from Figs. 4 and 5, the amplitude of the vertical movement involved depends on the type of masticatory activity in progress. This movement is predominantly vertical, with at the most a very small lateral component (Fig. 3) and a slight retraction occurring while the mouth is still wide open (Fig. 2). At the completion of the preparatory stroke, the dentary and its post-canine teeth on the active side are below the corresponding upper teeth, with their lateral surfaces vertically aligned.

Figure 2. Compound profiles from cinefluorographic recordings of mastication taken in the lateral projection. Top row: superimposed tracings showing the position of the marker in sequential frames of a single chewing cycle; bottom left, to show the path of movement of the marker during a puncture-crushing cycle; and bottom right, in a chewing cycle. (The position of the marker at the beginning of each stroke is shown.)

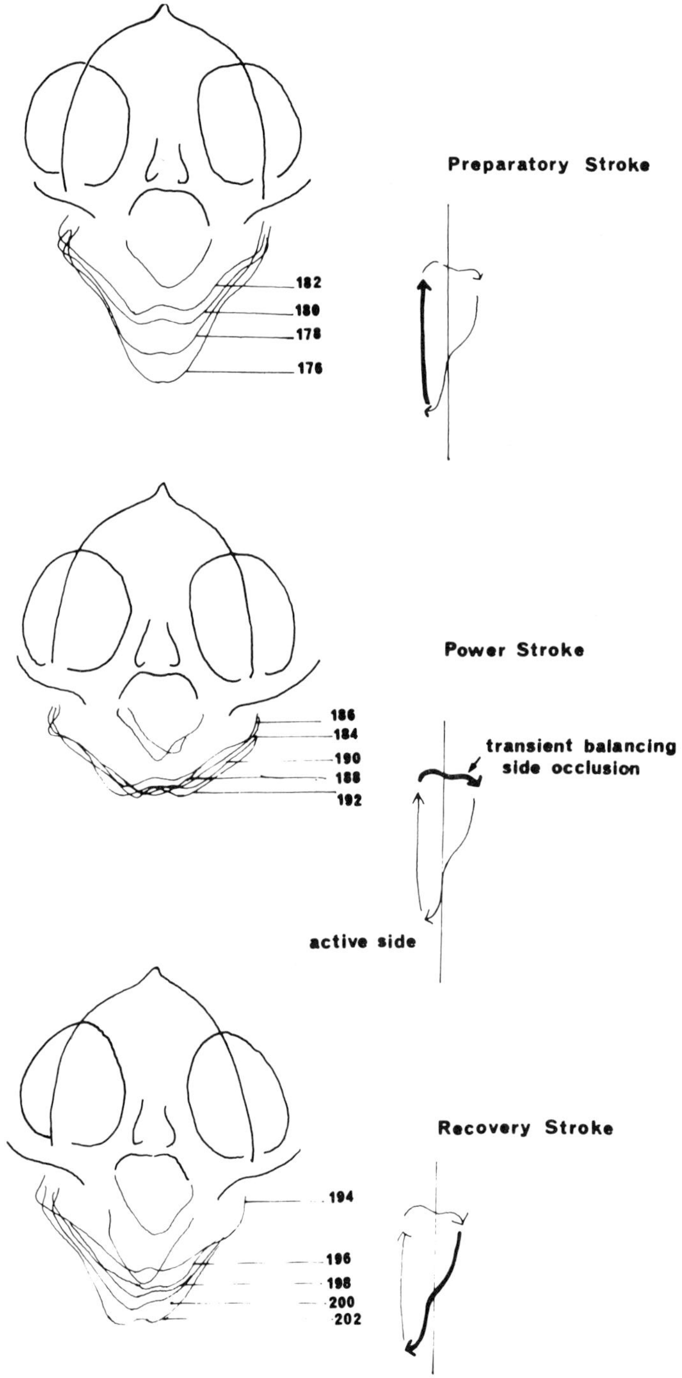

Figure 3. Compound profiles showing the chewing cycle in the antero-posterior (frontal) projection. Alternate frames have been used (as indicated by the numbering). The 'profile' of symphyseal and marker movement is shown on the right, somewhat enlarged.

2. The power stroke

The exact path of the movement of the dentary on the active depends on the type of mastication occurring.

In *ingestion and puncture-crushing* cycles, following separation of the bite, the power stroke continues as an extension of the preparatory stroke, but at a much reduced rate: there is a very slow upwards movement (Fig. 4) in which the food is compressed between the teeth. This is followed by a short period in which the jaw is 'held' at the minimum gape for that cycle. It can be argued that the food is being compressed during this time by largely isometric contraction of the jaw muscles. The dorso-ventral and antero-posterior recordings show that there is some medial movement of the dentary on the active side to bring it into, or just beyond, centric relation before the recovery stroke commences.

The movements of the dentary on the active side in the power strokes of *chewing* cycles are more complex. Viewed in lateral projection, the dentary is seen to come into occlusion, to move upwards and occasionally very slightly forwards and then to move slowly downwards until a clear freeway space is visible. In dorso-ventral and antero-posterior cinefluorographs, the movement of both dentaries in the horizontal and transverse planes can be seen. Initially the dentary on the active side moves slowly upwards, medially and slightly forwards until centric occlusion is reached. This is phase I. The dentary on the active side then continues to move medially through phase II, the symphysis crossing the midline (Fig. 3) to bring the balancing side into what appears from the cineradiographs to be a transistory occlusion. Both dentaries then drop and move into the recovery stroke. This long medial excursion of the active side dentary towards the balancing side lengthens the apparent duration of the power stroke.

3. The recovery stroke

The recovery stroke begins very slowly with the mandibular symphysis to the balancing side of the midline. When mandibular depression has reached a gape of about 8-10° (a large freeway space) the downward movement stops for a short time and the lower jaw is protruded (see Figs. 2 and 3). From the protruded position the lower jaw continues its downward and lateral swing towards the active side. The symphysis appears to cross the midline from active to balancing sides about halfway through the stroke (Fig. 3). The stroke is completed when the jaw reaches the maximum gape. At this point the dentary on the active side is normally well lateral to the midline.

The movements of the mandibular condyles on the glenoid fossae and of the symphysis cannot be clearly seen in all the radiographs.

However, the following movements appear to be occurring:

(*a*) *Preparatory stroke.* The jaw is rotating smoothly upwards with a little retraction (posterior translation) in the first stage. If the condyle on the active side has not reached a lateral position during the last phases of the recovery stroke, it moves laterally on the glenoid fossa during the early part of the preparatory stroke.

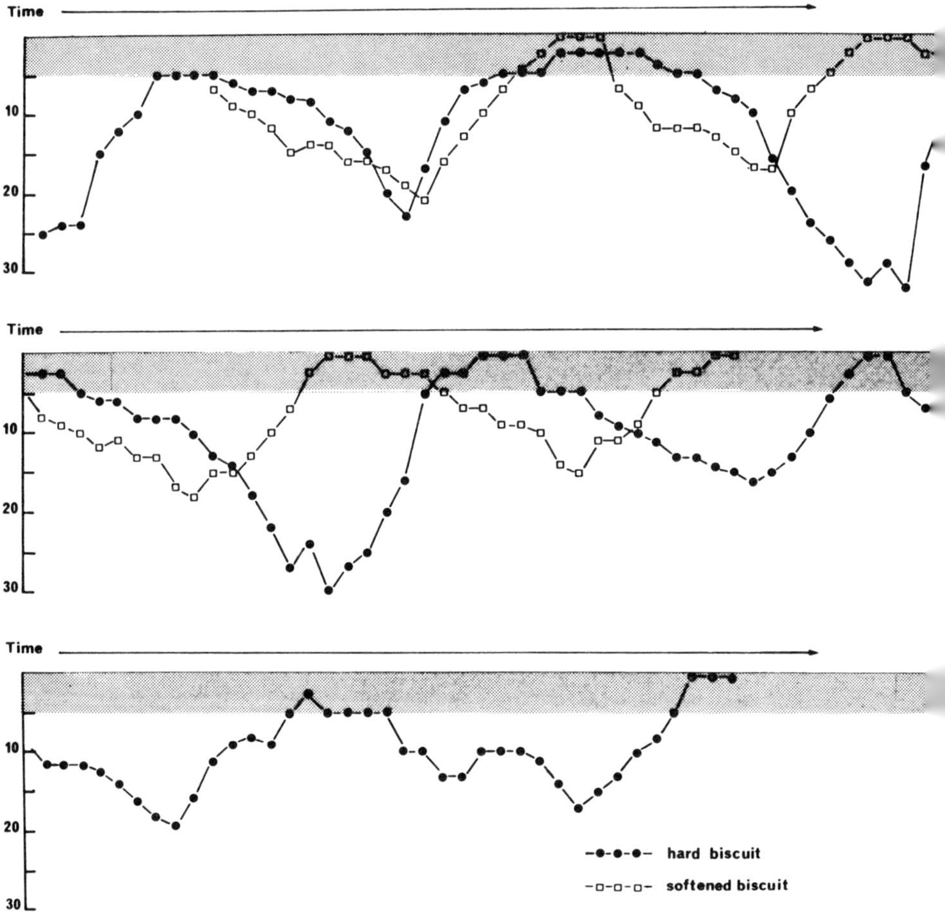

Figure 4. Plot showing the *gape* (in degrees) in lateral projection plotted against time for successive single frames during mastication on hard and very soft biscuit. Each horizontal line represents a total elapsed time of 0.83 secs. The shaded area represents the range of gapes, defined as 'wide freeway space', 'narrow freeway space' and into occlusion (i.e. equivalent to 5, 2.5, and 0 degrees of gape). The first three cycles for hard biscuit show the amplitude of movement associated with puncture-crushing mastication, and the second three that for chewing. Softened biscuit is not puncture-crushed and the maximum gape for the cycles shown is similar to that for the chewing cycles for the hard biscuit.

(*b*) *Power stroke.* During the last stages of the power stroke the condyle slips medially on the glenoid fossa, a Bennett movement. At the same time some slip occurs at the symphysis, the active dentary apparently moving slightly further forwards than the one on the

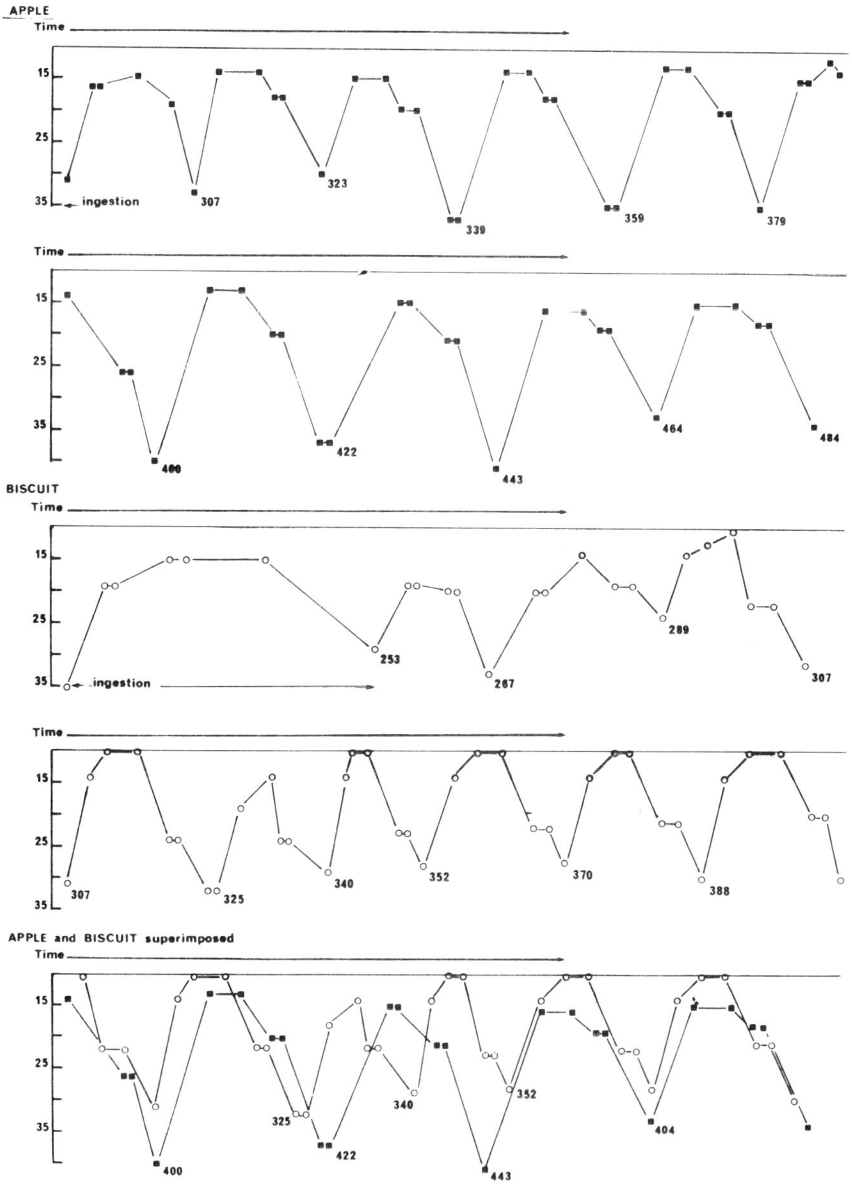

Figure 5. Plot showing successive cycles, beginning with ingestion, for feeding on apple and hard biscuit. Frames at the beginning of each stroke only have been traced and measured. The time elapsed for each line is 1.66 secs. Frame numbers are included for comparison of durations and to demonstrate rhythmicity.

balancing side. The behaviour of the mandibular condyles during the two phases of the power stroke is discussed below.

(*c*) *Recovery stroke.* After the initial slow opening phase, the jaw is then protruded (a forward translation of the condyle) and then depressed by condylar rotation on the glenoid fossa. The lateral movements of the jaw from the balancing to the active side will involve some condylar rotation about a vertical axis. The relative duration and the great regularity of the various types of masticatory cycle recorded is indicated in Figs. 4 and 5. The details of the time course for mastication in *Galago* will be discussed elsewhere,[23] but it is clear from the figures that a puncture-crushing cycle takes about 10% longer than a cycle involving chewing or lapping.

Occlusion

The cinefluorographic study confirms that *Galago* uses two different actions to triturate its food: puncture-crushing and chewing. Each produces a different type of wear on the teeth. In puncture-crushing the teeth do not come into contact and are used in a fashion analogous to a meat tenderiser; this blunts or flattens the tips of the cusps. In chewing the teeth contact in a regular fashion, their relative movement is controlled by the configuration of the cusps and ridges on their crowns. These movements produce clearly delineated matching wear facets on both upper and lower teeth. Examination of such wear facets allows the occlusal relationships between contacting teeth to be determined.

1. Molar occlusion in Galago

The upper and lower molars of *Galago* and *Palenochtha* (Fig. 6) are compared in Fig. 7. Fig. 6 illustrates the terminology used.[2] The molar morphology of *Palenochtha* from the middle Palaeocene is only slightly modified from the tribosphenic pattern which is likely to have characterised the Cretaceous ancestors of primates. A comparison between the molars of the two genera places the functional interpretation of *Galago* mastication in an evolutionary perspective. In both *Galago* and *Palenochtha* the protocone, which is displaced posteriorly, the paracone and metacone are all well developed on the upper molars. The stylar area, with para- and metastyles present in *Palenochtha*, is absent in *Galago*. The crests which primitively run antero- and postero-externally from the paracone and metacone respectively (paracrista and metacrista) are much abbreviated and unsupported by stylar cusps in the latter. The paraconule, metaconule and the crests (pre- and post-paraconule crista, premetaconule and post-metaconule crista) are also very much reduced or absent. The large hypocone on the postero-internal

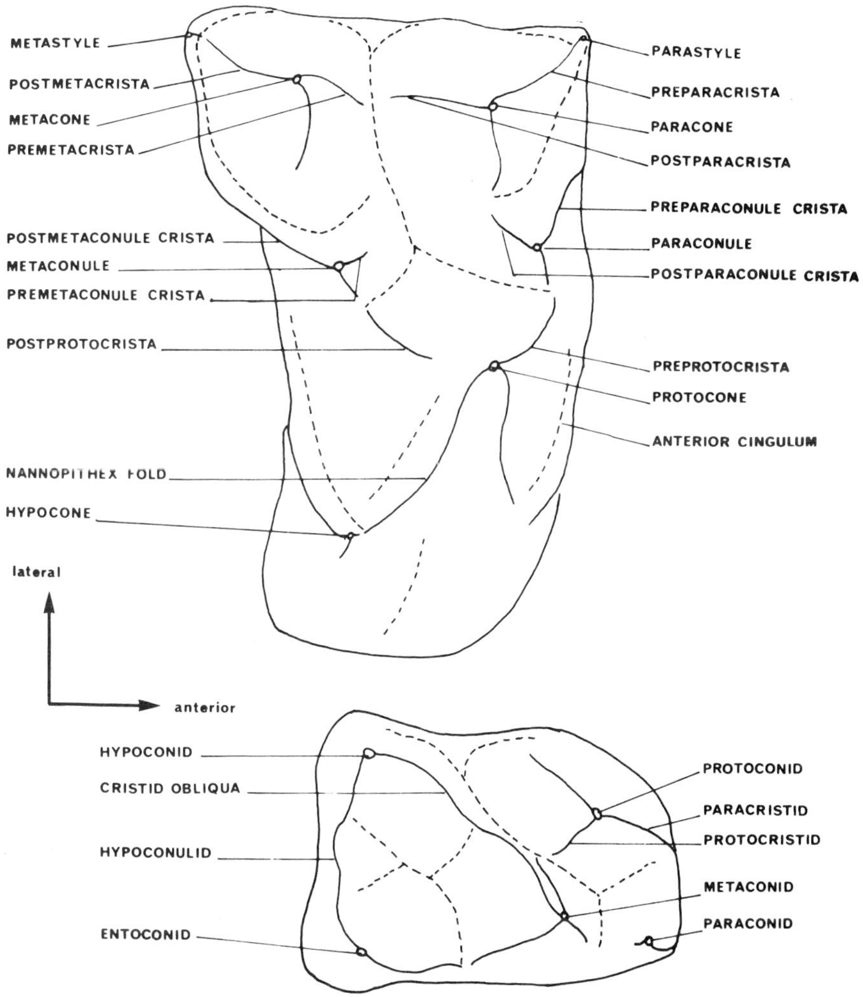

Figure 6. Outline drawings of upper and lower first molars of *Palenochtha minor* (AMNH 35443 and AMNH 35450 respectively) to show the nomenclature of the major features on their crowns.

cingulum is separated from the metacone by a distinctive notch in *Galago*. A cuspule is present in the same position on the first and second molars of *Palenochtha*.

The lower molars of *Galago* are modified by comparison with those of *Palenochtha*: the trigonid is reduced with the loss of the paraconid and is only slightly raised above the talonid. The metaconid has moved slightly posterior to the protoconid. The expanded and fully enclosed talonid basin is bordered by an enlarged hypoconid and a slightly smaller entoconid, whilst that of *Palen-*

ochtha is slightly open internally. The hypoconulid is reduced and virtually absent.

Consequent on these changes in gross morphology, the wear facets seen in *Galago* (although they are found, with two exceptions, in *Palenochtha*) have altered in size and relative importance. Those associated with transversely orientated embrasure shear in the primitive tooth (Facets 1*a*, 1*b*, 2*a*, 2*b* – see Crompton,[24] and also for the numerology and terminology used) are reduced from the sharply curvilinear and duplicated facets seen in the upper molars of

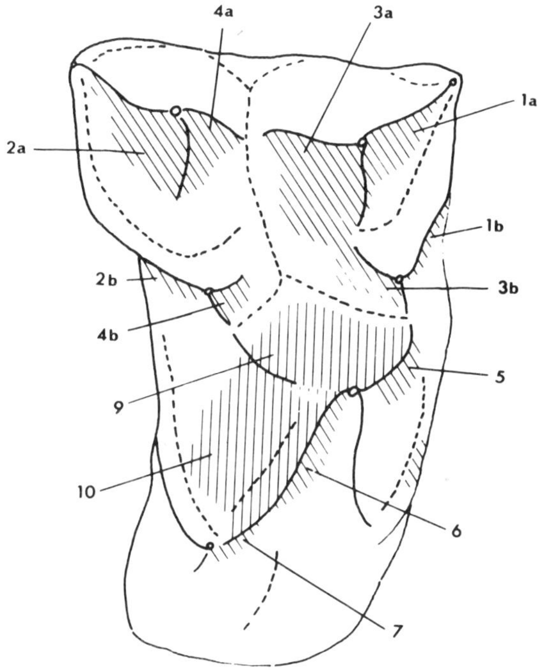

A

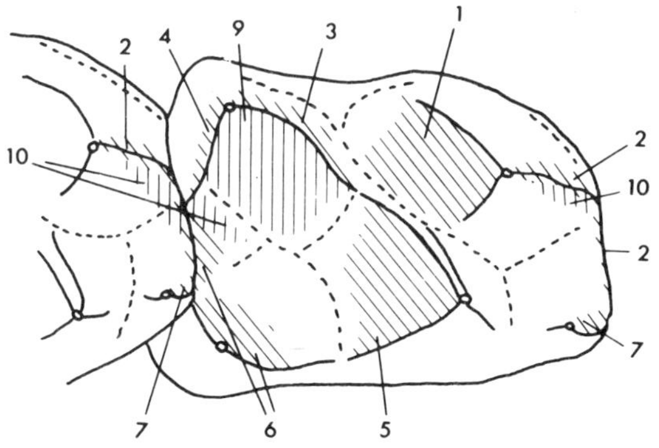

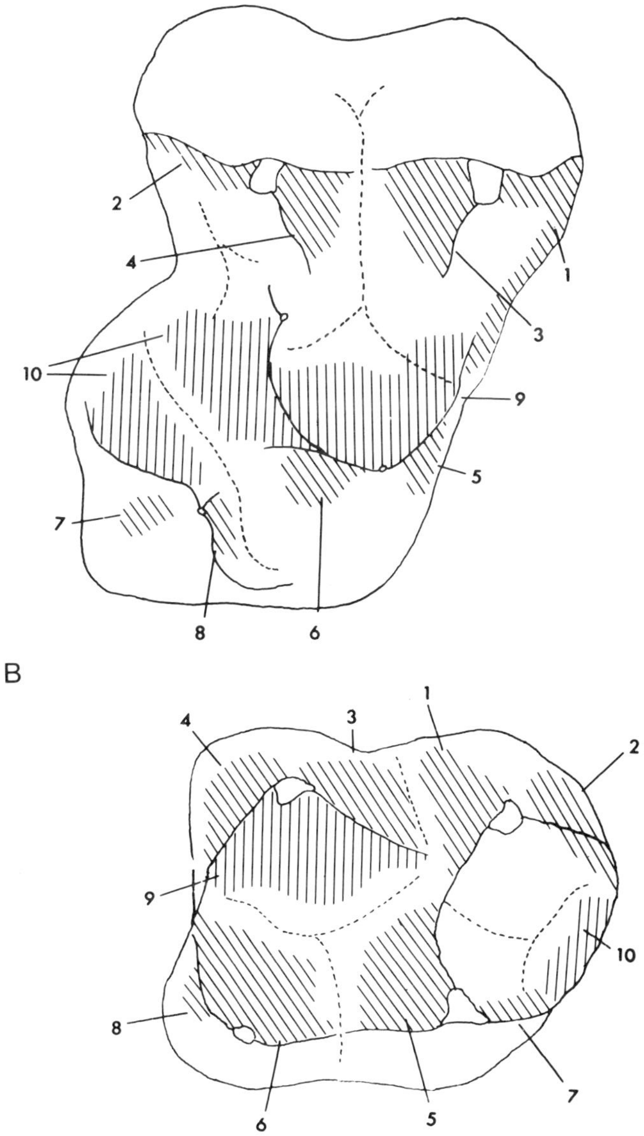

Figure 7. Upper and lower second molars of (*A*) *Palenochtha* and (*B*) *Galago* orientated as in Fig. 6 and shaded to show the positions and areas of the wear facets. The numbers show corresponding facets, but the cross-hatching does not indicate the direction of the striations. The molars of *Palenochtha* are from the Gidley Quarry, Montana; the specimen of *Galago crassicaudatus lasiotis* (MCZ 39414) was collected on the 'Tanganyika' Amboni Estate near Tanga. The teeth are shown at magnifications of approximately x 25 and x 15, respectively.

Palenochtha to comparatively small and much less transverse areas in *Galago*. The corresponding facets on the lower molars of *Galago* are greatly abbreviated: facet 1 no longer extends medially to the tip of the metaconid; facet 2 is rotated antero-externally, reducing its transverse orientation, and no longer extends far towards the metaconid; facets 3 and 4 are present in *Galago* on the postparacrista and the premetacrista as in *Palenochtha*, but facets 3*b* and 4*b* on the internal slopes of the postparaconulecrista and premetaconulecrista are absent in *Galago*. The wear facets found on the internal slopes of the talonid basin (facets 5, 6 and 9), an area associated with grinding, are much more extensive in *Galago*. Facets 5 and 6 are produced by contact between the preprotocrista and postprotocrista with the internal surfaces of the metaconid and entoconid. Facet 5 has so expanded in *Galago* that it reaches laterally almost to the protoconid. In *Palenochtha* facet 6 extends upwards and posteriorly along the nannopithex fold, supported at one end by a hypocone-like cingulum cusp. Wear facet 6 on the lower molars extends along the crest running anteriorly from the entoconid and posteriorly from that cusp onto the paraconid of the tooth behind it. The posterior cingulum cuspule occludes onto the paraconid. The corresponding facet in *Galago* does not reach the hypocone. The enlarged facet 9 seen in *Galago* results from contact between the outer surface of the protocone and the internal slope of the hypoconid. A partially enclosed basin has developed between adjacent teeth in *Galago* and is bordered by the entoconid of the anterior molar with the protoconid and metaconid of the posterior teeth. Wear facets 7, 8 and 10 are formed on the slopes of this basin by the hypocone. This arrangement, absent in *Palenochtha*, reflects an increased emphasis on crushing and grinding during the power stroke.

The major differences between the wear facets on the molars of *Palenochtha* and *Galago* demonstrate a shift in the emphasis placed on the leading edges of the shearing blades, as opposed to the wear surfaces supported by the cusps used for grinding the food. *Palenochtha* molars are designed to cut the food between a succession of shearing blades involving accentuated vertical and transverse movements in the first stage of the power stroke, whereas *Galago* molars are designed initially to cut and then to grind the food between a cusp and its related basin by a combination of shearing and compression. That this is the case in *Galago* can be readily demonstrated by considering the order in which the facets come into contact.

As the lower molar of *Galago* moves upwards and medially in phase I of the power stroke, facets 1, 2, 3 and 4 all contact simultaneously, followed successively by facets 5 and then 6. In *Palenochtha* the sequence is: facets 1*a*, 2*a*; 3*a* and 4*a*; 1*b*, 2*b*, 3*b*, and 4*b*, 5 and 6 to bring the teeth into centric occlusion (*sensu*

Crompton and Hiiemae).[14] As centric occlusion is reached in *Galago*, facets 7 and 8 (absent in *Palenochtha*) come into contact on the postero-medial slope of the hypocone and the posterior surface of the entoconid. The movement from the lateral position through a series of shearing contacts and into centric occlusion (phase 1) effectively completes the power stroke in a primitive mammal such as *Didelphis* or the Late Cretaceous *Didelphodus*.[24] In contrast, phase 1 in both *Galago* and *Palenochtha* is followed by a second movement in which the lower molars continue to move anteriorly and medially towards the midline and in doing so facets 9 and 10 move past each other. This is phase II of the power stroke. The relative degree of development of these facets in *Palenochtha* and *Galago* illustrates the greater functional significance of phase II in the latter. (See also Figs. 9 & 10.)

2. *Geometrical analysis of occlusion*

A geometrical construction has been developed to allow the direction of movement between any contacting pair of teeth to be estimated. This construction is based on the following: the wear facets on the molars of *Galago* are, to a first approximation, planes; more than one pair of non-parallel wear planes are in simultaneous contact in any pair of actively occluding teeth during tooth-to-tooth contact; for simultaneous contact to be maintained between these non-parallel matching wear planes, movement between the teeth must be confined to one, and only one, axis. Two angles, commonly used in structural geology, 'dip' and 'strike', completely characterise the position of any plane in space relative to a reference plane and an arbitrary axis in the reference plane (see Fig. 8). *Dip* is defined here as the angle of intersection between a wear plane and the occlusal plane. *Strike* is the angle between an approximate antero-posterior axis in the occlusal plane of the tooth and a line formed by the intersection of a wear plane with the occlusal plane. A line in space may be completely characterised by two angles which describe its relation to the occlusal plane and to the antero-posterior axis in that plane: its 'orientation' is the angle between the antero-posterior axis and the projection of the line on the occlusal plane; its 'plunge' is the angle between the projection of the line onto the occlusal plane and the line itself. The orientation and plunge of the line of intersection of two non-parallel wear facets are equal to the orientation and plunge of movement between the occluding teeth at that stage of mastication. Given that the dip and strike for each of the two wear facets is known, then the plunge and orientation of their line of intersection can be calculated.[19] From these figures the orientation and plunge of movement during phase I of the power stroke in *Galago* have been calculated.

As shown in Fig. 8, the upward movement of the lower jaw into

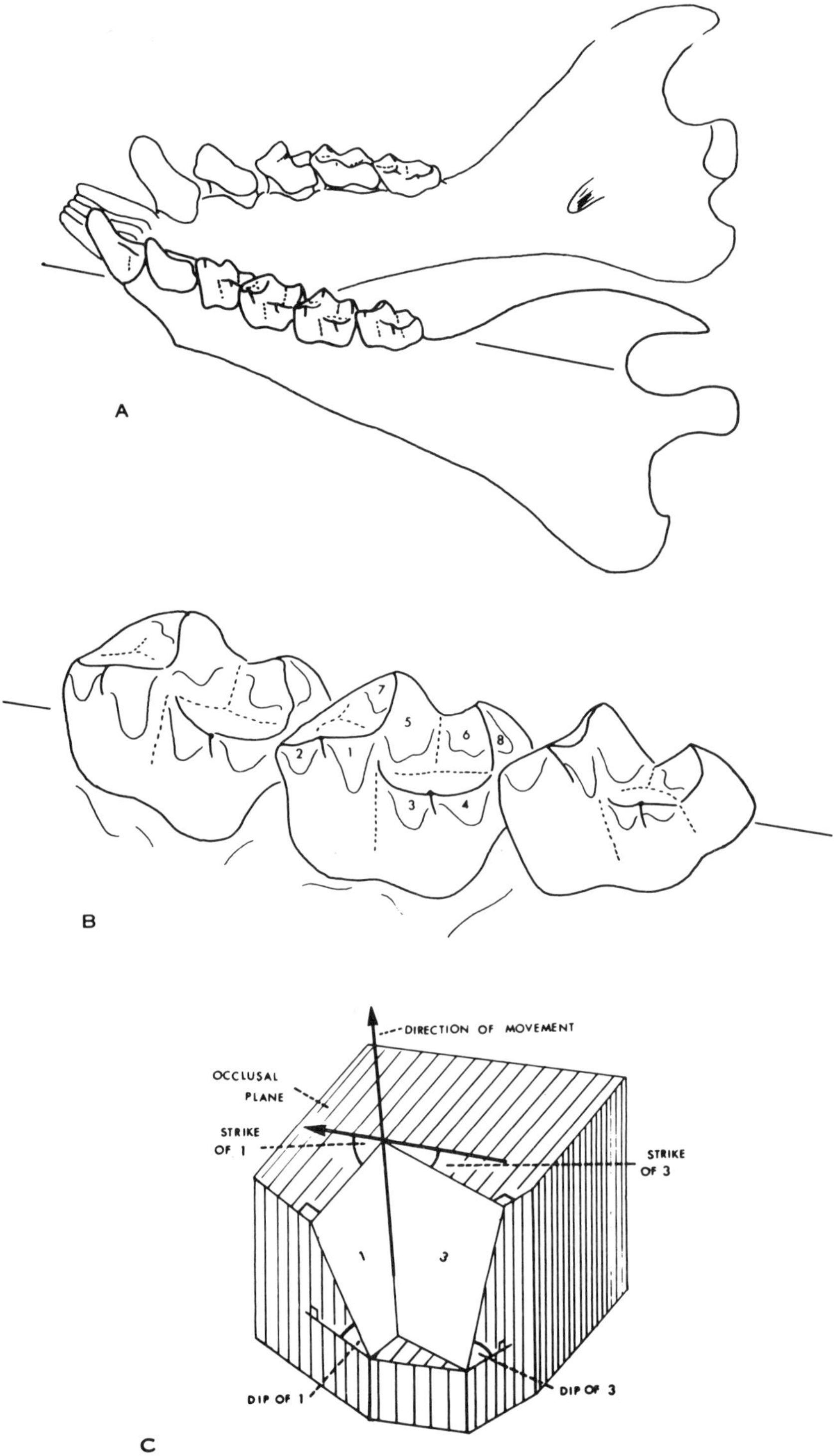
A
B
C
DIRECTION OF MOVEMENT
OCCLUSAL PLANE
STRIKE OF 1
STRIKE OF 3
DIP OF 1
DIP OF 3

centric occlusion during tooth-tooth contact may be described as a rotation of the dentary on the active side about a point located antero-medial to the active condyle, posteromedial to the third molar on that side and above the level of the occlusal plane. This results in an increase in the antero-posterior (mesiodistal) component of movement backwards along the post-canine tooth row and an increase in the transverse component of movement anteriorly along the tooth row. As the centre of rotation is above the tooth row, this will also augment the anterior component of movement posteriorly along the tooth row. (These observations and their relation to the concept of 'condylar rotation' will be discussed below).

Phase II of the power stroke may be described as an anteromedial translation of the active side of the lower jaw downwards and anteromedially out of centric occlusion during which contact is maintained between the protocone and talonid basin and the hypocone and trigonid basin. Since wear facets 9 and 10 on each lower molar are sub-parallel, an accurate estimate of orientation and plunge by the method using the intersection of planes is extremely difficult. However, visual observation suggests that the plunge of this movement is of the order of 50-55 degrees and is probably less steep than in *Palenochtha.* Kay shows that the movement of phase II in *Saimiri* is not a rotation about the condyle.[13] This appears also to be the case in *Galago*, a finding substantiated by the cinefluorographic observation that anteromedial condylar translation occurs in the later stages of the power stroke. The angle between the plunge of phase I and that of phase II is probably more oblique in *Galago* than in primitive primates.[23]

Fig. 10 is a scanning electron-micrograph showing the slope of wear facet 5 and the striations on its surface. At the bottom of this facet the striations continue to move without interruption into facet 9. This gradual change of direction supports the conclusion that phase I and phase II are parts of a single continuous power stroke on the active side and not the result of movement on the balancing side as suggested by Mills.[6] This conclusion is further substantiated by the observation that in *Tupaia* where phase II striations are seen, there is no medial excursion of the mandible such as to bring the balancing side into functional occlusion during the power stroke.[21] Moreover it has been shown that the orientations of phase II movements are not consistent with bilateral simultaneous contact in *Saimiri*, which has a fused symphysis.

A detailed view of facet 7 is shown in Fig. 9, demonstrating the

Figure 8. (*A*) Paraocclusal view of the lower jaws of *Galago* showing the arbitrary antero-posterior axis used. (*B*) The wear facets outlined in *A* enlarged. (*C*) Diagram showing the relationships of dip, strike and direction of movement to the occlusal plane, and the antero-posterior axis for wear planes 1 and 3.

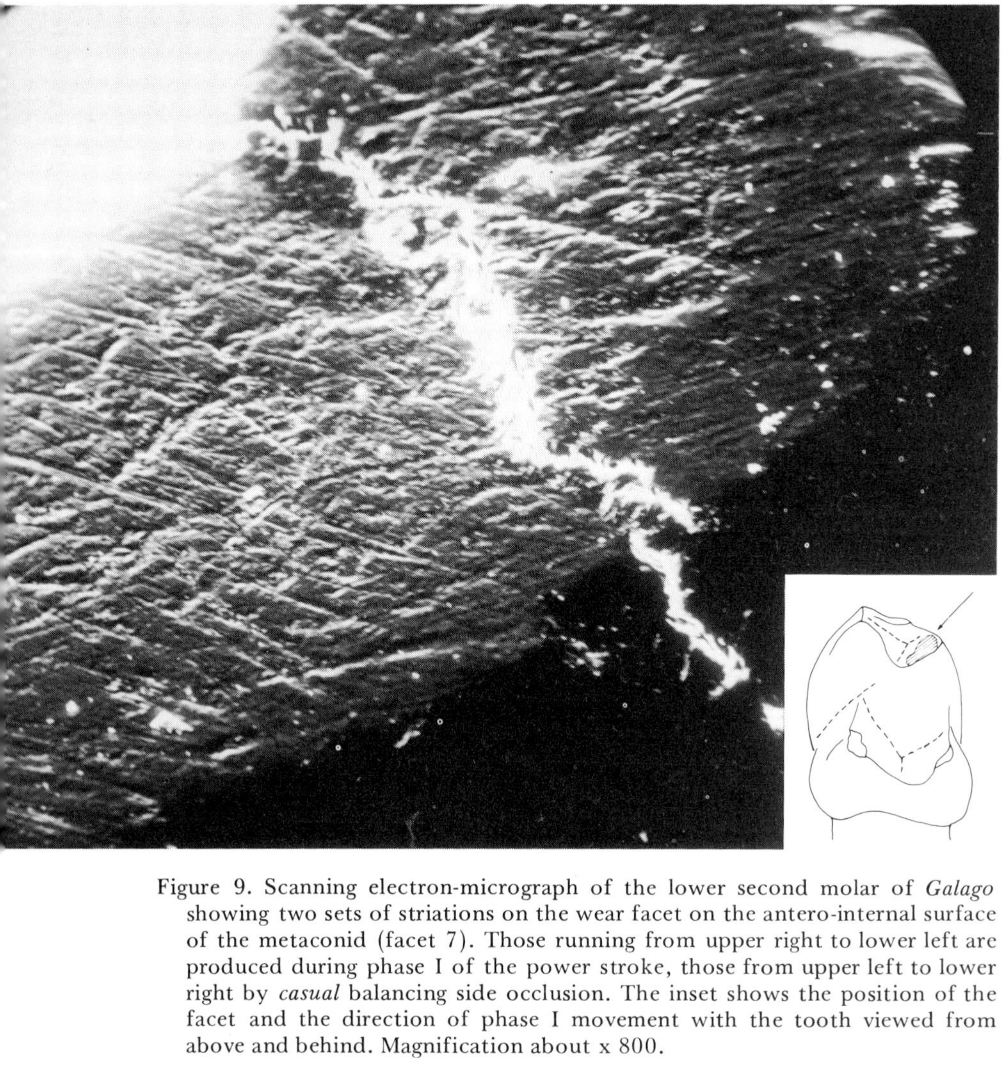

Figure 9. Scanning electron-micrograph of the lower second molar of *Galago* showing two sets of striations on the wear facet on the antero-internal surface of the metaconid (facet 7). Those running from upper right to lower left are produced during phase I of the power stroke, those from upper left to lower right by *casual* balancing side occlusion. The inset shows the position of the facet and the direction of phase I movement with the tooth viewed from above and behind. Magnification about x 800.

presence of two groups of striations running virtually at right angles to each other. The group crossing the field from upper right to lower left are parallel to the striations formed during phase I of the power stroke. The second group, passing from upper left to lower right, are only found on facet 7. Their orientation appears consistent with the effect of movement downwards and laterally by the molars on the balancing side following phase II contact on the active side. It can, of course, be argued that these striations are produced by upwards and backwards movement of the lower teeth on the active side as has been postulated from the position of wear facets and the orientation

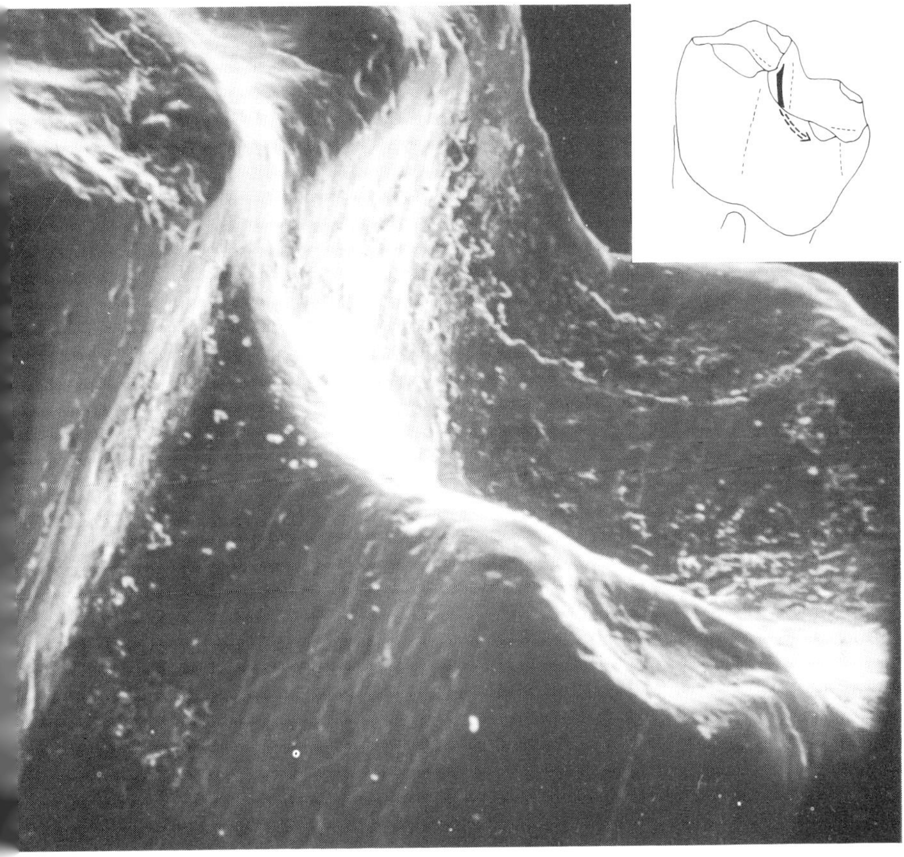

Figure 10. Scanning electron-micrograph showing the posterior margin of the trigonid and part of the talonid of the lower second molar of *Galago*. The protoconid (upper left) and the hypoconid (lower right) have been worn and dentine exposed during puncture-crushing mastication. The striations produced during shearing-grinding power strokes can be clearly seen on facet 3 (foreground) and on facet 5 (centre) where they curve continuously from phase I into phase II as indicated by the arrow in the inset. Magnification approximately x 100.

of the striations crossing them.[25] However, to date, in none of the assorted mammals studied has such a movement pattern been observed.

Discussion

This study has confirmed that in *Galago*, as in *Didelphis*,[12,14] only a very small proportion of the time spent triturating the food involves actual contact between upper and lower teeth, and that the movements of the whole lower jaw may not be predictable from the striations on the wear facets produced by such contact. The behaviour of active and balancing sides will not necessarily parallel each other where the mandibular symphysis is mobile and each dentary can rotate about its long (antero-posterior) axis. Although puncture-crushing and chewing power strokes utilise different parts of the teeth and produce different types of wear, the overall pattern of the masticatory cycle and the direction of the power stroke are similar. Since the shape of the teeth only guides and controls the relative movement of upper and lower teeth on a 'chewing' power stroke, this similarity of pattern suggests that the path of movement is almost certainly 'learned' at an early stage and is generally regulated by proprioceptive information emanating primarily from the muscles and the periodontal ligaments of the teeth, but also from the jaw joint. It follows that although direct extrapolation from one type to another, for example from *Galago* to the Miocene lorisiforms, is only wholly justifiable for the power stroke in chewing, it is likely that given occlusal and dietetic similarity, masticatory behaviour overall is likely to be comparable.

This view is to some extent substantiated by the observation that the incisors of *Galago* have a negligible masticatory function. Whilst this could be attributed to the presence of the 'dental comb' and therefore a consequence of this specialisation, behavioural studies of *Tupaia* and *Didelphis* have shown that in these animals, the incisors are rarely used and then mainly for grasping. It is suggested that since the incisors in primitive mammals have little functional importance, the adaptations for non-feeding purposes, such as grooming, do not mitigate against masticatory efficiency. If this is the case, then the spatulate 'cutting' incisors of the anthropoids are themselves a specialisation, reflecting a shift of 'biting' function from the canine-premolar series towards the incisors. This shift is not complete, since *Ateles*,[22] *Saimiri*,[13] and even man will use the premolars for tough food.

Puncture-crushing is often used in ingestion (ingestion by mastication) and involves all the cheek teeth but particularly the premolars. As might be expected, the tougher the food, the longer it is pulped in this fashion. In *Galago*, as in *Didelphis*, the gape between cycles is greatest during this stage of mastication. The explanation for the wide gape almost certainly lies in the behaviour of the muscles of mastication, as does the reason for the 'hold' seen in the recovery stroke. A suppression of downward movement while protrusion occurs, and at a gape slightly in excess of that seen in the rest position, is not unique to *Galago*. The same phenomenon occurs in *Tupaia, Didelphis, Ateles* and *Saimiri*. It is possible that protrusion following on limited opening, triggered by the external pterygoid, is necessary to give some greater mechanical advantage (in the widest sense) to the anterior suprahyoid muscles as they come into action as mandibular depressors.

The masticatory cycle in *Galago* conforms in its general pattern to that seen in all the other mammals studied using combined occlusal and cinefluorographic techniques. The essential difference between the primitive condition, as represented by *Tupaia* and *Didelphis*, is in the development of a consistent second phase in the power stroke. This is associated with the appearance of a 'grinding' type of trituration and is reflected in the changing distribution, extent and orientation of the wear facets. The facets seen in *Palenochtha* and *Galago* molars are generally very similar, but the gradual reduction and disappearance of facets 2*b*, 3*b* and 4*b*, the coalescence of facets 1*a* and 1*b*, together with the appearance of a functional hypocone with its associated facets 7 and 8, reflect a positive functional trend. The evolution of facets 1-6 in early mammalian molars has been traced from the Late Triassic to the Late Cretaceous by Crompton.[24] Their appearance and subsequent development can be related, at least in part, to the evolution of embrasure shearing as the mechanism of trituration. It involved a progressive increase in the number of transverse and obliquely orientated shearing blades coming into successive contact during a predominantly upwards and medially directed power stroke. The loss in *Galago* of many of the facets associated with this first phase of the power stroke, the reorientation of the remainder to give a more antero-posterior direction and the shift from sequential to near simultaneous contact reflects a reduction of simple embrasure shear with the new addition of a grinding action between areas contacting at centric occlusion and through the second phase of the power stroke.

In primitive mammals such as *Didelphis* there is little or no contact as the teeth move out of the power stroke, the recovery stroke involves a simple downwards movement from centric occlusion. In *Palenochtha*, as in all known Palaeocene Primates, a functional contact between the lateral slope of the protocone and the medial slope of the hypoconid and between the posterior slope of the protocone and the medial section of the paracristid can be observed. In these genera, the stylar areas, large in Late Cretaceous placentals, are much abbreviated, leading to a shorter transverse component of movement in phase I of the power stroke. This change, with the expansion of the contact areas of phase II, characterises the earliest primates, amongst other groups, appearing in the Caenozoic. *Galago* shows both these trends. The dip of many of the wear facets in *Galago* is much lower than for the corresponding facets in *Palenochtha* and the height difference between the trigonid and talonid basins is reduced. Both these features can be related to selection for an increasing component of occlusal force oriented normal to these surfaces, an adaptation for grinding, rather than parallel to the contacting blades as in shearing.

A two-stage power stroke was described by Mills for mammals, including primates: a 'buccal phase' producing wear facets with striations orientated transversely to the tooth row and a 'lingual phase' with wear facets striated obliquely.[6,16,17] Furthermore, he stated that the buccal phase on one side is coincident with the lingual phase on the other, producing a balanced occlusion, and suggested that this may help reduce the stress on the mandibular joint.[16] In the primitive mammal *Didelphis*, no lingual phase, and therefore no balanced occlusion were observed. The situation in *Galago* is somewhat different, since a phase II has developed. To that extent, phase I as described here can be equated with Mills' 'buccal phase' and phase II with his 'lingual'. However, this equation presents some difficulty in so far as the definitions given above carry the explicit statement that a phase on the active side has a simultaneous phase on the balancing side, whereas there is no evidence that phase I in *Galago* involves any tooth contact on the balancing side. Similarly, although the 'lingual phase' or phase II on the active side is, in *Galago*, followed by a delayed transitory occlusion on the balancing side, there is again no evidence that such contact is in any way functional.

Occlusion in *Galago* cannot therefore be regarded as 'balanced' in the sense Mills describes. Presumably a 'balanced' occlusion is one in which the force distribution about the jaws is such that no particular joint is unduly loaded. To achieve this in function would, in the extreme case, either require no food between the teeth on the active side, so eliminating any tendency for the lower jaw to 'rock' about a longitudinal axis, or comparable amounts of food on both sides and an effectively alternating power stroke. Neither of these conditions has been seen in *Galago*. Moreover there is a discrete and observable time lag between the completion of phase I with the onset of phase II and the transitory occlusion on the balancing side. The rapidity with which this contact occurs and is broken (assuming it to be true contact rather than simple vertical superimposition seen on the dorso-ventral projection radiographs) also mitigates against it having any real masticatory function. The terms 'buccal' and 'lingual' are therefore avoided in the present paper on the grounds that their connotations, as defined, cannot be justified on the evidence available.

In an earlier paper, Butler and Mills stated that 'the oblique scratches were produced (on certain wear facets) during rotation (of the mandible) about the contralateral condyle when the lower teeth were to the lingual side of the centric position'.[26] The transverse striations of the buccal phase were said to result from movement about the ipsilateral condyle. The free mobility of the mandibular condyles on the glenoid fossae has been observed in all the animals studied cinefluorographically and the actual range of movement occurring could not have been predicted from occlusal analysis.[13,14] Even so, a geometrical construction based on the occlusal analysis demonstrates that in *Galago*, as in *Saimiri*,[13] rotation is not occurring in the fashion described by Mills. Fig. 11 shows this construction: two lines, one from the lateral and one from the medial edge of the condyle, are drawn at a tangent to the arc of phase I movement to be expected at the second molar, given that this depends on condylar rotation. A third line is drawn showing the actual observed orientation of phase I movement. Expressing the orientation of each line as an angle between it and an arbitrary mesiodistal axis in the occlusal plane, it is clearly seen that the orientation of actual movement does not fall within the expected range for condylar rotation. This hypothesis may, therefore, be rejected. The smooth continuity between phase I and II on the active side and the medial shift of the condyle during the power stroke, suggests that the movements involved are smooth and unilateral. The mandibular joints may limit the possible range of movements in mammals including *Galago*, but do not appear to guide them.

It follows that since phase I movement is not a function of ipsilateral condylar rotation and that the angles of orientation and plunge of this movement change along the tooth row, no generalisations as to the direction and angulation of the power stroke in any mammal, let alone in fossil forms, can be made without certain precautions. Strict comparability in respect of the tooth or teeth used is essential. The movement described, since it represents a line in space, has to be characterised by relating that line to the occlusal plane or a prescribed axis in that plane. The power stroke in chewing in all mammals so far studied (including *Galago*) has upwards, medial and forwards components. It is the relative proportions of these three that delineate the distinctive nature of the power stroke in each animal. Evolutionary dental trends are normally reflected by the changing orientations of the power stroke. These cannot be adequately described in terms of either the vertical or horizontal plane alone.

This study has shown that in terms of general masticatory behaviour *Galago* conforms to the pattern found in primitive mammals such as *Didelphis* and *Tupaia* and in two other primates, *Saimiri* and *Ateles*. The implications of this observation will be discussed elsewhere.[23] However, *Galago* differs from the primitive mammals so far studied in that the power stroke has developed from a single movement carrying the lower teeth into centric occlusion (phase I) with the introduction of a crushing-grinding second phase (phase II). This appears to mark the first stage in a definite shift in primates and herbivores from simple embrasure shearing to a grinding type of power stroke.

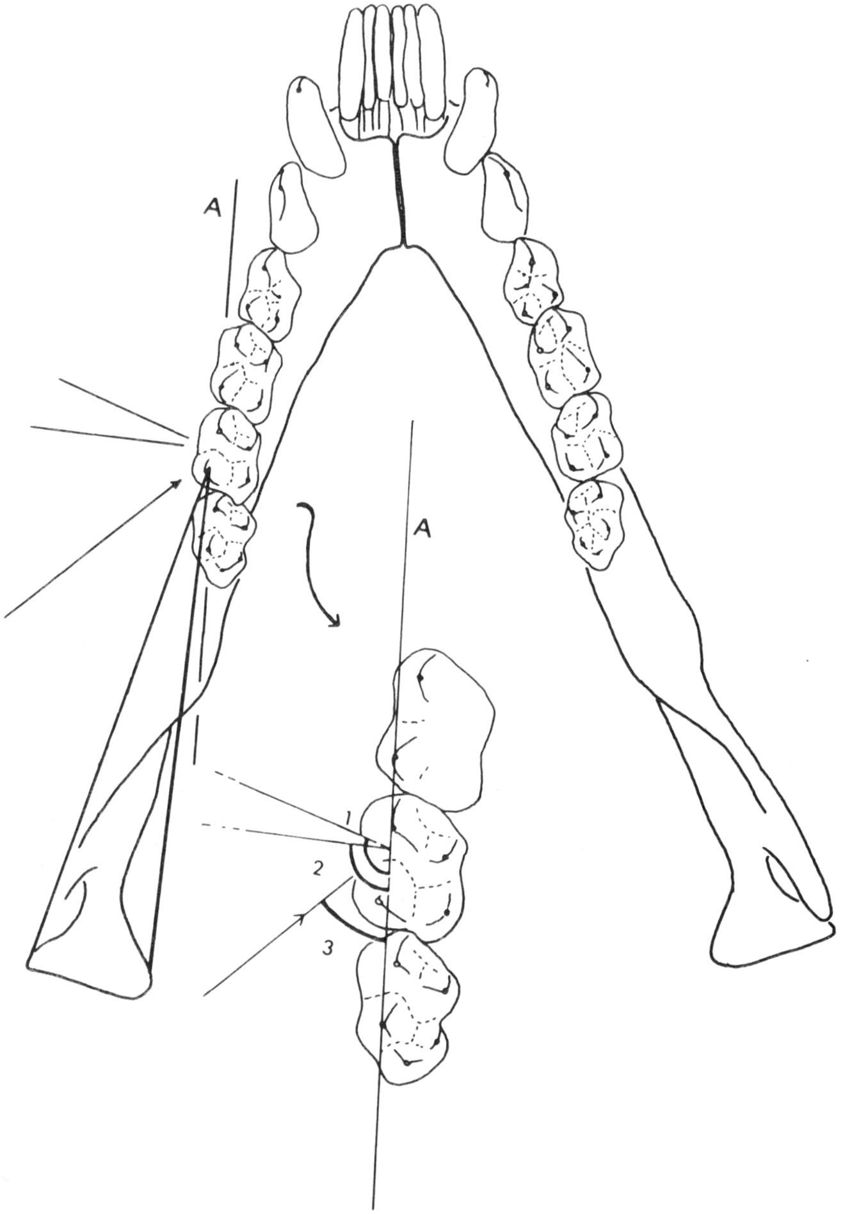

Figure 11. Occlusal view of the lower jaw and (inset) of M_{1-3} of *Galago*. The line A represents an arbitrary antero-posterior axis. Lines are shown projected from the lateral and medial borders of the condyle onto the hypoconid of M_2. On the assumption that the condyle is the centre of rotation for phase I of the power stroke, two lines (to the left of the complete tooth row) have been drawn at a tangent to the two axes of the condyle and represent the maximum range of orientations expected for the null hypothesis that condylar rotation occurs. The arrow to the left of the complete tooth row shows the observed orientations of the striations on the second molar produced in phase I. The angles 1 and 2 on the inset diagram show the angles between the antero-posterior axis and the range of angles subtended for condylar rotation. Angle 3 shows the observed angle between the orientation of phase I and the antero-posterior axis. The observed angle does not agree with that predicted for condylar rotation, confirming that such rotation does not occur during phase I of the power stroke.

Acknowledgments

The authors would like to express their great indebtedness to Dr B. Trum and especially to Dr F. Garcia of the New England Regional Primate Center for making available the specimen of *Galago crassicaudatus* used in this study, for their technical assistance and the use of laboratory facilities. One of us (R.F.K.) would like to acknowledge the support of the Wenner Gren Foundation for Anthropological Research Grant no. 2735, and the other (K.M.H.) USPHS Grant no. DE 02648. Mrs Gaby Dundon, Mrs Janet Lock and Mrs Pat Elson kindly typed and retyped the manuscript.

NOTES

1 Simpson, G.G. (1967), 'The Tertiary lorisiform primates of Africa', *Bull.Mus.Comp.Zool.* 136, 39-61.

2 Walker, A.C. (1969), 'True affinities of *Propotto leakeyi* Simpson 1967', *Nature* 223, 647-8.

3 Lewis, G.E. (1933), 'Preliminary notice of a new genus of lemuroid from the Siwaliks', *Amer.J.Sci.* 26, 134-8.

4 Simons, E.L., personal communication.

5 Walker, A.C., this volume.

6 Mills, J.R.E. (1955), 'Ideal dental occlusion in the Primates', *Dent.Pract.* 6, 47-61.

7 Ansell, W.H.F. (1960), *Mammals of Northern Rhodesia*, Lusaka.

8 Bishop, A. (1964), 'Use of the hand in lower primates', in Buettner-Janusch, J. (ed.), *Evolutionary and Genetic Biology of Primates*, vol. 2, New York and London.

9 Møller E. (1967), 'The chewing apparatus: an electromyograph study of the action of the muscles of mastication and its correlation with facial morphology', *Acta Physiol. Scand.* 69, suppl. 280.

10 Hiiemae, K.M. (1967), 'Masticatory function in the mammals', *J.Dent.Res.* 46, 883-93.

11 Hiiemae, K.M. and Ardran, G.M. (1968), 'A cinefluorographic study of mandibular movement during feeding in the rat (*Rattus norvegicus*)', *J.Zool.Lond.* 154, 139-54.

12 Hiiemae, K.M. and Crompton, A.W. (1971), 'A cinefluorographic study of feeding in the American oppossum, *Didelphis marsupialis*', in Dahlberg, A.A. (ed.), *Dental Morphology and Evolution*, Chicago.

13 Kay, R.F. (1972), 'Mastication in the squirrel monkey (*Saimiri sciureus*) and primate molar occlusion', unpublished MS.

14 Crompton, A.W. and Hiiemae, K.M. (1970), 'Molar occlusion and mandibular movements during occlusion in the American opossum *Didelphis marsupialis*', *Zool.J.Linn.Soc.* 49, 21-47.

15 Kraus, B.S., Jordan, R.E. and Abrams, L. (1969), *Dental Anatomy and Occlusion*, Baltimore.

16 Mills, J.R.E. (1963), 'Occlusion and malocclusion in the teeth of primates', in Brothwell, D.R. (ed.), *Dental Anthropology*, Oxford.

17 Mills, J.R.E. (1967), 'A comparison of lateral jaw movements in some mammals from wear facets on the teeth', *Archs.Oral.Biol.* 12, 645-61.

18 Manufactured by Ralston Purina Co., Saint Louis, MO 63188.

19 Lahee, F.H. (1957), *Field Geology*, 6th ed., New York.

20 *Palenochtha minor* was originally described by Gidley (1923) as '*Palaeochthon minor*', but Simpson (1937) transferred this species to a new genus *Palenochtha.* See Gidley, J.W. (1923), 'Palaeocene primates of the Fort Union, with discussion of relationships of Eocene primates', *Proc.U.S.Nat.Mus.* 63, 1-38; Simpson, G.G. (1937), 'The Fort Union of the Crazy Mountain Field, Montana and its mammalian faunas', *Bull.U.S.Nat.Mus.* 169, 1-287.

21 Hiiemae, K.M., unpublished observations.

22 Hiiemae, K.M. and Kay, R.F., unpublished observations.

23 Hiiemae, K.M. and Kay, R.F. (1973), 'Evolutionary trends in the dynamics of primate mastication', *Proc. Int. Congr. Primat.*, Vol. 3, pp. 28-69. Kay, R.F. and Hiiemae, K.M. (1974), 'Jaw movement and tooth use in recent and fossil primates', *Amer. J. Phys. Anthrop.*, 40, 227-56.

24 Crompton, A.W. (1971), 'The origin of the tribosphenic molar', in Kermack, D.M. and Kermack, K.A. (eds.), *Early Mammals*, London.

25 Gingerich, P., in press.

26 Butler, P.M. and Mills, J.R.E. (1959), 'A contribution to the odontology of *Oreopithecus*', *Bull.Brit.Mus.Nat.Hist.Geol.* 4, 1-26.

P. GINGERICH

Dental function in the Palaeocene primate Plesiadapis

Introduction

The main function of mammalian teeth is to reduce food matter to a size and consistency that can be swallowed and digested. The wide range of dental types seen in living mammals is the result of successive evolutionary radiations of animals adapted to masticating specific diets. In each radiation the dental morphology and adaptation are modified from those of an ancestral species to produce descendant species with a range of dental adaptations. Thus within any radiation it is usually possible to identify a number of dental types, each derived from the ancestral morphology and retaining features of it. For example, the Palaeocene primates *Plesiadapis, Phenacolemur, Carpolestes*, and *Palaechthon* exhibit four rather different adaptive modifications of the ancestral primate molar morphology. They also share many features which presumably were inherited from a common ancestor. The adaptive significance of the morphological differences seen in the teeth of these early primates can only be determined by a detailed consideration of how the teeth function.

Functional occlusion produces matching striated wear facets on upper and lower teeth. Much of the chewing behaviour of an animal can be reconstructed by studying these striated wear facets. Butler and Mills were the first systematically to map wear facets and classify them according to the direction of their striations.[1, 2] Mills distinguished two sets of wear facets on primate molars. One set is the result of an upward, medial, and slightly forward movement of the lower jaw into centric occlusion on the active side. This phase of occlusion Mills referred to as the 'buccal' phase. From centric occlusion the mandible moves forward, medially, and slightly downward on the active side, producing the second set of wear facets. This phase of occlusion Mills termed the 'lingual' phase.

Recent important cineradiographic studies of mastication in the opossum by Crompton and Hiiemae,[3] and in *Galago crassicaudatus* by Kay and Hiiemae,[4] have clarified several aspects of mandibular movement during molar occlusion. Fortunately, striated wear facets are as well preserved on the teeth of fossil mammals as on those of living animals. Thus it has been possible to identify both buccal and lingual phase facets on molars of the Eocene primate *Adapis*, as well as a third 'orthal retraction' set of facets indicating an upward and backward movement of the mandible during one stage of chewing.[5]

In this paper the wear facets on the molars and incisors of specimens of *Plesiadapis rex* (Gidley)[6] from the early Late Palaeocene Cedar Point quarry in northwestern Wyoming are described. This description will form the basis for a later comprehensive study of the evolution of dental function in the Plesiadapidae.

Molar morphology and function

Upper and lower molars of species of *Plesiadapis* have been illustrated and described in detail by Matthew,[7] Jepsen,[8] Simpson,[9] and Russell,[10] among others. The terminology used here in describing *Plesiadapis* molars is illustrated in Fig. 1. The lower molars are roughly rectangular; the protoconid and metaconid are joined by a strongly developed protocristid; the metaconid and paraconid are connate (as Matthew suggested, possibly what is here called the paraconid is really the metaconid, and what is here called the metaconid should really be considered a metastyle);[7] the paracristid runs forward to join the postcristid of the preceding molar; the cristid obliqua forms a strong shearing crest connected with the postcristid at the hypoconid; lingually the postcristid is supported by a well developed entoconid; a buccal cingulum is well developed on all molars, and a very weak lingual cingulum is suggested on some.

The upper molars are roughly triangular; the metacone and paracone support a continuous series of shearing crests: the paracrista, centrocrista, and metacrista; the paraconule and metaconule are joined to the protocone by the preprotocrista and postprotocrista respectively; the '*Nannopithex* fold'[11] is invariably a well developed crest running posteriorly from the protocone and turning abruptly buccally at the posterior margin of the tooth to form a post-cingulum; a cingulum borders the molars buccally, and a lingual cingulum is developed along the anteromedial border of the molars. Both upper and lower molars are low in profile and bear bulbous cusps.

Fig. 2 illustrates upper and lower second molars of *Plesiadapis rex* and indicates the wear facets formed during function. Each of the major crests connecting the cusps on the lower molars supports a buccal phase wear facet on its buccal surface. For descriptive

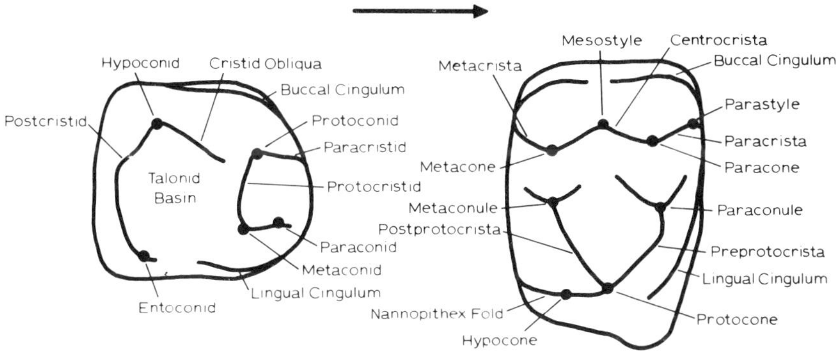

Figure 1. Terms used to describe molars of *Plesiadapis.* Lower molar is on the left, upper molar on the right. Heavy arrow is buccal to both molars and points anteriorly.

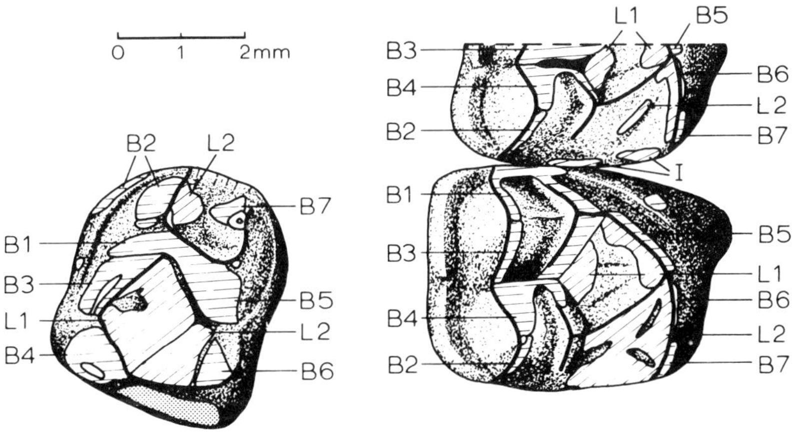

Figure 2. Wear facets on upper and lower molars of *Plesiadapis rex*, seen in occlusal view. On the left is a lower left second molar, on the right is part of an upper right first molar and an upper right second molar. Matching buccal (B) and lingual (L) phase facets are correspondingly numbered. Dotted area is an interstitial facet. Lines on the wear facets indicate the actual direction of striations and movement.

purposes, these are numbered B1-B6, following the numbering scheme recently introduced by Crompton.[12] An additional facet, B7, is present on the buccal side of the paraconid. Each buccal phase facet on the lower molars is matched by a buccal phase facet on the lingual side of the crests on the upper molars, correspondingly numbered B1-B7. The striations on all buccal phase facets are parallel, indicating that they were formed by a single jaw movement.

A second set of striated wear facets on the lower molars is formed mainly in the talonid basins. These are numbered L1 and L2 in

Fig. 2. The facet L2 generally continues posteriorly from the talonid of one molar onto the trigonid of the following molar. Thus the lower molar illustrated in Fig. 2 has a second L2 facet lingual to the protoconid, which is the posterior extension of the L2 facet on the preceeding molar. As the teeth become more heavily worn, this L2 facet on the trigonid may obliterate the neighbouring B7 facet. The L1 and L2 facets on the lower molars match facets developed on the upper molars, correspondingly numbered L1 and L2. Again, the striations on the lingual phase facets are parallel, indicating that they were formed during a single jaw movement. The two facets marked I on the upper first molar, and the corresponding area of L2 on the following molar have striation directions intermediate between those of the buccal and lingual phase facets. Matching intermediate facets are developed on the trigonid of the lower molars in some specimens. This confluency of buccal and lingual phase facets indicates that the buccal and lingual phases in *Plesiadapis* were consecutive components of a single transverse ectental jaw movement.

The buccal phase performs primarily a cutting function, with crests on the lower molars shearing past those on the upper molars. To better serve this function, the number of cutting edges on the molars was increased by two means. Cutting edges were added by developing new crests parallel to the previously existing ones. Thus the B1 protocristid cutting edge on the lower molars sheared past first the paracrista, then the preparaconule crista, and finally past the lingual cingulum on the corresponding upper molar. As the postmetaconule crista on the upper molars was limited in development by the L2 facet, a second B2 cutting edge was developed from the buccal cingulum of the lower molars. The second means of adding cutting edges is illustrated by both the B3 and B4 facets on the lower molars. As is shown in Fig. 2, here the enamel has been worn through relatively early, resulting in two resistant enamel cutting edges on a single crest. On heavily worn teeth almost all the crests bear two cutting edges as a result of wear having penetrated the enamel.

The molar teeth of *Plesiadapis* occluded in such a way that they produced on the active side a series of three compression chambers for expressing juices. At the beginning of the buccal phase the B3, B4, and L1 facets on the upper molars, and the L1, L2, B5, and B6 facets on the lower molars completely enclosed a space, the volume of which was progressively reduced as the teeth approached centric occlusion. The pulp remaining after this compression was then ground during the lingual phase.

The lingual phase performs primarily a grinding function. Food is ground between opposing planar areas of the matching lingual phase facets. In *Plesiadapis rex* the relatively rough texture of the enamel in the areas where lingual phase facets developed, illustrated in Fig. 2, caused windows in the facets until they became heavily worn. These windows functioned as cutting edges, suggesting that in *Plesiadapis* the lingual phase performed a cutting as well as a grinding function, at least in younger individuals.

Incisor morphology and function

The type species of *Plesiadapis, P. tricuspidens*, was so named because of its enlarged and distinctive tricuspid upper incisors.[13] The morphology of the lower incisors of *Plesiadapis* are also unique to the Plesiadapidae. These enlarged procumbent upper and lower incisors have been figured and described by Matthew and by Russell.[7,10] Both are illustrated in occlusal aspect in Fig. 3. The upper incisor (Fig. 3*a*) has a large anterior central cusp flanked by smaller medial and lateral cusps. Four ridges connect these three anterior cusps with a posterior cusp. These ridges became worn successively, beginning with the medial ridge, until the whole occlusal surface of the tooth anterior to the posterior cusp was excavated. A large interstitial facet developed where the right and left upper incisors contacted each other. With age, the large central anterior cusp and the medial and lateral cusps became heavily worn and blunted.

The lower incisor (Fig. 3*b*) has a slightly convex dorsal surface bordered laterally by a ridge of enamel running roughly parallel to the large interstitial facet on the medial side. The only cusp was small and developed at the posterior end of the lateral enamel ridge. With age, the tip of the incisor was gradually worn back. The tip was never sharply pointed. In some specimens the enamel is worn off on a large area of the central dorsal surface of the incisor crown; however, on other very heavily worn specimens this enamel covering is still complete.

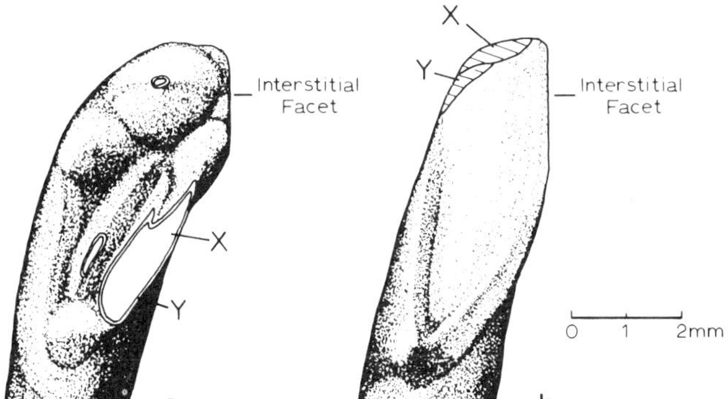

Figure 3. Wear facets on upper and lower incisors of *Plesiadapis rex*, seen in occlusal view. (*a*) right upper incisor; (*b*) left lower incisor. X and Y are matching facets on the upper and lower incisors.

The reconstructions of the skull of *Plesiadapis tricuspidens* by Simons,[14] Russell,[10] and Szalay[15,16] show that the function of these enlarged incisors was not fully understood. On the slightly worn specimens illustrated in Fig. 3 it is possible to distinguish two separate wear facets. The striated X facet on the lower incisors indicates an upwards, backwards, and slightly medial movement of the tip of the lower incisor across the ridges of the occlusal surface of the upper incisor. As the tip of the lower incisor approached the posterior cusp of the upper incisor it was forced more medially, and the leading edge of the Y facet on the lower incisor sheared past the enamel edge Y on the posteromedial side of the X facet on the upper incisor. The left and right lower incisors are just wide enough to have passed between the posterior cusps of the left and right upper incisors. From the wear on the incisors it is clear that the tips of the lower incisors functioned predominantly by shearing across the ridges (particularly the most medial ones) located just anterior to the posterior cusp on the upper incisors. The way the tip of the lower incisor closed the space in front of the posterior cusp on the upper incisor, shearing against the medial side of the posterior cusp, can best be seen in lateral view (Fig. 4). The result was a cutting device which would function most effectively in cutting stems of soft vegetation.

The tips of the upper and lower incisors could be opposed in *Plesiadapis*, but the tips of both the upper and the lower incisors became blunted with use. They therefore cannot have functioned as do the incisors of living rodents and lagomorphs, and would not have been effective in taking bites from a larger mass of food. Possibly the most similar incisor mechanism among living mammals is that found in the marsupial macropodines.[17] From striations on the interstitial facets of some of the lower incisors it is apparent that the

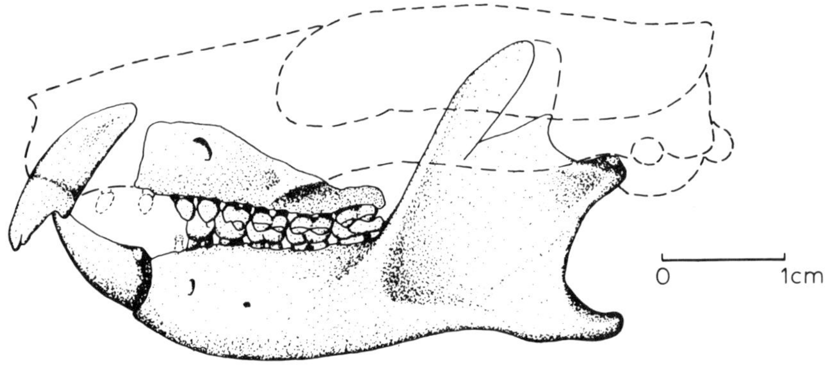

Figure 4. Reconstruction of centric occlusion in *Plesiadapis rex*, seen in lateral view. Stippled areas are actual specimens, dashed lines are hypothetical. Based on PU 21300, 21246, and an isolated upper incisor, all from the Cedar Point Quarry.

mandibular symphysis of *Plesiadapis* was highly mobile, as is true also of macropodines. The direction of movement of the lower incisors during function was upward and backward, the force for which was generated chiefly by the temporalis muscles.[18]

From this analysis of the function of the enlarged incisors of *Plesiadapis* it is apparent that when the molars were in occlusion, the incisors must have occupied approximately the position shown in Fig. 4. The lower incisors rested between and sufficiently far behind the upper incisors not to contact them during the buccal and lingual phases of molar occlusion. For the incisors to function it was necessary for the lower jaws to translate forward. With this forward translation, the mandibular condyle rode up on the articular tubercle and separated the upper and lower cheek teeth sufficiently for the incisors to function independently.

Discussion

Any attempt to determine the diet of an extinct mammal must consider the morphology of all the teeth: incisors, canines, premolars, and molars. The molars of *Plesiadapis* show relatively great development of buccal phase cutting edges, as do the molars of living grazing herbivores. The incisor mechanism of *Plesiadapis* indicates a cropping mode of food acquisition. The reduction and loss of canines and premolars in *Plesiadapis*, leaving a large diastema between the incisors and the remaining premolars and molars, is also consistent with a cropping incisor mechanism. The canine and premolar teeth of many mammals are used to shear and puncture large bites of food when they are first ingested. Reduction and loss of canines and premolars, with the resulting large diastema, is correlated with initial ingestion into the mouth of small pieces of food. Grazing mammals have a large diastema because they feed on food that is by its nature already in small pieces. Rodents have a large diastema because they gnaw a series of small pieces from a larger piece of food rather than ingesting the larger piece.[19] The great reduction of canines and premolars in *Plesiadapis*, compared for example with the large pointed premolars of *Phenacolemur*,[20] thus suggests that *Plesiadapis* also ingested food in small pieces. The shape of the cutting edges on the incisors of *Plesiadapis* further suggests that the original food mass was predominantly in the form of stems.

Teilhard de Chardin was the first to suggest that *Plesiadapis* may well have been a terrestrial animal, based on the relative abundance of *Plesiadapis* remains at Cernay.[21] In the Princeton collection from the Cedar Point quarry approximately 140 out of 400 catalogued placental mammal specimens are *Plesiadapis*, further supporting Teilhard's observation of their abundance. Simons suggested that some species of *Plesiadapis* most probably were terrestrial to account for the wide geographic distribution of the genus in North America and Europe.[22] The predominantly herbaceous diet and grazing feeding mode inferred above for *Plesiadapis* also suggest a terrestrial habitat. Among living gregarious terrestrial herbivores, the post-cranial remains of *Plesiadapis tricuspidens* compare closely in size and many morphological features with the skeleton of a marmot (genus *Marmota*). The habits and appearance of *Plesiadapis* may have been similar to those of living marmots, although this hypothesis must be tested by more detailed comparative work on the postcranial anatomy of *Plesiadapis*.

Recently Charles-Dominique and Martin suggested that allocation of *Plesiadapis* to the order Primates should be reviewed.[23] *Plesiadapis* is included in Primates by palaeontologists for two reasons. The first reason was most clearly stated by Gidley and by Simpson.[6,9] The structure of the upper and lower molars of *Plesiadapis* is virtually identical to the structure of the molars of the undoubted primate *Pelycodus* of the early Eocene. To quote Simpson:

> As regards molar pattern, *Plesiadapis* resembles the primitive Notharctinae more closely than any other group. . . . The resemblance to *Pelycodus*, most primitive known notharctine, is really amazing and extends to the apparently most insignificant details. The upper molars are of almost identical structure throughout, differing only in details of the cingula and proportions such as may characterise species of one genus. In the lower molars, *Pelycodus* has the paraconids slightly more distinct, but the resemblance is equally striking and includes even such features as the minute grooving of the trigonid face of the metaconid and the exact structure of the complex grooving of the talonid face of the hypoconid and of the whole heel of M_3.[9]

As will be discussed below, molar morphology similar to that of *Plesiadapis* is known in living mammals only within the order Primates.

The second reason for including *Plesiadapis* in Primates concerns the presence and construction of the ossified auditory bulla. The bulla of *Plesiadapis* is apparently composed of an ossification completely continuous with the petrosal,[24] a characteristic of all known adult primates.[25]

Martin employed the term *synapomorph* to designate characters acquired during the separation of a new stock, distinguishing them from *symplesiomorph* characters which are present in the ancestral stock as well as the new stock.[26] Thus it is the synapomorph characters which distinguish and define any new stock. In his list of synapomorph characters of Primates, Martin included features of the orbit, auditory region, limbs, nervous system, and reproductive tract. No mention was made of the dentition, on which the palaeontologist must base much of his interpretation. In connection with the manner of elongation of M_3, the position and mode of disappearance of the paraconid, the structure of the paracristid, and the development of the '*Nannopithex* fold' to form a posterior basin on the upper molars (all of which are associated with increased propalinal mandibular movement and encroachment of the lingual phase facets from the talonid basin of one molar onto the trigonid of the following molar), Gidley stated:

> In fact, this peculiar development of the upper and lower cheek teeth apparently constitutes a distinctively primate characteristic, which while not found in all families of the order, seems to have been repeated over and over again, with slight variations, in several related or unrelated groups, and, so far as I am aware, is not found in any other order of mammals.[6]

As molar morphology similar to that of *Plesiadapis* and *Pelycodus* (in the manner of elongation of M_3, molar trigonid construction, and development of the '*Nannopithex* fold') is found only among living and fossil Primates, this morphology apparently constitutes an important synapomorph character of the order. Later radiations have modified this basic primate molar pattern, but it must, as it is present in all early members of the order, closely approximate the morphology developed by early primates during their separation from the ancestral mammalian stock. The fact that most Eocene tarsioid primates have enlarged central incisors similar to those of primitive plesiadapoids is additional evidence that the ancestry of Primates is to be sought in these archaic forms.

In conclusion, the Plesiadapidae are best considered an early, probably terrestrial, herbivorous radiation which shared little except common ancestry with the contemporary, presumably arboreal, primates that gave rise to Eocene and later radiations. The trends toward dental reduction and specialisation seen in plesiadapid lineages preclude their having had any close relationship with living prosimians, but the Plesiadapidae are nevertheless important in that further study should reveal much about the origin and evolution of the earliest primates.

Acknowledgments

I am indebted to Professor Glenn L. Jepsen, Princeton University, for my introduction to fossil primates, and the opportunity to work with him at the Cedar Point quarry for three summers. I thank Vincent Maglio and Donald Baird for allowing me to study specimens of *Plesiadapis* in the Princeton collections. David Parris of Princeton calculated the relative number of *Plesiadapis* specimens in the Cedar

Point collection and identified them as *P. rex.* I also thank Professor J.P. Lehman and D.E. Russell of the Institut de Paléontologie, Paris, for access to their collection of *Plesiadapis tricuspidens.* I have profited greatly from discussions with Elwyn Simons and David Pilbeam, Yale University. This work was supported by a travel grant from the US National Science Foundation and by a grant-in-aid of research from the Society of Sigma Xi.

NOTES

1 Butler, P.M. (1952), 'The milk molars of the Perissodactyla, with remarks on molar occlusion', *Proc. zool. Soc. Lond.* 121, 777-817.

2 Mills, J.R.E. (1955), 'Ideal dental occlusion in the Primates', *Dent. Pract. Bristol*, 6, 47-61.

3 Crompton, A.W. and Hiiemae, K. (1970), 'Molar occlusion and mandibular movements during occlusion in the American opossum, *Didelphis marsupialis* L.', *Zool. J. Lin. Soc.* 49, 21-47.

4 Kay, R.F. and Hiiemae, K.M., this volume.

5 Gingerich, P.D. (1972), 'Molar occlusion and jaw mechanics of the Eocene primate *Adapis*', *Am. J. phys. Anthrop.*, 36, 359-68.

6 Gidley, J.W. (1923), 'Paleocene primates of the Fort Union, with discussion of relationships of Eocene primates', *Proc. U.S. Nat. Mus.* 63, 1-35.

7 Matthew, W.D. (1917), 'The dentition of *Nothodectes*', *Bull. Am. Mus. nat. Hist.* 37, 831-9.

8 Jepsen, G.L. (1930), 'Stratigraphy and paleontology of northeastern Park County, Wyoming', *Proc. Am. Phil. Soc.* 69, 463-528.

9 Simpson, G.G. (1935), 'The Tiffany fauna, Upper Paleocene: 2. Structure and relationships of *Plesiadapis*', *Am. Mus. Novit.* 816, 1-30.

10 Russell, D.E. (1964), 'Les mammifères paléocènes d'Europe', *Mém. Mus. nat. Hist. nat. Paris*, sér. C, 13, 1-324.

11 See footnote in Simpson, G.G. (1955), 'The Phenacolemuridae, new family of early primates', *Bull. Am. Mus. Nat. Hist.* 105, 435. This crest or shelf is probably better termed a postprotocingulum.

12 Crompton, A.W. (1971), 'The origin of the tribosphenic molar', in Kermack, D.M. and Kermack, K.A. (eds.), *Early Mammals, Zool. J. Linn. Soc.* 50, Suppl. 1.

13 Gervais, P. (1877), 'Quelques ossements d'animaux vertébrés recueillis aux environs de Reims', *J. Zool. Paris*, 6, 74-9.

14 Simons, E.L. (1960), 'New fossil primates: a review of the past decade', *Am. Scient.* 48, 179-92.

15 Szalay, F.S. (1971), 'Cranium of the Late Paleocene primate *Plesiadapis tricuspidens*', *Nature* 230, 324-5.

16 Gingerich, P.D. (1971), 'Cranium of *Plesiadapis*', *Nature* 232, 566.

17 Ride, W.D.L. (1959), 'Mastication and taxonomy in the macropodine skull', *Systematics Association Publication* 3, 33-59.

18 Gingerich, P.D. (1971), 'Functional significance of mandibular translation in vertebrate jaw mechanics', *Postilla* 152, 1-10.

19 Hiiemae, K.M. and Ardran, G.M. (1968), 'A cine-fluorographic study of mandibular movement during feeding in the rat (*Rattus norvegicus*)', *J. Zool. Lond.* 154, 139-54.

20 A description of dental wear and function in *Phenacolemur* is in preparation.

21 Teilhard de Chardin, P. (1922), 'Les mammifères de l'éocène inférieure francais et leurs gisements', *Ann. Paléont.* 10, 169-76, and 11, 1-108.

22 Simons, E.L. (1967), 'Fossil primates and the evolution of some primate locomotor systems', *Am. J. phys. Anthrop.* 6, 241-54.

23 Charles-Dominique, P. and Martin, R.D. (1970), 'Evolution of lorises and lemurs', *Nature* 227, 257-60.

24 Russell, D.E. (1959), 'Le crâne de *Plesiadapis*', *Bull. Soc. Géol. Fr.*, sér. 7, 1, 312-14.

25 McKenna, M.C. (1966), 'Paleontology and the origin of the primates', *Folia primat.* 4, 1-25.

26 Martin, R.D. (1968), 'Towards a new definition of Primates', *Man* 3, 377-401.

D. SELIGSOHN and F. S. SZALAY

Dental occlusion and the masticatory apparatus in Lemur *and* Varecia: *their bearing on the systematics of living and fossil primates*

Introduction

Fossil primates are usually characterised by their dental and mandibular parts, primarily because teeth and jaw fragments are the only remnants in most cases. Diagnoses are based on these fragments, and more often than not it is the characters of the dentition, particularly of the molars, which serve to delineate the fossil taxa from one another or from living relatives.

Apart from the taxonomic importance of the dentition, the teeth, along with the jaws and skull, when preserved, provide the only basis for deciphering the role of each of the fossil species in their extinct ecosystem. Yet, without a clear appreciation of the dietary factors which molded the feeding mechanism of living species, and of their inherited morphology, there is little hope of understanding the palaeobiology of past radiations. Recent outstanding work by Charles-Dominique,[1,2] Hladik and Hladik,[3] Hladik et al.,[4] Petter and Hladik,[5] Petter and Peyrieras,[6] Sussman,[7] and others on the natural history of several primate species increasingly demonstrates that most living primates have a relatively species-specific dietary regime, in a feeding zone or area of the forest equally characteristic for most species. These fundamental works in primate natural history are of extreme value and supply the foundations upon which morphologists may begin causal analyses of the dental, mandibular, and cranial morphology. Most living primates are still very poorly understood in this respect, and consequently their extinct relatives fare no better.

We believe that the general practice of using dental criteria to separate fossil species and genera is well tested and sound. Yet in mammalian systematics the time is approaching when the diagnostic differences between species or at least genera should be causally explained.

The range of dental modifications in the Recent lemurs of Madagascar (the subfossils included) surpasses the spectrum of diversity shown by either the known catarrhines, platyrrhines, or the once very abundant tarsiiforms, the omomyids, microchoerids, and the extant tarsiids. They are therefore an ideal living group for the study of phylogenetic modifications of the masticatory system. The dentitions of the *Lemur-Varecia* group are highly derived ones, apart from the obvious specialisations of the tooth comb and associated modifications of second lower premolars. Because many features of their anatomy point to a closer common ancestry between *Lemur* spp. and *Varecia* spp. among known lemurids than any of these with either species of *Lepilemur* or *Hapalemur*, the divergence in their feeding mechanisms is of particular interest.

We have selected these two species groups of lemuriforms because they have been treated, as a result of no critical evaluation, under one genus, yet clearly fall into two very distinct genera according to the morphological criteria of the feeding mechanism, the dentition in particular. We have consequently analysed the differences in morphology, occlusion and other function and attempted to correlate those differences with what is known of the diet of these groups.

Although the extant *Varecia variegatus* and the extinct *V. jullyi* and *V. insignis* have been often categorically lumped in the genus *Lemur* with the *Lemur catta, L. rubriventer, L. mongoz,* and *L. macaco* group, the glaring differences in their dentition immediately signify an evolutionary divergence of generic level to students of fossil mammals. Based on what we know of the total morphology of *Lemur* and *Varecia*, it appears that they have evolved from a common ancestor. The divergence in the dentition and the musculoskeletal part of the feeding mechanism can be explained, we believe, by the differences of their dietary adaptations.

We hope that this study will achieve a better understanding of the phylogenetic modifications of the feeding mechanism in these genera, demonstrate the derived generic differences, point to some of the possible causes for these differences, as well as point out the necessity for rigorous research on as many living species as possible, to understand mammalian feeding mechanisms. Without scrutiny of diet, dental, mandibular and cranial morphology, and masticatory function in the living species, the morphology of fossils, rich with information, will supply us only with a small part of the precious knowledge of past adaptations.

Materials and methods

Numerous skulls of *Lemur catta, L. mongoz, L. macaco, L. rubriventer, Varecia variegatus, V. jullyi,* and *V. insignis* were examined primarily in the American Museum of Natural History, New York; the British Museum (Natural History), London; the Muséum National d'Histoire Naturelle, Paris; and the Academie Malagache, Tananarive. With the aid of a metric grid, line drawings were made of the upper left and lower right molars of *Lemur macaco* (AMNH no. 170764, Figs. 1 and 2) and *Varecia variegatus* (AMNH no. 77792, Figs. 3 and 4). Transparencies were then made of the line drawings of the lower right molars, and the masticatory cycle was simulated by moving the tracings of the lower molars across the upper ones. Figures were consequently prepared (see later – Fig. 6) from the superimposed drawings, controlled by rechecking the occluding dentitions.

The masticatory power strokes of both *Lemur macaco* and *Varecia variegatus* were then simulated by hand, using the mandibles and crania. Implicit in this simulation was the assumption that the lower jaws are naturally 'guided' through the masticatory stroke by virtue of the fact that the morphology of the upper and lower tooth rows complement each other.

For reference points, the tips of the protoconids of M_2 and M_3 were observed as they traversed the upper molars during numerous instances of simulated mastication (Fig. 6). These points on the transparencies, i.e. the tips of the protoconids of M_2 and M_3, were then plotted on the line drawings of the upper left molars indicating the successive points through which the protoconids passed during the initial (occlusal), centric occlusal and, where appropriate, terminal phases of the masticatory stroke. The main axes of the upper and lower tooth rows are also shown on Fig. 6. The axis of the upper tooth row is represented by a line connecting a point at the central base of the upper canine with a point at the centre of the glenoid fossa. The axis of the lower tooth row is represented by a line connecting a point at the central base of the lower P_2 with a point lying medially along the transverse axis of the mandibular condyle.

With the aid of these reference markers, the transparencies of the lower right molars were then superimposed on the line drawings of the upper left molars. The former were then moved over the latter with the appropriate dots on each coinciding in each successive phase of the simulated power stroke.

Additional information on dental, mandibular, and cranial morphology was gathered and analysed.

Molar morphology

1. *Lemur macaco*

The upper and lower molars of *Lemur macaco*, as in *L. catta* or *L. mongoz*, demonstrate relatively little relief in their crown structure (see Figs. 1 and 2). The upper molars possess low, sub-hemispherical paracones and metacones with edges, the former being more prominent than the latter. The protocones are broad, flattened, U-shaped shelves which tilt in a buccal and slightly posterior direction. This results in trigon basins (central foveae) which are relatively shallow and ill-defined. Projecting from the protocones of M^1 and M^2 are a pericone and a hypocone, mesio-lingually and disto-lingually located, respectively. The more mesial pericones are more lingually projecting and lie relatively flush with the palate. Viewing the palate ventrally, the hypocones are more prominently elevated above the palate than the pericones.

The lower molars are constructed in a complementary manner. The trigonid basins (anterior foveae) are but moderately elevated above the talonid basins (posterior foveae), while the relief of the cusps which surround these basins is minimised or eliminated. The trigonid and talonid basins are thus actually flattish planes, lingually down-tilted from the buccal shearing edges. The protoconids and metaconids of all three lower molars are separated by relatively restricted anterior foveae. There is a significant difference between the configurations of the anterior fovea of M_1 and M_2. The anterior

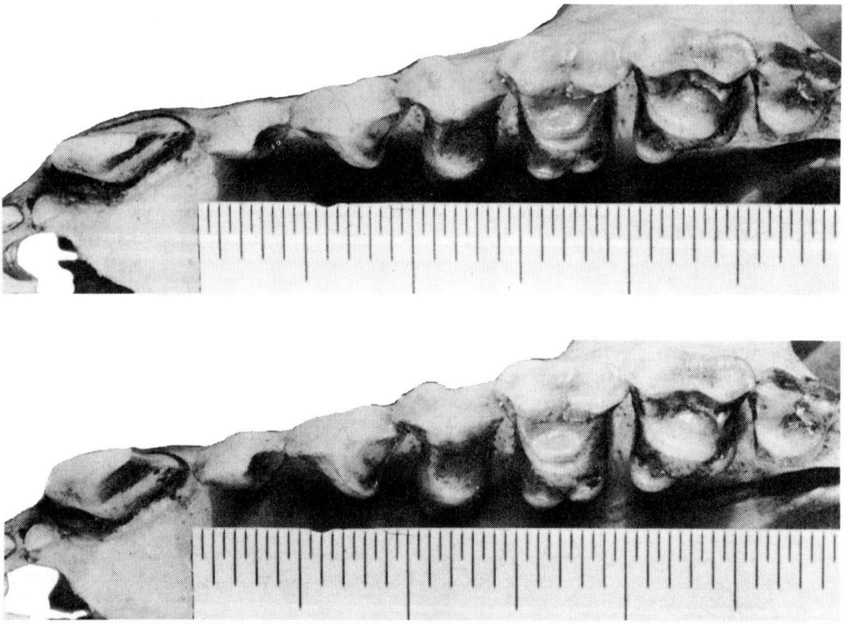

Figure 1. *Lemur macaco*, AMNH no. 41264, occlusal view of left upper dentition.

fovea of M_1 is 'closed off' anteriorly by a paracristid projecting antero-lingually from the protoconid. The anterior fovea of M_2, however, is open mesially. The lower molars are also distinctive in that the trigonid and talonid basins are displaced lingually on the occlusal surfaces of the crowns to a relatively great degree. The buccal faces of the lower molar crowns thus appear to slope downward in a markedly diagonal fashion from the rims of the trigonid and talonid basins. The talonid basin is completely closed off distally and lingually by a long postcristid.

Both M^1 and M^2 are oriented similarly with respect to the axis of the upper tooth row, with only a moderate rotation of M^3. This results in a relatively continuous system of ectolophic crests. With respect to M^1 and M^2, M^3 is only moderately reduced in size.

The lower molars are similarly oriented relative to the axis of the lower tooth row; there is relatively little tooth rotation.

2. Varecia variegatus

The molar teeth of *Varecia variegatus*, like those of *V. insignis* and *V. jullyi*, display more relief than those of *Lemur* spp. (see Figs. 3 and 4). The paracones and metacones of the upper molars are higher, steeper and mesio-distally more elongated than those of *Lemur*. As in

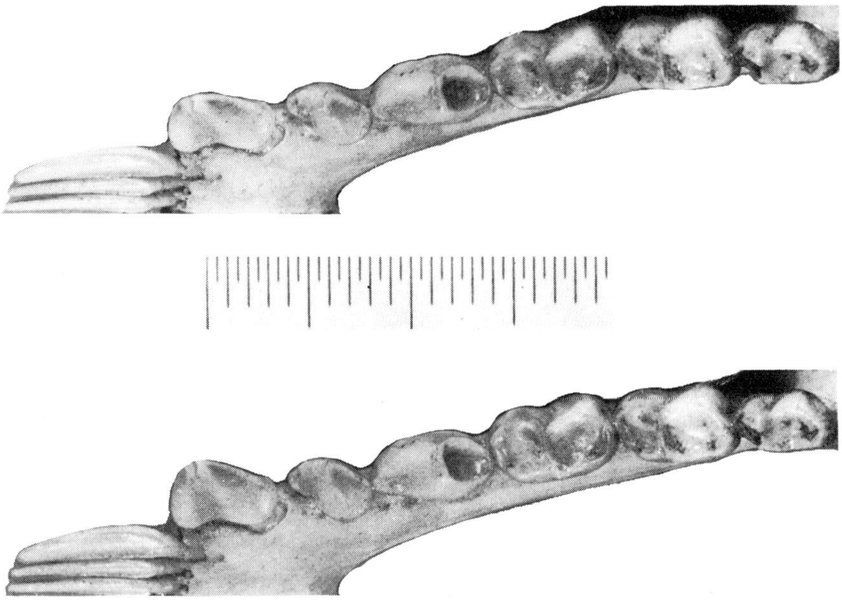

Figure 2. *Lemur macaco*, AMNH no. 41264, occlusal view of right lower dentition.

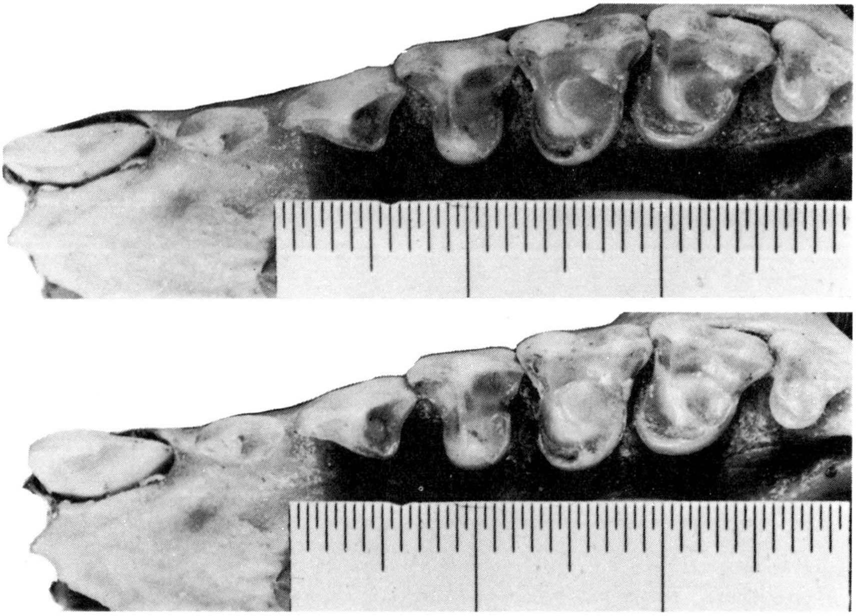

Figure 3. *Varecia variegatus*, AMNH no. 77792, occlusal view of left upper dentition.

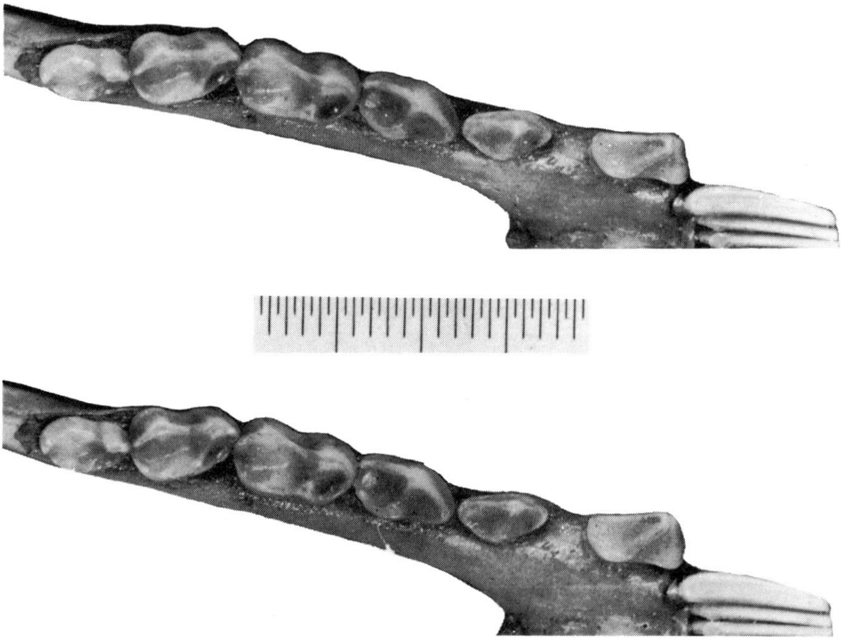

Figure 4. *Varecia variegatus*, AMNH no. 77792, occlusal view of left lower dentition.

Lemur, the paracones are more prominent than the metacones. The protocones of M^1 and M^2 are more restricted and conical in configuration, and are rotated greatly on the occlusal surface so as to dip virtually distally, and only slightly buccally. The basic bucco-lingual orientation of the protocone crests is thus different from the mesio-distal one of *Lemur*. The resulting trigon basins are relatively deeper and more defined, but are shifted from a central to a more distal position as compared to *Lemur*. There are extensive, shelf-like lingual cingula which project mesio-lingually from the base of the protocones of M^1 and M^2. These cingula 'dip' palatally as they extend mesially. The M^1 lingual cingulum usually possesses a small cuspule on the central portion of its lingual rim.

The lower molars of *Varecia variegatus*, *V. insignis*, and *V. jullyi* are similarly distinguished from those of *Lemur* spp. The trigonid and talonid basins are basically large excavations in the lower molar tooth crowns, enclosed by narrow, cusp-studded crests. The cusps and crests forming the buccal rims of the trigonid and talonid basins are far more prominent than those forming the lingual rims. The 'slopes' of the talonid basins thus complement the 'slopes' of the corresponding protocones. It should be noted that the buccal faces of the lower molar tooth crowns are more nearly vertically disposed.

The trigonid basin of M_2 tends to be more open anteriorly than that of M_1 and the talonid basins in general run in a mesio-buccal-disto-lingual direction. Because of their very low entoconids and postcristids, the talonid basins appear to extend through disto-lingually situated 'gaps' in their basin rims.

The upper molars, M^2 and M^3 in particular, tend to be rotated; M^3 is relatively and absolutely more diminutive than M^3 in *Lemur macaco*. Whenever rotation is present, it results in an interrupted ectolophic ridge system.

The second lower molar, as M_3, tends to be rotated, corresponding to the condition of M^2. The M_3, like M^3, is markedly reduced.

Molar wear

Because of the low relief of the lower molars of *Lemur macaco*, the cutting edges on them, resulting from tooth-on-tooth and tooth-food-tooth contact, are rather extensive and elaborate (Figs. 5*A* and *B*). The enamel cutting edges in the trigonid basins form large, modified figure 8s with extensive exposures of protoconid and metaconid dentine within the loops of these 8s. Occlusal wear around the talonid basins results in extensive exposures of dentine. The cutting edges generated around these exposures of denture run along the cristid obliqua and postcristid. On M_1 especially, the enamel cutting edges of both the trigonid and talonid regions form one continuous system.

In the upper molars extensive dentine exposures are evident along the crests of the ectolophs (the continuous crests of the paracone and metacone), especially on M^1, and along the crescentic rims of the protocones. Striated wear facets on the upper molars are evident, especially along the buccal rim of the crests of the protocones and along the centrocristas of M^1 and M^2 (see Fig. 5*A*). The striations on all these facets run in a lingual and moderately mesial direction. Striations on wear facets in the lower molars also run in a mesiolingual direction. These striated facets run along the lingual edge of the cristid obliqua, as well as along the buccal face of the crown below the protoconid and cristid obliqua.

Dentine exposures in the lower molars of *Varecia* (Fig. 5*D*) usually appear to be less extensive than those of *Lemur*. They are restricted to the narrow crests defining the anterior foveae and talonid basins and are best developed on the tips of the protoconids and hypoconids of M_1, M_2 and M_3. As in the lower molars, dentine exposures in the upper molars of *Varecia variegatus* (Fig.5*C*) are usually poorly developed, compared to those of *Lemur* spp., being essentially restricted in a comparable stage of wear to the very narrow ridges along the ectolophs and protocones.[8] The wear gradient in *Lemur macaco* appears to be drastically steeper than in *Varecia variegatus*. Faint striations, running essentially antero-posteriorly, were detected on M^1 and M^2 in the polished 'valleys' created at the junctures of the steep-sided paracones and preproto-cristas. The upper molar occlusal surfaces are, however, pre-ponderantly nondescript or mottled in appearance, with mottling being very evident within the trigon basins and along the ectoloph crests. The crest of the protocone of M^2 demonstrates a more pitted appearance.

Striations were detected on the buccal portions of the postcristids of M_1 and M_2, running in a markedly mesial and slightly lingual direction. As with the upper molars, however, the lower molar basins and crests are basically mottled or pitted, the pitting being most evident on the tops of the hypoconids of M_2 and M_3.

Occlusion

In studying the dentitions and the two types of wear on the molars of the two genera, two occlusal patterns became apparent. One pattern is associated with an inferred, primarily crushing-piercing type of chewing, as in *Varecia*, and the other with one in which cutting is of primary importance, as in *Lemur*. In the first type the cusps become abraded by the pounding-crushing contacts, whereas in the second type, in forms with extensive edge-to-edge shear, distinct and (depending on the diet) sometimes polished, sometimes grooved shear facets are especially characteristic of the molars. There

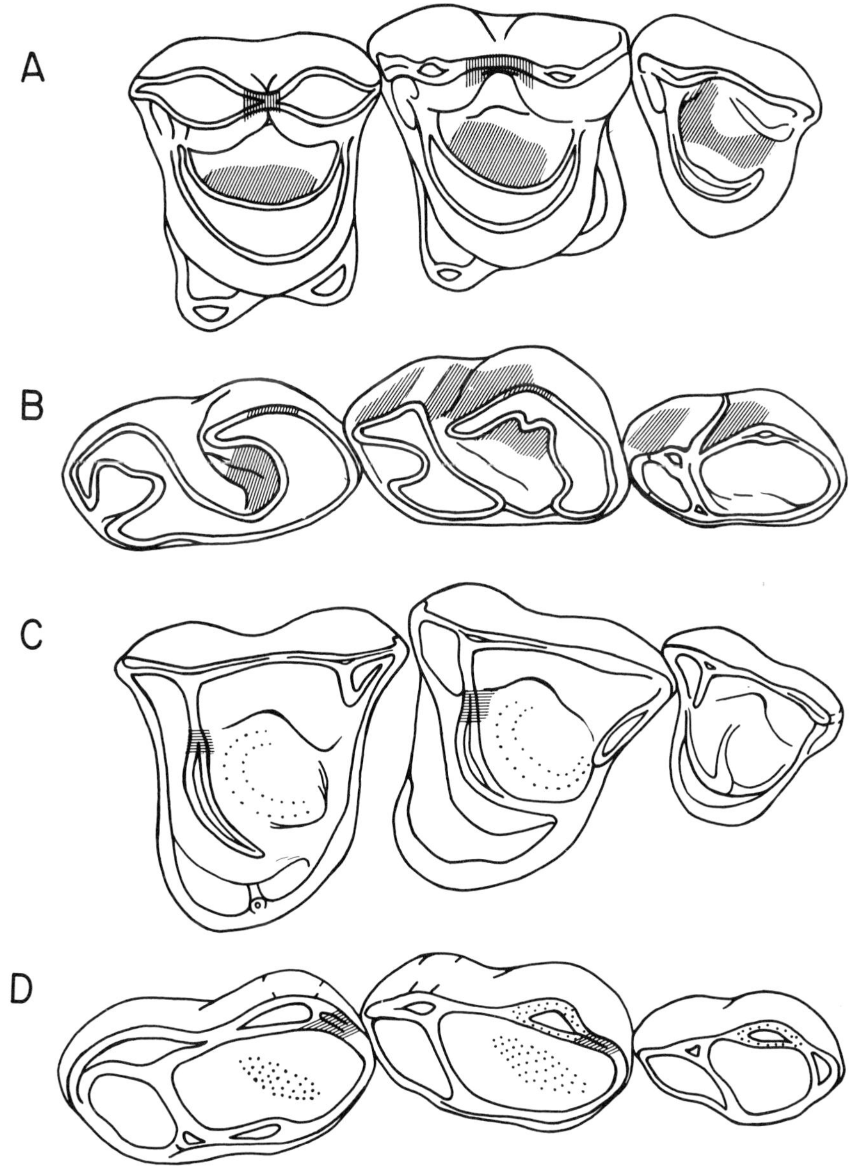

Figure 5. Contrast of molar wear in *Lemur macaco* and *Varecia variegatus*. *A* and *B* represent occlusal views of the upper and lower molars, respectively, of *Lemur macaco*, while *C* and *D* represent the occlusal views of the upper and lower molars, respectively, of *Varecia variegatus*.

The closely spaced parallel lines in *A*, *B*, *C*, and *D* represent patterns of striations in regions where they were observed. In *C* and *D*, dots in the trigon and talonid basins schematically indicate a mottled surface. Dots on the hypoconids of M_2 and M_3 indicate a more pitted surface.

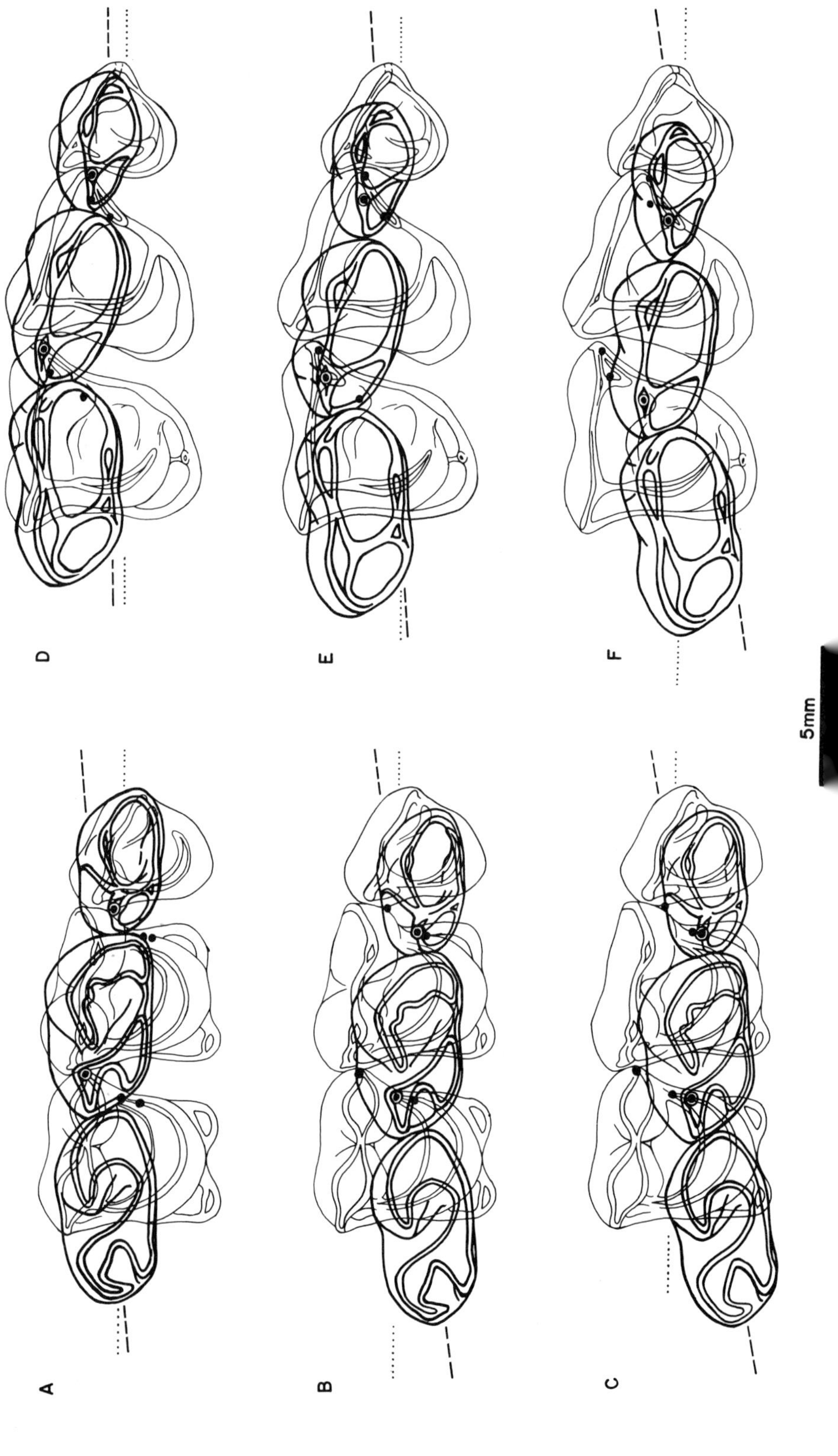
A
B
C
D
E
F
5mm

are some indications, according to the hitherto unappreciated discoveries of Every,[9,10] that cutting edges destroyed by the crushing-puncturing action are restored, and sharpened during non-chewing masticatory activities, though this is difficult to prove. The pioneering studies of Butler and subsequent works by Mills[11-16] have shown that grooves and striations, whenever present on contact facets, are particularly excellent indicators of the direction of jaw movements.

The patterns of molar occlusion for both *Lemur macaco* and *Varecia variegatus* are illustrated in occlusal view (Fig. 6). As mentioned previously, the protoconids of M_2 and M_3 are being used for reference. The figures indicate the paths taken by the lower molars as they move across the upper molars during the power stroke. In addition, the illustrations demonstrate the fact that the molar morphology of the *Lemur macaco* and *Varecia variegatus* differs considerably and that food is processed in distinctive ways.

The major 'theme' of the *Lemur* species group, as shown by the occlusion of *Lemur macaco*, appears to be primarily transverse cutting. The process is continuous but can be conveniently divided into two phases.

The first phase of the power stroke features the mesiolingual and moderately dorsal passage of the paracristid, cristid obliqua and postcristid of each lower molar across the appropriate ectoloph on the upper molars and premolars. Immediately following this, the diagonally sloping buccal faces of the lower molar crowns occlude with the ectolophs. This process is also mesiolingual and moderately dorsal. The entire first phase, when viewed in the coronal plane,

Figure 6. Occlusal diagrams of the right molars in *Lemur macaco* (*A,B,C*) and *Varecia variegatus* (*D,E,F*). The encircled dots indicate the tips of the protoconids of M_2 and M_3, while the three solid dots on M_1 and M_2 indicate the successive (and arbitrarily selected) positions through which the M_2 and M_3 protoconids pass during mastication.

The dotted lines represent the longitudinal axes of the upper tooth rows, which are defined here to connect a point at the central base of the upper canine with a point in the centre of the glenoid fossa. The axes of the upper tooth rows of both genera are aligned here in continuous and parallel fashion in *A* and *D*, *B* and *E*, *C* and *F*. The dashed lines represent the longitudinal axes of the lower tooth rows, which are defined here to connect a point at the central base of P_2 with a point medially situated on the transverse axis of the madibular condyle.

The three solid dots on M^1 and M^2 on *Lemur macaco* (*A*, *B* and *C*) indicate the positions of the M_2 and M_3 protoconids during the occlusal (*A*), centric occlusal (*B*), and terminal phases (*C*) of the masticatory stroke. The three solid dots on M^1 and M^2 in *Varecia variegatus* (*D*, *E* and *F*) represent the positions of the M_2 and M_3 protoconids as they move from the occlusal (*D*) to the centric occlusal phase (*F*) of the masticatory stroke.

A, *B* and *C* demonstrate that transverse motion is relatively greatly emphasised in *Lemur macaco*, while *D*, *E* and *F* indicate a relatively greater anterior component in the mandibular movement in *Varecia variegatus*.

describes a relatively low trajectory, being far more ectental (transverse) than orthal (vertical). This permits, as can be seen, the very extensive apposition and contact of cutting edges.

The conclusion of the first phase merges into the second phase, that of centric occlusion, followed by a final, short, mesio-ventro-lingual movement of the talonid basins across the protocones. Centric occlusion possibly results in the pulverisation of the previously sliced vegetable material as the plane-like talonid basins firmly engage with the shelf-like protocones. The terminal phase results in one last shearing movement: the cutting edges along the buccal rims of the talonid basins approach those along the protocone crests and probably rasp the food pressed in between. The striation patterns on the wear surfaces of the molars appear to reflect the primarily transverse cutting action of these teeth.

During the power stroke, the lower molars are displaced lingually to a relatively great extent.

The anterior fovea of M_2 seems to function, quite interestingly, as a sort of funnel or scoop. As the anterior fovea of M_2 lingually traverses the postprotocrista of M^1, it can conceivably catch food and channel it buccally toward the protoconid of M_2. This food would then be sheared as the cutting edges of the M_2 protoconid moved across the M^1 postprotocrista. The functional significance of the morphological differences seen in the M_1 and M_2 anterior foveae becomes readily apparent when one realises that the trigonid of the M_2 occludes with the postprotocrista of a preceding upper molar, whereas that of M_1 does not.

The function of the lingual cingulum and the pericones, so well developed in *Lemur* and many other primates, is difficult to explain. No pericones come into tooth-on-tooth contact with the lower molars until the crowns of the upper molars are well worn. Only then does the disto-lingual pericone of M^2 actually occlude with the M_2 postcristid, thus serving as an accessory cutting edge.

The basic functions of the cheek tooth row in *Varecia*, as displayed in *Varecia variegatus* (Fig. 6*D, E, F*), are significantly different from *Lemur* spp. The power stroke is comprised of two rather distinct phases. In the first phase, the lower molars are displaced upwards and mesially. The protoconids, hypoconids, and associated crests are thus 'slid' upward and mesially past the lingual rims of the upper molar ectolophs. In the second phase, the lower molars continue on their upward course, but are deflected lingually so that their movement in the occlusal plane is now mesio-lingual. With this movement, the molars ultimately achieve centric occlusion, as follows: (i) the protoconids of M_2 and M_3 engage with M^1 and M^2 lingual to the metacones; (ii) the protocones of M^1, M^2 and M^3 and the hypoconids of M_1, M_2 and M_3 are thrust respectively into the talonid basins of M_1, M_2 and M_3 and the trigon basins of M^1, M^2 and M^3.

The first phase of the masticatory stroke emphasises the sliding of sharp crests past each other. This situation seems well suited for rolling and cracking solid, particulate plant material. The second phase features the firm engagement of cusps with basins in mortar and pestle fashion. This occlusal pattern, it appears, excellently equips the molars to crush and mash stems or hard, discrete food objects once they have been cracked and splintered during the first phase. The wear of the teeth seems to corroborate this interpretation.

The movement of the lower jaw of *Varecia variegatus*, when viewed in the sagittal plane, emphasises both vertical and particularly mesial components, in contrast to the homologous jaw movement of *Lemur macaco*. The lower jaw of *Varecia variegatus* is thus displaced mesially to a relatively greater degree than it is displaced transversely, whereas the reverse is true for *Lemur macaco*.

The anterior foveae of both M_2 and M_3 appear to be quite functional. They are extensive and are posteriorly juxtaposed to the preceding talonid basins. During mastication, food is probably mashed within the talonid basins. Some of the processed food may be forced out of these basins, being carried outward primarily in a disto-lingual direction through the characteristic spout-like 'outlets' in the disto-lingual rim of the talonid basin described previously. A similar spout-like outlet characterises the lower molars of the Palaeocene picrodontids.[17] It would appear that the anterior foveae may be adapted to catch and retain this food-overflow. The retaining action of the anterior foveae and the extensive mesial travel of the cheek tooth row would insure that a maximum mass of food is available to the protocone-talonid and trigonid basin mashing apparatus. The anterior foveae of M_2 and M_3 can then be interpreted as functional extensions of the preceding talonid basins.

The extensive lingual cingula of M^1 and M^2 may partially function as broad retention areas for the masticated food carried mesio-lingually, although this is uncertain. We do not believe that extensive contralateral contact occurs or is responsible for the development of the lingual shelves in *Lemur* or *Varecia*. It is not necessary here to discuss the evidence for or against a 'lingual phase', *sensu* Mills.[15,16] It may be noted, however, that one of us (Szalay), after having re-examined the Palaeocene evidence, concluded that contralateral contact of molars is not responsible for several facets on the occluding hypoconid-protocone and the hypocone shelf-paracristid in Palaeocene primates. Every's studies should eventually reflect on this subject.

From the molar evidence, we have established a model for mastication in *Lemur macaco*, which emphasises transverse cutting. The premolars also seem to adhere to this model (see Figs. 1 and 2). The lingual elements of both P^3 and P^4 are relatively well developed, while the metaconid of P^4 is prominent. These features result in bucco-lingually elaborated cutting edges on both the mesial and distal slopes of these teeth. Running in parallel down the buccal and lingual margins of both the mesial and distal slopes of P_4 are two cutting edges. Because P_4 tightly interlocks with P^3 and P^4 during occlusion, it is naturally guided lingually between these teeth. The cutting edges on the mesial and distal slopes of P_4 thus make firm contact with the lingually extended cutting edges of P^3 and P^4. This situation seems to be conducive to efficient cutting and slicing, and appears to be consistent with the occlusal pattern of the molars outlined above.

The situation in *Varecia*, exemplified by the extant *V. variegatus*, is quite different (see Figs. 3 and 4). The lingual elements of P^3 and P^4 are weakly developed. The lingual portion of P^4 is distinctly more cuspate, while P^3 appears more caniniform. The lower fourth premolar has a low metaconid, with fossae replacing cutting edges on both the mesial and distal slopes. The lower fourth premolar is also rotated mesio-lingually. Occlusion, therefore, between it and P^3 and P^4 is altered. Because of its rotation, P_4 slides mesio-lingually between P^3 and P^4. This movement is facilitated by the fact that two relatively wide diastemata exist between the upper canine and P^2 and between P_2 and P_3, which permit the lower premolars to slide mesio-lingually between the upper premolars without interlocking. The morphology and occlusion of the premolars of *Varecia variegatus* thus minimise the possibility of efficient transverse cutting and slicing, and suggest instead a functional emphasis on the grasping, rolling and cracking of solid, particulate vegetable material. This reflects the functional theme seen in the occlusion of the molars.

Mandibular and cranial morphology

The lower jaw of *Varecia variegatus* displays marked thickening immediately beneath and medial to P_4-M_2. This thickening, when viewed ventrally, appears to align with the centrically occluded protocones and talonid basins. This thickening may possibly be a means by which the lower jaw is buttressed against stresses as they are generated ventrally through the medial half of the body of the mandible by the crushing function of the molars. *Lemur macaco* shows no such bucco-lingual thickening in the mandible.

The distinctive patterns of mastication in *Lemur* and *Varecia* seem also to be reflected in the osteology of the crania. In *Lemur macaco*, there is a tendency for the zygomatic arches (when viewed in a coronal plane) to be oriented ventro-medially, in some cases to a great extent (Fig. 7*A*). This may suggest a relatively more oblique line of action for the masseter in the coronal plane. At the same time, the pterygoid plates tend to be relatively more medially situated with respect to the upper tooth rows, with the lateral pterygoid plates projecting ventro-laterally from the basicranium. This might indicate that the medial pterygoid muscles in *L. macaco*, like the masseters, possess a relatively more oblique line of action when viewed in the coronal plane. Because the lateral pterygoid plates tend to be situated relatively more directly medial to the mandibular condyles, the medial pterygoid muscles, when viewed ventrally, would probably tend to have a predominatly medio-lateral orientation. The masseters and the medial pterygoids, then, would seem to be disposed so as to facilitate primarily transverse mandibular movement.

The cranium of *Varecia variegatus*, when similarly viewed, contrasts with that of *Lemur macaco*. In *Varecia variegatus* the zygomatic arches are not oriented ventro-medially, but are essentially vertical (Fig. 7*B*). A more vertical line of action is suggested for the masseter in *Varecia*. Similarly, the location of the pterygoid plates is relatively more lateral with respect to the upper tooth rows, while the pterygoid plates are also more vertically disposed. This suggests a more vertical line of action for the medial pterygoid muscles. In contrast to *Lemur macaco*, the medial pterygoid muscles may possibly facilitate more anterior mandibular movement, as the lateral pterygoid plates lie more anterior to the mandibular condyles.

In *Varecia variegatus*, then, the masseters and medial pterygoids appear to be better suited for producing relatively more orthal (vertical) movement in the lower jaw. The medial pterygoids also appear to have a greater ability to protract the mandible than in *Lemur macaco*. The evidence from the cranium apparently agrees with that of the molars in suggesting a more vertically and more anteriorly directed lower jaw.

Discussion and conclusions

Studies on tooth wear of extant and extinct taxa certainly contribute toward the understanding of functional significance of tooth morphology in a given species. On the other hand, meticulous studies on the natural history of living species, assessment of the dietary regime, and an understanding of the path of phylogeny and evolutionary dental morphology are of no less significance. It is only with a judicious combination of all these three approaches, and not by wear studies alone, that one can hope to understand the feeding mechanism (the teeth in particular) of any living or fossil species of mammal.

One of the greatest weaknesses of our pilot study is the still insufficient information about the dietary habits of the species in question in their natural habitats. It was the consensus at the conference among several students of prosimian ecology that most

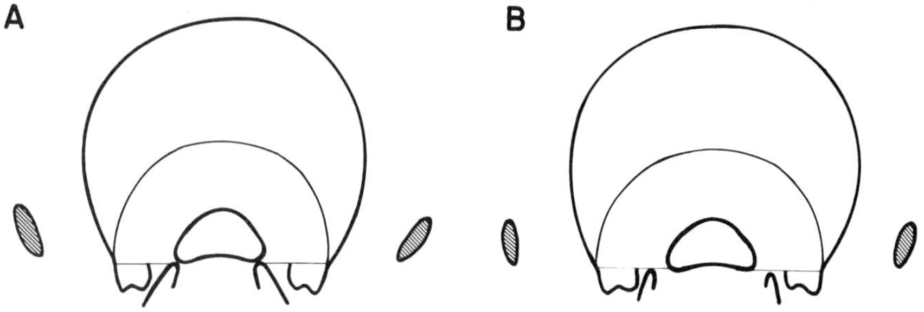

Figure 7. Comparative, schematic views of the posterior ends of the crania in *Lemur macaco* (*A*) and *Varecia variegatus* (*B*) to show the relative orientations of the zygomatic arches and pterygoid plates. In *L. macaco* the zygoma tends to be oriented ventro-medially, whereas in *V. variegatus* it is essentially dorso-ventral in orientation. In the former the pterygoid plates are distinctly more medial to the tooth row and closer to the articular condyles than in the latter. The cranial evidence indicates that in *L. macaco* both the masseters and pterygoid muscles were oriented in a relatively more diagonal fashion. In contrast, the masseter in *V. variegatus* lacks this relatively marked lateral component, and the pterygoid muscles appear disposed to pull more orthally and propalinally, rather than transversely.

of our knowledge on diet is derived from studies conducted during the dry seasons. This obviously has the drawback of not providing information on the diversity of the diet during the whole year. It should be remembered, however, that dry season studies might be indicative of the restrictive factors in the diet of a species, as food supply during the wet seasons is probably more abundant and diverse.

Petter and Jolly indicated that the natural diet of *Lemur macaco* consists chiefly of leaves and fruits.[18,19] Jolly lists *Harungana madagascariensis, Tristemma virusanum*, and *Trema orientalis* as species of 'lush green plants' which are particularly prominent in the diet of *Lemur macaco collaris.*[19]

The nature and consistency of most leaves and fruits are such as to warrant the dental morphology and masticatory jaw movements outlined above for *Lemur macaco*. As with those of *Lemur macaco, Lemur* spp. dentitions feature upper and lower tooth rows which: (*a*) tightly interlock during occlusion, (*b*) relatively evenly distribute cutting and shearing function throughout their premolar and molar regions, and (*c*) move transversely (ectentally) rather than mesiodistally (propalinally) with respect to one another during mastication. It would seem, then, that such dentitions have been selected to best process a diet of pliable, fibrous, vegetable material, such as leaves, given the hitherto unknown ancestral morphology.

The diet in the wild of *Varecia variegatus* is not well known. As J.-J. Petter remarked at this Research Seminar, study of this animal is extremely difficult as it is usually found at the very top of the canopy. R. Sussman (personal communication) has informed us of one report in which *Varecia variegatus* was observed in its habitat to have eaten large quantities of hard, spherical fruits or nuts, a specimen of which was collected by him. If this report is accurate, and does indeed reflect the food preference of *Varecia variegatus*, then some significant correlations can be made with the dental, mandibular and cranial morphology. It was noted above that mastication in *Varecia variegatus* seemed to stress relatively orthal puncturing, crushing and mashing. Such chewing would appear to be well adapted for processing stems as well as hard, spherical fruits, seed pods, or nuts. The relatively pointed and blade-like premolars, which occlude in the manner indicated above, combined with the extensive diastemata in the premolar region, all suggest that the premolars function primarily to grab and puncture rather than to cut or shear.

Similarly, some of the peculiar features of the molar region may be adaptations to process stems or solid, compact, tough-shelled nuts or fruits or seed-pods, presumably eaten by *Varecia variegatus.* Stems or fruits or nuts of the kind mentioned above would not require processing by an extensive portion of the upper and lower tooth rows, and could instead be efficiently cracked or mashed by just two adjacent pairs of upper and lower molars. It may be for this reason that in *Varecia variegatus*, M^3 and M_3 are reduced while M^2 and M_2 are often rotated. These rotations appear to have shifted the M^2 protocone and M_2 talonid basin mesially, thus functionally bringing the M^2 and M_2 mashing apparatus closer to that of M^1 and M_1. This phenomenon in the upper molars has resulted in the interuption of the primitively continuous M^1-M^2-M^3 ectolophs. For a leaf-eater, such a situation could be maladaptive. In *Varecia variegatus*, however, the sharply crested ectolophs, though oriented in a quasi-parallel rather than continuous fashion, can efficiently function to hold, squeeze, and splinter as the lower jaw is moved upward and anteriorly.

Summary

We have attempted to demonstrate the importance of evolutionary dental morphology and function in understanding the feeding mechanism in living and fossil species. We have indicated that for extant groups such studies facilitate interpretations of phylogenetic relationships, and that at the same time they shed light on the possible selective forces operating on the dentitions of closely related fossil taxa.

Because of the nature of the mammalian fossil record, students of mammalian palaeontology are very often constrained to resort to dental evidence in their efforts to elucidate the ecological role of extinct species. The study of dental function in extant groups (drawing as it does upon evidence which is not available in fossils) enables us to sharpen the interpretive tools that are needed to explain the (very often dental) fossil remains of extinct species. Two current taxonomically problematical groups of lemurines, *Lemur* and *Varecia*, were compared in this study with respect to dental morphology and dental function. We proceeded to demonstrate that the 'themes' of molar occlusion in each of these two groups differ dramatically, at a level expressed by generic separation; often mastication in *Lemur* emphasises transverse slicing and cutting, and that in *Varecia* features relatively orthal cracking and mashing. With this comparison, we hope to have contributed towards a clarification of the taxonomic status of these two groups, and towards some understanding of the forces of selection that produced them.

Acknowledgments

We are grateful to Dr Pierre Vérin of the Musée d'Art et Archéologie of Tananarive for his aid to Szalay while in Madagascar. The illustrations were skilfully prepared by Anita J. Cleary. The research was supported by NSF Grant GB 20085, and a grant for African studies by the Wenner-Gren Foundation for Anthropological Research, both to Szalay.

NOTES

1 Charles-Dominique, P. (1966), 'Analyse de contenus stomacaux d'*Arctocebus calabarensis, Galago alleni, Galago elegantulus, Galago demidoffi*', *Biol. Gabon.* 2, 347-53.

2 Charles-Dominique, P. (1971), 'Eco-éthologie des prosimiens du Gabon', *Biol. Gabon.* 7, 124-228.

3 Hladik, A. and Hladik, C.M. (1969), 'Rapports trophiques entre végétation et primates dans la forêt de Barro Colorado (Panama)', *Terre et Vie* 23, 25-117.

4 Hladik, C.M., Charles-Dominique, P., Valdebouze, P., Delort-Laval, J. and Flanzy, J. (1971), 'La coecotrophie chez un Primate phyllophage du genre *Lepilemur* et les corrélations avec les particularités de son appareil digestif', *C.R. Acad. Sci. Paris* 272, 3191-4.

5 Petter, J.-J. and Hladik, C.M. (1970), 'Observations sur le domaine vital, et la densité de population de *Loris tardigradus* dans les forêts de Ceylan', *Mammalia* 35, 394-409.

6 Petter, J.-J. and Peyriéras, A. (1970), 'Nouvelle contribution à l'etude d'un lémurien malagache, le aye-aye (*Daubentonia madagascariensis* E. Geoffroy)', *Mammalia* 35, 167-93.

7 Sussman, R., this volume.

8 Due to lack of data on the individual age of museum specimens and lack of information on the time of eruption of the molars, the statement 'comparable stage of wear' is a tenuous one indeed. Quite arbitrarily we assume that the third

molars erupted at comparable ages in these closely related taxa, and use the similarity in their wear as 'comparable stage' for the specimens compared.

9 Every, R.G. (1960), 'The significance of extreme mandibular movements', *Lancet* 2, 37-9.

10 Every, R.G. (1970), 'Sharpness of teeth in man and other primates', *Postilla* 143, 1-30.

11 Butler, P.M. (1937), 'Studies of the mammalian dentition: 1. The teeth of *Centetes ecaudatus* and its allies', *Proc. Zool. Soc. Lond.* 107, 103-32.

12 Butler, P.M. (1952), 'The milk-molars of Perissodactyla, with remarks on molar occlusion', *Proc. Zool. Soc. Lond.* 121, 777-817.

13 Butler, P.M. (1952), 'Molarization of the premolars in the Perissodactyla', *Proc. Zool. Soc. Lond.* 121, 819-43.

14 Butler, P.M. (1961), 'Relationships between upper and lower molar patterns', International Colloquium on the Evolution of Lower and Non-specialised Mammals, Brussels. *K. vlaamse Acad. Wetens. Belgie*, 117-126.

15 Mills, J.R.E. (1954), 'Ideal dental occlusion in the Primates', *Dent. Pract.* 6, 47-61.

16 Mills, J.R.E. (1966), 'The functional occlusion of the teeth of Insectivora', *J. Linn. Soc. Zool.* 47, 1-25.

17 Szalay, F.S. (1968), 'A family of early primates, the Picrodontidae', *Amer. Mus. Novitates* 2329, 1-55.

18 Petter, J.-J. (1962), 'Recherches sur l'écologie et l'éthologie des lémuriens malagaches', *Mém. Mus. nat. Hist. nat.*, sér. A, 27, 1-146.

19 Jolly, A. (1966), *Lemur Behavior*, Chicago.

I. TATTERSALL

Facial structure and mandibular mechanics in Archaeolemur

Introduction

Archaeolemurinae, the indriid subfamily containing the species *Archaeolemur majori* (Fig. 1), *A. edwardsi* (Fig. 2) and *Hadropithecus stenognathus*, has long been considered, particularly among English scholars, to be characterised by a variety of features linking it to the higher primates. This point of view is especially evident in the work of Major and Standing,[1,2] although, at about the same time as these two authors were writing, Grandidier expressed the opinion that 'Il est d'abord hors de doute qu'il faut les ranger parmi les lémuriens'.[3]

A recent reappraisal of the cranial anatomy of the archaeolemurines has fully confirmed Grandidier's views.[4] In all respects except, to a limited extent, their masticatory apparatus, the archaeolemurine lemurs remain extremely close to the living indriids; they were not advanced, for instance, over the indriine level in their neural or peripheral sensory organisation. Walker has recently shown that *Archaeolemur* possesses in its postcranial skeleton a variety of adaptations characteristic of the most terrestrial of the cercopithecoid monkeys,[5] but nonetheless from a phylogenetic point of view the postcranial anatomy of these animals bears an unmistakably lemuroid stamp.[6] Those features which led Major to the opinion that *Archaeolemur* might best be classified as a monkey are, in the words of Grandidier, 'les caractères acquis par une similitude de vie'.

The characters used by Major, and later by Standing, in suggesting that the archaeolemurines were more 'advanced' than the indriines reside primarily in the dentition, and, to a lesser extent, in the other components of their masticatory apparatus. This fact prompted the present investigation of the archaeolemurine masticatory system, and led to the generation of a model of masticatory function in these animals.

The mandible as a bent lever

Most previously proposed models of mandibular function have depended on the jaw's operation by means of independently-acting bent lever systems, usually comprising the temporalis and the masseter-internal pterygoid complexes.[7] Such models require that these levers produce rotation around the temporomandibular joint, which therefore acts as the fulcrum of the system.

The functioning of the temporomandibular joint as the fulcrum in a lever system is open to question on several counts. First, there is the inherent improbability of a system in which, for instance, depending on the position of the bite-point, only between about 30%-50% of the available power exerted by the temporalis and the masseter-internal pterygoid complexes is usefully expended at the dentition in *Archaeolemur edwardsi* (Fig. 3*a*). More important, however, is the fact that the remainder of the muscles' available effort goes to produce a reaction force at the jaw joint. Examination of the morphology, internal and external, of the condylar neck and

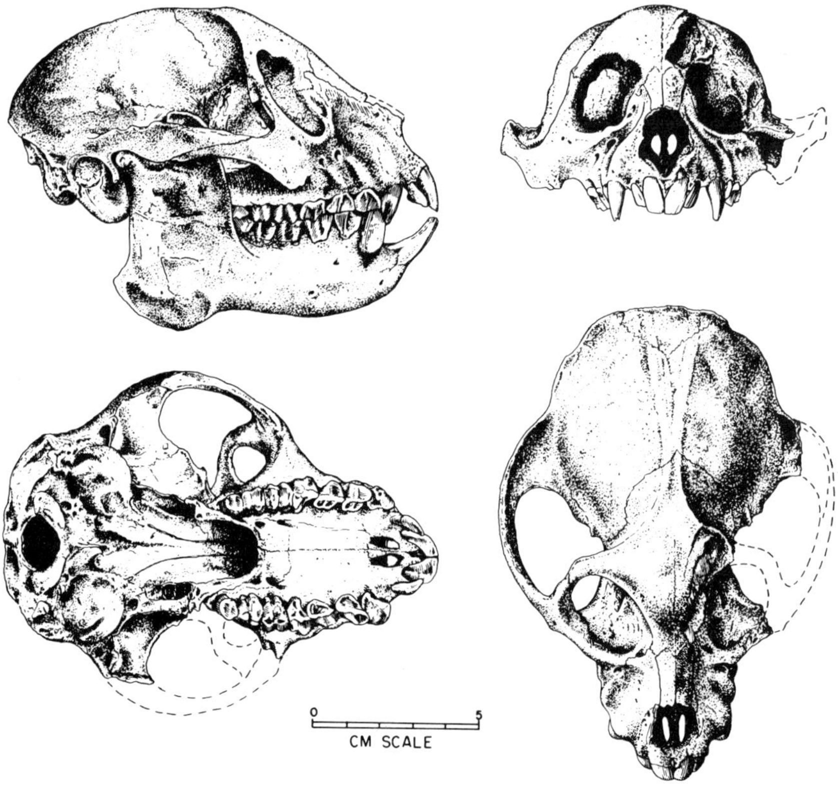

Figure 1. Cranium and mandible of *Archaeolemur majori* (BM M 7374).[21]

the glenoid fossa, of the low articular eminence and of the gracile postglenoid process in all the archaeolemurines, suggests very strongly that these structures could not have withstood the forces which would have been imposed upon them in a lever system. In addition, the functional surfaces of the temporomandibular joint in modern forms are lined with fibrous connective tissue ('fibro-cartilage'), rather than with the hyaline cartilage characteristic of the load-bearing joints of the body, such as the knee. Fibrocartilage, while providing an excellent sliding surface, is unsuited to withstand the large compressive forces which would inevitably be present if the jaw worked as a lever around the temporomandibular joint.

A further argument against a simple lever operation of the jaw is provided by the fact that in such an arrangement the resultant force at the dentition would be oriented tangential to the arc of adduction

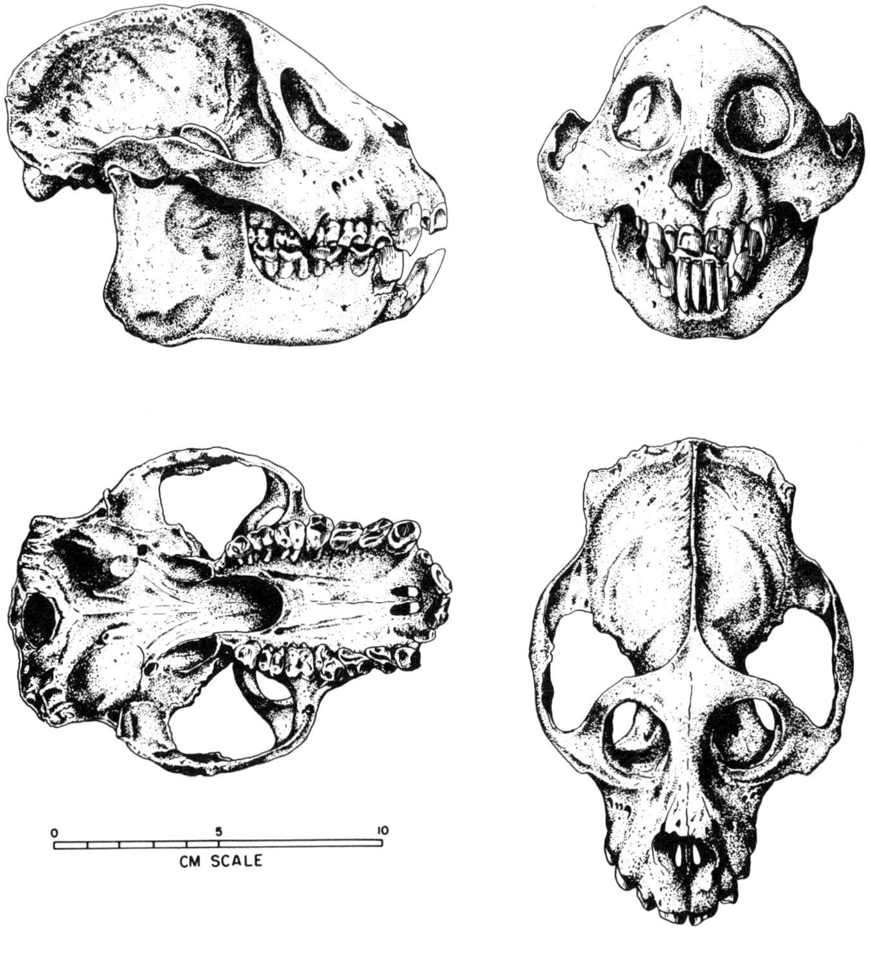

Figure 2. Cranium and mandible of *Archaeolemur edwardsi* (BM M 9909-9910).[21]

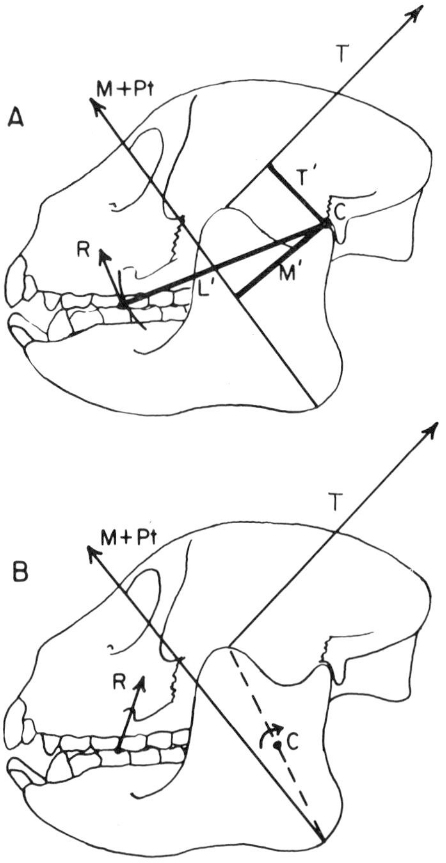

Figure 3. Mechanics of mandibular elevation in *Archaeolemur edwardsi.* (*A*) viewed as a bent lever system operating around the temporomandibular joint; (*B*) as a couple system producing torque around the mandibular foramen.
Abbreviations: C, point of rotation; M+Pt, vector of masseter-internal pterygoid complex; M', lever arm of masseter-internal pterygoid; T, mean vector of temporalis complex; T', lever arm of temporalis complex; R, resultant force at dentition; L', load arm.

of the jaw, i.e. orthogonal to the axis connecting the fulcrum and bite-point. (Some authors have indicated their belief that the occlusal force resulting from the rotation of the mandible about the jaw joint would be oriented vertically with respect to the tooth-row; this is emphatically not the case except where, as among Carnivora, the fulcrum and the bite-point lie in the same plane).[15] The architecture of the face in the archaeolemurines (and, indeed, in mammals in general) is totally inappropriate to the resolution of such anteriorly-directed resultant forces. A simple parallelogram of forces is sufficient to show that invariably a component vector will be directed out of the facial structure, placing the posterior part of the palate under considerable tension, and producing a bending movement in the face. Bone, while extremely strong in compression, is weaker in tension, and skeletal architecture is invariably designed to reduce, or even eliminate, tensile forces.[8]

All this is not to say that no forces ever occur at the temporomandibular joint; small locating forces necessarily occur, since the joint must, after all, function as a normal diarthrosis; but their magnitude is insignificant compared with those which would be present were the jaw joint a fulcrum. The fact that human patients with bilateral resection of the condyles are able to chew efficiently makes it sufficiently evident that no significantly large forces occur at this joint.[9]

Davis has proposed an alternative model of jaw function, based on the hypothesis that the temporalis and the masseter-internal pterygoid (anterior adductors) act not as separate lever forces but as a couple.[10] As Davis notes, this provides the advantage of reducing to zero the forces acting at the fulcrum. But Davis concludes that to the extent that the power exerted by the temporalis (an extremely large muscle in the bear, the animal under study) exceeds that exerted by the masseter, the jaw works as a bent (modified class I) lever. The misapprehension in Davis' scheme, even as it applies to carnivores, lies in the fact that the large size of the temporalis does not necessarily mean that it always exerts a force greater than that generated by the masseter, even though it may be capable of doing so. Moreover, and more importantly, the production of greater power by one muscle of the couple may in fact be *required* if a reaction force at the temporomandibular joint is to be eliminated.

More recently Turnbull, in a very useful work, has adopted Davis' scheme and generalised it to include mammals other than carnivores.[11] This is, however, unfortunate for a variety of reasons. For instance, although in some of the carnivores the condyle lies about midway between the mean points of insertion of the temporalis and anterior adductor complexes, with the result that it does approximate to (although it never precisely coincides with) the centre of rotation of the system,[12] in more herbivorous animals such as the primates, in which the temporomandibular articulation is elevated considerably above the plane of the tooth-row, the jaw joint and the centre of rotation of the jaw are far from coincident. It should also be noted that the couple action does not necessarily produce the vertical resultant force at the dentition that Davis illustrates, or the anterior one shown by Turnbull; one of the major advantages of a couple system lies in that the orientation of this force is variable along the tooth-row.

Facial structure

It has already been suggested that an anteriorly-directed resultant force would be inappropriate to the facial structure of *Archaeolemur* and other mammals.[13] What kind of resultant force, then, is the facial skeleton of *Archaeolemur* designed to resolve? Herein lies one of the keys to the operation of the archaeolemurine masticatory system. In essence, the facial region is designed as a bilateral tripod or pyramid, its base on either side formed by the robust maxillae. The apex of this structure lies at the most anterior and superior point on the frontal bone. The three legs of the tripod are formed by the facial profile, by the very strongly developed post-orbital bar/anterior zygomatic root complex, and by a strong trabecular structure, clearly visible in radiographs, running from the apex of the pyramid to the area of the palatine hamulus. The whole structure is reinforced by the bony palate and the medial septa of the nasal cavity and frontal sinus.

A structure of this sort is ideally designed to resolve forces directed towards its apex. The anteriorly-directed occlusal force in the face produced by a mandible rotating around the temporo-mandibular joint, then, would make functional anatomical nonsense of the facial architecture of *Archaeolemur.* Clearly, though, if masticatory stresses in the face of *Archaeolemur* were oriented towards the apex of the facial pyramid, the occlusal force generated by the masticatory musculature must have been variable in orientation according to the position of the bite-point along the tooth-row. Just such a vector is provided by the model of mandibular mechanics in *Archaeolemur* discussed below.

Mandibular movements

Several authors, particularly Mills,[14] have used the facets formed by wear on the occlusal surfaces of the teeth in reconstructing mandibular movements during mastication. This procedure is somewhat dubious in those cases where the dentaries are united in a mobile syndesmosis,[15] but is generally reliable when the mandibular symphysis is fused in an immovable synostosis, as in *Archaeolemur.*

Determination of the direction of jaw movements is normally made through examination of the orientation of the minute striae which are commonly found on the dental wear facets. In the case of *Archaeolemur*, although incisal and premolar wear is usually marked, the molars were subject to relatively little attrition and abrasion, and the molar wear facets in any single individual are inadequate for the complete reconstruction of mandibular movements. However, a composite pattern can be reconstructed from examination of molar wear in a variety of individuals. This question, and others related to it, are discussed at length elsewhere.[4]

In *Archaeolemur* there were three primary masticatory functions: incisal (nipping); premolar (shearing); and molar (crushing/grinding). The movement of the lower incisors against those of the upper jaw was the result of a direct posterior movement, combined with elevation. The condyles moved forward bilaterally as the mouth opened, and biting took place by means of a strong bilateral retraction of the condyles as the jaw moved upwards.

The premolar series of *Archaeolemur* is highly specialised as a continuous shearing blade. In young adults, at least, the shearing facets of the premolars, buccal in the lower series, lingual in the upper, are inclined at an angle of about 45° or less to the sagittal plane, and invariably bear clearly-marked striae which run almost directly transversely. The mandibular movements involved in this type of shear were probably as follows: as the jaw opened, both condyles slid forwards. After this, the ipsilateral condyle moved posteriorly on its articular surface at some point during elevation of the mandible to the point of tooth-food-tooth contact, or of dental approximation. A transverse movement then occurred, during which the ipsilateral condyle rotated medially; simultaneously, the contralateral condyle translated posteriorly. The mandibular movements involved in molar occlusion appear to have been substantially similar to those used to produce premolar shear.

Mechanics of mastication

That the basic mammalian pattern of jaw musculature is designed to produce an appropriate resultant force while eliminating any reaction at the temporomandibular articulation has been demonstrated in the case of *Archaeolemur majori* by construction of a simple two-dimensional static model. It is fully recognised that such a model of the jaw in elevation is highly oversimplified; however, it does serve to illustrate the possibilities inherent in a mechanism of this type.

The model consists of a plywood cutout of the mandible of *A. majori* suspended against a lateral outline of the cranium. The mandible is held in static equilibrium by means of springs of known stretching constant which represent the vectors of the major muscle groups (anterior adductors; anterior temporalis; posterior temporalis) and the resultant occlusal force. The model forcibly demonstrates that static equilibrium (as occurs during biting of a resistant object)[16] can only be achieved when the resultant force is of the type described earlier. Only in the presence of such a resultant do the available forces at the bite-point sum to zero. This is not to say that the resultant occlusal force equals all those muscle forces working to produce it; not only is a small amount of energy (variable, depending

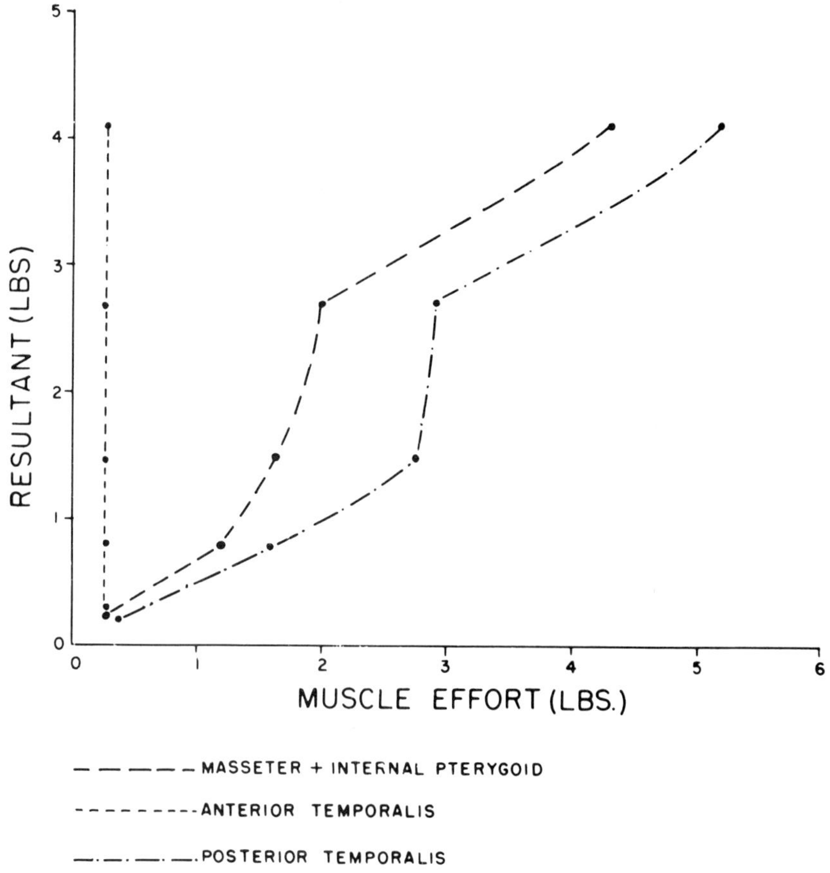

Figure 4. Muscular efforts required to produce increasing resultant forces at a bite-point corresponding to the anterior premolar.

on the bite-point) used in balancing the jaw, but, since muscles exert equal force at their origins and insertions, much effort is inevitably expended in producing forces on the neurocranium. This dome-like structure is well-adapted to resolve such forces; even those from the masseter are transferred from the zygomatic arch to the braincase by means of the temporalis fascia.[1 7]

Essentially, the mandible is held in equilibrium during biting by a couple action (Fig. 3) between the posterior temporalis and the anterior adductor muscles (particularly the superficial masseter, the dominant masseteric component, which originates very powerfully anteriorly via the anterior masseteric tendon). This posterior temporalis-anterior adductor couple produces an anti-clockwise torque around a point intermediate between the insertions of these two muscle groups; this coincides with the position of the mandibular

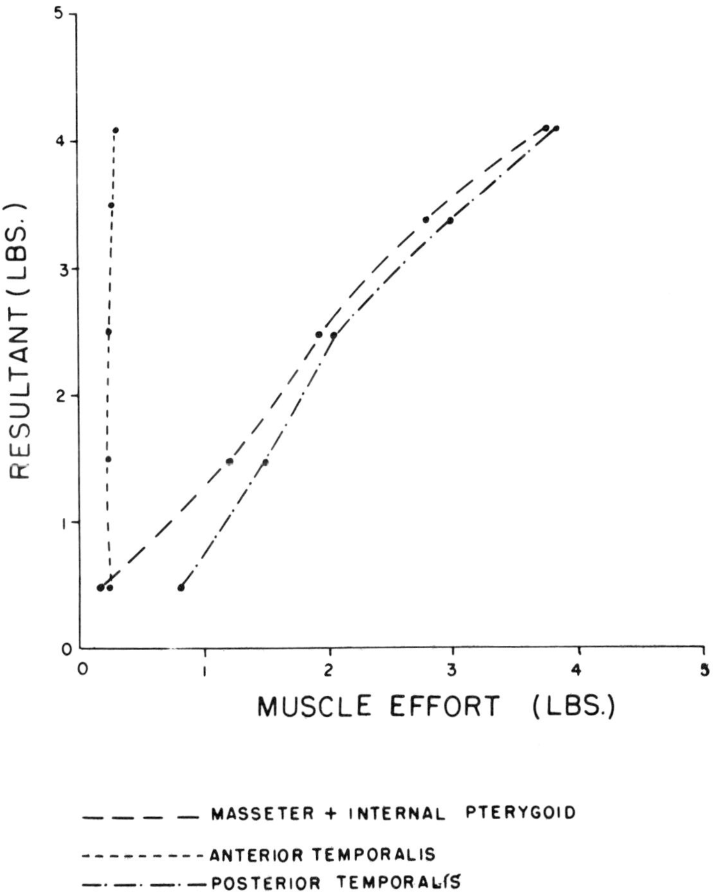

Figure 5. Muscular efforts required to produce increasing resultant forces at a bite-point corresponding to P4-M1.

lingula, the point of attachment of the sphenomandibular ligament. This ligament assists in the stabilisation of the jaw around this point during elevation. Since the condyle lies posterior to the lingula, the effect of the couple is to rotate the condyle ventrally. Such rotation is resisted primarily by the resultant occlusal force (i.e. the resistance at the teeth), and to a lesser extent by the antagonistic pull of the anterior temporalis.

Manipulation of the model described above can serve to provide an approximate idea as to the various muscular combinations which act to provide the appropriately-oriented masticatory force at different bite-points along the tooth row. Figs. 4 and 5 illustrate the muscle combinations which produce increasing resultant forces at two different bite-points; the anterior premolar and P4-MI, where wear indicates that much of the masticatory activity took place.

During incisal biting, at a 10° gape, the anterior temporalis contributes only a small, constant force to the system; the bulk of the effort is provided by the posterior temporalis working in the couple. Since the masseter produces what is essentially a balancing force at this bite-point, the effort of the posterior temporalis increases far faster than that of the masseter-internal pterygoid complex as bite-force is increased. At a wider gape, the anterior adductors play a greater role, although the posterior temporalis still predominates. The posterior temporalis must therefore produce a disproportionate amount of effort at this anterior bite-point, since it both balances the jaw and provides almost all the bite-force, while not in a particularly mechanical advantageous position to do the latter. This presumably explains the enormous size of the posterior temporalis and the diminution of the anterior moiety of this muscle in the extremely long-faced lemur *Megaladapis*, together with the fact that far more pressure may be exerted at the posterior than at the anterior teeth.

The retraction of the mandible during incisal biting results from additional pull by the posterior temporalis; the condyles translate posteriorly along a plane parallel to the major axis of action of the posterior temporalis, and therefore not resistant to their movement.

At a more posterior bite-point, corresponding to the middle of the premolar series, the anterior temporalis continues to play a minor role, while the anterior adductors are more strongly involved in the production of the resultant force, although their contribution is still less than that of the posterior temporalis. When biting at a 10° gape takes place at P_4^4-M_1^1, the anterior adductors and the posterior temporalis play more equal roles, although the effort expended by the anterior adductors only equals that of the posterior temporalis at high masticatory pressures. Under some conditions the contribution at this bite-point of the anterior temporalis may be rather greater than Fig. 5 indicates. During chewing at the most posterior teeth the couple muscles do little more than balance the jaw, i.e. their contribution, particularly that of the posterior temporalis, does not increase greatly as the bite force increases. Most of the occlusal force is provided by the anterior temporalis; indeed, at wide gapes, which bring the bite-point almost in line with the mean axis of action of this muscle, the force exerted by the anterior temporalis increases almost proportionately with the resultant force. Nonetheless, only a limited amount of masticatory activity took place at this point, as shown by the relatively low degree of wear on the teeth and by their small size; this was presumably due to the restricted amount of lateral motion possible.

Clinical evidence such as that advanced by Moss[18] substantiates the hypothesis of the mandible's rotation around the lingula. Although Moss' discussion is limited to the human temporo-mandibular joint, which possesses an articular eminence far more pronounced than that of *Archaeolemur*, the argument appears equally applicable in the case of the latter. During the opening of the jaw, the condyle rides forward on the articular eminence, while the angle of the mandible moves back. The functional surface of the temporal portion of the jaw joint is not comprised by the glenoid fossa, but by the more anteriorly positioned eminence, as can be seen from the distribution of fibrocartilage in the joint (Fig. 6). The glenoid fossa in man merely serves as a receptacle for the condyle when the jaw is at rest.

During motion of the jaw the mandibular foramen moves virtually not at all; this is necessarily so, since this foramen, adjacent to the lingula, is the point at which the inferior alveolar neurovascular bundle enters the mandible. Movement here during opening and closing of the jaw, as in a lever system, would tend to rupture the neurovascular triad, particularly the inferior alveolar nerve. It should also be pointed out that in the case of a mandible rotating around the temporomandibular joint, the amount of posterior movement of

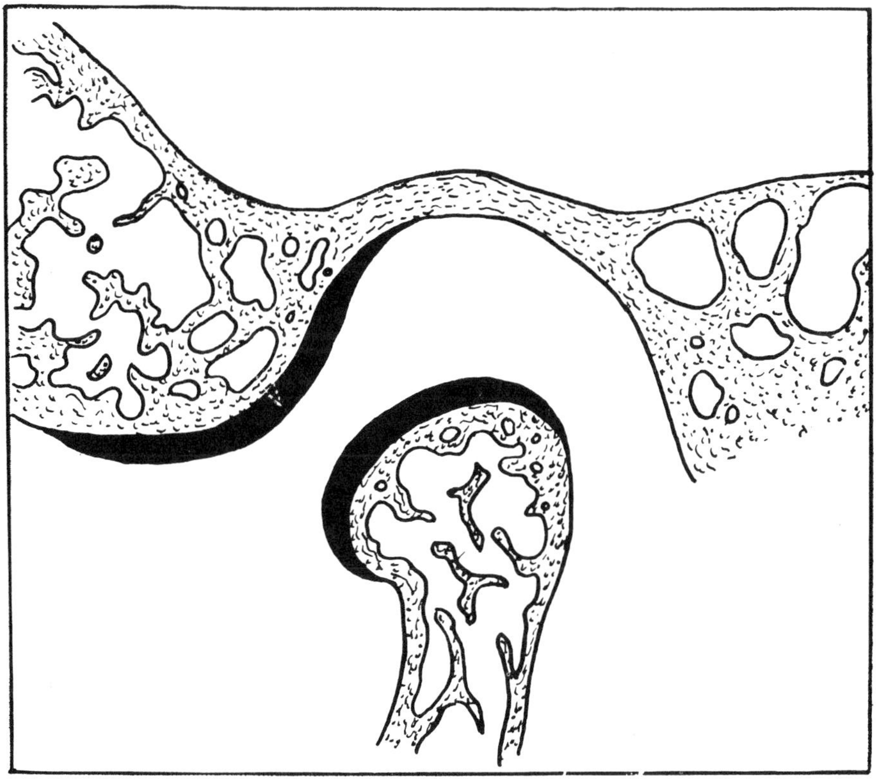

Figure 6. Diagram showing those areas (black) of the condylar head and the articular eminence in man which are covered with fibrocartilage; these represent the functional surfaces of the joint. Redrawn after Moss.[18]

the mandibular angle involved would be such as to endanger the pharyngeal structures of the throat.

In *Archaeolemur* lateral motion of the mandible during mastication was apparently achieved through contralateral rotation of the ipsilateral condyle; simultaneously the contralateral condyle translated posteriorly. Ipsilateral condylar rotation could have taken place at any phase of elevation, but presumably occurred generally during close approximation of the tooth-rows, when the ipsilateral condyle was in a posterior position. Translation of the contralateral condyle, and hence the prime component of lateral motion, was evidently largely effected by the almost directly posterior pull of the posterior temporalis; the movement would have been stabilised by the antagonistic action of the contralateral external pterygoid.

Since during lateral motion the ipsilateral condyle was effectively fixed, rotating, in this plane at least, around a point within itself, it would appear at first sight to have acted as the fulcrum of a lever. The geometry of the system, however, is such that the point of resistance, i.e. the food, would have tended to become the centre of rotation, causing the ipsilateral condyle to move anteriorly. The ipsilateral posterior temporalis would have acted to eliminate such movement. In the absence of a tendency of this sort, any reaction force in this plane would have been nullified by the anterior, antagonistic, pull of the ipsilateral lateral pterygoid.

Derivation of the masticatory apparatus

In most of its morphological details, as well as in its basic plan, the archaeolemurine masticatory apparatus is very close to that of the living indriines (Fig. 7). Apart from precise dental morphology, the most striking differences between the archaeolemurine and indriine chewing mechanisms lie in the form of the symphyseal region of the mandible, the robustness and proportions of the mandibular corpus, and in the manner in which the dentaries are joined. The lack of symphyseal fusion in the indriines is a particularly significant distinction; Crompton and Hiiemae have shown that a mobile symphysis implies a totally different pattern of mastication from that obtaining in mammals possessing synostosed dentaries: one in which the two dentaries may move independently of each other.[15]

In a living indriine such as *Propithecus*, given the presence of a mobile symphyseal syndesmosis, it is evident that the contralateral temporalis cannot function to produce internal movement of the ipsilateral dentary. Other, entirely ipsilateral, muscular combinations must be operative to produce jaw movements not dissimilar to those in *Archaeolemur*. As Smith and Savage have pointed out,[7] internal movement of the mandible by contraction of the temporalis of the

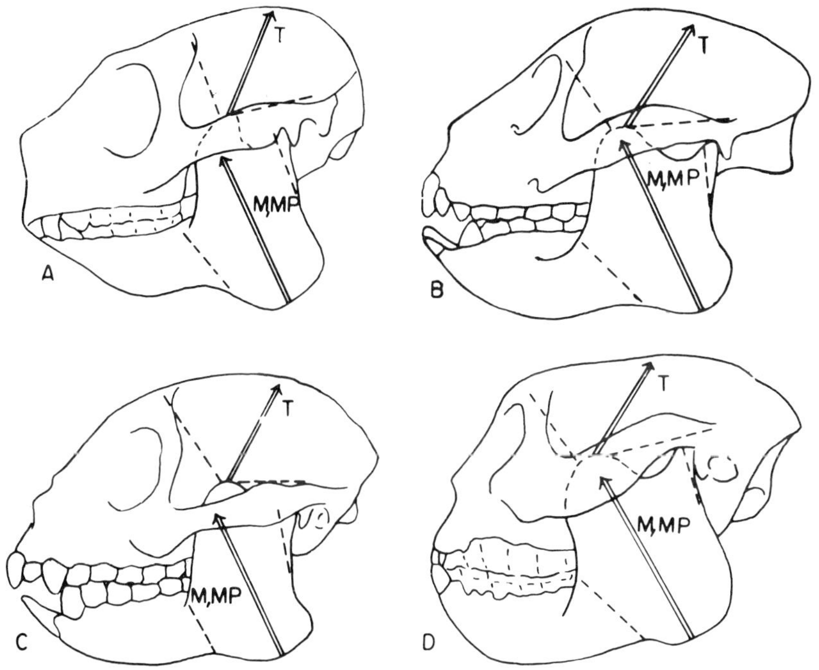

Figure 7. Comparison of the orientations in lateral view of the major masticatory muscles in four indriid species: *A, Propithecus verreauxi; B, Archaeolemur edwardsi; C, Archaeolemur majori; D, Hadropithecus stenognathus.* Abbreviations as in Fig. 3.

contralateral side would involve extreme stressing of the lower jaw; these stresses would be transmitted anteriorly, around the symphysis, and back again along the ipsilateral corpus. Despite this disadvantage, such a situation is precisely what we find among the archaeolemurines, together with massive jaws and an extremely strongly buttressed symphysis, characterised in particular by the possession of a strong inferior transverse torus.

Goodman has recently discussed the relation between form and function in the pongid symphysis.[19] By stressing both a photoelastic model and pongid mandibles coated with strain-sensitive lacquer, Goodman established that the inferior transverse torus of the pongid symphysis represents an adaptation for withstanding laterally-directed forces, i.e., those which would tend to spread the dentaries apart. This torus is well-developed in the mandible of *Archaeolemur*, while the forces developed in this area during contralateral temporalis action would have been very much of the sort Goodman describes.

It is clear, then, that the masticatory system of *Archaeolemur* could have been very simply derived from an ancestral pattern quite similar to that seen in the extant indriines, particularly *Propithecus*, by fusion of the mandibular symphysis. Other modifications, such as the dominance of the contralateral temporalis in lateral movements, change in symphyseal form and the increased robustness of the mandible, are all correlates of this fundamental shift, and presumably reflect a change towards a more powerful masticatory pattern, possibly accompanying the initial increase in size of the archaeolemurine ancestors. Presumably the change took the shape it did, with minimal modification of the pre-existing apparatus, because the common ancestor of Indriinae and Archaeolemurinae was already fairly highly specialised towards the modern indriine condition. In short, although the archaeolemurine masticatory apparatus resembles in some morphological respects (e.g. bilophodont molar teeth, possession of inferior transverse torus) that of various catarrhine primates, and was evidently the response to similar selective pressures,[4,20] in its general plan and in its origin it was radically different: an analogous variation on an altogether different theme.

Acknowledgments

The study of which this report is an outgrowth was supported by the Wenner-Gren Foundation for Anthropological Research Inc., by the Boise Fund of Oxford University and by various Fellowships of Yale University. Completion of this paper was assisted by a Grant-in-Aid of the Society of Sigma Xi. I should like to acknowledge here stimulating discussion of this topic with David Roberts, Department of Geology and Geophysics, Yale University. The illustrations were prepared by Mr Nicholas Amorosi, Department of Anthropology, American Museum of Natural History.

NOTES

1 Major, C.I.F. (1896), 'Preliminary notice on fossil monkeys from Madagascar', *Geol. Mag.* (decade 4), 3, 433-6; (1900), 'A summary of our present knowledge of extinct primates from Madagascar', *Geol. Mag.* (decade 4), 7, 492-9; (1901), 'Some characters of the skull in the lemurs and monkeys', *Proc. Zool. Soc. Lond.* 1901, 129-52.

2 Standing, H.F. (1908), 'On recently discovered subfossil primates from Madagascar', *Trans. Zool. Soc. Lond.* 18, 69-162.

3 Grandidier, G. (1905), 'Recherches sur les lémuriens disparus et en particulier ceux qui vivaient à Madagascar', *Nouv. Arch. Mus. Hist. Nat. Paris* 4 sér., 7, 1-142.

4 Tattersall, I. (in press), 'Cranial anatomy of Archaeolemurinae (Lemuroidea, Primates)', *Anthrop. Pap. Amer. Mus. Nat. Hist.*

5 Walker, A.C. (1967), 'Locomotor adaptation in recent and subfossil Madagascan lemurs'. PhD Thesis, University of London.

6 Jouffroy, F.-K. (1963), 'Contribution à la connaissance du genre *Archaeolemur* Filhol, 1895', Ann. Paléont. 49, 129-55.

7 e.g. Smith, J.M. and Savage, R.J.G. (1955), 'The mechanics of mammalian jaws', *School Sci. Rev.* 1955, 289-301; Ostrom, J.H. (1964), 'A functional analysis

of jaw mechanics in the dinosaur *Triceratops*', *Postilla* 88, 1-35; Crompton, A.W. and Hiiemae, K. (1969), 'How mammalian molar teeth work', *Discovery* 5, 23-4.

8 Oxnard, C.E. (1971), 'Tensile forces in skeletal structures', *J. Morphol.* 134, 425-35.

9 Moss, M.L. and Rankow, R.M. (1968), 'The role of the functional matrix in mandibular growth', *Angle Orthodontist* 38 (2), 95-103.

10 Davis, D. (1955), 'Masticatory apparatus in the spectacled bear *Tremarctos ornatus*', *Fieldiana: Zool.* 37, 25-46.

11 Turnbull, W.D. (1970), 'Mammalian masticatory apparatus', *Fieldiana: Geol.* 18 (2), 149-356.

12 The closeness of the jaw joint to the centre of rotation of the mandible may form part of the explanation for the tight, restricted condyle-glenoid relationship in carnivores, as opposed to the much looser temporomandibular articulation among herbivores; however, the point is far from clear, and cinefluorographic studies of movement at the carnivore jaw joint during chewing have yet to be made.

13 For the purposes of this discussion the dentitions and masticatory musculatures of the two species of *Archaeolemur* are virtually identical, the only major difference lying in the greater size of *A. edwardsi.* Unless otherwise stated, then, the discussion throughout applies equally to both species.

14 Mills, J.R.E. (1967). A comparison of lateral jaw movements in some mammals from wear facets on the teeth. *Archs. Oral. Biol.* 12, 645-61.

15 Crompton, A.W. and Hiiemae, K. (1969), 'Functional occlusion in tribosphenic molars', *Nature* 222, 678-9. Crompton, A.W. and Hiiemae, K. (1970), 'Molar occlusion and mandibular movement during occlusion in the American opossum', *Zool. J. Linn. Soc.* 49, 21-47.

16 Because, at least for the purposes of a simplified argument, the jaw in motion can be visualised as passing through an infinite series of static equilibria, it is permissible to generalise, within limits, from the static to the dynamic state.

17 Eisenberg, N.A. and Brodie, A.G. (1965), 'Antagonism of temporal fascia to masseteric contraction', *Anat. Rec.* 152, 185-92.

18 Moss, M.L. (1959), 'Functional anatomy of the temporomandibular joint', *New York J. Dentistry*, 29 (11), 315-19.

19 Goodman, M.C. (1968), 'Structure and function in the symphyseal region of the pongid mandible', Dissertation, Yale.

20 Jolly, C.J. (1970), '*Hadropithecus*, a lemuroid small-object feeder', *Man*, n.s., 5, 619-26.

21 B.M. = British Museum (Natural History), London.

R.G. EVERY

Thegosis in prosimians

Introduction

An analysis of thegosis in prosimians demonstrates how knowledge of this phenomenon has introduced a new dimension to the study of mammals. Many hitherto unsuspected questions can now be asked. In the area of primatology concerned with basic primate origins these questions are most pertinent. For, if basic theory proves fallacious, the implications are relevant to *Purgatorius ceratops*, the earliest known primate, and to *Homo sapiens*, the primate that concerns us most.

Knowledge of the origins of primates depends almost exclusively on teeth. Interpretation of the selective forces that produced teeth is therefore of crucial importance. The prime taxonomic feature for the earliest primates is the enlargement of the talonid and trigon basins, concurrent with the loss of the paraconid and reduction of the trigonid. The virtually universal interpretation of this phenomenon is that the evolution of primate teeth is dominated by a 'rubbing', 'grinding', 'mortar-and-pestle', 'surface-shearing' advantage of the enlarged basins.

From a study of thegosis this fundamental interpretation, basic to all primates, is now seen as fallacious.

Thegosis is a phylogenetically derived behaviour.[1] The term comes from Homer's use of the Greek *thego*: to whet, sharpen. In the *Iliad*, Homer accurately describes not only the boar's anatomical preparation of his formidable dental weapons (the lower canines) but also the animal's physiological preparation, the emotional state that basically determines his immediate behaviour. The term thegosis, therefore, includes the metaphorical connotation inherent in the original Greek *thego*: to excite, provoke.

Since all the author's basic concepts of thegosis emerged during the course of a psychiatrically-orientated dental practice, the initial experimental animal used was, of course, man himself. Most of these experiments, which began over thirty years ago, were possible only on man and cannot be duplicated on any other animal. In the study of any other animal one must, therefore, lean heavily on the evidence gained from man. Although this area of evidence is too extensive even to summarise here, it nonetheless forms the basis of this present study of prosimians.[2]

The earliest mammals represent another area of evidence that is also significant to this present study, but can only be briefly introduced.[3] When discussing this aspect of mammalian evolution, Simpson states in his introduction: 'The discovery of the ancestral type would introduce into the study of dentitions the principle of homology and would supply the background that would knit together knowledge of teeth and transform it from a mere morphological catalogue to a true and potent science. . . . The point of view adopted is principally functional and dynamic.'[4]

It seems that Simpson's prediction of a true and potent science may have been vindicated as far as general homology is concerned, but functionally it has remained virtually as he himself stated it in 1936. For example: 'The primitive upper molar did not consist simply of three cusps and nothing more. [Footnote] This fact, well understood by Cope and Osborn and by all their successors, has from time to time been hailed as a discovery and supposed to invalidate the Cope-Osborn theory.'

When one realises that Cope preceded Simpson by 53 years, it is sobering to read what was known and understood. Simpson says:

> The protocone, internal main tubercle, is crescentic and from it two sharp crests run externally (labially) along anterior and posterior sides. In occlusion these are shearing crests. . . . The main external cusps, paracone and metacone, are sub-equal, and it appears to have been a primitive character for them to bear crests descending anteriorly and posteriorly from each apex, to some degree uniting the two cusps. Aside from these crests, the cusps are sometimes nearly conical, sometimes compressed or lancelate, and sometimes strongly crescentic. The crescentic character was probably not generalised but was a modification that early arose within certain lines of descent . . . and greatly affected the subsequent development of their dentitions.[4]

Simpson was describing what he in that paper termed the 'tribosphenic molar':

> Its principal advance is the further development of opposition, first by the widening and basining of the talonid and second by the development of the hypoconid which occludes against the

central basin of the upper molar and between and internal to the main outer trigon cusps, now two in number, although not completely separated. In addition to this double opposition, the elements of alternation and shear are preserved.

It is undeniable that the advent of the reciprocally opposing, cusps-to-basins feature was an important evolutionary event. It allowed the division of exogenous substances that eluded the action of the cutting blades. It also had the advantage of squashing soft pulpy food and mixing it and other divided food with saliva, the first digestive juice of the alimentary canal. Odontologists, especially primatologists, have in a number of ways tended to be confused by this, however, and have postulated theories that now appear as fallacious. 'Preventing overclosure of the jaw' is one, 'grinding', 'mortar-and-pestle action', 'lingual phase of occlusion', 'lessening importance of shearing blades', and the 'shearing surface' as constituting the 'shearing' agent, are others. Since it is undeniable that the prime taxonomic feature diagnostic of the earliest primates is at present the enlargement and raising of the talonid relative to the trigonid, concomitant with corresponding changes in upper molars, whether such arguments are valid or not becomes of prime importance. For if they prove fallacious, Simpson's criterion for a 'true and potent science', that is, '. . . principally functional and dynamic', has not been vindicated; and, furthermore, knowledge of teeth, especially primate teeth, has remained virtually a 'morphological catalogue'.

In the hindsight knowledge of the phenomenon of thegosis, one may now detect some of the inherent difficulties. Apart from the misconceptions regarding phylogenetically derived behaviour and learning (the nature-nurture problem), which undoubtedly delayed the perception of thegosis as a fundamental mammalian characteristic, I believe functional analysis of teeth was, and still is, crippled by its functionless terminology, as the following examples show:—

'Crest' (crista) depicts a location, not a function; it means the summit of anything, as a comb or tuft on the head of a bird, roof-ridge, hill or wave; the only other anatomical use is a ridge along the surface of a bone.

'Selenodont' (Greek *selene*, a moon) and 'crescent' (Latin *crescere*, to grow) analogise a tooth-feature to a waxing moon.

'Trigon' is a triangle.

'Talon' (Latin *talus*, the heel) connotes, not a function, but a spatial relationship.

The suffix 'cone' has an inherent connotation of a solid figure with a circular base tapering to a point.

'Loph' is a crest (Greek *lophos*).

'Cingulum' is a girdle.

Apart from the apex of the cone, none of these terms connotes a function; none, especially not the cone, connotes a feature that has a functionally sharp cutting edge. 'Crest', in fact, merely serves to emphasise the (fallacious) importance of the so-called 'shearing surface' which it merely surmounts.

Simpson's own term, 'tribosphenic', is also suspect. 'Tribo-' (Greek *tribein*, to rub, grind, or pound), and '-sphenic' (Greek *sphenos*, a wedge) are, as Simpson explains, 'suggestive of the mortar-and-pestle, opposing action of protocone and talonid and of the wedge-like, alternating and shearing action of trigon and trigonid'.[4] The prefix depicting the talonid is functional; the suffix depicting the trigonid, unfortunately, is inconsistently morphological.

That Simpson's use of 'wedge-like' is merely morphological is clearly indicated in his discussion of symmetrodonts: 'These teeth have become triangular and in occlusion their opposed apices alternate and the teeth fit into each other like opposed series of blunt wedges. The crests have become sharp and pronounced and they add the important occlusal factor of shearing.'[4]

For Simpson to associate a wedge with a dividing action when the dividing agent is, in fact, not the wedge's thin edge but two blunt side edges, is particularly unfortunate. Although 'wedge' is used for anything more or less like a wedge, as a large piece of cake or the flying formation of geese, the basic connotation remains an instrument that has a function of splitting, forcing apart, or fixing tightly, etc. Its shape is characteristically thick at one end sloping to a thin (often sharp) edge at the other. In engineering, the inherent principle of any cutting instrument is described as the production of a crack that is propagated. Anyone who has used a wedge to split open a log is familiar with this action. If there is no existing crack to propagate, one is created. The wedge is hammered lightly into the grain of the wood until it grips; it is then hammered forcibly, propagating the crack. In this action, not only is the thin edge of the wedge functional, but also the sloping side surfaces. In fact, for a material that splits easily, the crack, once initiated, may be propagated by the sloping sides alone, i.e. surface against surface.

This, then, seems an appropriate place to discuss the common but fallacious concept of the so-called 'shearing surface', i.e., the idea that cutting and dividing exogenous material is achieved by grinding it between two opposed surfaces, especially in the manner of a mortar and pestle. A mortar-and-pestle unit is essentially two opposed surfaces, one concaved at a greater radius than the other's convexity. Since Butler's publication on the 'facet' and Mills' interpretations of 'buccal and lingual phases of occlusion',[5,6] the 'shearing surface' concept has, even more fallaciously, been argued to be the facet's inherent function. So much so that the facet is now identified as the 'shearing surface'; this has become the basic unit in the study of dentitions. It is the vogue to detect a 'shearing surface', to find its counterpart and its homology, to give it a number and analyse its efficiency and importance according to size. 'Shearing surfaces' are found to shear 'up' and 'down', 'across' and 'against' other 'shearing surfaces'. Occasionally (and often ambivalently) the functional significance of a 'crest' is noticed and mentioned.

The term 'shear', itself, needs analysis. Although 'shears' in engineering have come to apply to a larger instrument of similar kind to scissors, the term 'shearing', itself, is basically the dividing, cutting action of a single blade: a sickle, knife, chisel, lathe tool, bow or propeller of a ship, provide examples.

'Shearing', then, is a valid term for the initial actions of mastication – the action of Simpson's 'Opposition . . . of positive elements, cusps, crests, against other positive elements';[4] also for the first phase of the 'power stroke', described (for *Galago*) by Kay and Hiiemae in this volume,[7] by Crompton and Hiiemae (for *Didelphis*),[8] and commonly observed in man and other anthropoids. These constitute a number of relatively vertical strokes (made on a bolus of food) that terminate appreciably before the cusps and blades themselves come into apposition.

When the term 'shearing' is used for the action of 'shears' comparably to that of 'scissors', there is no confusion about the relationship of the blades. The efficiency of the cutting action requires the close apposition of the cutting edges. When a material is interposed between the blades' surfaces (i.e., between the facets of teeth-blades), the instrument is rendered useless.

Despite this overt mechanism, the identical term 'shearing' is used by some odontologists to apply precisely to the above situation that renders shears and scissors useless. Shearing is supposed to result when the material is interposed between two blade-surfaces (i.e., their facets) that move parallel but opposite to each other. This is the 'grinding', 'mortar-and-pestle', 'surface-shearing' action. To function at all, the surfaces would have to be considerably roughened in order to grip the substances being 'shorn', or the substance would have to possess considerable adhesiveness to grip the surfaces and, in addition, the forces perpendicular to the surfaces would have to be excessive. The reality is that the facet, compared to its adjacent tooth-surfaces, is characteristically polished, shiny and smooth.

'Shearing' is a term that not only may have different connotations for different individuals, but also may have one connotation for the same individual who uses it for a number of entirely different mechanisms.

Full discussion of this confusing term and its degradation of odontology is beyond the scope of this paper.[9] The situation, however, is so chaotic that I am suggesting a temporary – if not permament – suspension of its use.

Terminology

Should the phenomenon of thegosis eventually be generally accepted as a phylogenetically derived behaviour which was crucial to both the origin and the subsequent radiation of mammals, the magnitude of the inevitable change in our thinking can hardly be overrated. It would become apparent that the present discussion among odontologists about different notions of tooth function, and the terms used to label them, is highly significant. Even before thegotics has succeeded in making any great impact, there are signs of changes taking place. Odontology faces a considerable readjustment of its basic principles.

While odontology is in this state of flux, it would hardly be conducive to a flexible approach to argue that tradition justified the retention of confusing or inadequate terms, let alone those that are both morphologically and functionally erroneous. It is the very chronicity of a lesion that ultimately demands its excision. And it would hardly be conducive to the efficiency of the operation were the surgeon to discuss the functional properties of his lance in terms of 'crest' or 'cone'. When discussing the ancestral mammal, the theories and the theoreticans, Simpson comments: 'In this field there are and always have been workers whose opinions are not much influenced by fact.'[4]

Since I have the task of defining the phenomenon of thegosis and therefore the new science of thegotics, I am forced to use my own language if I am to facilitate the discussion. Fortunately, the changes are minimal and the introductions not extensive. In combination, however, they build to a complicated tooth-formula that at first sight appears rather formidable. However, the chemical formula for a complicated molecule appears to be even more complex at first sight. Yet there is no complaint that the chemical formula confuses our understanding rather than elucidates it. Nor is there any difficulty in devising an abbreviation, though this can only be introduced after its full components are known. This terminology is planned to suggest the selective forces in the evolutionary process. The aim is to simplify the conception of both the evolutionary process (the selective forces that produced the dentition) and the functional spatial relationships of the teeth themselves.

Although my professional as well as research experience has involved the articulation of artificial replacements for natural (human) teeth, and I have never delegated this aspect of the 'technical work' to others, and although I am familiar with working on a target viewed in a mirror, I nonetheless cannot achieve the feat, frequently demanded of me, of conceiving the spatial relationships of teeth as they are depicted in much of the current literature. The frequent practice of depicting right upper molar teeth alongside left lower teeth is particularly objectionable. In order to perceive the dynamics of function, I have first to convert (in my mind) the lower occlusal contours to their mirror image (i.e., to the right side) before I can procede to articulate in my mind the appropriate contours to their appropriate counterparts – as I am accustomed to do with the concrete objects. This defies my capacity.

The practice of depicting right upper molars with left lowers may have some advantage in presenting a 'radiographic profile' of a static occlusal relationship. Yet even this facility is doubtful; why, for example, was it not detected earlier that, in this view, the articulating blades of the 'tribosphenic' molars are mutually convex?[1,3]

If workers cannot perceive static relationships, how can they perceive dynamic ones? And if workers cannot perceive dynamic spatial relationships in one direction (transverse) how can they perceive these in two (transverse followed by oblique)?

The magnitude of this problem is exemplified by my colleagues in dentistry who, when fabricating artificial replacements, endeavour to articulate the teeth so that they achieve a so-called 'balanced occlusion'. This is an extremely difficult task to achieve visually, and is seldom even attempted. If a 'balanced occlusion' is ever achieved it is through the agency of a mechanical device: the 'grinding in' on an anatomical articulator using carborundum powder.

That scientists seldom appreciate the spatial relationships involved, despite their day-to-day involvement with the problem, is suggested by the consistent failure of authors to publish illustrations that depict the valid (frontal) cross-sectional relationship of teeth in lateral occlusion. Invariably the jaw (of a transverse masticating animal) is depicted, not as rotating about an axis near the postglenoid tubercle, but translating bodily in a transverse direction, which is fallacious.[10]

The new terminology not only suggests precise morphological and functional characteristics, but also retains the connotation of phylogenesis despite an adaptive variation where, for example, a transverse blade becomes virtually longitudinal, or a feature becomes amalgamated with another, or absorbed in a wall or base, or lost.

When the primate dentition is analysed in the light of thegosis, its striking characteristic is not, as has been generally supposed, the enlargement of the basins, but the modification and elaboration of cutting blades. Contrary to current opinion, bluntness, although elaborated, has not become the dominant characteristic. Sharpness (the cusp-blade unit) remains the prime advantage; its apparent depreciation is a result of the overt shortening of some of the blades that in the ancestral alpha dentition dominated the occlusal surface.[3] The remainder of the blades, however, are elongated. Apparently overlooked is the concomitant of a large basin: the elongation of its peripheral blades. Moreover, and significantly, a new cusp-blade unit has appeared.

Cingular blades, also, are elaborated. Not only are they raised higher on the crown, but new cingular blades and cusps are frequently present. Cingular blades are subsidiary (reserve) blades which function in the gerontic stage of the tooth's life-cycle. They are, therefore, appropriately oriented to the primary (occlusal surface) blades they eventually replace.[3]

The overall effect is that, because 'basined' surfaces have been enlarged and heightened on the crown, the cutting blades, instead of being spread out over the occlusal surface as in the ancestral condition, are now featured more on the periphery. This therefore ensures a greater ontogenetic continuity of cutting blades generally, an advantage – not a sacrifice – to the greater ontogenetic continuity of the 'basined' surfaces of which blades form an integral part.

These and other significant changes will be discussed later when using the facility of the new terminology.

The prime evolutionary advantage of teeth is their sharpness; the secondary advantage is bluntness. It is the sharp conical apex and blunt conical walls of our original aquatic ancestors' teeth that allowed their original grasping advantage. In the original mammalian molar it is the sharp thegosed apex of the cusp and the sharp thegosed blade that runs from it that allowed the additional advantage of cutting in scissor-like action. In the later mammalian molar it is the blunt, opposable, platform-surface bounded by sharp cusps and blades that allowed the additional advantage of cutting, cracking, splitting, crushing and mashing by striking action.

These three basic features of the characteristic (basic) mammalian dentition – thegosis, sharpness, and opposable sharpness and bluntness – can therefore be used to build a functional terminology.

Apart from *thegosis* itself, the terminology is virtually built up around four basic new terms:

1. *Scissorial:* Dividing exogenous material by point-cutting in a scissor-like action.
2. *Incusive* (Latin *incuso*, to hammer; *incus*, anvil): Dividing exogenous material by striking when held by a surface as an anvil or chopping-block.
3. *Akis* (Greek *akis*, small pointed object, splinter, arrow-head, point of a chisel): A cusp with a sharp edge running from the apex as an arrow-head or point of a chisel or sickle.
4. *Drepanon* (Greek *drepane*, a sickle): A concave blade with the sharpness on the edge of the concavity as a sickle.

There are further terms; some of these are mentioned below; all (including the above basic units) are presented in detail elsewhere.[3]

Types of dentitions

Dentitions can be categorised according to the location of the thegosis-facet relative to the direction of the enamel prisms (or rods). The cutting blades of primitive teeth before the Cretaceous and (for advanced teeth) those of typical insectivores and carnivores, for example, are thegosed on the surface of a plate of enamel, that is (predominantly) across the ends of the enamel rods. This is *alpha thegosis.*

The cutting blades of the typical herbivores, for example, are thegosed transversely across a plate of enamel including the amelo-dentinal junction (the line of demarcation between enamel and dentine), that is (predominantly) on the sides of the enamel rods. This is *beta thegosis.*

The present study of fossil and extant Primates shows clearly that this Order is characterised by cutting blades that function for a part (sometimes a considerable part) of the animal's lifetime with alpha thegosis before beta thegosis eventually appears.

Types of blades

1. Scissorial blades

'Scissorial' is a general term to include the basic action common to all paired blades, whether vertical or horizontal.

Scissorial action is virtually the first category of Simpson's definition: 'Shearing: (*a*) The parallel passage of crests sharing equally in shear.'[4] Although Simpson later explains that 'the crests shear past each other, a truly scissor-like, crest-past-crest shearing mechanism',[4] his use of 'parallel' in the definition is misleading, if not erroneous. Parts of the juxtaposing blade-surfaces may be regarded as parallel, but the blades' functional cutting edges are far from being so. Simpson's use of 'parallel' would be appreciated by those who, despite the defining 'crest', conceive of the parallel surfaces as the cutting agents.

The essence of scissorial action is that the diagonally opposing blades, while confining an interposed exogenous material to their cutting actions, concentrate the dividing force at a point, thus utilising the principle of producing a crack which is propagated: point-cutting. Were the blades in direct opposition, i.e., parallel, there would be even more efficient confining, but it would be at the cost of cutting along a line, i.e., the force required for such an action would be multiplied. There is, therefore, an optimum angle for efficient point-cutting compromised with the requirements for efficient confining. The difference between point-cutting and line-cutting actions is comparable to that between guillotine and axe.

In the evolution of early mammalian dentitions there was an interesting switch in the mechanism of cutting blades.[3] The first evidence we have of this is in the extinct dryolestids and paurodonts. Its occurrence in the ancestral mainstream, however, first appears in the Cretaceous therian *Pappotherium*, i.e. concurrent with the appearance of opposable basins and their peripheral blades and cusps.

In the primitive cusps-in-a-row and symmetrodontoid conditions there is a relatively simple action of blades with cusp-to-valley interdigitation. In the advanced condition, there is a cusp-to-cusp opposition with the blades vertically concave (i.e. the concavity of one blade opposes that of the other) and horizontally convexed (i.e. the convexity of one blade opposes that of the other). This complicated system occurred to the advantage of ensuring automatic confining of exogenous material to the dividing, cutting action of the blades. It is an essential characteristic of advanced alpha dentition.

In beta dentition, however, with the transfer from a predominantly vertical blade action to a predominantly horizontal one, the advantage of the opposing-convexity system has given place to another. The presence of blade-concavities perpendicular to the line of cutting action remains, however, very much an advantage. Alpha/beta dentition is intermediate.

Because primate dentition is intermediate, the horizontal-opposing-convexity system is somewhat depreciated. For this reason, and because the spatial relationships are somewhat difficult to describe, it is only mentioned here. Nonetheless, this feature has had important bearing on the characteristics of primate dentition.

In the primitive dentition with simple scissor-like (in some aspects more guillotine-like) blades, point-cutting begins at one end of the blade-pair and proceeds along the whole course of the cutting edges to terminate at the other end. The blades are zig-zagged (somewhat in the manner of tailors' pinking-shears), which aids, primitively, the confining of exogenous material to the cutting action. In the advanced alpha dentition (Simpson's 'tribosphenic' dentition, and the dentition basic to my own terminology) there is a profound change. Alphascissorial blades are now consistently drepanoid (sickle-like), and at both ends of the blade there is an *akis* (in an upper blade) or *akid* (in a lower),thus forming a *diakidrepanon* (upper) or *diakidrepanid* (lower). In function, each akid of the (dynamic) diakidrepanid approximates to and juxtaposes its counterpart akis of the (static) diakidrepanon; the concavity of the diakidrepanid, therefore, also approximates to and juxtaposes its counterpart concavity of the diakidrepanon. Such a morphological arrangement allows functional point-penetration at each juxtaposing akid-akis extremity, followed by point-cutting as the blades juxtapose. Interposed exogenous material is thereby confined to the dividing action of the blades. (It is to be emphasised that the horizontal-opposing-convexity feature has an important bearing on this confining action; it and other relevant features of the advanced condition are presented elsewhere.)[3]

In the basic (advanced) alpha dentition, each diakidrepanoid blade is combined with another, each blade sharing a cusp of the other; the combination therefore becomes a *triakididrepanon*, or *triakididrepanid*. In occlusal view the triakididrepanoid unit has the appearance of an asymmetrical Gothic arch; both blades are (horizontally) convexed, one is oriented (predominantly) transversely across the occlusal plane, the other obliquely. As Simpson has emphasised, Cope and Osborn were aware of the blades that ran from the three cusps of the 'trigon'.[4] Had they observed that there were two profoundly different blade systems present in early mammals (i.e. cusp interdigitation and cusp apposition) they might have realised that the blades at each cusp of their 'trigon' form a unit 'trigon' that, in fact, is homologous with the 'trigonid'.

These terms are abbreviated in the following way. With, for example, the prefix *para*:

Para-Triakis for Para-Triakididrepanon (PaT)
Paratransversis for Para-diakidrepanon-transversa (Pat)
Paraobliquis for Para-diakidrepanon-obliqua (Pao).

Because lower transverse blades articulate with upper transverse blades, and oblique with oblique, the terms allow easy appreciation of occlusal relations.

In the primate (alpha/beta) dentition, the angle between the triakididrepanoid blades progressively becomes obtuse. Despite such adaptive variation (discussed below), where both transverse and oblique blades progressively become longitudinal, the terminology, by retaining the phylogenetic connotation, also retains the connotation of occlusal relationships.

The Palaeocene primate *Plesiadapis rex* (Fig. 1*a*) illustrates the characteristic widening of the angle of the triakididrepanoid blades; the Paratransversis and Metatransversis have virtually become as obliquely orientated mesially as the Paraobliquis and Metaobliquis are distally. The Eocene primate *Adapis magnus* (Fig. 3*b*) illustrates further widening; the angle has now become obtuse.

In *Plesiadapis* the paraobliquakis and metatransversakis, although immediately adjacent, are virtually discrete entities; whereas in *Adapis* they are completely amalgamated. In *Plesiadapis* the Paraobliquis and Metatransversis each share an akis with their respective cingular blades; whereas in *Adapis* they are widely separated from all contact. The parastylo-Cingulis and metastylo-Cingulis themselves are now combined at a shared (pseudo-) akis.

Thegosis characteristically occurs as the teeth erupt and come into occlusion. The behaviour is adapted to facilitate the alignment of the scissorial blades which are thus brought into accurate functional apposition. A further advantage of this facility is the alignment that

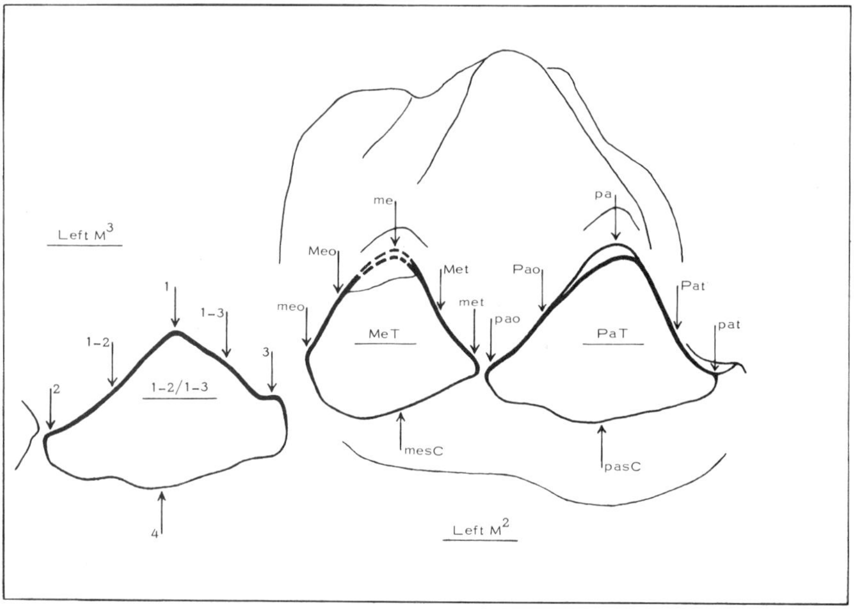

Figure 1(*a*). *Plesiadapis rex.* Left M^2 and part of M^3. Buccocclusal view.
(Figs. 1*a*, *b* and *c* are from the same specimen.)
1, 2 & 3 Akis (a cusp with a sharp blade running from the apex).
1-2 Oblique diakidrepanon (a sickle-like blade with an akis at either end).
1-3 Transverse diakidrepanon (phylogenetically orientated transversely across the occlusal plane).
1-2/1-3 Triakididrepanon (a combination of two diakidrepana).

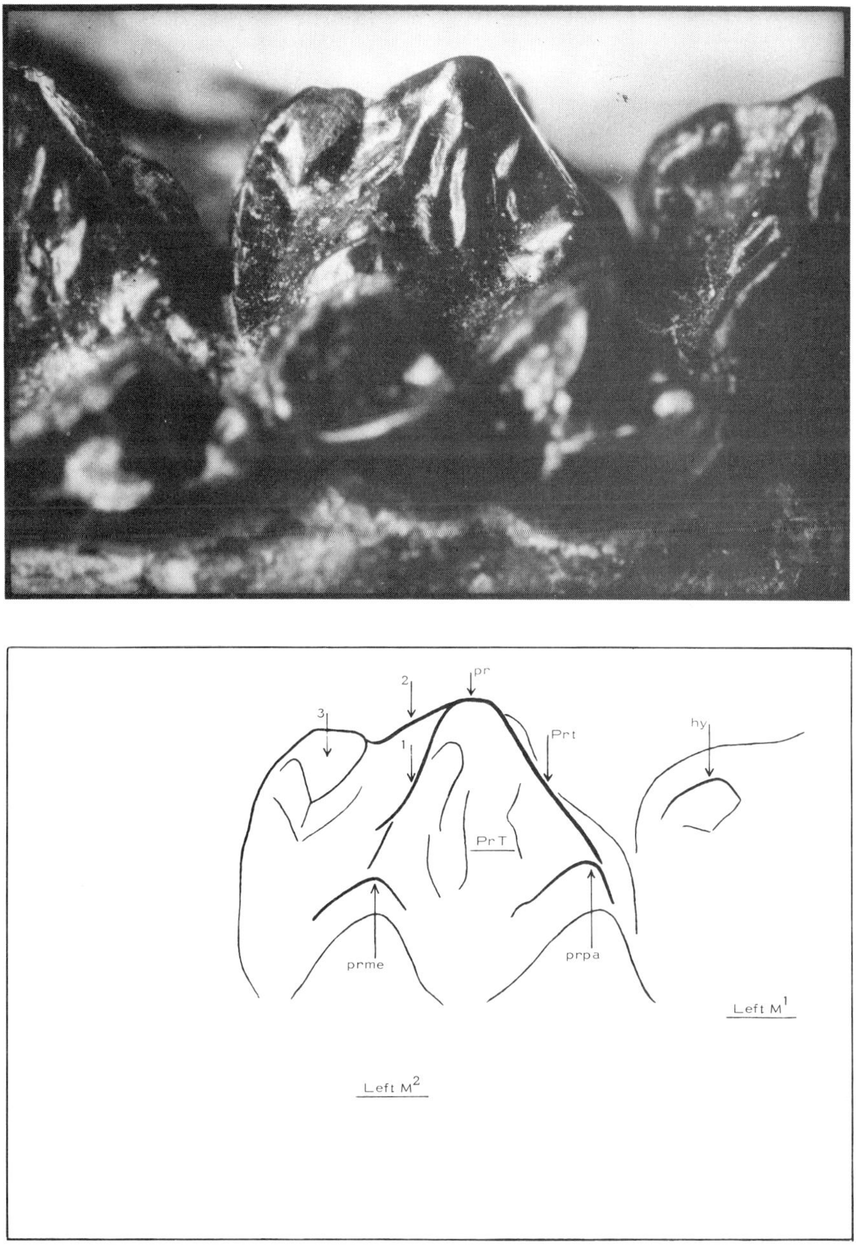

Figure 1(*b*). *Plesiadapis rex.* Left M^2 and part of M^1. Buccocclusal view.

1 The (relict) Protobliquis (Pro).

2 The (pseudo-) Protobliquis, i.e., functioning with the blades from the entoakid.

3 Facet on the (pseudo-) Hypo-Triakis, with abrasion scratches superimposed.

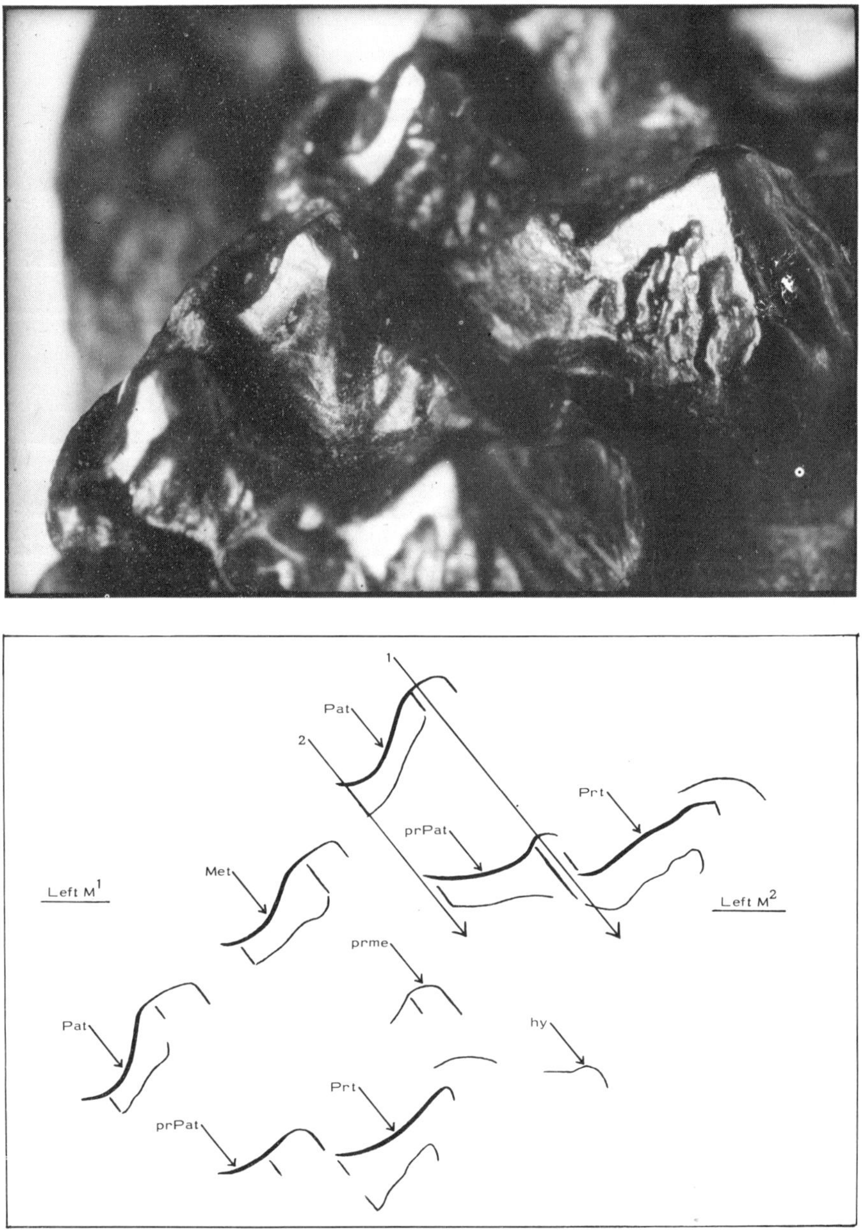

Figure 1(c). *Pleisiadapis rex.* Left M^1 and M^2. Mesiolingual view.
1 & 2 Direction of scissorial action of the Hypotransversid.

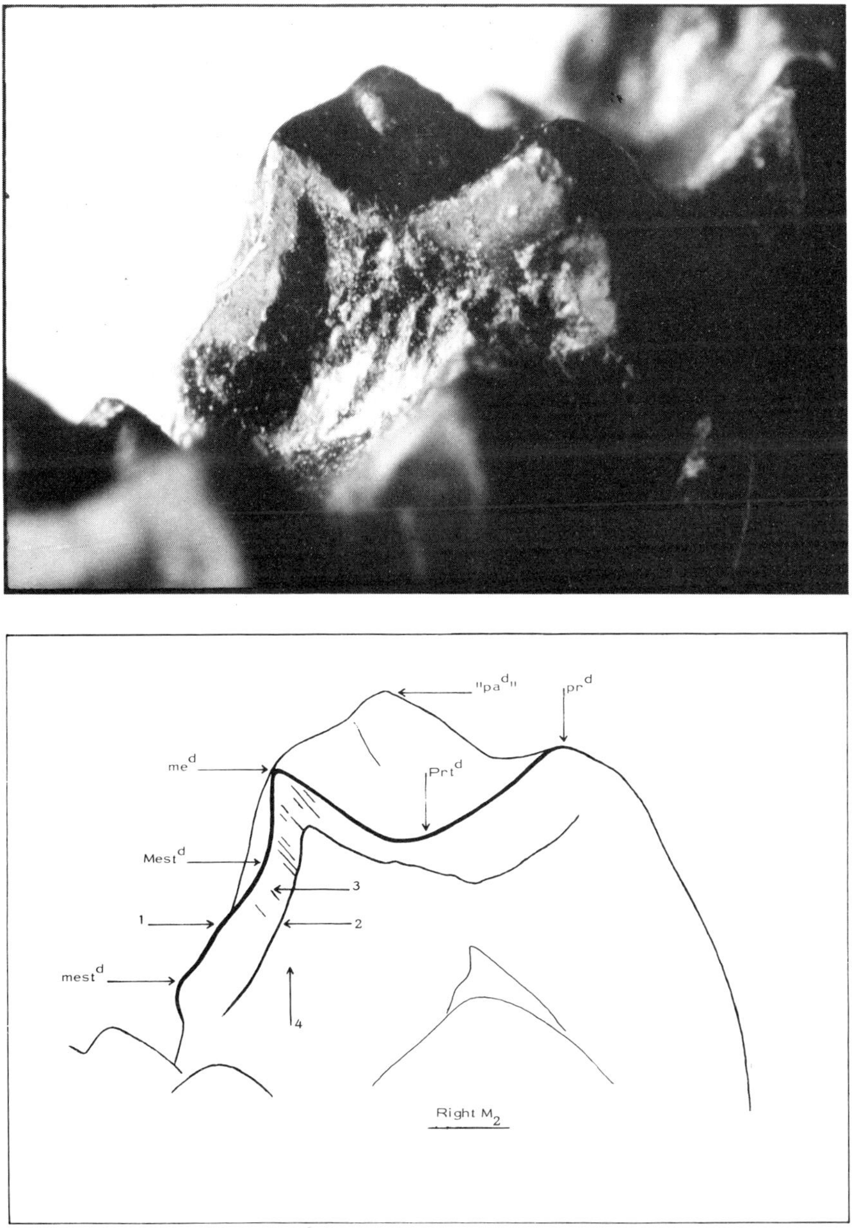

Figure 2 (*a*). *Plesiadapis rex.* Right M_2. Distal (slightly buccocclusal) view.
1 Leading edge of the Metastylotransversid.
2 Trailing edge of the Mest^d.
3 Thegosis-facet forming a narrow strip at the scissorial edge.
4 Primary (developmental) hollowed-out blade-surface.

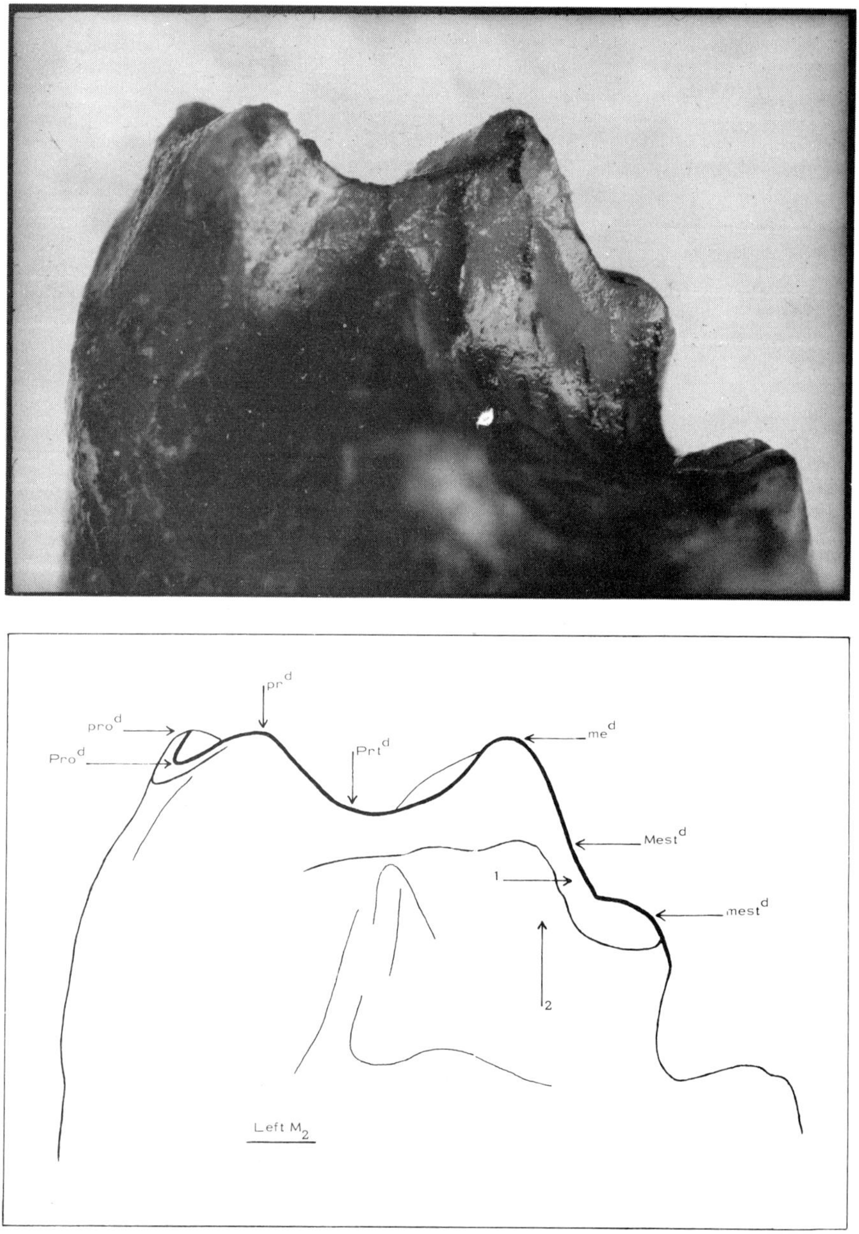

Figure 2(b). *Adapis magnus.* Left M_2. Distal (slightly buccal) view.
(Figs. 2 *b* and *c* are from the same tooth).
pro^d The protobliquakid, the pseudo-parakid, i.e. functionally homologous to the original parakid.
Pro^d The functional Protobliquid.

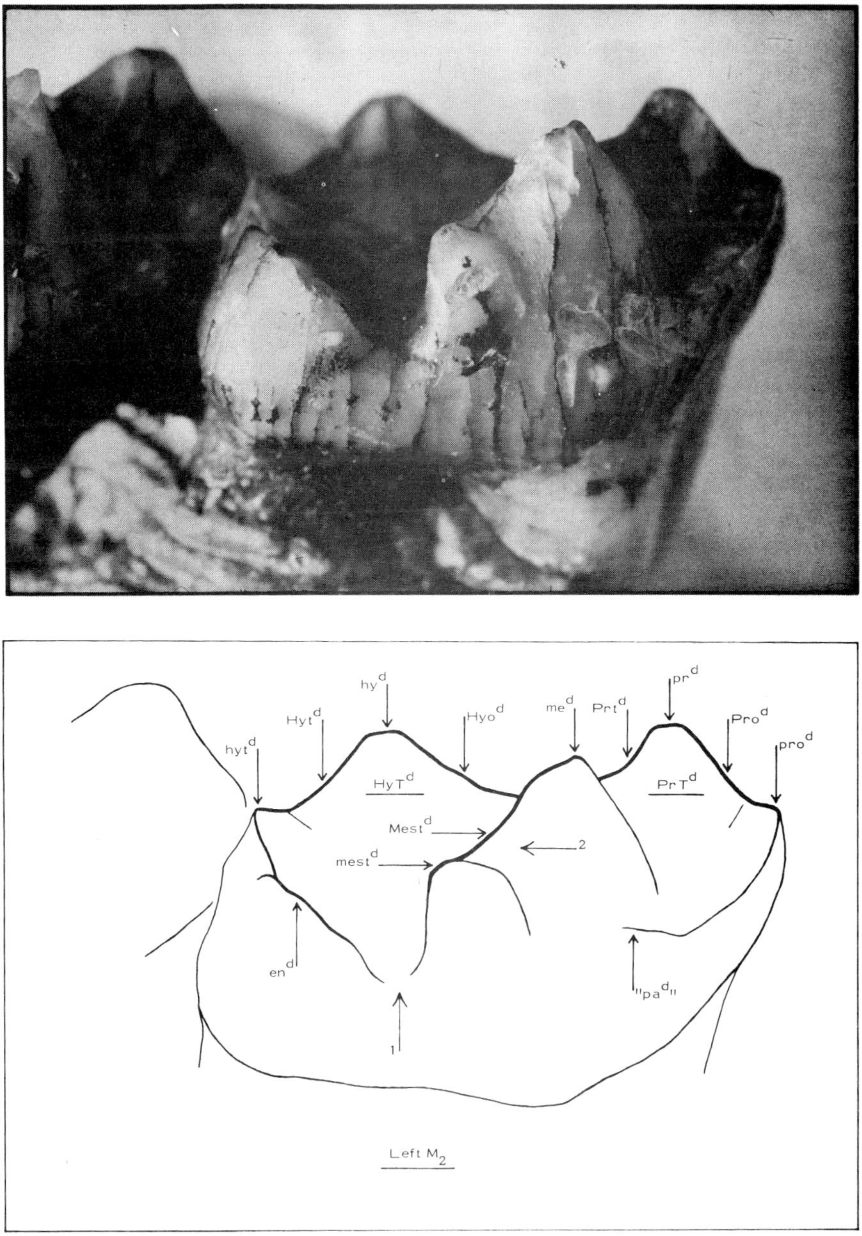

Figure 2(*c*). *Adaptis magnus*. Left M_2. Linguocclusal view.
(NB The lingual surface of this specimen has a number of preparation artifacts.)
1 Talonid escapement.
2 Metastylotransversid escapement.

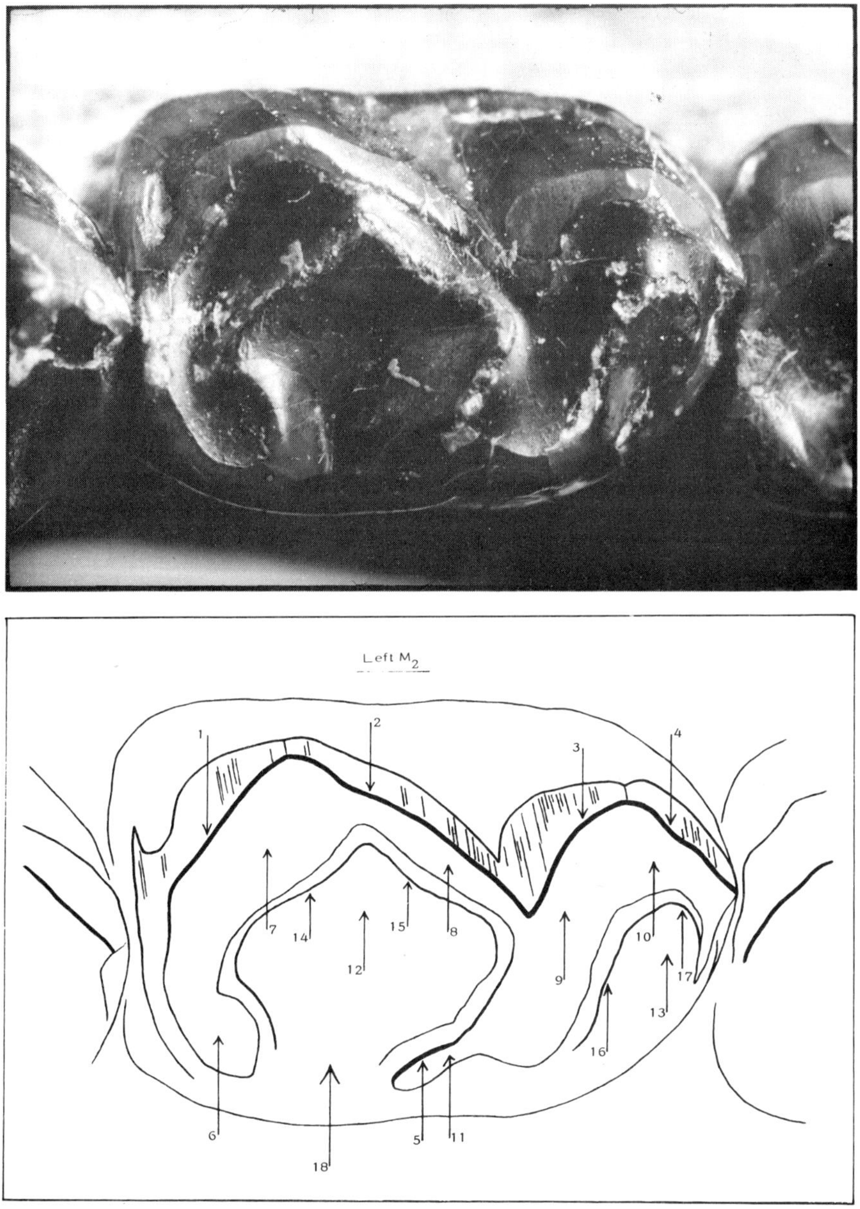

Figure 3 (*a*). *Adapis magnus.* Left M_2. Occlusal view.

1-2-3-4 & 5 Leading edges of secondary, scissorio-incusive blades, now the leading blades of the tooth, i.e. Hypotransversid, Hypobliquid, Prototransversid, Protobliquid, and Metastylotransversid.

6-7-8-9-10-11 Secondary (abrasive) hollowing to the secondary blades. Secondary hollowing, also, now forms secondary (dentinal) incusive basins coalesced into one continuous basin.

12 & 13 Remnants of primary (enamel) incusive basins of Hypo-Triakid and Proto-Triakid.

14-15 & 16-17 Remnants of primary (scissorio-incusive) blades of HyT^d and PrT^d, now incusive only.

18 Talonid escapement.

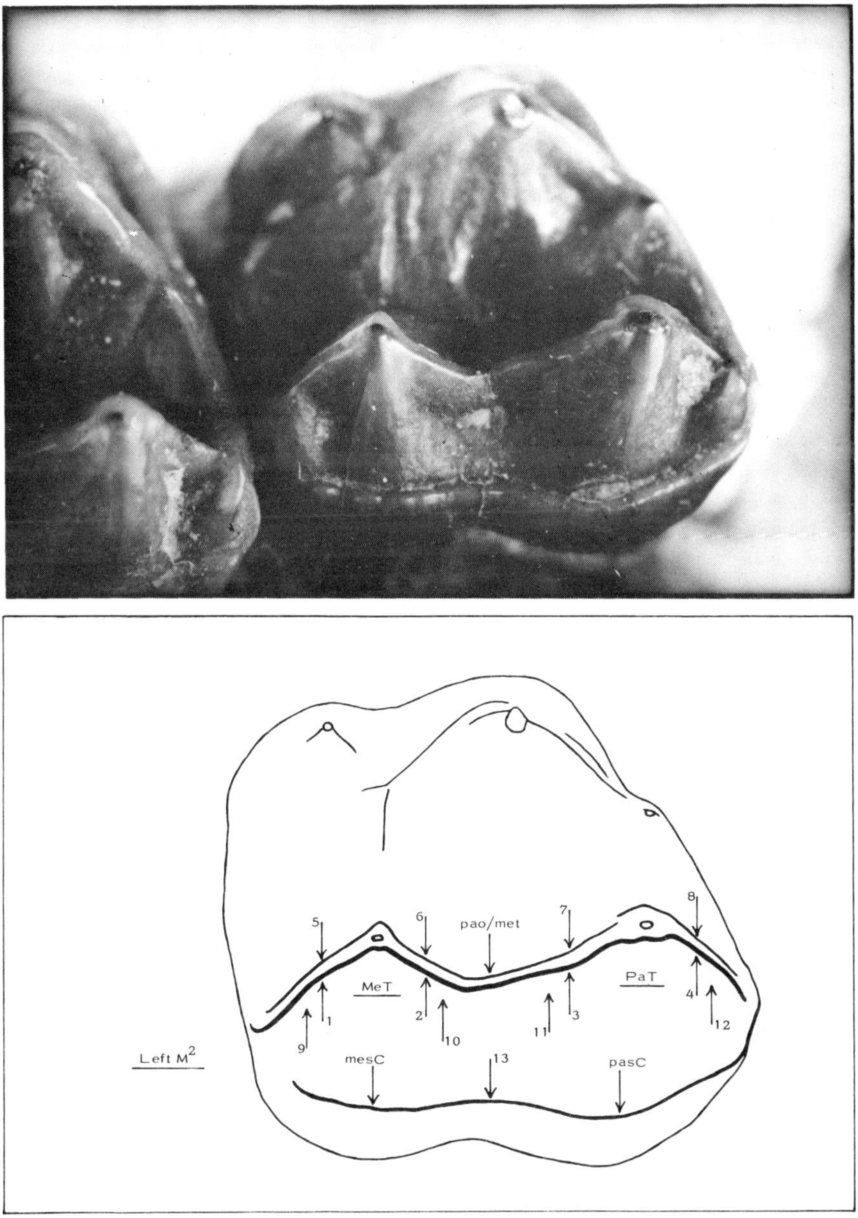

Figure 3(*b*). *Adapis magnus.* Left M^2. Buccocclusal view.
(Figs. 3, *b* and *c* and Figs. 4, *a*, *b* and *c* are from the same tooth.)
1-2 & 3-4 Leading edges of the blades of the Meta-Triakis and Para-Triakis.
5-6 & 7-8 Trailing edges of the blades of the MeT and PaT.
9 & 10 Primary hollowed-out surfaces to the leading edges of the MeT.
11 & 12 Similar surfaces to the PaT.
13 Shared (pseudo-) akis of the parastylo-Cingulis and metastylo-Cingulis.
pao/met Amalgamation of paraobliquakis and metatransversakis.

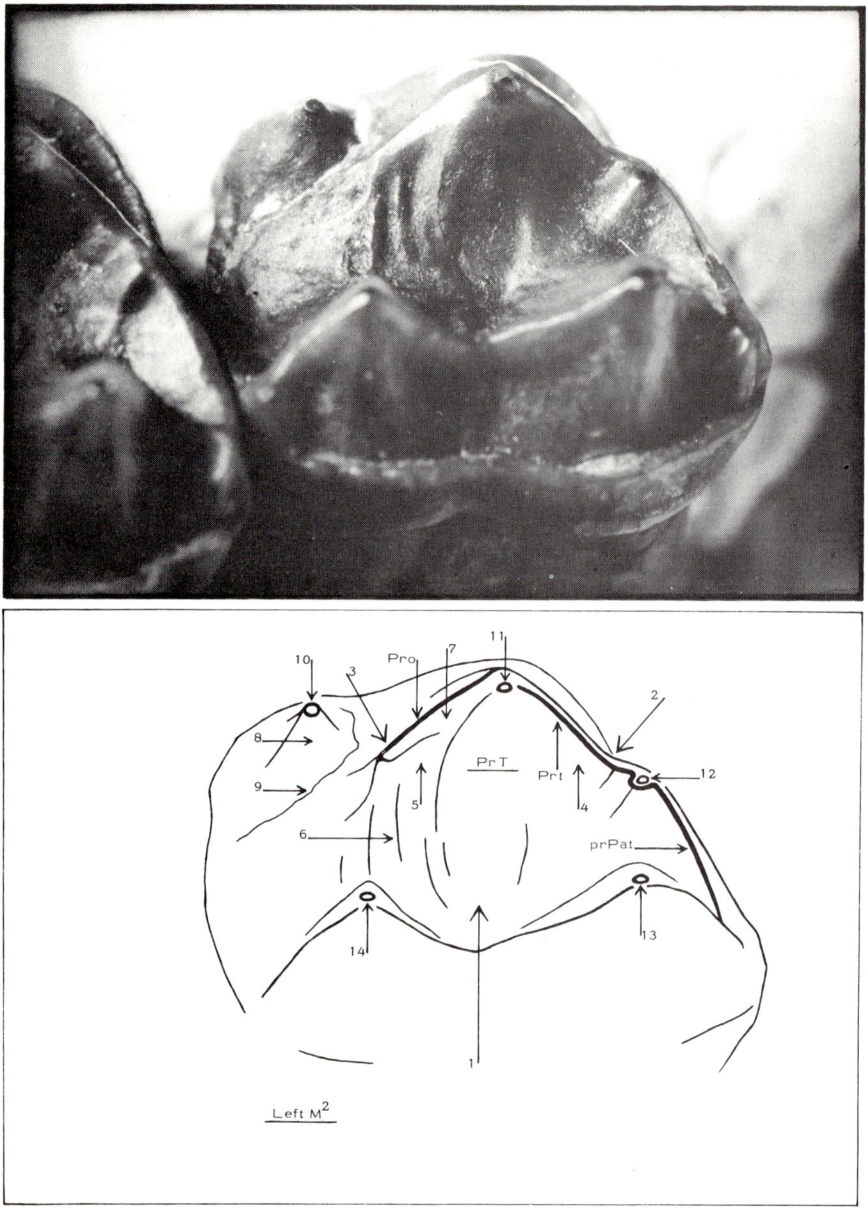

Figure 3(*c*). *Adapis magnus.* Left M^2. Buccocclusal view.

1 Direction of scissorio-incusive action; terminal occlusion of the hypoakid.

2 Point where the Hypobliquid in terminal occlusion crosses the Prototransversis.

3 Point where the Hypotransversid in terminal occlusion crosses the Protobliquis.

4 & 5 Primary hollowed-out incusive surfaces to the leading edges of the Proto-Triakis.

6 Ridges concaved towards the centre of the incusive action.

7 Facet caused by oblique thegosis of the Hypotransversid with the Protobliquis.

8 Exposed dentinal core of the (pseudo-) Hypo-Triakis; the enamel cap has fractured in preparation (9).

10 Functionally abraded apex of the dentinal core.

11, 12, 13 & 14 Corresponding dentinal abrasion of the protoakis, protoparakis, parakis and metakis.

can be achieved over an appreciable range of morphological variation. It is a feature, moreover, that offers considerable opportunity for biological progress.

To facilitate this and subsequent action, the juxtaposing surfaces of scissorial blades are hollowed out[3] (or are oriented to produce the equivalent effect). In the alpha stage (e.g. Fig. 3*b*; Fig. 4*a*) the hollowing is primary (developmental); in the beta stage (Fig. 5*c*) it is part primary and part secondary (abrasive). This therefore allows:

1. Close juxtapositioning of the cutting edges.

Advantage:
- (a) Obviates the jamming of an interposed exogenous material between, and therefore separating, the cutting edges.
- (b) Facilitates division of exogenous material.

2. The restriction of thegosis to a narrow strip at the cutting edge. This is particularly evident in the beta condition where the facet is limited by the width of the enamel plate itself.

Advantage:
- (a) Facilitates the thegotic process through an economy of time; a minimum amount of tooth-substance has to be removed.
- (b) Ensures the ontogenetic continuity of thegosis through an economy of tooth-material, especially the crucial edge-forming enamel.
- (c) Further facilitates division.

The origin of the widened angles of primate triakididrepanoid units, both upper and lower, is a significant aspect of the primate dentition. To understand this, additional features of the ancestral alpha dentition need some discussion.

When describing the basic 'tribosphenic' lower molar, Simpson gives the morphology of an overt feature of this and subsequent alpha dentitions (where it often becomes even more overtly adapted to appear – additionally – in the upper molars), but he gives it no functional significance: 'The external cusp, protoconid, is crescentic, and from its apex run two sharp, shearing crests each of which sinks, or is distinctly notched, near the middle and then rises again to a distinct cusp.'[4]

Although the presence of this 'notch' on the ancestral Protobliquid and Prototransversid is occasionally mentioned in the literature, the absence of any attempt to give it functional significance is as conspicuous as is the absence of the feature in the descendent primate dentition.

One of the advantages of the scissorial diakidrepanoid blade is that it allows an extended length of point-cutting edge. It also capitalises on the horizontal-opposing-convexity principle mentioned above.

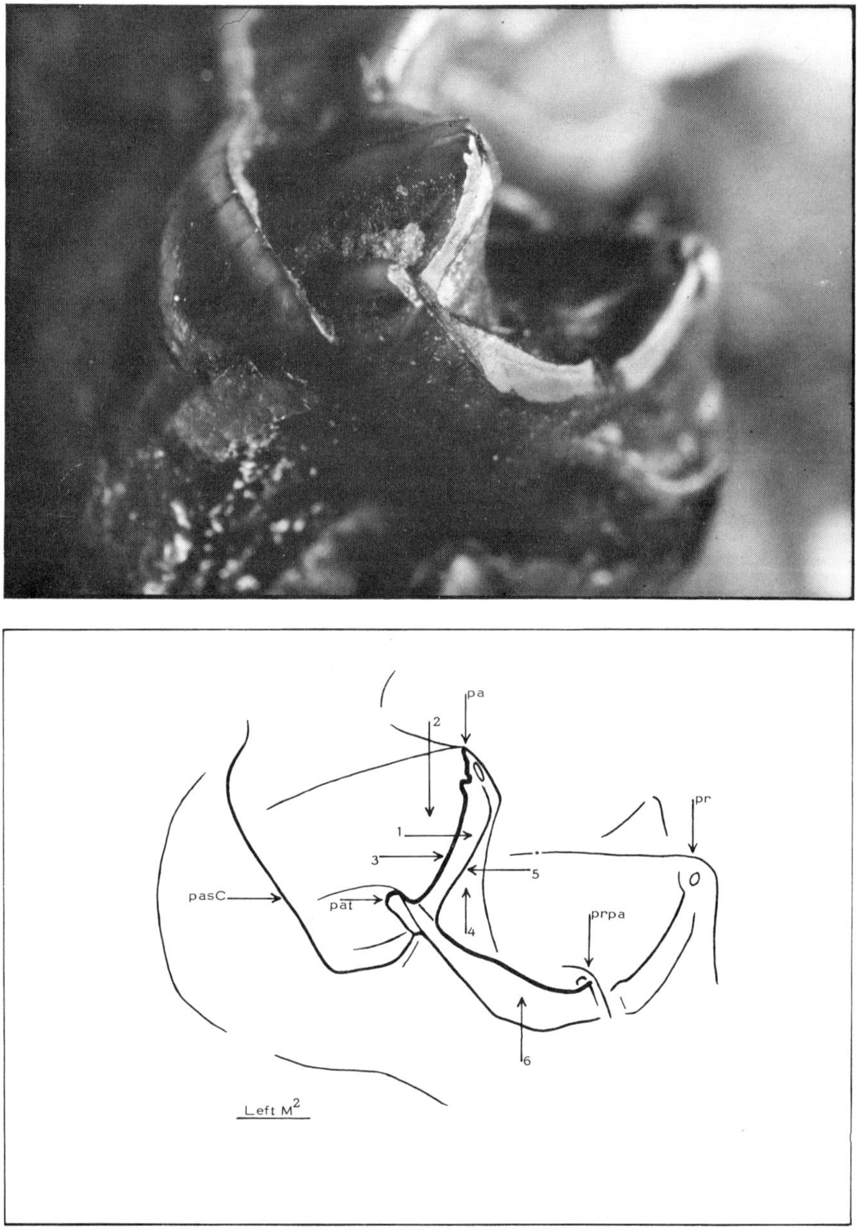

Figure 4(*a*). *Adapis magnus.* Left M^2. Mesiobuccocclusal view.

1 Alpha-thegosis of the Paratransversis; there is primary hollowing (2) to the leading edge (3), also primary hollowing (4) to the trailing edge (5).

6 Alpha-thegosis of the proto-Paratransversis, with corresponding characteristics.

pasC The parastylo-Cingulis, an ontogenetic, reserve, leading blade of the crown.

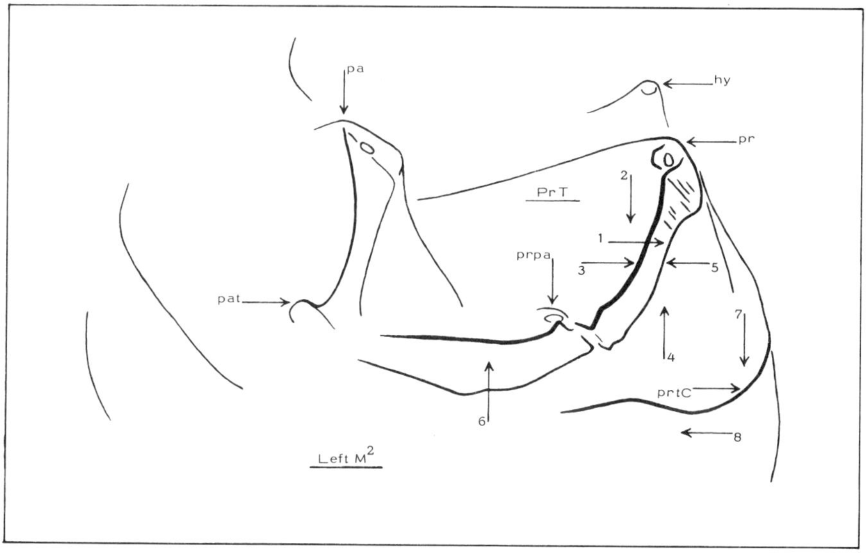

Figure 4(*b*). *Adapis magnus.* Left M^2. Mesiocclusal view.

1 Alpha-thegosis of the Prototransversis. The primary hollowing (2) to the leading edge (3) also forms part of the incusive basin of the Proto-Triakis; the primary hollowing (4) to the trailing edge (5) forms part of the mesiobuccal wall of the crown.

6 Alpha-thegosis of the proto-Paratransversis, with corresponding features.

7 & 8 Primary hollowing to the prototransverso-Cingulis.

prtC The prototransverso-Cingulis, an ontogenetic, reserve, trailing blade of the crown.

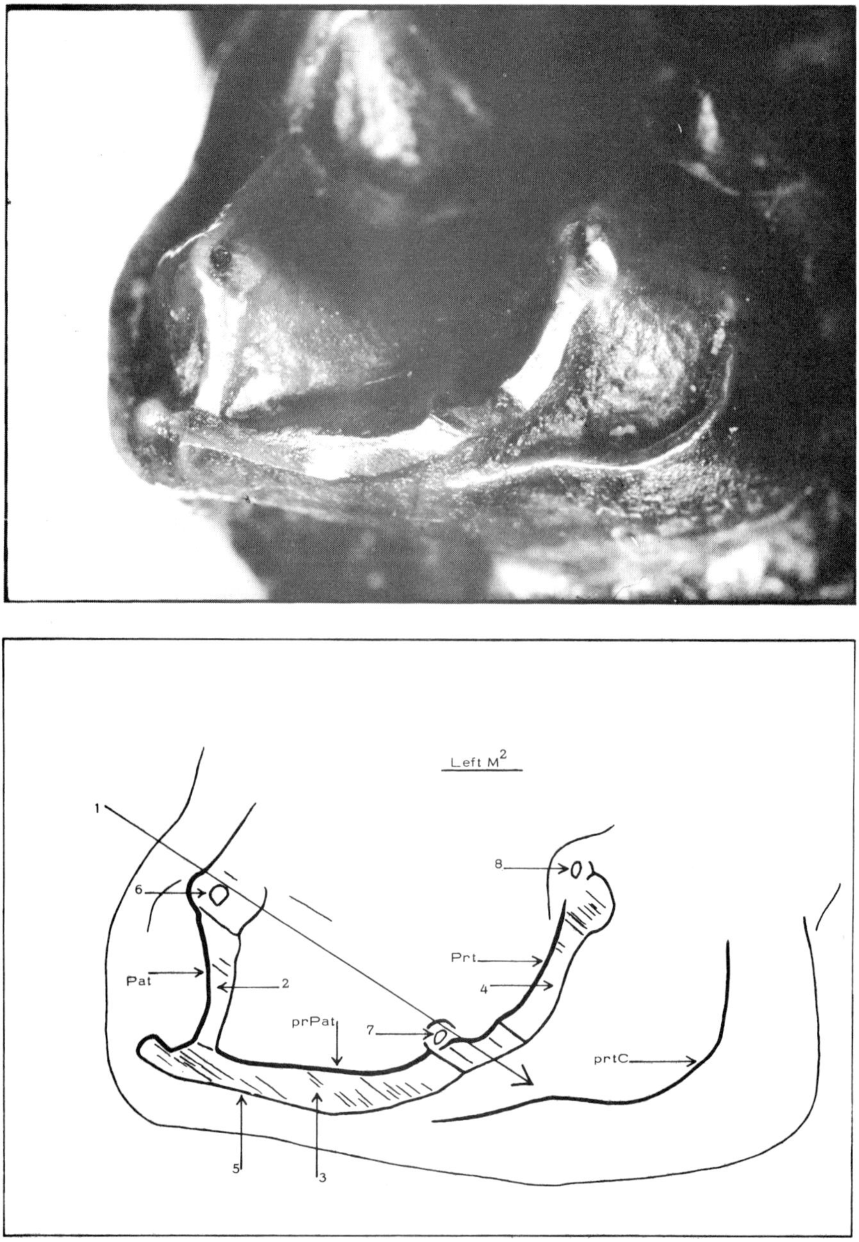

Figure 4(*c*). *Adapis magnus.* Left M^2. Mesiocclusal view.

1 Direction of scissorial action.

2, 3 & 4 Narrow strips of thegosis faceting of the Paratransversis, proto-Paratransversis and Prototransversis.

6, 7 & 8 Beginnings of secondary hollowing at the apices of the parakis, protoparakis and protoakis.

Simple diakidrepanoid blades have an inherent disadvantage that becomes especially significant when blades are elongated: when the central concavities of two blades approximate, point-cutting becomes line-cutting. It is a disadvantage, however, that is obviated by the morphological *fissuring* of one or both the blades;[3] point-cutting is thus maintained.

In the (advanced) alpha dentition the Protobliquid is the dominant blade of the lower molar and is characteristically fissured. Moreover, it overshadows the basin of the adjacent mesial tooth, on the distal margin of which is the Hypotransversid and entoakid. Inevitably, then, when the (adjacent) basin is progressively raised, the function of the Protobliquid is affected. When the two structures (i.e. the Protobliquid and the Hypotransversid) eventually become level, the portion of the original Protobliquid lingual to the interproximal contact-point is rendered functionless and is, in fact, vestigial or lost (Figs. 2*c*; 6*b*, *c*). Significantly, the characteristic scissorial, diakidrepanoid blade, although greatly shortened, has remained intact (Figs. 2*b*, *c*; 5*a*, *b*). In some primates the relict of the parakid is apparent, in others its former location is represented by a void.

The question arises, since there is still the characteristic akid at the free (lingual) end of the remaining blade: has the parakid in some primates 'migrated', or is the existent akid invariably a new evolutionary feature? Since in many species an obvious remnant parakid occurs, I therefore propose the functional term *protobliqakid*. The blade remains the Protobliquid.

The new blade, mentioned above, that occurs as a result of the enlargement of the incusive surfaces, runs distolingually from the metakid (Figs. 2*a*, *b*, *c*; 5*a*, *b*). A blade in this position also occurs in many dentitions other than those of the primates, particularly alpha/beta and beta dentitions. It even occurs, but in a primitive form, in alpha dentitions where there is a quasi-enlargement of the basin of the Proto-Triakis as a result of a secondary reduction of the Para-Triakis (e.g. *Didelphis*).

Crompton (whose system of facet numbering is increasingly used) depicts the facet of this blade as 'shearing surface 5'.[11] He seems more concerned, however, with the size and 'efficiency' of this and other surfaces than he is with their cutting edges. Although Crompton recognises my personal communications with him regarding such features,[11] he has yet to recognise the phenomenon of thegosis upon which all this is based. He recognises, for instance, that 'the food to be sheared was trapped in an ellipsoid space which decreased in size as the leading edges passed by one another. Once the leading edges had passed one another, no further shearing was possible'. Despite this, Crompton seems ambivalently preoccupied with the 'shearing surface', and his analysis of this feature becomes the dominant basis for his discussion of tooth function generally, for instance:

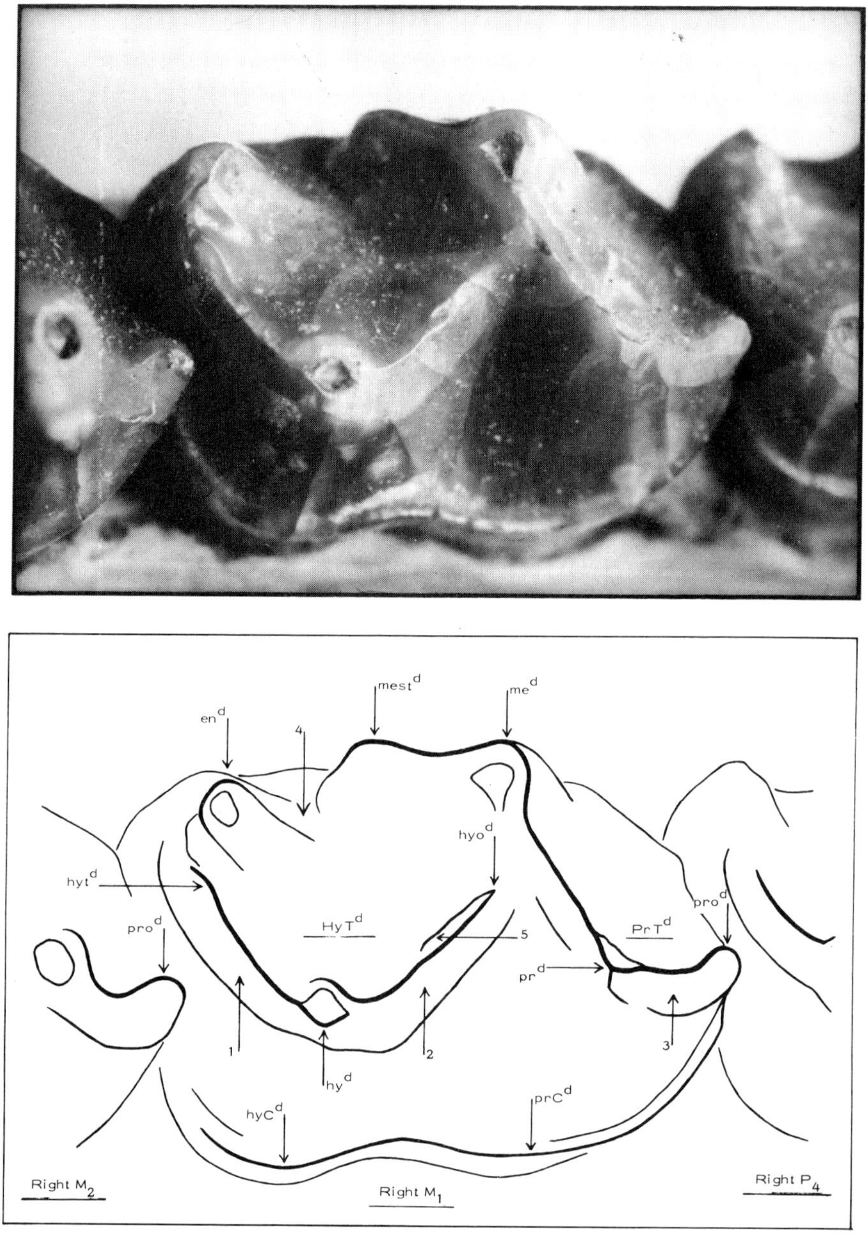

Figure 5(*a*). *Adapis parisiensis.* Right M_1 and parts of P_4, M_2. Buccoclusal (slightly distal) view.

(Figs. 5, *a*, *b* and *c* are from the same specimen.)

1, 2 & 3 Alpha-thegosis of the Hypotransversid, Hypobliquid and Protobliquid.

4 Talonid escapement.

5 Facet caused by oblique thegosis of the Hypobliquid with the Prototransversis.

hyo^d (Pseudo-) hypobliquakid, absorbed in the wall of the Prototransversid.

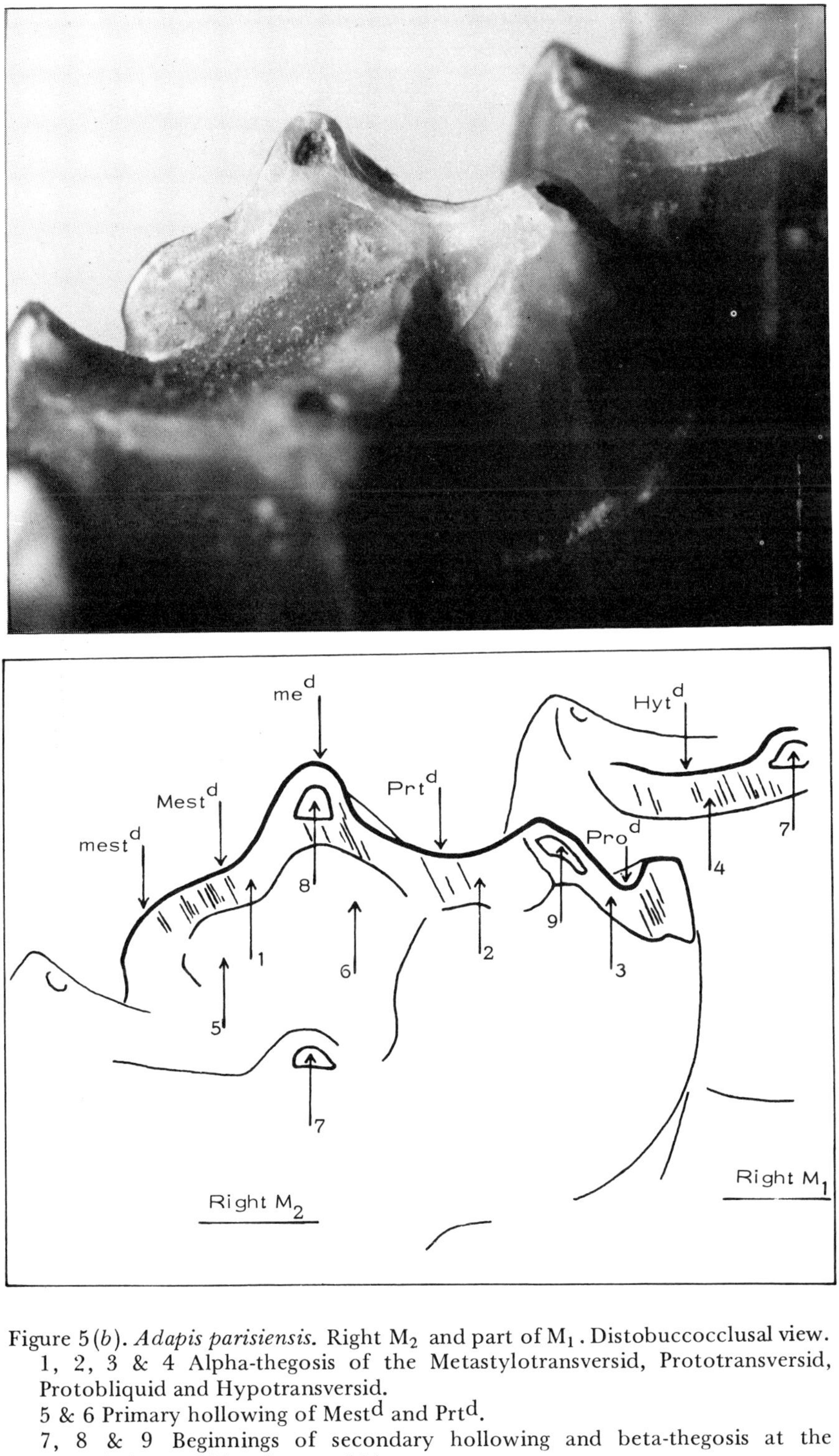

Figure 5(*b*). *Adapis parisiensis.* Right M_2 and part of M_1. Distobuccocclusal view. 1, 2, 3 & 4 Alpha-thegosis of the Metastylotransversid, Prototransversid, Protobliquid and Hypotransversid.
5 & 6 Primary hollowing of $Mest^d$ and Prt^d.
7, 8 & 9 Beginnings of secondary hollowing and beta-thegosis at the hypoakid, metakid and protoakid.

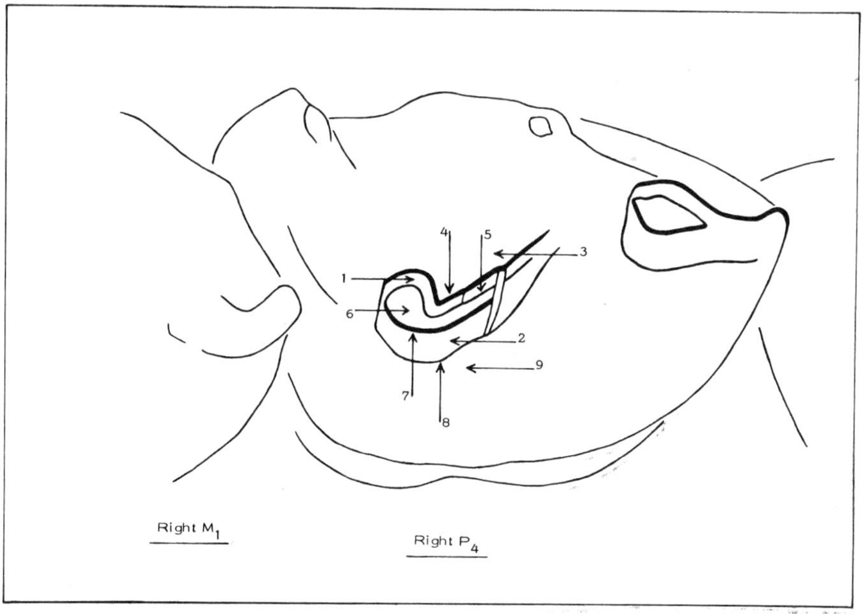

Figure 5(*c*). *Adapis parisiensis.* Right P_4. Buccocclusal (slightly distal) view.

1 Beta-thegosis of the primary Hypobliquid blade; there is primary hollowing (3) to the leading edge (4) and secondary hollowing (6) to the trailing edge (5).

2 Beta-thegosis of the secondary Hypobliquid blade; there is secondary hollowing (6) to the leading edge (7) and primary hollowing (9) to the trailing edge (8).

The position, orientation, and relative size of the shearing surfaces in *Didelphodus* were determined by the direction of mandibular movement and the type of food to be broken down. The teeth are constructed so that some of the shearing surfaces were used simultaneously while others were used successively. . . . The evolution of the tribosphenic molar from those of a form such as *Kuehneotherium* consisted essentially of increasing the transverse component of mandibular movement, increasing the size and efficiency of some of the shearing surfaces present in the early therians, and adding shearing surfaces that were used successively.

In prosimians generally the new blade is seen to have become highly developed. The blade is diakidrepanoid, it has a characteristically concave *escapement* area (Fig. 2*c*),[3] and, contrary to Crompton's suggestion of 'shearing efficiency' of a large facet, the juxtaposing surface of this scissorial blade is characteristically hollowed: the facet, therefore, characteristically forms a narrow strip at the cutting edge.

Its counterpart, the Prototransversis (with which the Prototransversid of the ancestral alpha dentition orginally articulated), also shows the scissorial diakidrepanoid characteristic (Figs. 1*c*; 4*b*, *c*). It has the narrow strip of thegosis-faceting; the hollowing (present on both sides of the blade) is particularly evident within the basin itself (Figs. 1*b*; 3*c*; 4*b*, *c*).

In some prosimians, e.g. *Palaeopropithecus* and *Lepilemur* (Fig. 6*a*, *b*, *c*), the blade is so extensive that the entoakid has entirely disappeared. That this is a loss and not an amalgamation is evident from the fact that the extremity of the blade (i.e. the metastylotransversakid) is separated from the extremity of the Hypotransversid (i.e. the hypotransversakid) by the escapement valley that phylogenetically separates it from the entoakid. Since this new scissorial blade is virtually an extension, at the metakid, of the transverse blade of the Proto-Triakid, I therefore propose the term *Metastylo-diakidrepanid-transversa*, abbreviated to *Metastylotransversid*.

With respect to two disputed primates *Tupaia* and *Plesiadapis*, the absence of the Metastylotransversid in the former and presence in the latter confirms that *Tupaia* should be placed clearly in the Insectivora and *Plesiadapis* equally clearly in the Primates (Fig. 2*a*).

2. *Scissorio-incusive blades*

Despite Simpson's statement, when discussing the origin of the basining of the talonid and central basin of the upper molar, that 'in addition to this double opposition, the elements of alternation and shear are preserved',[4] his preoccupation with the 'tribo-' function for the basins has emphasised the importance of the central bluntness rather than the peripheral sharpness of these structures. Simpson

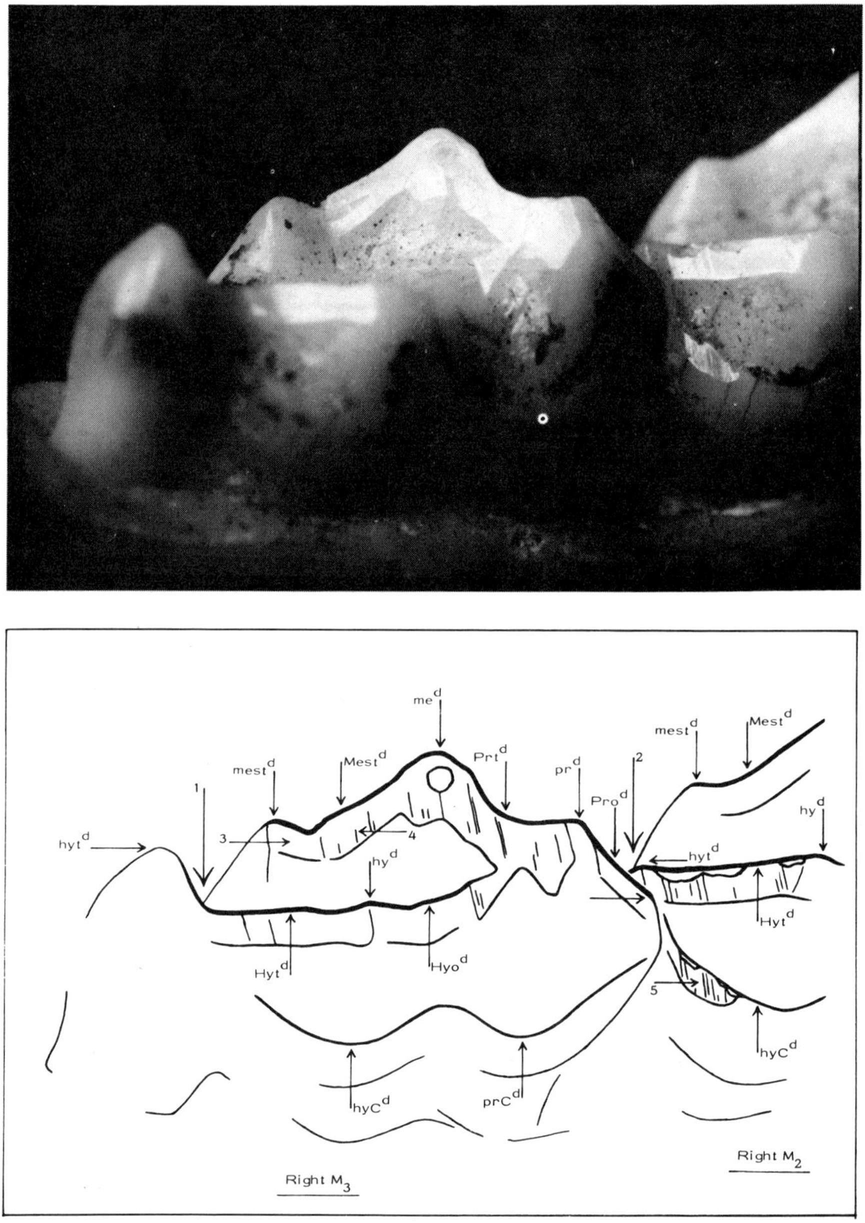

Figure 6(*a*). *Lepilemur mustelinus.* Right M_3. Buccocclusal (slightly distal) view.
1 & 2 Talonid escapement, now between the metastylotransversakid and the hypotransversakid (phylogenetically between the metastylotransversakid and the entoakid), i.e. the entoakid is lost.
3 & 4 Alpha-thegosis of the Metastylotransversid (facet 3 is in the shade).
5 Alpha-thegosis of the hypo-Cingulid.

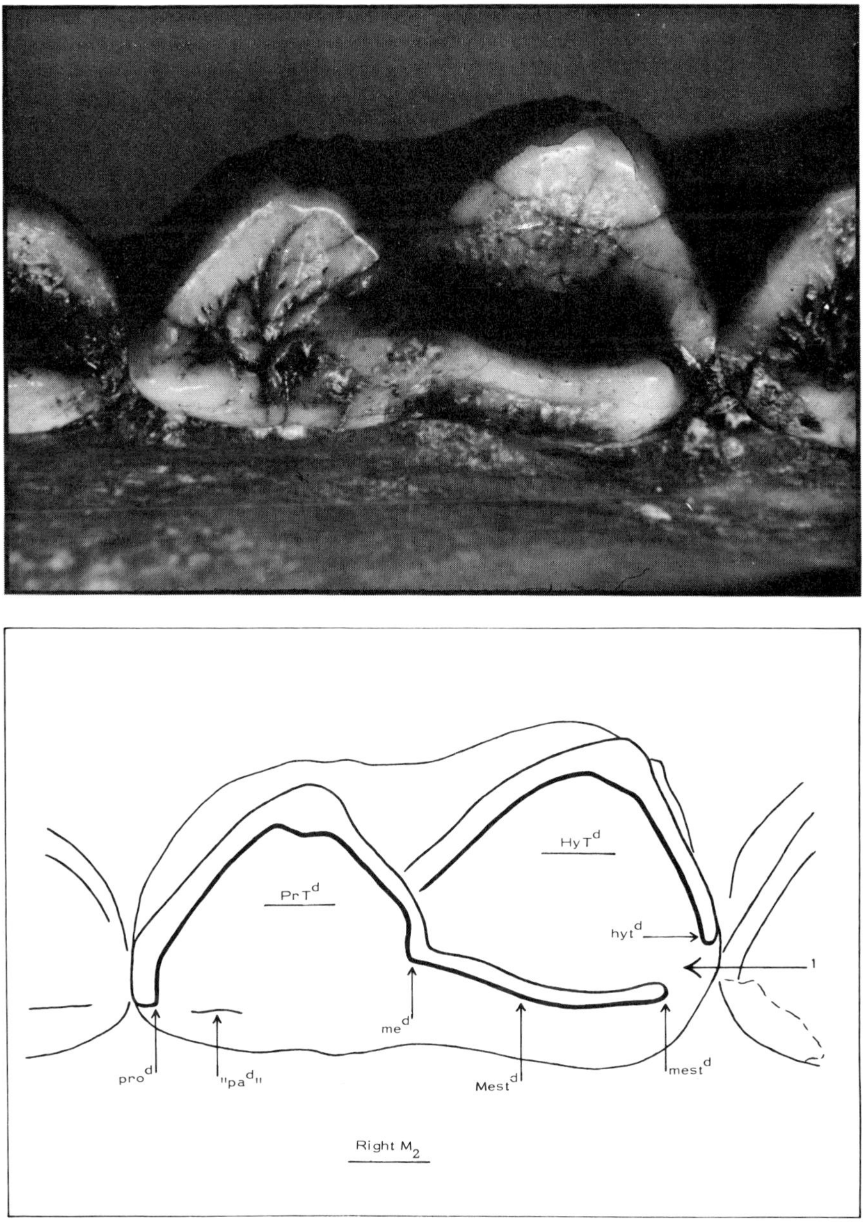

Figure 6(b). *Palaeopropithecus maximus*. Right M_2. Occlusal view.
1 Talonid escapement, as in *Lepilemur*.

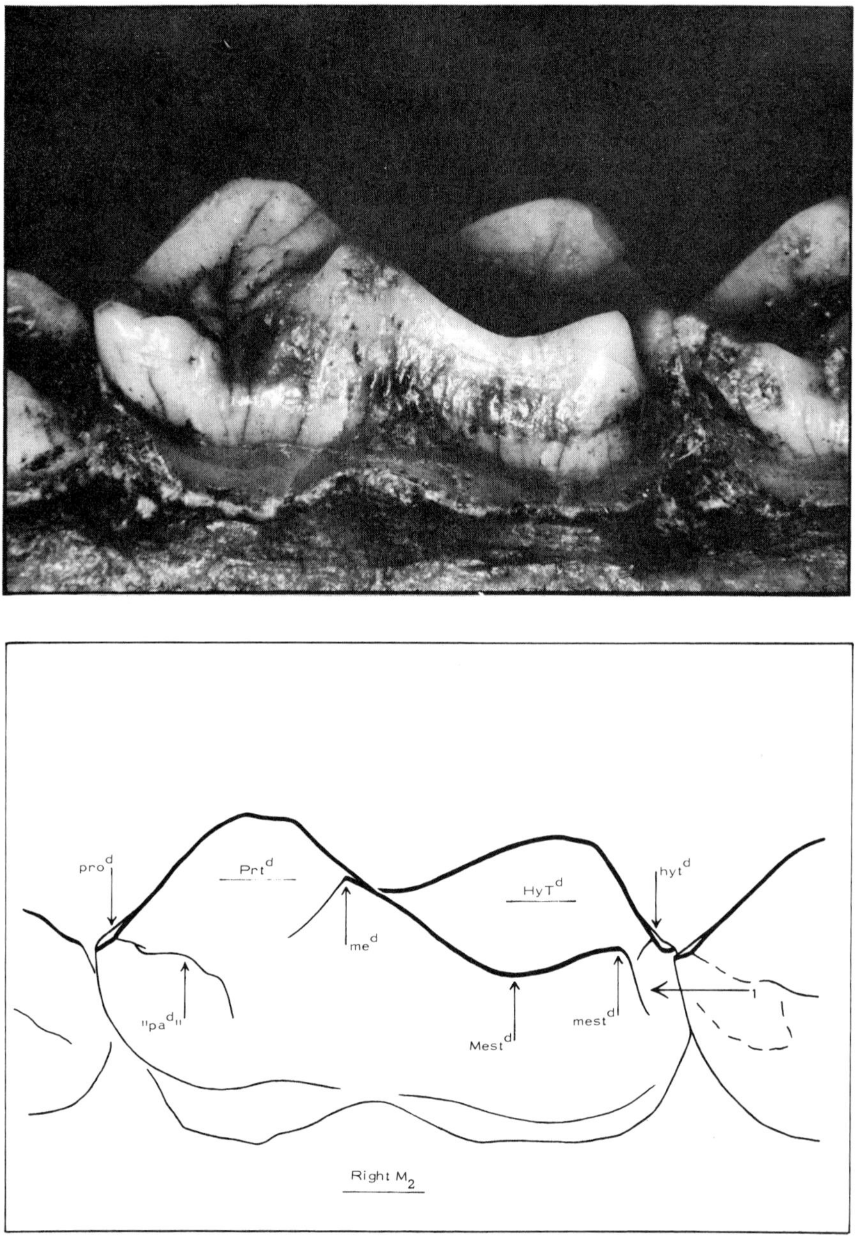

Figure 6(*c*). *Palaeopropithecus maximus.* Right M_2. Linguocclusal view.
1 Talonid escapement.

relates the origin of these structures (including their peripheral sharpness, which he, ironically, did not entirely overlook) to a development from a 'protribosphenic' Middle Jurassic molar whose talonid showed 'simple opposition', i.e. an initial structure with bluntness but no sharpness.

This argument now proves fallacious. The talonid of the Middle Jurassic pantotheres functioned as a scissorial blade that protected the vulnerable interproximal mucous membrane from pathogenic impaction of exogenous material. There is no (blunt) opposability of the talonid.[3]

It can hardly be overemphasised that the prime evolutionary advantage of teeth is their sharpness, the secondary advantage is bluntness. For although most triakididrepanoid blades function purely scissorially, those of the talonid double in function, acting both scissorially and incusively.

The transverse blade of the Hypo-Triakid occludes first scissorially with the transverse blade of the Meta-Triakis, then incusively with the oblique blade of the Proto-Triakis; whereas the oblique blade occludes, first scissorially with the oblique blade of the Para-Triakis, then inclusively with the transverse blade of the Proto-Triakis.

There are, therefore, two fundamentally different occlusal relations:

1. Scissorial: the blades are in point-cutting relationship, i.e. transverse blade occludes with transverse blade, and oblique with oblique.
2. Incusive: the blades are crossed diagonally in striking relationship, i.e. transverse blade occludes with oblique blade, and oblique with transverse.

The dominant function of the incusive (basin) surface is to confine exogenous material to the striking (dividing) action of the incusive blade-and-cusp. Occasionally, however, exogenous material will elude the blade-and-cusp, yet be divided (cracked, split, mashed) by the mutual striking action of the incusive surfaces alone.

With a double (scissorio-incusive) function for the feature there was, therefore, selective advantage for a greater reserve bulk of tooth substance for this segment of the tooth; this allowed greater ontogenetic viability for the crucial scissorio-incusive blade-and-cusp. Hence the Hypo-Triakid/Proto-Triakid levelling, and the inevitable extension of incusive functions to the Proto-Triakid.

With a history of general acceptance of the dominant 'tribo-' function for the pre-primate basin it is hardly likely that the enlarged primate basin, or even the 'lingual-phase-of-occlusion' concept, should raise any question. From the knowledge of the phenomenon of thegosis a number of pertinent questions can now be asked. One of these is fundamental to an understanding of the whole primate dentition: is the increased bulk of the talonid, for example, primarily adapted to the increased functional wear of the basin or to that of its peripheral blades and cusps?

Perhaps the answer can be found in two characteristic features: the differential hardness and the differential morphology of enamel and dentine, the two basic tooth-structures.

Fig. 3*c* shows the dentinal core of the hypoakis of *Adapis magnus* left M^2; the enamel cap has fractured off in preparation. The apex of the dentinal core has been worn flat by functional abrasion, as has that of the protoakis, protoparakis, parakis and metakis; all are equally worn (see also Fig. 4*c*). Fig. 5*a, b* show the same worn dentinal apices of *Adapis parisiensis* right M_1. Fig. 5*c* (P_4 of the same specimen) shows the extension of dentinal abrasion along the (now) beta scissorial blades. With continued functional abrasion the dentinal areas widen, lengthen, and eventually coalesce. They thus form a strip along the entire blade system, i.e. the Protobliquid, Prototransversid, Metastylotransversid, Protobliquid, Hypotransversid, and entoakid (Fig. 3*a*).

At this stage, the primary blades progressively lose their thegotic contacts and, therefore, progressively lose their scissorial function; their incusive function remains.

The scissorial function of the secondary blades (now the leading scissorial blades) are, however, enhanced. This is a result of the widening and deepening of the secondary hollowing, which now becomes a secondary incusive basin.

Finally, i.e. in the gerontic condition, the secondary (dentinal) basins take over from the primary (enamel) basins, which disappear. The occlusal surface now becomes one large, secondary, dentinal basin, encircled by peripheral (secondary) blades that are still subject to beta-thegosis, although now ontogenetically (i.e. adaptively) attenuated (Fig. 7*a, b, c*).

Even with a superficial observation of this phenomenon, it is surprising that the overt breaching of enamel at the cusp apex, the

Figure 7(*a*). *Galago crassicaudatus.* Left M^1, M^2, M^3. Occlusal view. Showing abrasion (scratch and pit marks) on the secondary (dentinal) incusive surfaces. The enamel in this specimen is clearly defined, being slightly decalcified in preparation.

1 & 2 Remnants of the primary (enamel) incusive surfaces of the Proto-Triakises of M^1 and M^2.

3, 4 & 5 Secondary (dentinal) incusive basins of the remnant Proto-Triakises of M^1, M^2 and M^3.

6, 7 & 8 Remnants of the primary incusive basins of the (pseudo-) Hypo-Triakises of M^1, M^2 and M^3.

9 & 10 Secondary incusive basins of the remnant (pseudo-) Hypo-Triakises of M^1 and M^2.

11 Coalescence of secondary incusive basins of M^1.

12 Tangent to the general direction of mandibular movement in the scissorio-incusive action, i.e. mastication. This is the direction of the so-called 'buccal phase of occlusion'.

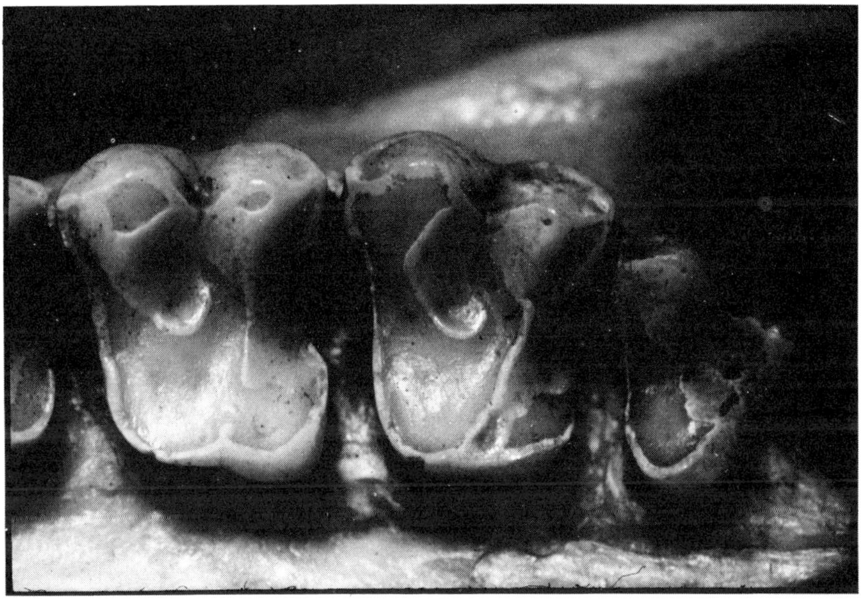

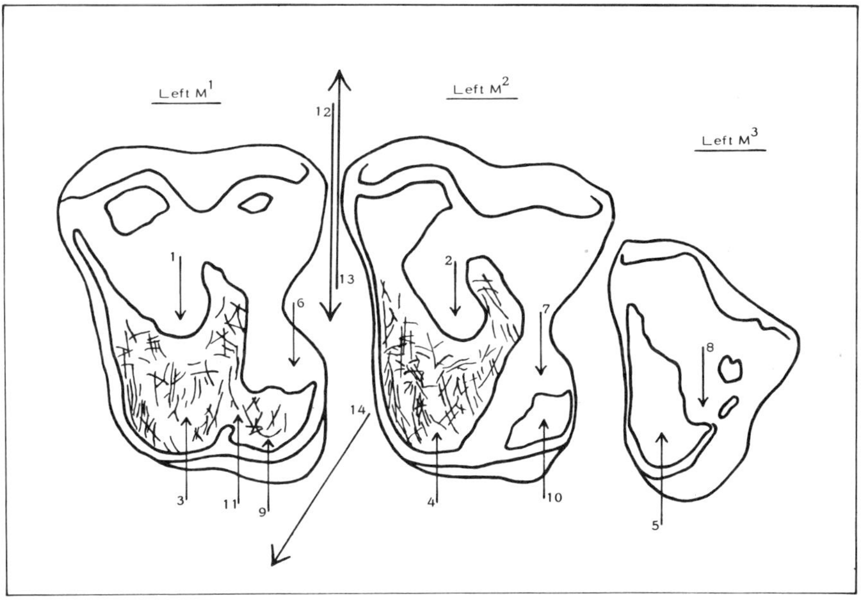

13 Tangent to the general direction of the (dominant) transverse thegotic action (as shown in young adult *Galago* specimens).
14 Tangent to the general direction of the (dominant) oblique thegotic action (from young adult specimens). This is the direction of the so-called 'lingual phase of occlusion'. The evidence of functional (masticatory) action, however, is not the matching parallel striations on facets but the curved, divergent, scratch, pit and gouge marks of abrasion. The evidence of multiple incusive basins and abrasion, alone, falsifies the concepts of a 'mortar-and-pestle' action and a 'lingual phase of occlusion'.

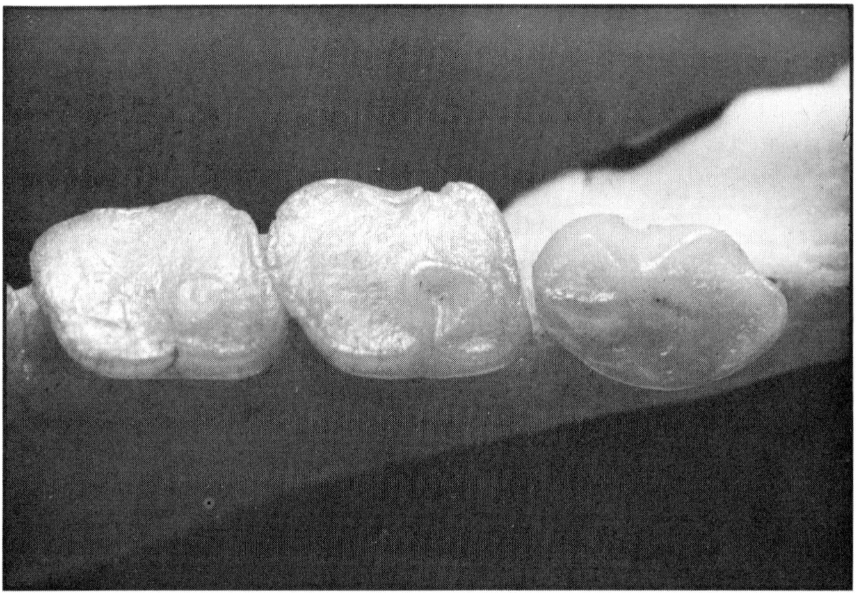

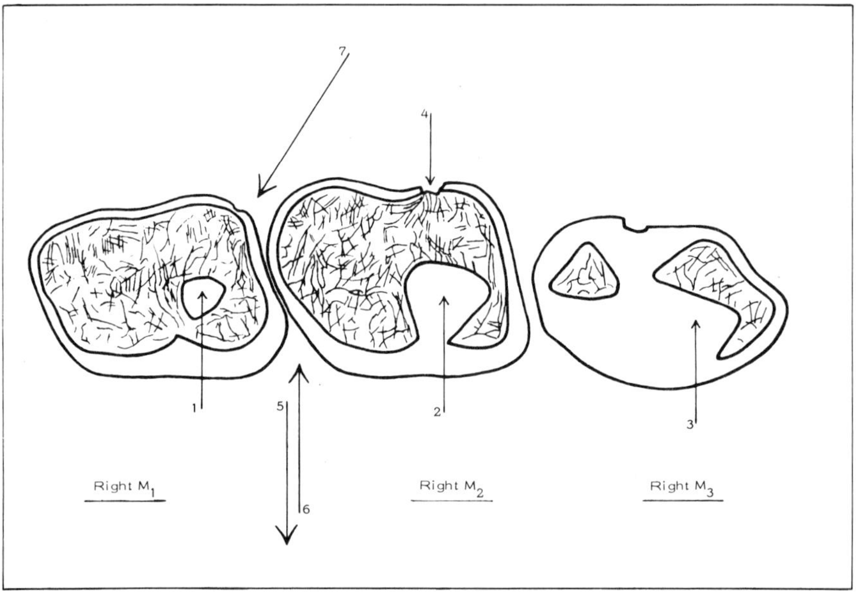

Figure 7(*b*). *Galago crassicaudatus.* Right M_1, M_2, M_3. Occlusal view. Showing abrasion on secondary incusive surfaces. All secondary incusive basins of M_1 and M_2 are coalesced.

(Figs. 7*b* and *c* are from the same animal.)

1, 2 & 3 Remnants of primary incusive basins of the Hypo-Triakises of M_1, M_2 and M_3.

4 Small fracture of enamel periphecy, with abrasion channelled into the opening.

5 Tangent to the general direction of scissorio-incusive action.

6 Tangent to the general direction of transverse thegotic action.

7 Tangent to the general direction of oblique thegotic action.

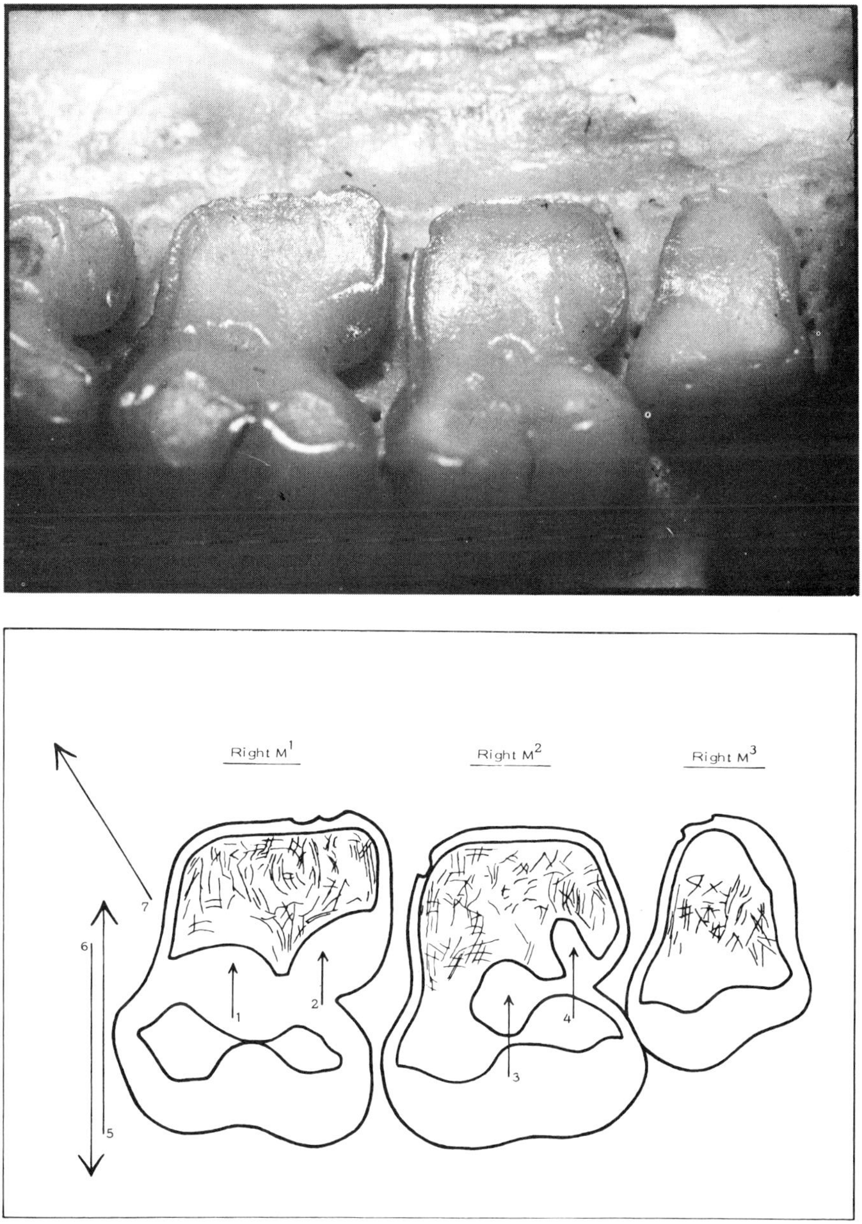

Figure 7(*c*). *Galago crássicaudatus.* Right M^1, M^2, M^3. Occusal view, showing abrasion on secondary incusive surfaces; all basins are now coalesced.
1, 2, 3 & 4 Remnants of primary incusive basins.
5 Tangent to the general direction of scissorio-incusive action.
6 Tangent to the general direction of transverse thegotic action.
7 Tangent to the general direction of oblique thegotic action.

exposure and rapid wear of the dentine along the blade system, has been interpreted as evidence of dominant 'tribo-' functions, or, as Kay and Hiiemae suggest in this volume, that 'puncture crushing' before the teeth 'come into contact' 'blunts or flattens the tips of the cusps'.[7]

Three further questions arise for the evolutionists:

1. What is the adaptive significance of a finely tapered dentinal core, especially that of a 'pestle' cusp, when its effect is merely to jeopardise the cusp's viability?
2. What is the adaptive significance of a 'pestle' cusp, especially when 'blunted', when its function is not only jeopardised by its shortening, i.e. removal of contact with the 'mortar' but by the creation of a second 'mortar' in addition to the (reduced) first, the second even deeper, moreover, than the first?
3. What is the adaptive significance of a 'lingual phase of occlusion' when no cutting blade is oriented in this direction and, in fact, the phase produces no evidence of (functional) abrasion that is in any way different from that produced by the buccal phase (Figs. 1*b*; 7*a, b, c*)?

As is to be expected, the morphological characteristics of dentine when analysed are seen to be adapted, as is its differential hardness, to facilitate the ontogenetic viability of the functional features of the enamel which the dentine supports. They are adaptations in terms of the prime evolutionary advantage of teeth: their sharpness.

Evolutionary control of tooth-wear

One feature of anthropoid thegosis that may prove crucial to our understanding of prosimian phylogenies is the suggestion that in primates the direction of the final phase of the (transverse) masticatory action is opposite to that of the dominant thegotic action. The two actions are discrete, not only directionally, but also temporally – occurring at quite different times.[1,2,10,13]

The phylogenesis of upper canine thegosis is an important aspect of this evidence and relevant to thegosis in prosimians, but it is excluded here, partly because my studies of this aspect of the evidence have been limited by the paucity of fossil canines, and partly because I have concentrated on molar thegosis.

As the idea gains further ground that the sharp boundary produced by faceting is the functional cutting agent, it can be anticipated that those who find the thegosis concept unacceptable will attempt alternative arguments of 'self-sharpening' mechanisms for various teeth (i.e. automatic sharpening, concomitant with the masticatory action itself).

The self-sharpening idea occurred to me some 25 years ago. My electric shaver at that time had rather soft stainless-steel blades that, too often, wore out and had to be replaced. Although the metal was soft, the blade edges were always sharp. It was immediately clear, however, that there was no parallel between this action and that of mastication. The action of an ordinary pair of scissors, where virtually each stroke is a cutting one, could more reasonably be considered as a parallel. Unlike the shaver, scissors characteristically become blunt, not sharp, with use. To resharpen a scissor blade effectively one has to remove a whole layer off one surface, evenly and precisely, so as to maintain the optimum angle of bevel to the leading cutting edge. It is better done by an expert who knows what is the optimum angle, what force to use in the action, and when to stop.

A fundamental characteristic of phylogenetically-derived behaviour is the inherent mechanism that starts (evokes) the behaviour and stops (attenuates, terminates) it. This mechanism is species-specific: it is as characteristic and stereotyped as the animal's morphology.

Considering the strong selective forces that must be operating on brachydont teeth to conserve their irreplaceable substances, evolutionary processes have doubtless operated to produce a mechanism to resharpen enamel blades in a controlled manner. With fine control of both the occurrence and the extent of the action there would be selective advantage; without it, there would be no way of balancing the need for blades to be resharpened with the need to avoid over-sharpening, i.e. unnecessary wastage of tooth-substance. The structures of the oral cavity, particularly the peridontal membranes, are richly supplied with proprioceptive pathways; it would be surprising were they unable to cope with a mechanism guiding and eventually terminating a masticatory stroke just short of collision of tooth-substances, a collision that would bring discomfort to the mouth and no advantage whatsoever to the division of exogenous material. The jarring discomfort of chewing on an undetected spicule of bone, even a grain of sand, illustrates the argument.

This paper is in a way a preliminary report as I have only recently had access to prosimian material. The exposure, however, has shown many avenues that I hope to pursue.

Acknowledgments

I thank Mrs Margaret Black of Kent, England, who not only financed the entire trip from London's antipodes but also most of several months' expenses (including the photography) while I studied the relevant material and wrote this paper. There were many others who contributed, chiefly Frank Canaday through the Explorers' Club, New York, and my own family. I also thank all those individuals and institutions who gave me access to their material.

NOTES

1 Every, R.G. (1970), 'Sharpness of teeth in man and other primates', *Postilla* (Yale) 143, 1-30.

2 Every, R.G. (1960), 'The significance of extreme mandibular movements', *Lancet* 2, 37-9.

3 Every, R.G. (1972), *A New Terminology for Mammalian Teeth: Founded on the Phenomenon of Thegosis*, Christchurch, N.Z.

4 Simpson, G.G. (1936), 'Studies of the earliest mammalian dentitions', *Dent. Cosmos* 78, 791-800.

5 Butler, P.M. (1952), 'The milk molars of the Perissodactyla, with remarks on molar occlusion', *Proc. Zool. Soc. Lond.* 121, 777-817.

6 Mills, J.R.E. (1955), 'Ideal dental occlusion in the Primates', *Dent. Pract.* 6, 47-61.

7 Kay, R.F. and Hiiemae, K.M., this volume.

8 Crompton, A.W. and Hiiemae, K.M. (1970), 'Molar occlusion and mandibular movements during occlusion in the American opossum, *Didelphis marsupialis*', *Zool. J. Linn. Soc.* 49, 21-47.

9 Every, R.G., in preparation.

10 Every, R.G. (1965), 'The teeth as weapons: their influence on behaviour', *Lancet* 1, 685-8.

11 Crompton, A.W. (1971), 'The origin of the tribosphenic molar', in Kermack, D.M., and Kermack, K.A. (eds.), *Early Mammals, Zool. J. Linn. Soc.* 50, Suppl. 1, 65-87.

12 Gingerich, P.D., this volume.

13 Every, R.G., in preparation.

List of abbreviations

en^{d}	entoakid
hy	hypoakis
hyC^{d}	hypo-Cingulid (Hypo-cingulo-diakidrepanid)
hy^{d}	hypoakid
hyo^{d}	hypobliquakid
Hyo^{d}	Hypobliquid (Hypo-diakidrepanid-obliqua)
hyt^{d}	hypotransversakid
Hyt^{d}	Hypotransversid (Hypo-diakidrepanid-transversa)
HyT^{d}	Hypo-Triakid (Hypo-Triakididrepanid)
me	metakis
me^{d}	metakid
meo	metaobliquakis
Meo	Metaobliquis (Meta-diakidrepanon-obliqua)
mesC	metastylo-Cingulis (Metastylo-cingulo-diakidrepanon)
mest^{d}	metastylotransversakid
Mest^{d}	Metastylotransversid (Metastylo-diakidrepanid-transversa)
met	metatransversakis
Met	Metatransversis (Meta-diakidrepanon-transversa)
MeT	Meta-Triakis (Meta-Triakididrepanon)
pa	parakis

'pa^{d}'	(relict) parakid
pao	paraobliquakis
Pao	Paraobliquis (Para-diakidrepanon-obliqua)
pao/met	amalgamation of paraobliquakis and metatransversakis
pasC	parastylo-Cingulis (Parastylo-cingulo-diakidrepanon)
pat	paratransversakis
Pat	Paratransversis (Para-diakidrepanon-transversa)
PaT	Para-Triakis (Para-Triakididrepanon)
pr	protoakis
prC^{d}	proto-Cingulid (Proto-cingulo-diakidrepanid)
pr^{d}	protoakid
prme	protometakis
Pro	Protobliquis (Proto-diakidrepanon-obliqua)
pro^{d}	protobliquakid (pseudo-parakid, i.e. functionally homologous to the original parakid)
Pro^{d}	Protobliquid (Proto-diakidrepanid-obliqua)
prpa	protoparakis
prPat	proto-Paratransversis (proto-Para-diakidrepanon-transversa)
Prt	Prototransversis (Proto-diakidrepanon-transversa)
PrT	Proto-Triakis (Proto-Triakididrepanon)
prtC	prototransverso-Cingulis (Proto-cingulo-diakidrepanon-transversa)
Prt^{d}	Prototransversid (Proto-diakidrepanid-transversa)
PrT^{d}	Proto-Triakid (Proto-Triakididrepanid)

J. H. SCHWARTZ

Premolar loss in the primates: a re-investigation

Introduction

From his study of fossil and living lemuroids, Gregory derived a theory of premolar loss during primate evolution:

> The indrisine lemurs (Indrisidae) afford an instructive example of the reduction in the number of the premolars from four to two on each side, by the loss of the anterior two, probably conditioned by the shortening of the face and by the crowding of the dentition due to the increase in the anteroposterior diameter of the remaining cheek-teeth.
>
> In the final stage of this evolution [referring to the differentiation of the premolars], the first and second premolar disappear entirely and the remaining premolars are homologous with the third and fourth of the primitive placental dentition.
>
> In all primates the elimination of the two anterior premolars is probably correlated with the shortening of the face, with the marked increase in size of P^3 and P^4, and often with the anteroposterior lengthening of the molars.[1]

It was these three statements (the only three which Gregory appears to have made in regard to premolar loss in the primates) which led the author to reconsider premolar loss. There did not seem to be a sufficient basis for such a generalised theory. Therefore, the purpose of this paper is to re-evaluate this theory, and to suggest that the second lower premolar lost during primate evolution may have been the homologue of the fourth premolar of the primitive placental dentition.

Palaeocene primates

Among the Palaeocene primates one finds evidence of various trends of specialisation. The succession in the Carpolestidae (*Elphidotarsius-Carpodaptes-Carpolestes*) demonstrates a reduction in size and number of the anterior premolars with a concomitant increase in size of a specialised posterior premolar.[2] Although this can serve as supportive evidence for Gregory's theory of premolar loss, it must be remembered that the Carpolestidae were a specialised evolutionary branch of primates which left no descendants,[2] and that this specific pattern of premolar loss in the family was the result of evolutionary pressures towards specialisation involving not only morphology and function of each individual tooth, but the interaction of morphology and function of the dentitions of both upper and lower jaws.

The Plesiadapidae (Fig. 1), further shown to be primates by Gingerich,[3] also display a trend for reduction in size and number of the anterior premolars as the posterior premolar increased in size and complexity.[4,5] Further specialisations in dentition, such as the lack of the lower canine, reduction of the upper canine, and the functional significance of the enlarged and procumbent upper and lower incisors,[3] contribute to the consideration that this family was an aberrant branch in the evolution of the Primates and cannot, therefore, be used as evidence for premolar loss.

The Paromomyidae (Paromomyinae[6]) were not as specialised in the premolar region as the Carpolestidae or Plesiadapidae; the posterior lower premolar was about as high as M_1 and either similarly or less molariform than the same tooth in the Eocene subfamily

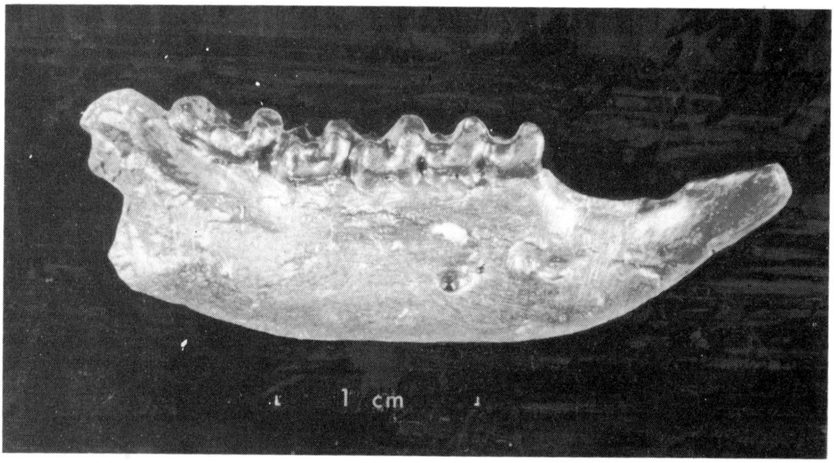

Figure 1. *Plesiadapis fodinatus*, B.M. (N.H.) 13922. Right dentary, showing complete tooth-row.

Omomyinae. The members of this family were specialised in the direction of greatly enlarged lower front teeth (either an incisor or a canine), a condition which, in comparison with succeeding Eocene primates, would exclude them from direct relationship to the evolution of later primates.

Eocene primates

The Adapidae, including the North American subfamily Notharctinae and the European subfamily Adapinae, retain the original placental mammal number of premolars.[7,8] The first premolar of both the upper and lower dentitions is greatly reduced in size and, as with the Palaeocene primates, can be regarded as evidence of eventual evolutionary loss of this tooth. Starting with the second premolar and proceeding posteriorly, there is a gradient of increasing size, and incorporation of gross morphological characteristics such that the spatulate shape seen in P^2_2 grades into the molariform P^4_4 and does not create a break in the premolar-molar morphological continuum (Figs. 2 and 3). Unlike the Palaeocene primates, the Adapidae do not display specialisations which would exclude them from relationship to the evolution of later primates. Both the Adapinae and Notharctinae have peg-like upper incisors with a definite gap between the central incisors. (Le Gros Clark describes the upper incisors of *Adapis* as being 'of moderate size and furnished with a chisel-like cutting margin', and says that 'in neither genus, however, are the upper incisors separated by a conspicuous median gap' (p.93).[4] This is not so, as can clearly be seen in specimens of *Adapis parisiensis* and *A. magnus* at the British Museum of Natural History (B.M.N.H., M.1633, M.1345, M.4487, M.7504) and *A. parisiensis* (M.583) in the Watson Collection, Cambridge (Fig. 4)). Although the Adapidae differ from extant primates in dental morphology, the presence of a median gap between peg-like upper incisors relates them to the strepsirhines (i.e. prosimians excluding *Tarsius*), which have similarly shaped upper incisors and a median gap for the passage of a rhinarium.[9] Further relationship of the Adapidae to strepsirhine evolution is seen in the similarity of the ectotympanic region of the basicranium.[10]

The family Anaptomorphidae is currently generally considered to contain the subfamilies Anaptomorphinae and Omomyinae,[2] as had been proposed by Simpson and *contra* Gazin.[6,11] Description of the differences in dental morphology between these two subfamilies is given by Simons and Simpson.[2,6] One particular characteristic of the Anaptomorphinae is that P^4_4 is generally very enlarged and is noticeably higher than the crowns of adjacent teeth, while that of the Omomyinae is generally not elevated above adjacent teeth, and P_4 is incipiently or nearly molariform, similar to that of the

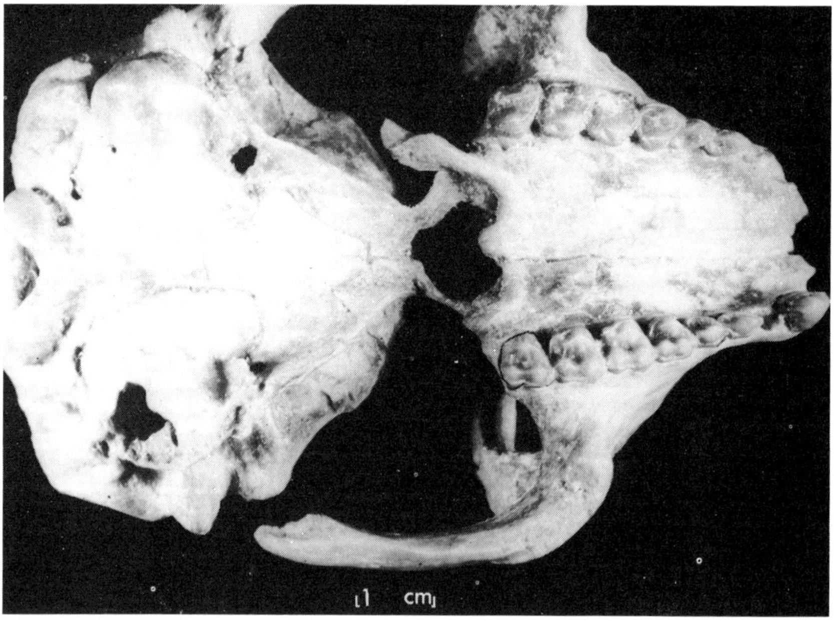

Figure 2. *Adapis parisiensis*, B.M. (N.H.) M.1345. Ventral view of skull, showing upper dentition. Note smooth transition from premolars to molars.

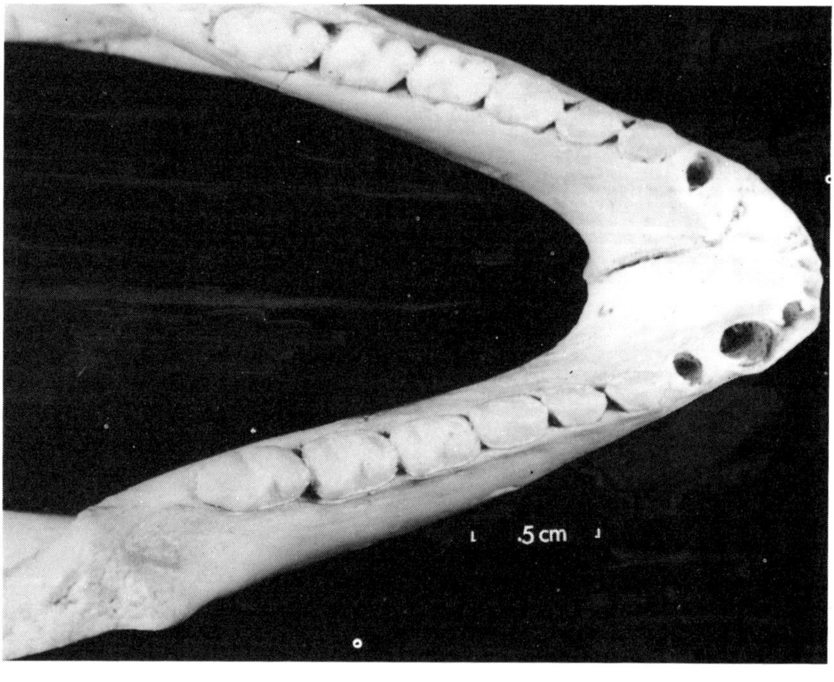

Figure 3. *Adapis parisiensis*, B.M. (N.H.) M.1634. Lower jaw, showing incomplete dentition. Note smooth transition from premolars to molars.

Figure 4. *Adapis parisiensis*, B.M. (N.H.) M.1633 (left) and *Adapis magnus*, B.M. (N.H.) M.4487 (right). Frontal view of snout, showing anterior upper dentition.

Microchoerinae (= Necrolemurinae). The lower dental formula of the Omomyinae is typically 2.1.3.3., but that of *Teilhardina belgica*, morphologically similar to *Omomys*, is 2?.1.4.3. It has been suggested that the Anaptomorphinae, Omomyinae, and Microchoerinae were derived from a form similar to or prior to *Teilhardina*, giving all three subfamilies common tarsioid affinities.[2,10,12] As with the Carpolestidae, reduction in the number of premolars can be seen in successive species of several genera of the Anaptomorphidae: (1) *Teilhardina*, the basal form, which has a greatly reduced P_1; and (2) the anaptomorphine genera *Tetonius* and *Absarokius*. Although anaptomorphid molar and upper and lower dental arcade morphology indicate tarsioid affinities, relationships to the evolution of *Tarsius* are not as close as are those of the Microchoerinae.

Although members of the microchoerine genera *Microchoerus* and *Necrolemur* show specialisations in their cusp morphology which remove them from direct relation to tarsioid evolution, members of the genera *Pseudoloris* and *Nannopithex* are generalised so as to indicate more direct tarsier affinities of the subfamily.[10] In general, the Microchoerinae are characterised by one enlarged lower anterior tooth; by a P_4 which is not elevated but nearly molariform, with a short heel; and by $P^{3,4}$ which are low and oval in shape (Figs. 5 and 6). (For molar morphology see Stehlin,[8] Simons,[10] and Simpson[6].)

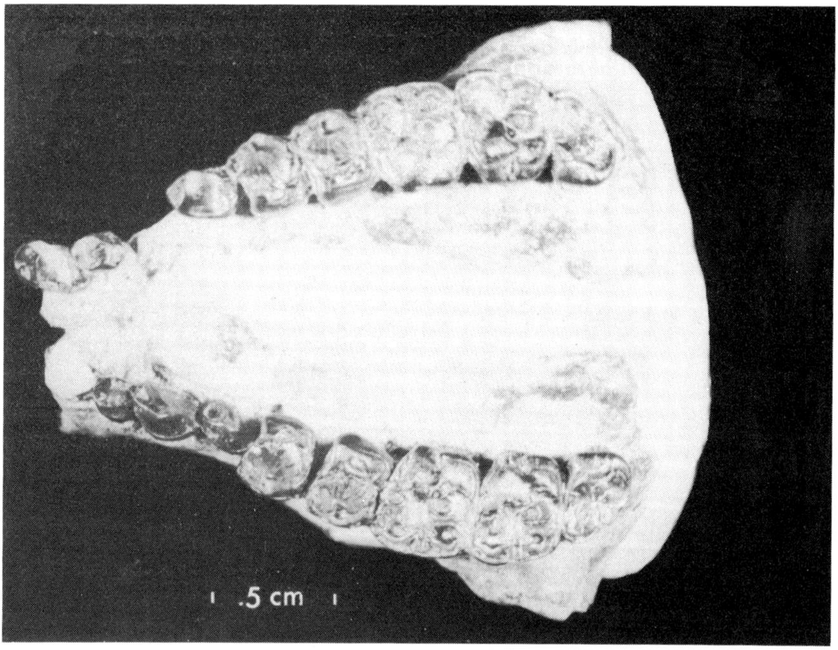

Figure 5. *Microchoerus erinaceus*, B.M. (N.H.) 25.229. View of palate, showing upper dentition. Note the relatively smooth transition from premolars to molars.

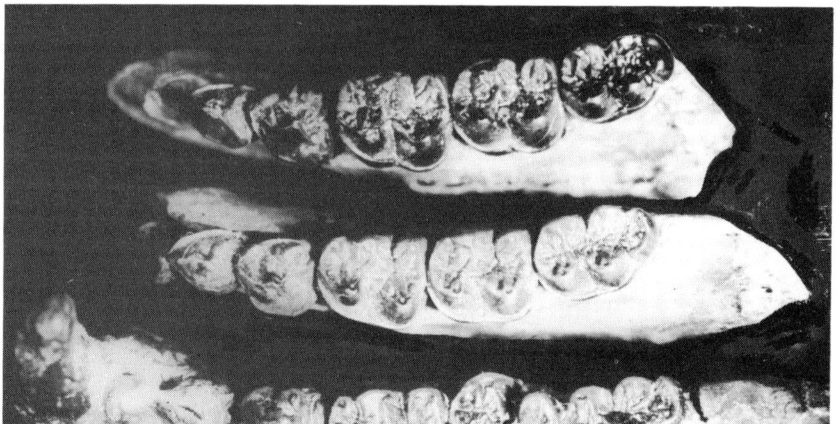

Figure 6. *Microchoerus erinaceus*; left mandibles, showing cheek-teeth.
Top: Sedgwick Museum, Cambridge C9670.
Middle: Sedgwick Museum, Cambridge C9681.
Bottom: B.M. (N.H.) 25.229.

The lower dental formula for all four genera is 0.1.4.3. and the upper is 2.1.3.3., except for *Nannopithex raabi* which is 2.1.3 *or* 4.3.[10] Recently it has been proposed that a new interpretation of the lower dentition of Microchoerines shows that the dental formula is 2.1.2.3.[2] Studies and x-rays made by the author on specimens from the British Museum (M.H.) and the Zoological Museum (Cambridge) indicate that the morphology of the tooth now proposed as a canine is more like that of a premolar; this is further indicated by comparison of x-rays of microchoerines and adapids, whose lower premolar regions are similar. As with the Adapidae and some of the Anaptomorphidae, the anterior premolar is greatly reduced in size.

Oligocene primates

A few members of the Eocene tarsioid families persist into the Oligocene (e.g. *Rooneyia*,[13] *Microchoerus, Macrotarsius*), but the majority of the fossils are catarrhine. (This does not include the Parapithecidae (*Parapithecus* and *Apidium*), which the author feels are not related to catarrhines, but represent, in comparison with other Oligocene primates, specialised and aberrant evolutionary forms.)[2,14,15] The Oligocene catarrhines have a lower dental formula of 2.1.2.3., and show at least incipient catarrhine characteristics of an enlarged canine and a sectorial anterior premolar. This is most definitely seen in the earliest catarrhine fossil, *Oligopithecus savagei*, where the enlarged anterior premolar even shows wear which would necessitate a 'sizeable upper canine',[16] indicating the typical catarrhine characteristic of a honing triad (i.e. large upper canine and lower canine and anterior premolars).[17] Subsequent forms, *Aegyptopithecus* and *Propliopithecus*, show the same condition. Furthermore, as opposed to the Adapidae, Omomyinae, Microchoerinae, and Parapithecidae, which show a gradient from the premolars through the molar set, the catarrhines of the Oligocene show a distinct break between the two sets of teeth. The primates of the Miocene and upwards, then, were to undergo further specialisations based on these Oligocene foundations.

Recent primates

Although the fossil record may be lacking in direct evidence of evolutionary divergence, major changes in the evolution of gross dental morphology had been established by the end of the Oligocene: strepsirhines and haplorhines can be differentiated on the basis of dental, palatal, cranial and post-cranial morphology and dental formulae.[9] (Unfortunately there is a dearth of platyrrhine material until the Miocene with *Homunculus, Cebupithecia*, and *Neosamiri*, which are very similar to modern types.[18] In dental formula and premolar-molar gradients, the platyrrhines are similar to many prosimians.) Since it would appear from the fossil evidence that haplorhines (more specifically, catarrhines) were evolving differently from the strepsirhines (except for the Indriidae) in the differentiation of the premolars from the molars, at least in the lower jaw, it should be possible to see this in the dentitions of extant primates.

The following notes are based on specimens studied at the British Museum (Natural History), London and supplemented by reference to prior publications:

The Lemuridae (e.g. *Lemur catta, Varecia variegatus, Hapalemur griseus, Cheirogaleus major, Microcebus murinus, Phaner furcifer*) Lorisidae (Lorisinae: e.g. *Loris tardigradus, Nycticebus coucang, Arctocebus calabarensis, Perodicticus potto*) and Galaginae (e.g. *Galago senegalensis, G. crassicaudatus, G. demidovii, Euoticus elegantulus*) all have the dental formula $\frac{2.1.3.3}{2.1.3.3}$, a tooth-comb (generally considered to contain the lower incisors and canines), and a caniniform lower anterior premolar. Both upper and lower jaws show a gradient from the premolar set, through the molariform last premolar, and on through the molars; however, the gradient in the lower jaw does not generally include the caniniform premolar (Fig. 7). *Lepilemur mustelinus* differs only in its dental formula, $\frac{0.1.3.3}{2.1.3.3}$. (Because of its unique specialisations, the aye-aye will not be considered in this discussion.) Thus, this characteristic of a dental gradient is generally maintained from the Eocene to modern strepsirhines.

Tarsius presents some difficulty in interpretation. The upper dental formula is clearly 2.1.3.3., but the anterior premolar tends to be relatively smaller than in other primates with three premolars, so that the dental gradient is generally seen to begin with the second premolar. Although the lower dental formula is commonly accepted as being 1.1.3.3., it can equally well be argued that it is 2.1.2.3. The last premolar may vary in its degree of molarisation from specimen to specimen, such that one might see a gradient in one animal, but some degree of differentiation between the premolar and molar sets in another. Qualitatively, it appeared that there was a correlation between the size of the larger of the anterior teeth and the degree of molarisation of the last premolar (i.e. the larger that tooth, the greater the differentiation between the premolar and molar sets.

The Cebidae (Fig. 8) (dental formula $\frac{2.1.3.3}{2.1.3.3}$) show wide intra-familial variation such that the post-canine dentition of the upper and lower jaws of *Aotus trivirgatus* and *Cacajao calvus* tends to be graded; that of *Chiropotes satanus, Pithecia monachus* and *Cebus*

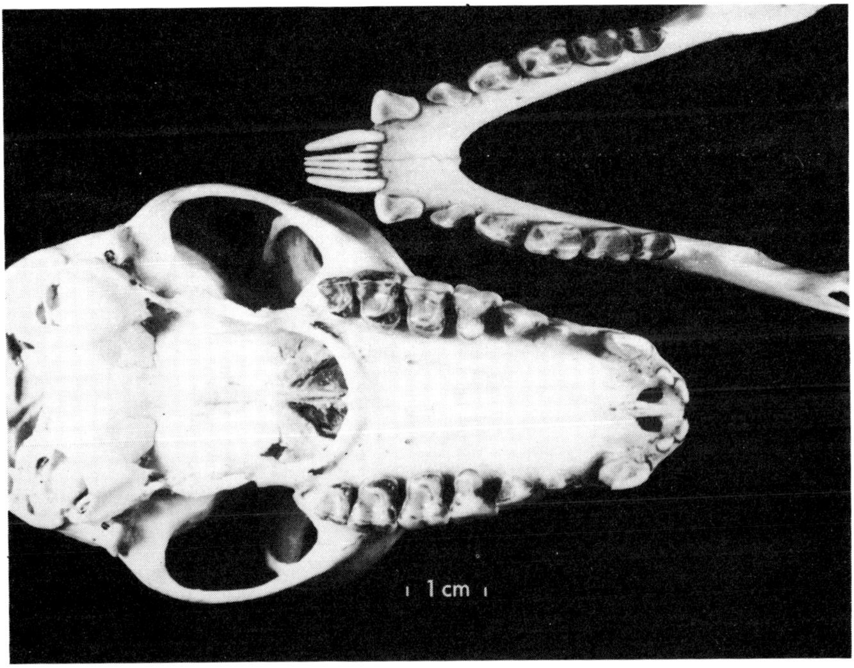

Figure 7. *Varecia variegata* (*'Lemur variegatus'*), B.M. (N.H.) 75.1.29.10 Skull and lower jaw, showing full dentition. Note the smooth transition from premolars to molars. (N.B. One of the incisors in the tooth-comb has been broken.)

apella less so; whereas that of *Callicebus torquatus, Alouatta seniculus, Saimiri sciureus, Ateles fusciceps, Brachyteles arachnoides, Lagothrix flavicauda* and *Callimico goeldii* presents a clearer distinction between the premolar and molar sets. Of the Callithricidae (dental formula $\frac{2.1.3.2}{2.1.3.2}$), *Saguinus midas* can be interpreted as having a dental gradient, but not *Callithrix flaviceps* or *Cebuella pygmaea*. In practically all specimens examined, if there was any degree of differentiation between the premolar and molar sets it was more so in the lower than upper jaw, the last premolar of which was relatively more molariform. As with the variability exhibited in the genus *Tarsius*, the Ceboids which showed less grading tended to have relatively larger upper and lower canines and slightly sectorial anterior lower premolars.

The catarrhines (Figs. 9 and 10) (dental formula $\frac{2.1.2.3}{2.1.2.3}$) differ from most extant primates not only in the number of premolars, but also in the anterior premolar of the lower jaw which – except in the Hominidae – is enlarged, elongated, and specialised sectorially so as

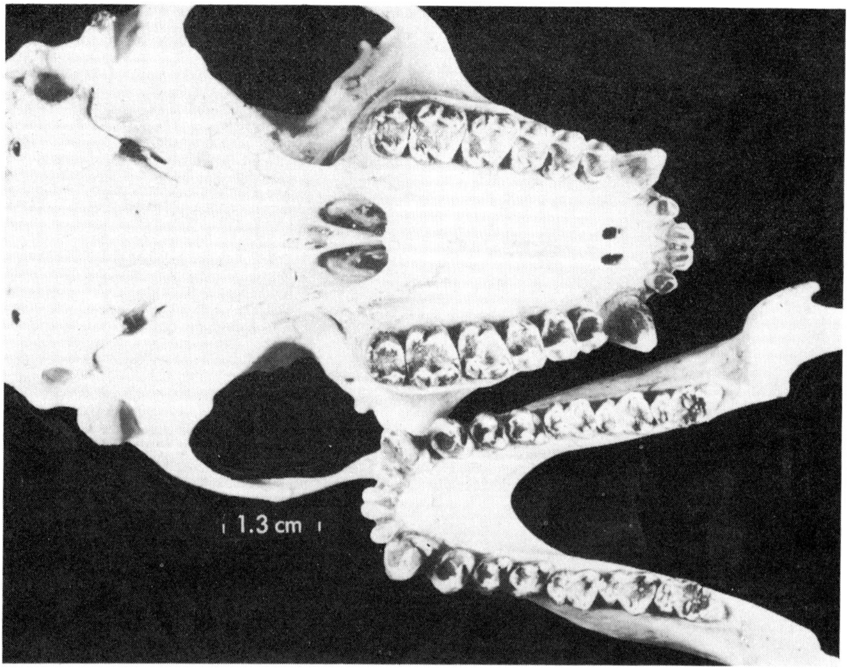

Figure 8. *Alouatta* sp., B.M. (N.H.) 3.7.11. Skull and lower jaw, showing full dentition. Note the fairly clear division between premolar and molar morphology, which is more obvious in the lower jaw.

to accommodate the large upper canine. Except for *Erythrocebus patas* and *Cercopithecus aethiops* (subfamily Cercopithecinae), in which the posterior premolar tends to be relatively larger and molariform compared with the other cercopithecines, the Cercopithecinae, Colobinae, Pongidae and Hylobatidae were seen to have relatively the same condition: clearly distinct sets of premolars and molars in the lower jaw, with the anterior premolar larger than the posterior and the upper jaw showing more of a tendency for grading between the two sets of cheek-teeth. Once again, the degree of differentiation or grading appeared to be correlated with the relative sizes of the upper and lower canines and lower anterior premolar, the components of the honing triad.

The Indriidae (Fig. 11) differ from the Lemuridae and Lorisidae in that they have fewer anterior teeth in the lower jaw and fewer premolars. The dental formula is $\frac{2.1.2.3}{1 \textit{ or } 2.1 \textit{ or } 0.2.3.}$’[19,20,21] but it is commonly accepted, by homology with the anterior dentitions of the lemurids and lorisids, that the lower dental formula is 1.1.2.3.[22] Indriids also differ from other prosimians in that the anterior lower premolar is not as much caniniform as it is sectorial, analogous in

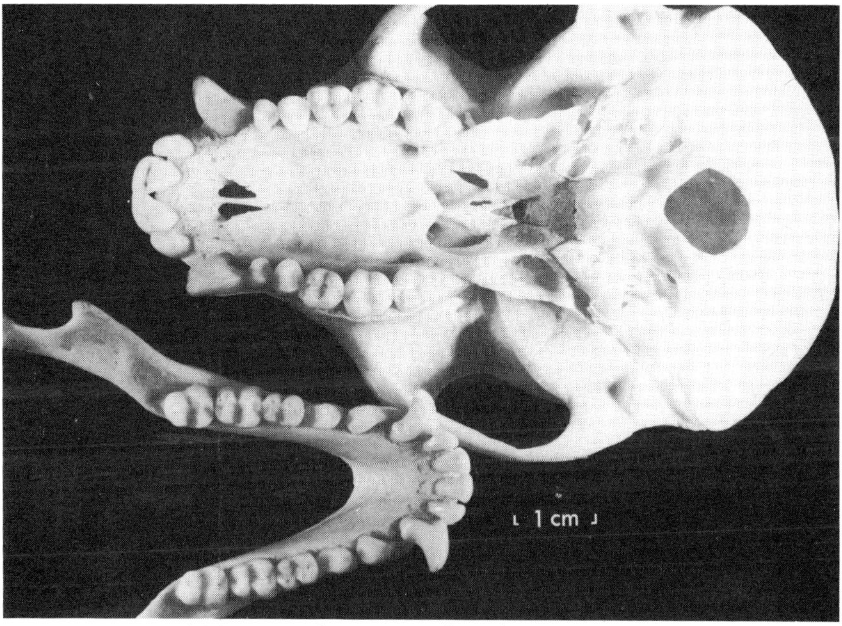

Figure 9. *Cercopithecus aethiops*, B.M. (N.H.) 66.504. Skull and lower jaw showing full dentition. Note the marked distinction in morphology between the last premolar and the first molar in both upper and lower tooth-rows.

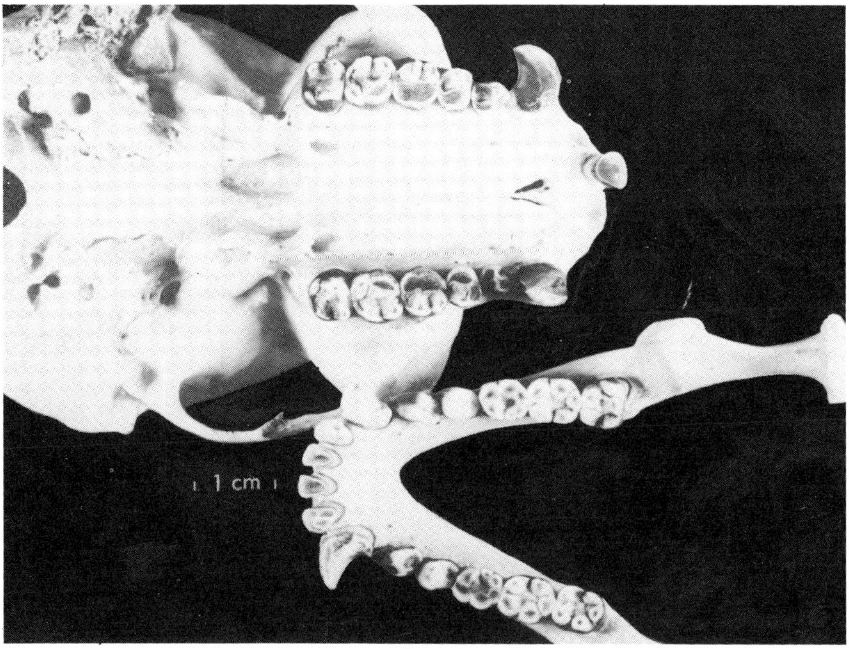

Figure 10. *Hylobates hoolock*, B.M. (N.H.) 33. Skull and lower jaw, showing most of the dentition (N.B. 3 incisors have been lost from the upper jaw). Note the marked distinction in morphology between the last premolar and the first molar in both upper and lower tooth-rows.

function and similar in shape to some catarrhines and those platyrrhines with relatively enlarged upper canines. The morphology of the lower jaw is similar to that of catarrhines, in that the anterior premolar is larger in height and length than the posterior premolar and there is distinct differentiation between the premolar and molar sets. This is more marked in *Propithecus* than in *Indri*, and is apparently correlated with *Propithecus* having a relatively larger upper canine than *Indri.* As with catarrhines and platyrrhines, there is a greater tendency for grading in the upper jaw. The Indriidae are unique among primates, though, in the fact that the number of deciduous molars is greater than the number of replacing premolars (Fig. 12): the formula for the deciduous dentition is said to be $\frac{2.1.3}{2.1.3}$, the same as lorisids and lemurids, except that *Lepilemur*[23] is $\frac{1.1.3}{2.1.3}$. Current research by the author indicates that the formula for the deciduous dentition of indriids is $\frac{2.1.2}{1.1.4}$ and that the replacing premolars in the lower jaw are not P_2 and P_4 or P_3 and P_4.[19,1]

Discussion

Comparison of the tooth rows and premolar regions of extant and fossil primates seems to support a separation of the strepsirhines and haplorhines. The strepsirhines, apart from the aye-aye and indriids, have lost one premolar on each side of the upper and lower jaws during their evolution and have maintained a dental gradient. The haplorhines, on the other hand, have gone through stages where one and then, in some cases, another premolar has been lost; the premolar region exhibits variability posteriorly and specialisation anteriorly, with the result that this region is, with few exceptions, distinguishable as a unit separate from that of the molars. The issue at hand is to try to understand what the premolar region represents and to what extent variability within it is significant.

Gregory interpreted the premolar region as 'intermediate in position and function between the canines and the molars'.[24] Butler adds agreement in the presentation of his Field Theory.[25,26] This theory states that there is a gradient of forms as one passes along the series of teeth indicating the '. . . presence in the embryonic jaw of morphogenetic fields that influence the tooth rudiments in different parts of the jaw to develop in divergent ways', and that the whole series of teeth is a '. . . range of forms into which a given type of tooth germ can develop when placed in the given range of environments'.[25]

The gradient is the 'field'; a tooth develops according to its position in the field; 'field genes' control regional differentiation; the

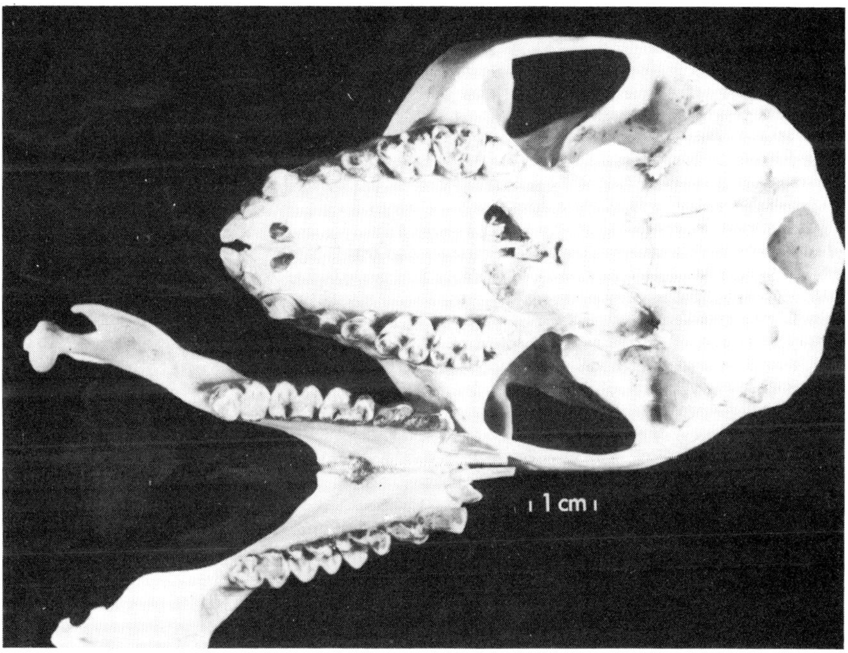

Figure 11. *Propithecus diadema edwardsi*, B.M. (N.H.) 1939.1204. Skull and lower jaw, showing the complete dentition (N.B. Most of the outer tooth of the tooth-comb has been lost from the right side of the lower jaw). Note that this indriid resembles the catarrhines in the marked distinction in morphology between the last premolar and the first molar in both upper and lower tooth-rows.

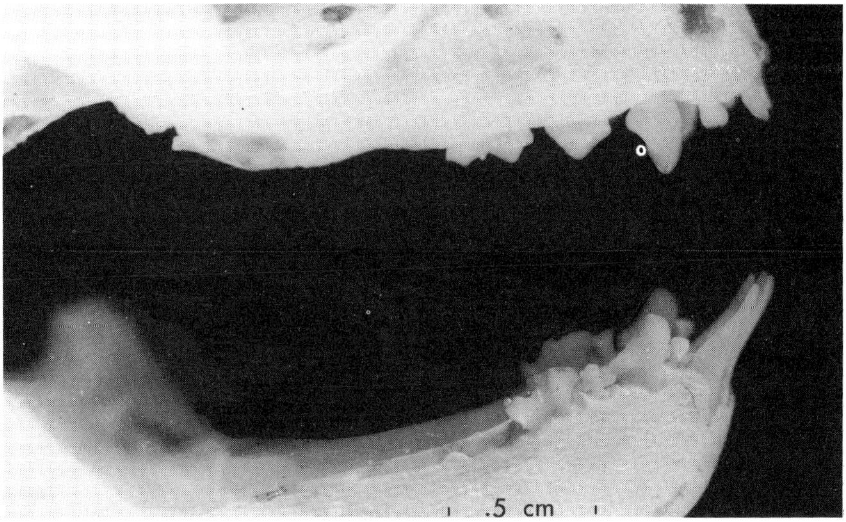

Figure 12. *Propithecus verreauxi*, B.M. (N.H.) 30.3.15.1. Upper and lower deciduous dentition (see text for explanation).

field is differentiated into incisor, canine, and molar regions; the premolar 'region' is the result of the combined affects of the canine and molar regions; and the morphology of each premolar is determined by the predominance of either the canine or molar region, or the equilibration of both (Fig. 13).[25,26] In terms of function, it would appear that the role of the premolar, whose morphology can range from being caniniform to molariform, is that of a potential resource which can be tapped by evolutionary pressures to enhance or expand the functional area of either the canine or the molars, or both. For example, the battery of cheek-teeth of *Equus* was expanded, during the course of evolution, by the molarisation of the premolars;[25] in lemur, the lower anterior premolar became extremely caniniform 'by assuming the function of the true canine, which . . . has been relegated to the

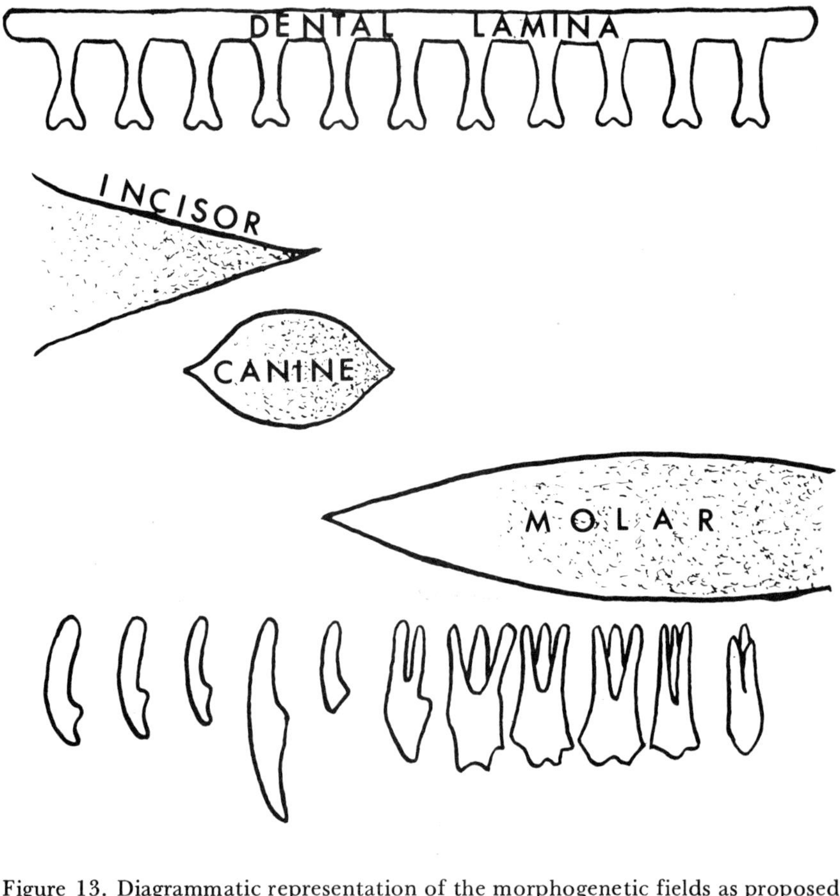

Figure 13. Diagrammatic representation of the morphogenetic fields as proposed by the Field Theory (Butler); [25,26] these three fields (incisor, canine and molar) are presented in relation to the dental lamina and the realised eutherian dentition. Note the overlapping influence of these fields in this instance.

incisor series' (Le Gros Clark, p.94).[4] Butler's Field Theory can be readily accepted at this level, considering only molarisation or caninisation; but the application of the concept of a morphogenetically patterned premolar region must also take into consideration the consequences of tooth loss.

According to the Field Theory, the characteristics of a premolar are determined by its position in the jaw in relation to the positions of the canine and molars. The area between the canine and molar fields is a gradient of the genetic potentials of each field. Evidence from the evolution of herbivores (e.g. the horse), where the length of the jaw and the number of molariform teeth were increased, indicates not only that the field can change so that there is a simultaneous modification of several teeth,[25] but also, more significantly, that the size, and thus genetic potential, of a field is influenced by the size of the jaw. If the potential of a field increases with the increase in the size of the jaw, it must also diminish in intensity with a decrease in the size of the jaw. Therefore, the reduction of the jaw and the process of tooth (i.e. premolar) loss, which occurred during the course of primate evolution, must have affected the magnitude of the potentials of both the canine and molar fields. In the light of this, the data presented earlier in this paper become critical in the interpretation of premolar loss in the primates, as to whether reduction has occurred consistently in an anteroposterior direction, or whether the second premolar to be lost was not, in fact, the numerically second premolar.

Subsequent to the Field Theory, other points of interpretation of dental gradients have been put forward. Patterson would have a premolar field as well as incisor, canine, and molar fields.[27] To postulate such a field seems to obscure the functional nature of premolars, which is potentially that of supplementing or approximating the function of the canine and/or molars. Van Valen proposed a model for a prepattern-gradient theory where 'all sites for prospective teeth have identical prepatterns, different parts of which are activated to different degrees by gradients of evocators', such that 'a prepattern is the competence to form a set of structures (a morphogenetic program for differentiating cusps, crests, etc. in a prescribed morphological relation to each other), the presence and relative development of each of these structures depending on the strength of an appropriate evocator'.[28] First of all, although Butler did not state it as such, it is obvious from the above quotations that the Field Theory incorporates this concept;[25] and, secondly, by eliminating morphogenetic fields, one is eliminating the control over the evocators, which would be analogous to eliminating the codon from the gene. In sum, the Field Theory, at least at this point, is the most direct means for interpreting differences in dental gradients.

The present consensus of opinion, that the upper and lower first premolar (as interpreted from its small size in the Adapidae and

Microchoerinae or its non-appearance in the evolution of the Carpolestidae, Plesiadapidae, and some Anaptomorphinae) was the first premolar lost in primate evolution, is in accordance with the Field Theory. The resultant reduction of the canine field did not eliminate the potential for influencing the formation of a caniniform premolar, as seen in the lemurids and lorisids, or of a sectorialised caniniform premolar, as seen in the indriids, some ceboids, and the catarrhines. However, it need not be the case that the anterior premolar is lost first, for 'in spite of the already reduced size of the first premolar in most eutherians . . . this position has not always been the initial one vacated in Tertiary therians', as is seen, for example, in the Palaearctic eutherian genus *Talpa*, where the upper and lower P2 were suppressed or lost, 'P_1 being retained because of its importance as a substitute canine'.[29] The diminution of the canine field in the primates must have been secondary in importance to the maintenance of the molar field and its molarising effect on neighbouring premolars, for the gradation of the premolar set into the molars, as seen in Eocene strepsirhines, to some degree similarly exists in the Lemuridae, the Lorisidae, *Tarsius* and, with some variation, the Ceboidea, the latter two groups having three premolars as well. As most of these living primates are herbivorous or frugivorous,[30,31,32] diet probably provided the pressure on early primates for the initial selection of the molar field potential over that of the canine field. The further process of reduction from three to two premolars is not so clear-cut, however.

It is not necessary that one should be able to observe the decrease in size and eventual loss of that tooth. The formation of the tooth follicle and subsequent development of the tooth germ before hard tissues are formed is greatly affected by the environment of the jaw, adjacent developing follicles, and already developed and erupted teeth.[33] Studies carried out by Osborn on the reptile *Lacerta vivipara* show that (1) a tooth germ may develop from the dental lamina but never fully develop and erupt; and (2) that the presence and proximity of young teeth can inhibit the development of a new tooth.[34,35] Although mammals differ from reptiles in the direction of budding of the anterior teeth and in the number of successions of teeth, the same principles of tooth formation or inhibition hold.[35] In most cases, the permanent premolars of non-hominid primates erupt after all but the canine and third molar of the permanent dentition have erupted.[36] Thus, the area in which the tooth germs of the premolars develop would be subjected to crowding and inhibition, which might even prevent a follicle from budding from the dental lamina, as would be the case with the Indriidae. Generally, during the course of the evolution of therian dentition, 'late eruption is directly correlated with late onset of ontogenetic development and, under certain circumstances, these manifestations may strongly indicate a trend toward evolutionary loss of the tooth involved', but this does not apply to the primate canine.[29]

Since the functional importance of the canine, and therefore the canine field, seems too great for this tooth to be lost, its late development would cause added crowding on the developing premolars. It is possible, then, that even though the deciduous molars had erupted, the permanent premolars might be subjected to such pressures that not only would one or more be distorted in shape, but eruption or tooth germ formation might not occur.[29] Again, one must consider the case of the Indriidae.

Oligopithecus is the first of the directly related fossil primates to exhibit a reduction in premolar number to two. As stated earlier, this haplorhine also shows at least an incipient catarrhine condition of specialisation of the lower anterior premolar towards being sectorial, as indicated by its relatively greater size and anteroposterior elongation. For example, the characteristics of a sectorial premolar (development of a robust (anterior) root and an increase in the height of the primary cusp) are characteristics which can be related to a canine rather than a molar field. The development of such a specialised tooth would increase its functional importance relative to the premolar, or premolars, posterior to it and, as a member of a late-developing group of teeth, would take precedence in the utilisation of the space available for development. It seems unlikely that the sectorial premolar could be the third premolar of the primitive placental dentition.

First of all, one must take into consideration the condition in living primates. Even assuming bias on the part of the author in his interpretation of the variability of grading versus distinctiveness of premolars and molars in the platyrrhines, one cannot deny the morphological differences which exist between the posterior premolar and first molar of the catarrhine and indriid dentitions. As outlined earlier in this paper, the strepsirhines excluding the indriids exhibit similar gradations of characteristics from the premolars through the molars. Since this gradation seems to be a function of the morphogenetic potential of the molar field, one would expect, with the reduction from three to two premolars, a relatively undisturbed premolar-molar gradient if the reduction had been the result of the loss of the second premolar. As shown in the above review, this is not the case with the catarrhines, fossil or extant. This would indicate that, in the early stages of catarrhine evolution, there was an alteration in the milieu of the posterior premolar; that there had been a diminution in the influence of the molar field over the development of a molariform premolar. Secondly, if the second premolar lost was the numerical second, the result would have been a further decrease in the morphogenetic potential of the canine field over the adjacent premolar. The characteristics of a sectorial premolar are too closely allied to those of a canine, with its strong, high primary cusp and robust root, to expect that there had been further reduction of the canine field. And, thirdly, with the shortening of the lower jaw and the selective pressures for a specialised and robust anterior premolar, the pressures for the non-eruption of a premolar would be directed toward the distal end of the premolar series, with the result that the area of development of the last premolar would be impinged upon, leading to the loss of that tooth and a decrease in the potential of the molar field. Therefore, it is suggested that, even though Gregory's interpretation of the lemuroids may be applicable to the loss of the first premolar of the lower jaw, the second premolar lost during the course of primate evolution was, in fact, the homologue of the fourth premolar of the primitive placental dentition – in other words, P_4. The pressures involving reduction in the upper jaw from three to two premolars were not the same.

The discussion of the interrelationships between the lower canine and the sectorial premolar must be expanded to include the function of these two teeth as a unit as well as the functional interactions of this unit with the upper dentition. Zingeser points out the basic mechanisms of the interaction between the lower canine and sectorial premolar and the upper canine: the mesiobuccal cusp of the sectorial premolar and the distal surface and cingulum of the lower canine form a V-shaped notch for the upper canine in the centric position, 'against which this tooth differentially wears to sharpen blade and cusp-tip during honing activities'. The upper canine, sectorial premolar and lower canine form a tongue-in-groove grinding and honing triad, whose primary function is to sharpen the upper canine.[17]

First of all, this adds weight to the previous discussion by emphasising the importance of maintaining the morphogenetic potential of the canine field during the loss of the second premolar as a mechanism for the eventual development of the lower unit of the honing triad. Secondly, it introduces the concept of the lower canine and sectorial premolar being a unit which must be responsive to the evolutionary changes occurring in the upper jaw, particularly those of the upper canine, thus implying a basis for a difference in the evolutionary processes affecting the upper and lower jaws.

Whereas in the lower jaw the combined shortening of the jaw and the increase in size of the anterior premolar would create pressures in a posterior direction, thus causing the loss of P_4, the pressures created in the upper jaw, by its shortening and by the increase in size of the canine would not be impeded by the added functional requirements of another tooth (i.e. an analogue to the sectorial premolar). Since it is the upper canine alone which is the basis of the honing triad of the catarrhines and the analogous condition of the Indriidae, there would not be a need to maintain its morphogenetic field, so that premolar loss could occur in an anteroposterior direction. As opposed to the lower jaw, where evolutionary pressures appear to have been acting to preserve the canine field, thereby differentiating the dental gradient, evolutionary pressures affecting the upper jaw might have been operating to preserve its molar field, as suggested by the higher incidence of gradation in the catarrhines and Indriidae; in which case P^2 would be lost.

In conclusion, it is suggested that differing demands on the canine field caused different modes of premolar loss in the upper and lower jaws during the course of primate evolution, resulting in the loss of the homologues of P^1 and P^2 and P_1 and P_4 of the primitive placental dentition.

Acknowledgements

I would like to thank R.D. Martin and J.W. Osborn, and Miss T. Molleson and members of the staff of the Department of Zoology in the British Museum (Natural History), London, without whose assistance and criticism this paper could not have been written.

NOTES

1 Gregory, W.K. (1922), *The Origin and Evolution of the Human Dentition*, Baltimore, pp. 146, 155.
2 Simons, E.L., this volume.
3 Gingerich, P., this volume.
4 Le Gros Clark, W.E. (1962), *The Antecedents of Man*. Edinburgh.
5 Simpson, G.G. (1935), 'The Tiffany fauna, Upper Palaeocene II: Structure and relationships of *Plesiadapis*', *Amer. Mus. Novit.* 816, 1-30.
6 Simpson, G.G. (1940), 'Studies on the earliest primates,' *Bull. Amer. Mus. Nat. Hist.* 77, 185-212.
7 Gregory, W.K. (1920), 'On the structure and relations of *Notharctus*, an American Eocene primate', *Mem. Amer. Mus. Nat. Hist. (n.s.)* 3, 49-243.
8 Stehlin, H.G. (1916), 'Die Säugetiere des schweizerischen Eocäns', Teil 7, 2 Hälfte, *Abhandl. schweiz. paläont. Ges.* 41, 1299-552.
9 Hill, W.C.O. (1953), *Primates*, vol. 1: *Strepsirhini*, Edinburgh.
10 Simons, E.L. (1961), 'Notes on Eocene tarsioids and a revision of some Necrolemurinae', *Bull. Brit. Mus. Nat. Hist. (Geol.)* 5, 45-69.
11 Gazin, G.L. (1958), 'A review of the middle and upper Eocene primates of North America', *Smithson. Misc. Coll.* 136, 1-112.
12 Simons, E.L. (1963), 'A critical reappraisal of Tertiary primates', in Buettner-Janusch, J. (ed.), *Evolutionary and Genetic Biology of Primates*, vol. 1, New York.
13 Wilson, J.A. (1966), 'A new primate from the earliest Oligocene, West Texas: preliminary report', *Folia primat.* 4, 227-48.
14 Simons, E.L. (1967), 'The earliest apes', *Sci. Amer.* 217, 28-35.
15 Simons, E.L. (1969), 'The origin and radiation of the primates', *Ann. N.Y. Acad. Sci.* 167, 319-31.
16 Simons, E.L. (1962), 'A new Eocene primate genus, *Cantius*, and a revision of some allied European lemuroids', *Bull. Brit. Mus. Nat. Hist. (Geol.)* 7, 1-36.
17 Zingeser, M.R. (1969), 'Cercopithecoid canine tooth honing mechanisms', *Amer. J. Phys. Anthrop.* 31, 205-13.

18 Stirton, R.A. (1951), 'Ceboid monkeys from the Miocene of Colombia', *Univ. Calif. Publ., Bull. Dep. Geol. Sci.* 28, 315-56.

19 Friant, M. (1948), 'Classification générique des lémuriens actuels, principalement basée sur l'ostéologie et la dentition', *Acta anat.* 6, 152-74.

20 James, W.W. (1960), *The Jaws and Teeth of Primates*, London.

21 Spreng, H. (1938), 'Zur Ontogenie des Indrisinengebisses', Dissertation, Bern University.

22 Mivart, St. G. (1867), 'On the skull of *Indris diadema*', *Proc. Zool. Soc. Lond.* 1867, 247-56.

23 Friant, M. (1947), 'L'état de la dentition d'un lémurien nouveau-né (*Lepilemur leucopus* F. Major)', *Rev. Stomat.* 48, 597-9.

24 Gregory, W.K. (1916), 'Studies on the evolution of the primates', *Bull. Amer. Mus. Nat. Hist.* 35, 239-355.

25 Butler, P.M. (1939), 'Studies of the mammalian dentition: differentiation of the post-canine dentition', *Proc. Zool. Soc. Lond.* 109, 1-36.

26 Butler, P.M. (1963), 'Tooth morphology and primate evolution', in Brothwell, D. (ed.), *Dental Anthropology*, Oxford.

27 Patterson, B. (1956), 'Early Cretaceous mammals and the evolution of mammalian molar teeth', *Fieldiana Geol.* 13, 1-105.

28 van Valen, L. (1970), 'An analysis of developmental fields', *Develop. Biol.* 23, 456-77.

29 Ziegler, A.C. (1971), 'A theory of the evolution of therian dental formulas and replacement patterns', *Quart. Rev. Biol.* 46, 226-49.

30 Chance, M.R.A. and Jolly, C.J. (1970), *Social Groups of Monkeys, Apes and Men*, London.

31 Jolly, A. (1966), *Lemur Behavior*, Chicago.

32 Napier, J.R. and Napier, P.H. (1970), *A Handbook of Living Primates*, London.

33 Butler, P.M. (1956), 'The ontogeny of molar pattern', *Biol. Rev.* 31, 30-70.

34 Osborn, J.W. (1970), 'New approach to *Zahnreihen*', *Nature* 225, 343-6.

35 Osborn, J.W. (1971), 'The ontogeny of tooth succession in *Lacerta vivipara* Jaquin (1787)', *Proc. Roy. Soc. Lond. (B)* 179, 261-89.

36 Schultz, A.H. (1935), 'Eruption and decay of the permanent teeth in primates', *Amer. J. Phys. Anthrop.* 19, 489-581.

A.C. BERRY

Non-metrical variation in the prosimian skull

Introduction

In the older descriptive works on anatomy, reference is frequently made to unusual ossicles, extra foramina and other minor variations in the bones of the skull. Initially these variations were regarded as nothing more than minor curiosities. Russel and Le Double noted the different incidences of these variants in different populations,[1,2,3] while Montagu used their varying incidences in human and other primates in evolutionary studies.[4,5]

Schultz reported variations in crania of several monkey species and carried out an intensive study on both metrical and non-metrical variation in large samples of *Alouatta, Cebus* and *Ateles.*[6,7,8] Wood Jones advanced human population studies,[9,10] based on the variation in incidence of a number of these non-metrical minor variants in the human skull (reviewed by Brothwell),[11] and this method has recently been considerably expanded (see below).

Before these variants can be critically harnessed for population comparisons, it is necessary to establish what factors control their occurrence. Genetical studies on mice by Grüneberg and co-workers (reviewed by Grüneberg),[12] discussed below, indicate that there is a considerable degree of genetical control, particularly if a wide range of variants is used.

Berry, using a range of skeletal variants, has been able to study genetical distances, based on differences in variant incidences, between different rodent populations.[13,14,15] He has also applied the method to samples of seals.[16] A range of 30 minor variants in the human cranium was described by Berry and Berry,[17] and genetical distances between human population samples have been calculated by Laughlin and Jørgensen for Eskimo,[18] by Kellock and Parsons for Australian Aborigines,[19,20] by De Villiers for South African negroes,[21] and by Berry for Northern Europeans.[22]

A pilot study on a number of non-human primates (Berry and Berry) showed considerable variation, both inter- and intra-specific, in the incidences of non-metrical variants.[23] Similar findings for *Aotus trivirgatus* were reported by Balan and Thorington.[24]

It appeared likely that variation of this type existed in the prosimian cranium. Blood samples for the investigation of serum protein and enzyme variation are hard to obtain for many prosimians, whereas a considerable amount of skeletal material is available in museums. Hence it seemed worthwhile to see whether non-metrical variants could be used as genetical indicators in the study of these species.

Materials

The present study has been limited to *Galago* spp. because of the good numbers available for study at the British Museum (Natural History). There is bound to be similar variation in other prosimian species. Although variants of the vertebral column are likely to exist and have been utilised in rodent studies (Berry and Searle),[25] the present investigation was confined to the skull simply because post-cranial material is not often preserved in collections. In general a sample of about 50 crania is desirable.

The following crania were examined.

G. crassicaudatus agisymbanus	31 from Zanzibar 28 from Pemba Island, about 40 miles away Total: 59 (mixed sexes)
G. senegalensis albipes	47 males 47 females from Uganda Total: 94
G. senegalensis moholi	51 (mixed sexes) from Zambia, S. Rhodesia, Malawi and Angola
G. demidovii	53 (mixed sexes) from various subspecies and localities

Method

1. Characters

Each skull was examined for the presence or absence of the following characters (many of which have been previously described and illustrated).[17,24,25] Some of these, common in man and higher primates, were absent in galagos or occurred in only one species.

1. *Asymmetry of nasal bones.* Usually the nasal bones articulate

with the frontal bones in a symmetrical manner. Occasionally, however, the articulation is asymmetrical. This variant was not scored in our previous primate study.

2. *Sutural bones.* Except for the variant below (No. 3), these were never seen in the present study.
3. *Bregmatic bones present.* Rarely an ossicle is found in the position of the anterior fontanelle. It occurs in man, larger primates and one of the *Galago* species studied.
4. *Asymmetrical bregma.* In larger primates, the suture between the frontal bones rarely persists into adult life. When it does the skull is said to be metopic. In the galagos studied the suture never fused, but usually joined the sagittal and coronal sutures to form a symmetrical junction at the bregma. Occasionally, however, the junction was asymmetrical. This variant is one in which scoring is to some extent subjective, as microscopic examination emphasises degrees of asymmetry unnoticed by the naked eye.
5. *Parietal foramina present.* The emissary foramina in the parietal bone that lie close to the sagittal suture were only seen in the larger *Galago crassicaudatus.*
6. *Stylomastoid foramen present.* There is a moderately large foramen lying in the groove between the bulbous mastoid part of the temporal bone and the occipital condyle. Posterior to this the much smaller stylomastoid foramen may be seen when present.
7. *Posterior condylar pits.* At the junction of the posterior part of the occipital condyles with the occipital bones there may be a small pit. This is difficult to score in galagos because of varying degrees of development in the pit, and its use was abandoned. However, it is a useful variant in larger primates.
8. *Anterior condylar canal double.* This canal, which carries the hypoglossal nerve, may be double. It is a widespread variant and occurs in rodents (where it is called *foramen hypoglossi*), man and other primates.
9. *Foramen paraovale divided.* This foramen lies close to the junction of the sphenoid with the temporal bone, immediately medial to the foramen ovale. Rarely it is divided by a bony septum.
10. *Pterygoid-sphenoid lamina incomplete.* The external pterygoid lamina stretches posteriorly and fuses with the postero-lateral corner of the sphenoid, arching over the foramen ovale. This variant has been previously used only in *Aotus trivirgatus.* (Thorington refers to it as 'external pterygoid lamina joined posteriorly to sphenoid').[24]

11&12. *Lateral or median sphenoidal foramina present.* The basal aspect of the body of the sphenoid bone is sometimes pierced by small foramina. These may be either lateral, in which case they are frequently paired, or a single median foramen may occur, usually more anterior. Rarely, a single lateral foramen may occur with a median one. In this case, both have been scored. There is considerable variation in the expression of these foramina and those that are merely a pin-prick have been scored as ½.

Occasionally in *G. crassicaudatus agisympanus* a single midline foramen was noted immediately anterior to the suture between the anterior and posterior parts of the sphenoid bone. Such a foramen was not scored. These foramina have not been noticed in other primates, but have been described in mice.

13. *Accessory palatine foramina present.* The foramina at the postero-lateral corners of the palatine bone were counted, and any in excess of one considered to be accessory palatine foramina.
14. *Asymmetry of palatine suture.* The suture between the palatine bones and the palatine processes of the maxilla usually crosses the midline suture of the hard palate symmetrically. Occasionally, however, the junction is asymmetrical.
15. *Triple orbital foramina.* In *G. crassicaudatus agisymbanus* the

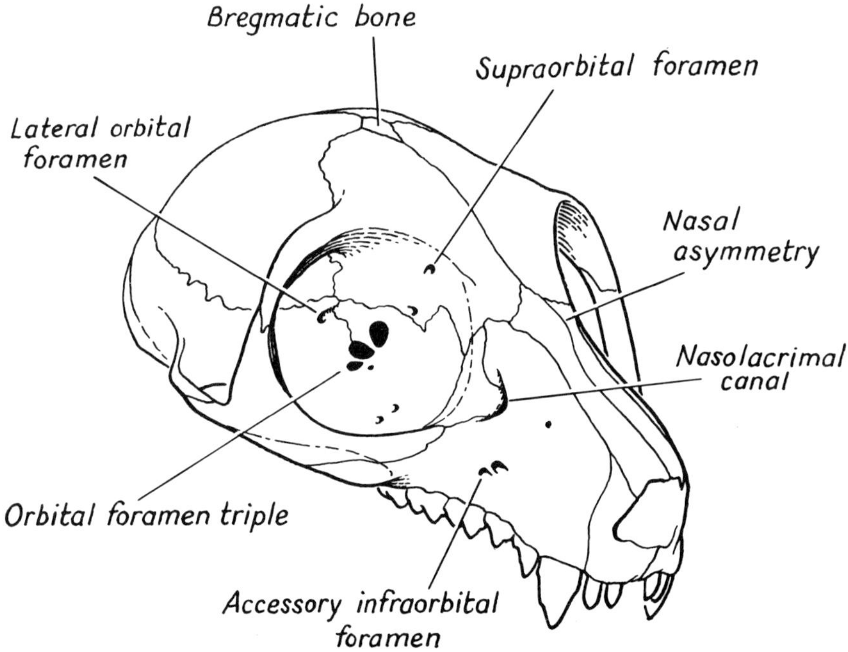

Figure 1. Anteriolateral view of a skull of *Galago* sp., to show the variants described in the text.

posterior wall of the orbit is perforated by two clearly defined foramina. Rarely, a third opening is clearly demarcated; but more frequently this third opening is only partially separated by an incomplete bony spicule from the other foramina (partial triple orbital foramen). This was scored as ½.

A tiny but distinct foramen sometimes occurred medial to the other foramina discussed above. This was not scored in the present study, but could well be used in future.

16. *Lateral orbital foramen absent.*

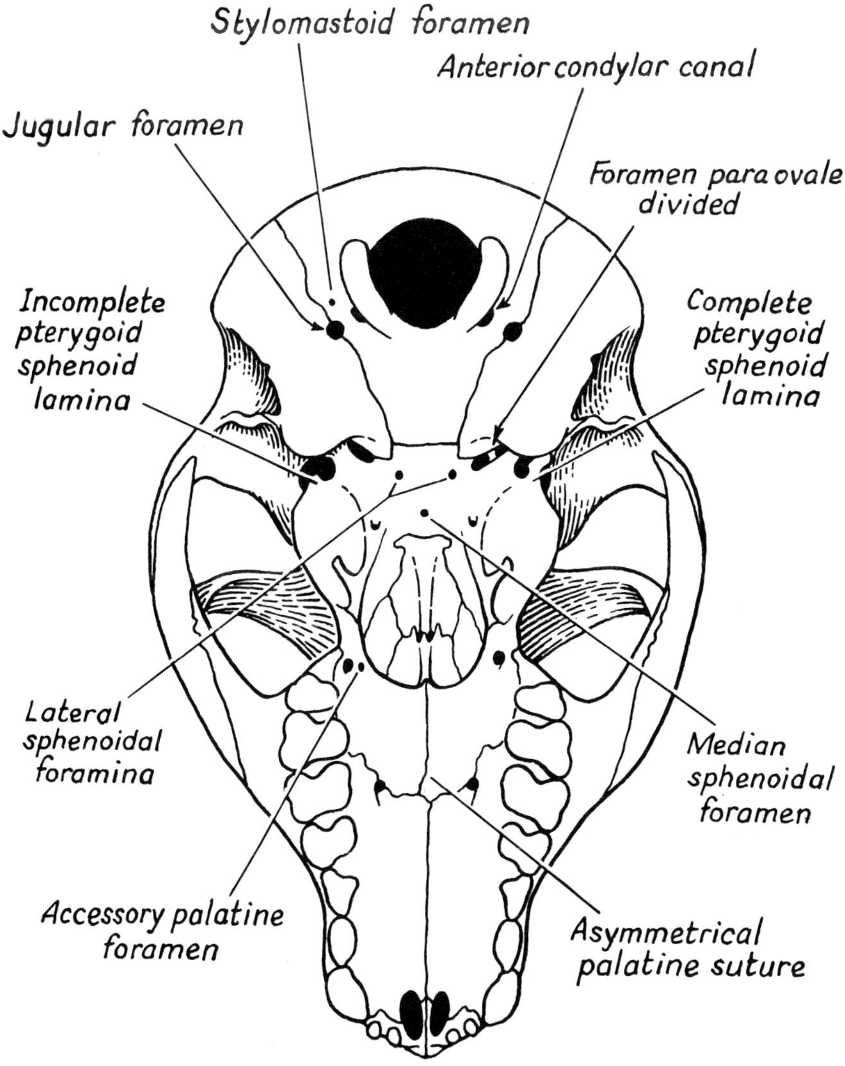

Figure 2. Ventral view of a skull of *Galago* sp.

17. *Lateral orbital foramen double.* A small foramen is usually found latero-superior to the main orbital foramina. The double foramen was only seen in *G. c. agisymbanus.*
18. *Supraorbital foramen absent.* The supraorbital foramen opens inside the orbit and is almost universally present. In rodents it has been named the frontal foramen.
19. *Accessory supraorbital foramen present.* Occasionally, further foramina were noted in the supraorbital region. These were scored as accessory supraorbital foramina.
20. *Accessory nasolacrimal foramen present.* Rarely the foramen immediately anterior to the orbit is double. This variant is called 'accessory pre-orbital foramen' in rodents.
21. *Accessory infraorbital foramen present.* The infraorbital canal runs from the orbit and emerges on the lateral surface of the maxilla. The emergent foramen may be double.
22. *Accessory mandibular foramen.* The anterior foramen on the external surface of the mandible may be double. In large primates, multiple foramina are frequently present.

Scoreable variation was found for 20 of these 22 characters in the three *Galago* species examined (see Figs. 1 and 2).

2. *Scoring difficulties*

Since the method entirely depends on accurate scoring of the variants described, it is important to consider the hindrances to accurate scoring.

(*a*) Observer subjectivity: inevitably, several of the variants will be sometimes scored as present by one observer and yet absent by another; ideally all material being compared should be scored by the same worker. Where this is not possible, careful agreement on the scoring methods and thresholds is essential.

(*b*) With smaller crania, the bone is very delicate and it is hard to know whether variants are truly present or merely artefacts from damage to the bone. The orbit is particularly vulnerable.

(*c*) Where the bone has been incompletely cleaned, foramina may be obscured and made unscoreable by fragments of soft tissue. Enzymatic cleaning as described by Searle is a great advantage, but is unpopular with museum curators since it results in disarticulation of some of the cranial elements in smaller species.[26]

(*d*) Foramina and sutures may be obscured by putty, glue, indian ink and other museum artefacts.

3. *Statistical procedure*

The incidences of the individual variants are calculated in each sample. As many of the variants are bilateral, the incidence for these are calculated for the number of sides scored. Certain variants are frequently not scoreable in some skulls, and consequently the total number scored differs for each variant. The incidences of the individual variants in different samples can then be compared with a χ^2 or similar test. A male sample of *G. senegalensis albipes* was compared in this way with a female sample from the same species.

For population comparisons, however, it is usually more profitable to calculate an 'estimate of divergence' between the samples. This calculation gives the root mean of the summed squared differences of transformed incidences of the individual variants. It is a measure of the genetical distance separating the two samples. The value of the method depends on the degree of correlation of variants with each other. Although no correlation study has been done on the *Galago* data, studies on other species have shown only very low levels of correlation.[17,27,28,29]

The calculation of the estimate of divergence, devised by Professor C.A.B. Smith, has been described in detail previously.[22,23] It is given by:

$$\text{Estimate of Divergence} = \sqrt{\frac{\Sigma\,[(\Theta_1 - \Theta_2)^2 - (\frac{1}{n_1} + \frac{1}{n_2})]}{\text{number of variants}}}$$

– where Θ is the angular transformation in radians of the percentage frequency (P) of the character, such that $\Theta = \text{Sin}^{-1}\,(1\text{-}2p)$; n_1 and n_2 represent the number of individuals in the sample being compared.

The variance of this function can easily be calculated.

Estimates of Divergence were calculated between *G. crassicaudatus agisymbanus* samples from Zanzibar and Pemba Island, and between the two *G. senegalensis* subspecies.

Results

1. *Incidences of variants.* Table 1 shows the percentage incidences of 20 variants in 5 samples.
2. *Comparison of incidences of variants* in male and female. *G. senegalensis albipes:* The only variant to show a difference in incidence between the sexes with a probability of <5% was 'Asymmetrical nasal bones'. This occurs more frequently in males than in females (p>0.01).
3. *Estimates of Divergence:*
   (i) Between *G. crassicaudatus agisymbanus* from Zanzibar and the same subspecies from Pemba Island: 0.084.

Table 1 Incidence of 20 minor skull variants in 5 samples of *Galago*

		G. crassicaudatus agisymbanus		*G. senegalensis albipes*		*moholi*	*G. demidovii*
		Zanzibar	*Pemba Is.*	♂	♀		
	Number of crania in sample:	*31*	*28*	*47*	*47*	*51*	*57*
	Variant	%	%	%	%	%	%
1.	Nasal asymmetry present	6.7	7.4	22.2	2.2	3.9	0
3.	Bregmatic ossicle present	0	25	0	0	0	0
4.	Asymmetrical bregma present	35.0	38.9	23.0	22.0	33.0	34
5.	Parietal foramina present	5.2	5.4	0	0	1.0	0
6.	Stylomastoid foramen present	13.8	12.7	71.8	57.5	11.7	32.5
8.	Anterior condylar canal double	9.8	3.6	8.9	7.9	8.7	28.9
9.	Foramen paraovale divided	0	0	14.1	5.7	0	0
10.	Pterygoid-sphenoid lamina incomplete	15.5	17.8	8.7	5.6	13.2	14.3
11.	Lateral sphenoidal foramen present	9.2	18.8	52.0	48.0	11.7	3.8
12.	Median sphenoidal foramen present	15.0	17.8	0	2.2	2.1	19.6
13.	Accessory palatine foramen present	28.2	31.5	17.9	6.6	4.2	3.9
14.	Asymmetrical palatine suture present	14.3	14.8	13.2	13.0	0	4.2
15.	Orbital foramen triple	0	0.9	36.1	23.5	0	3.7
16.	Lateral orbital foramen absent	0	0	10.5	38.4	36.9	20.4
17.	Accessory lateral orbital foramen present	3.7	7.7	0	0	0	0
18.	Supraorbital foramen absent	0	0	0	0	0	4.2
19.	Accessory supraorbital foramen present	8.8	18.9	3.2	1.1	2.1	0
20.	Accessory nasolacrimal foramen present	0	0	0	0	0	0
21.	Accessory infraorbital foramen present	1.8	0	1.0	0	2.0	1.0
22.	Accessory mandibular foramen present	40.3	35.1	3.2	9.6	6.0	4.0

(ii) Between *G. senegalensis albipes* and *G. senegalensis moholi:* 0.473.

Discussion

1. General conclusions

The results show that non-metrical minor variants do occur with varying incidences in *Galago* skull samples. Apart from the exception noted, the variant occurrence is unaffected by sex. Why asymmetrical nasal bones should occur more frequently in males than in females is hard to explain. It could be a chance finding. The low estimate of divergence between separated samples of the same subspecies is to be expected, but the dramatic difference in the incidence of bregmatic bone (Pemba 25%, Zanzibar 0%) is the sort of difference often found between island populations.[15,30] The large estimate of divergence found when the subspecies of *G. senegalensis* are compared is in reasonable accord with expectation. The amount of variation seems small compared with that found in other species. Not only is there a low number of traits found to exhibit variation in the samples examined, but even when variation does occur it frequently does so only rather rarely. This is a disadvantage to studies of the present type, as population incidences below about 10% affect distance estimates disproportionately.

Why the galagos should have this tendency to invariability is not obvious. It would be interesting to know if it was characteristic of other prosimian species and also whether galagos have a similar tendency when other systems such as coat colour or blood biochemistry are considered. In general, man seems to exhibit much more variation, though to some extent this is due to the frequent occurrence of sutural bones and tori in human crania. The large primates such as *Gorilla* and *Pongo* show considerably higher incidences of multiple foramina compared to the galagos. It is tempting to suggest that the number of accessory foramina might be related to skull size, since accessory foramina are considerably commoner in the larger species. This does seem to apply when the primates are considered as a whole or when the galagos alone are considered, but in rodents the opposite is the case – multiple foramina being commoner in the tiny harvest mouse than in the rat.

2. The validity of the method

Earlier studies have shown that the method gives apparently meaningful genetical distances between samples within a species.[13,15,17] However, once comparisons are attempted between samples quite unrelated to one another (such as between clearly separated species) confusing results emerge.

The comparability of the genetical distances calculated between samples depends on the assumption:

(*a*) that the variants occur independently of one another,
(*b*) that their occurrence is largely genetically controlled.

Correlation studies in the mouse and man all show little correlation between the variants.[17,27–9] Although no such study has been done on the present material, it is likely that the above results hold good.

The *genetical control* of the variants has been established in mice by Grüneberg, who showed them to be controlled multigenically, with occurrence or non-occurrence of a variant resulting from a final threshold effect.

Searle,[26,31] Deol and Truslove,[32] and Howe and Parsons[33] showed that maternal physiology and diet could exert a considerable effect on individual variants. These effects were largely related to birth-weight. If a large number of variants are used together, however, these maternal and dietary effects seem to cancel one another out and variation is largely under genetical control.[33] In general, the factors affecting non-metrical skeletal variants are thought to be some way from the primary site of gene action, so that they are highly sensitive to minor fluctuations in the developing foetus.[12] Hence the term 'epigenetic' was used by Berry and Searle to describe the causation of these non-metrical minor variants.[25]

3. The validity of the method when compared with other methods

The tools usually used in population studies tend to be morphological measurement, coat colour variation and variations in blood biochemistry. Comparison of metrical and non-metrical methods on Indian rats led Berry and Smith (unpublished) to believe that their distances based on non-metrical methods were more realistic than those based on metrical methods. With human material, the two methods tend to show similar relationships,[21,22,34] though a negative correlation was found in one study.[35]

Coat colour is a classical taxonomic character (for obvious reasons). In some circumstances it may be adaptive, but direct evidence for this is surprisingly slight. Certainly much of intraspecific colour variation is a consequence of the genetical history of the population: it may represent clinal variation or random assortment of genes, as in Hebridean races of *Apodemus sylvaticus.*[15,36]

Comparison of the results of biochemical studies with nonmetrical ones have produced unexpected results. Berry and Murphy, working on mice, found seasonal variations in both systems.[37] In humans, Berry found little correlation between distances calculated on blood groups and those from non-metrical variants;[22] but in general the latter distances tended to be larger than serological ones. The multi-factorial control of the minor skeletal variants means that a very large number of loci are represented when a wide range of variants is being considered. This should make the method a more delicate indicator of genetical variation than methods using serological markers which are represented (albeit very closely) by only a few loci.

Conclusion

This study was undertaken as a pilot project. It has established that variation does exist in prosimian crania and that it can be harnessed for use in population studies. Further variants could well be discovered by those more familiar with prosimian anatomy.

The present study has also entirely neglected the teeth. Minor dental variants occur in most animals and have been used in population studies of rodents,[38] gibbons,[39] and humans.[40] This would appear to be a potentially fruitful field.

Acknowledgment

My thanks are due to A.J. Lee for drawing the figures, and to the Keeper of Zoology, British Museum (Natural History), for giving me access to the material.

NOTES

1 Russel, F. (1900), 'Studies in cranial variation', *Am. Nat.* 34, 737-47.
2 Le Double, A.F. (1903), *Variations des os du crane*, Paris.
3 Le Double, A.F. (1906), *Variations des os de la face*, Paris.
4 Montagu, A.M.F. (1933), 'The anthropological significance of the pterion in primates', *Am. J. Phys. Anthrop.* 18, 159-336.
5 Montagu, A.M.F. (1937), 'The medio-frontal suture and the problem of metopism in the primates', *J.R. Anthrop. Inst.* 67, 157-201.
6 Schultz, A.H. (1923), 'Bregmatic fontanelle in mammals', *J. Mammal.* 4, 65-77.
7 Schultz, A.H. (1926), 'Studies on the variability of platyrrhine monkeys', *J. Mammal.* 7, 286-305.
8 Schultz, A.H. (1960), 'Age changes and variability in the skulls and teeth of the central American monkeys, *Alouatta, Cebus* and *Ateles*', *Proc. Zool. Soc. Lond.* 133, 337-90.
9 Wood-Jones, F. (1930-31), 'The non-metrical morphological characters of the skull as criteria for racial diagnosis', I, II and III, *J. Anat.* 65, 179-95, 368-78, 438-45.
10 Wood-Jones, F. (1933-34), 'The non-metrical morphological characters of the skull as criteria for racial diagnosis', IV, *J. Anat.* 68, 96-108.
11 Brothwell, D.R. (1958). The use of non-metrical characters in the skull in differentiating populations, *Dt. Ges. Anthrop.* 6, 103-9.
12 Grüneberg, H. (1963), *The Pathology of Development*, Oxford.

13 Berry, R.J. (1963), 'Epigenetic polymorphism in wild populations of *Mus musculus*', *Genet. Res.* 4, 193-220.

14 Berry, R.J. (1964), 'The evolution of an island population of the house mouse', *Evolution* 18, 468-83.

15 Berry, R.J. (1969), 'History in the evolution of *Apodemus sylvaticus* (Mammalia) at one edge of its range', *J. Zool. Soc. Lond.* 159, 311-28.

16 Berry, R.J. (1969), 'Non-metrical skull variation in two Scottish colonies of the Grey Seal', *J. Zool. Soc. Lond.* 157, 11-18.

17 Berry, A.C. and Berry, R.J. (1967), 'Epigenetic variation in the human cranium', *J. Anat.* 101, 361-79.

18 Laughlin, W.S. and Jørgensen, J.B. (1956), 'Isolate variation in Greenlandic Eskimo crania', *Acta Genet.* 6, 3-12.

19 Kellock, W.L. and Parsons, P.A. (1970*a*), 'Variations of minor non-metrical skeletal variants in Australian Aborigines', *Am. J. Phys. Anthrop.*, n.s. 32, 409-21.

20 Kellock, W.L. and Parsons, P.A. (1970*b*), 'A comparison of the incidence of non-metrical cranial variants in Australian Aborigines with those of Melanesia and Polynesia', *Am. J. Phys. Anthrop.*, n.s. 33, 235-9.

21 De Villiers, H. (1968), *The Skull of the South African Negro*, Johannesburg.

22 Berry, A.C. (1971), 'The use of minor skeletal variants in human population studies', PhD Thesis, University of London; and Berry, A.C. (1974), 'The use of non-metrical variations of the cranium in the study of Scandinavian population movements', *Am. J. Phys. Anthrop.*, in press.

23 Berry, A.C. and Berry, R.J. (1971), 'Epigenetic polymorphism in the primate skeleton', in Chiarelli, A.B. (ed), *Comparative Genetics in Monkeys, Apes and Man*, London and New York.

24 Balan, B.B. and Thorington, R.W. (1969), 'Discrete variation in *Aotus* skulls', Paper read at Annual Meeting of Am. Soc. Mammalogists.

25 Berry, R.J. and Searle, A.G. (1963), 'Epigenetic polymorphism of the rodent skeleton', *Proc. Zool. Soc. Lond.* 140, 577-615.

26 Searle, A.G. (1954*a*), 'Genetical studies on the skeleton of the mouse. IX: Causes of skeletal variation within pure lines', *J. Genet.* 52, 68-102.

27 Truslove, G.M. (1961), 'Genetical studies on the skeleton of the mouse. XXX: A search for correlations between some minor variants', *Genet. Res.* 2, 431-8.

28 Ossenberg, N.S. (1970), 'The influence of artificial cranial deformation on discontinuous, morphological traits', *Am. J. Phys. Anthrop.*, n.s. 33, 357-71.

29 Hertzog, K.P. (1968), 'Association between discontinuous cranial traits', *Am. J. Phys. Anthrop.*, n.s. 29, 397-404.

30 Berry, R.J. (1967), 'Genetical changes in mice and men', *Eug. Rev.* 59, 78-96.

31 Searle, A.G. (1954*b*), 'Genetical studies on the skeleton of the mouse. XI: The influence of diet on variation within pure lines', *J. Genet.* 52, 413-24.

32 Deol, M.S. and Truslove, G.M. (1957), 'Genetical studies of the skeleton of the mouse, XX: Maternal physiology and variation in the skeleton of C57BL mice', *J. Genet.* 55, 288-312.

33 Howe, W.L. and Parsons, P.A. (1967), 'Genotype and environment in the determination of minor skeletal variants and body weight in mice', *J. Embryol. Exp. Morph.* 17, 283-92.

34 Knip, A.S. (1970), 'Metrical and non-metrical measurements on the skeletal remains of Christian populations from two sites in Sudanese Nubia', I and II, *Proc. Kon. Ned. Akad. v. Wetensch. Amsterdam* C 73, 433-68.

35 Berry, A.C., Berry, R.J. and Ucko, P.J. (1967), 'Genetical change in ancient Egypt', *Man* 2, 551-68.

36 Delany, M.J., (1964), 'Variation in the long-tailed field-mouse (*Apodemus*

sylvaticus L.) in North West Scotland. I: Comparison of individual characters', *Proc. Roy. Soc. Lond. B* 161, 191-9.

37 Berry, R.J. and Murphy, H.M. (1970), 'The biochemical genetics of an island population of the house mouse', *Proc. Roy. Soc. Lond. B.* 176, 87-103.

38 Corbet, G.B. (1964), 'Regional variation in the bank vole *Clethrionomys glareolus* in the British Isles', *Proc. Zool. Soc. Lond.* 143, 191-219.

39 Frisch, J.E. (1963), 'Dental variation in a population of gibbons (*Hylobates lar*), in Brothwell, D.R., (ed.), *Dental Anthropology*, Oxford.

40 Hanihara, K. (1967), 'Racial characteristics of the dentition', *J. Dent. Res.* 46, 923-6.

M. CARTMILL

Daubentonia, Dactylopsila, *woodpeckers and klinorhynchy*

Introduction

Many insects, especially among the Lepidoptera and Coleoptera, feed as larvae on the cambium and woody tissues of the trunks and branches of trees. A great number of insect species have adopted this way of life; over 500 genera of wood-boring and cambium-eating insects, comprising many hundreds of species, infest the forests of the Congo region alone.[1] Wood-boring adaptations take many forms. Some larvae feed on cambium and phloem immediately under the bark; some can only feed on dead and partly decayed wood; some kill the tree or branch by tunnelling in the cambium and eat deeper into the underlying wood as it begins to decompose.[2,3] All these wood-boring adaptations have the effect of protecting the larvae and pupae from vertebrate predators to some degree.

The greater the specialisation of the larva for boring in wood, the greater the specialisation demanded of its predators. Larvae feeding directly under the bark of a tree may be pulled from crevices by the slim barbed tongues of titmice, nuthatches, woodhewers and other birds,[4,5] or fished out by the probing fingers of marmosets,[6] or of squirrel monkeys[7] and orangutans.[8] Insects boring in fallen branches or old tree stumps can be exposed and eaten by any predator strong enough to break open the rotten wood concealing them – by coatimundis, for example.[9] To recover insect larvac from their tunnels in the sound wood of living trees, however, requires special adaptations (*a*) for boring or cutting through the concealing wood and (*b*) for probing the exposed tunnel to catch the retreating insect.

In most continental forest communities, woodpeckers are the principal vertebrate predators on wood-boring insect larvae. Wood-pecking birds of the family Picidae inhabit forests in every part of the world except Madagascar, New Guinea and Australia, and various

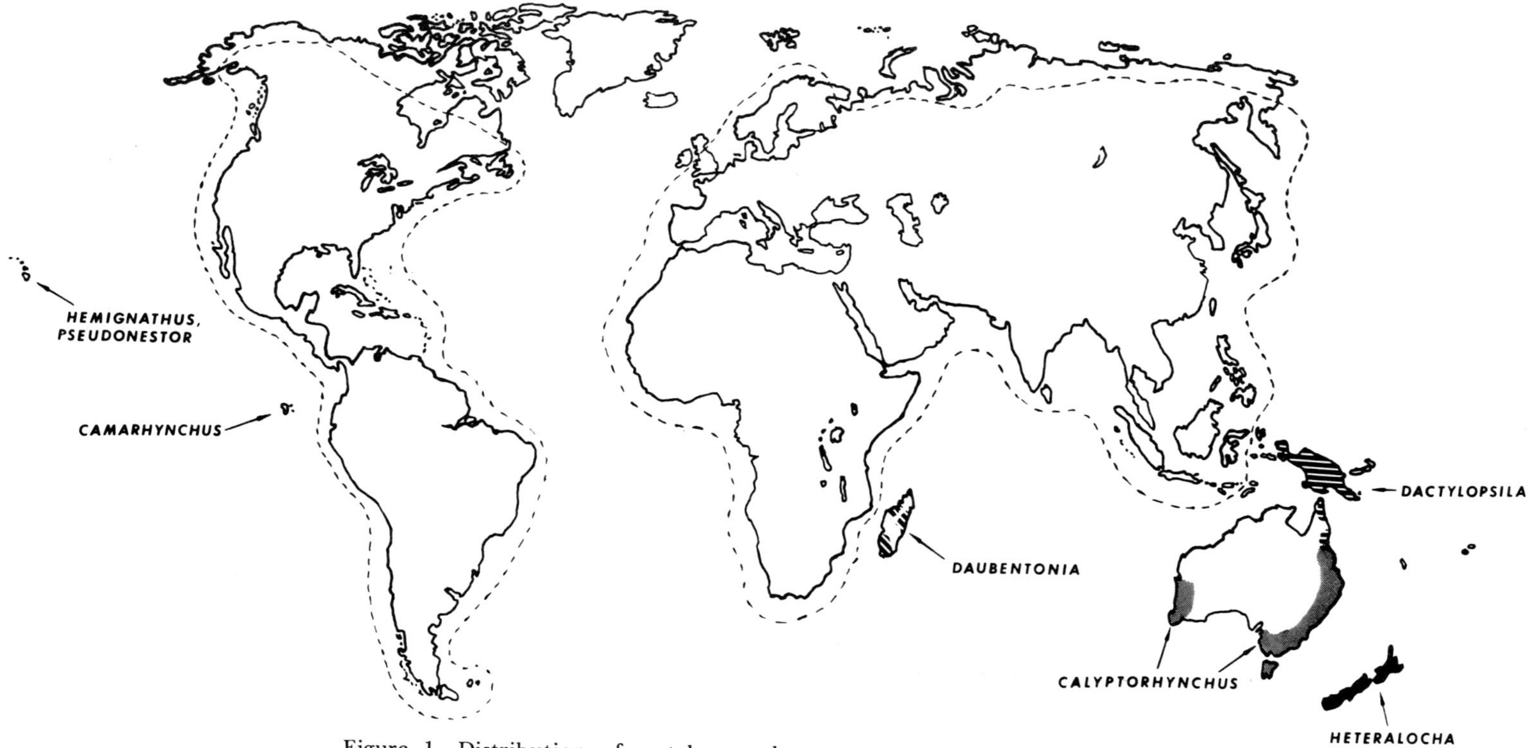

Figure 1. Distribution of vertebrates that prey on wood-boring insects (after various sources). The dashed line shows the limits of woodpecker distribution. The distribution shown for *Calyptorhynchus* includes only those species which feed on wood-boring insects. *Daubentonia* currently has a distribution more restricted than that shown here; the hatched areas represent all records of living or subfossil *Daubentonia*, including the extinct *D. robusta.*

oceanic islands (Fig. 1).[10] Although many woodpeckers get most of their food by picking insects from the surface of trees and from crevices in the bark,[11,12] a woodpecker can also expose a deeply imbedded insect by cutting into the wood with rapid blows of its chisel-like beak; the insect is then either picked up with the beak or extracted from its tunnel by the bird's long, barbed tongue.

On some of the oceanic islands never colonised by woodpeckers, other birds have developed adaptations analogous to the picid specialisations for tunnelling and probing in wood. On Hawaii, honeycreepers of the superspecies *Hemignathus lucidus* use their short, straight lower beaks to peck and excavate in rotting wood, and employ their long, slender and curved upper beaks to probe exposed tunnels.[13,14] The extinct huia *Heteralocha acutirostis* of New Zealand foraged for wood-boring insects using a sexual division of labour; the male bore a stout beak for excavating, and the female a long curving beak for probing.[15] The woodpecker finch *Camarhynchus pallidus* of the Galapagos Islands is well-known for its habit of using a cactus spine or twig as a tool for probing insect tunnels after they have been exposed by ripping off bark and wood with the beak.[16] In Australia, parrots of the genus *Calyptorhynchus*, especially *C. funereus* and *C. lathami*, bite and wrench chunks of wood from living trees to expose burrowing moth larvae; no probing adaptation seems to have been developed in these birds.[17,18] The Hawaiian honeycreeper *Pseudonestor* has a parrot-like beak and habits like those of *Calyptorhynchus.*[19]

New Guinea and Madagascar are the only extensive areas of the world which have not been colonised by woodpeckers or by any of the various woodpecker-like island birds. In these areas, the woodpecker niche is filled by specialised mammals, which have enlarged incisors that serve as wood-cutting organs and an elongated finger which is used as a probe.

The lemuriform primate *Daubentonia madagascariensis* forages for wood-boring grubs in the rain forest of the eastern coast of Madagascar. In feeding, it begins by tapping the surface of a branch, trunk, or fruit, turning its large ears toward the spot being investigated. Detecting an insect, it probes the crevices of the infested plant with a grotesquely attenuated third finger, exposing deep-burrowing larvae by biting and tearing at the wood with its rodent-like incisor teeth. Larvae are pulled out whole or piecemeal by the third finger and licked off.[20,21] Unlike the other Madagascar primates, *Daubentonia* possesses true claws on all its digits except the first toe.[22] These claws represent an adaptation for foraging on trunks, comparable to the specialised stiff tail which helps a foraging woodpecker to support its weight; they make movement and posturing on large trunks easier for *Daubentonia* than for a clawless animal of similar size.[23,24]

In New Guinea, similar morphology and foraging habits have been developed by diprotodont marsupials of the genus *Dactylopsila*. Much like *Daubentonia*, these animals feed by tapping and sniffing infested branches and trunks to locate prey, tearing off bark and wood with their enlarged incisors, and inserting an elongated finger – in this case, the fourth – into exposed tunnels to snare burrowing larvae.[25,26] Like *Daubentonia*, but unlike many of the other small arboreal diprotodonts, *Dactylopsila* has well-developed claws on all digits except the hallux; it differs from *Daubentonia* in being of much smaller size and in having rooted incisors. At least two species of *Dactylopsila, D. palpator* and *D. trivirgata*, inhabit New Guinea. *Dactylopsila palpator* has a longer and more attenuated fourth finger and larger wood-cutting incisors than *D. trivirgata*;[27,28] it is also distinguished from *D. trivirgata* by certain details of the basicranium,[29] and is placed by some in a separate genus *Dactylonax*. *Dactylopsila trivirgata* occurs in two areas on the eastern coast of the Cape York Peninsula, on the Australian mainland.[30] In the more southerly of these two areas, *D. trivirgata* is sympatric with *Calyptorhynchus funereus*;[17] there is presumably some competition between the two species in this small zone of sympatric occurrence.

Cranial morphology of Daubentonia *and* Dactylopsila

The skull of *Daubentonia* (Fig. 2) appears aberrant in a number of respects when compared to the other Madagascar lemurs. Its most obvious peculiarity is the great hypertrophy of its large, continually growing upper and lower incisors; these, together with the large post-incisor diastemata in both upper and lower jaws, give the skull a superficially rodent-like appearance. *Daubentonia*'s brain is globose and foreshortened, and its dorsal-ventral depth is unusually great relative to its size.[31–3] It is also unusually large; corrected for allometry, the brain/body-weight ratio of *Daubentonia* is greater than that of any other prosimian, and even exceeds that of many anthropoids.[34–6] Seen in midsagittal section, the facial skeleton of *Daubentonia* shows, like the brain, a pronounced decrease in anteroposterior length and increase in dorsal-ventral height by comparison with other prosimians.[32,37] The dental arcade is short and narrow. Klinorhynchy is marked in *Daubentonia*: that is, this lemur's facial skeleton is bent downward with respect to the cranial base, so that the obtuse angle between the floor of the nasal fossa and the clivus is about ten degrees less than in *Lemur*.[37]

The zygomatic arches of *Daubentonia* are exceptionally high, especially at their anterior ends,[37] and the zygomatic processes of the maxilla arise far rostrad, above P^4. In other extant lemuriforms, they ordinarily arise above M^2. Although *Daubentonia*'s orbits are of

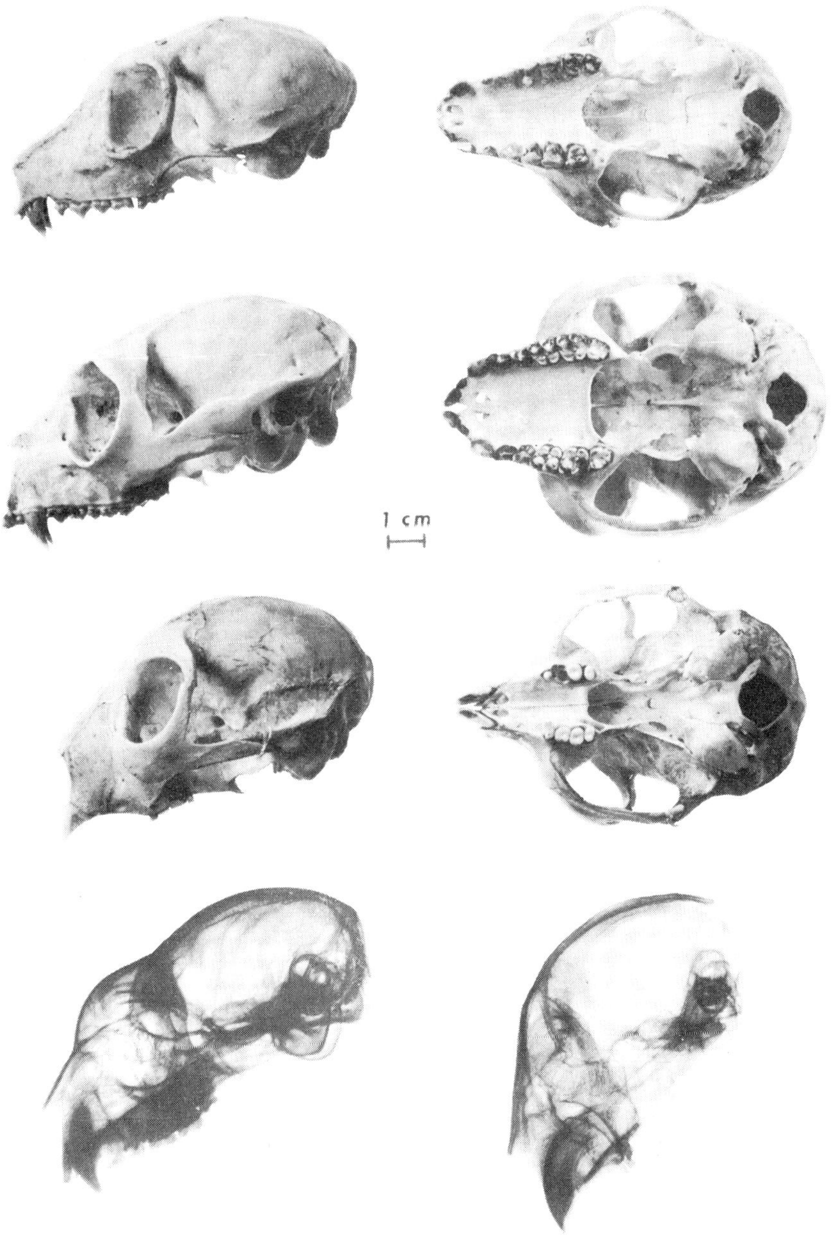

Figure 2. Skulls of prosimian primates. Top row, lateral and palatal views of *Lemur mongoz*, Am. Mus. Nat. Hist. no. 100539, showing relatively primitive configuration of zygomatic arch and pterygoid laminae. Second row, *Propithecus diadema*, A.M.N.H. no. 100557, showing indriid configuration. Third row, *Daubentonia madagascariensis*, A.M.N.H. no. 100632. Bottom row, lateral radiographs of *Propithecus verreauxi* (left), Field Mus. Nat. Hist. no. 8342, and *Daubentonia* (right), Yale Peabody Mus. no. 302, oriented on horizontal through clivus to show increased klinorhynchy in the latter; the *P. verreauxi* skull has been enlarged slightly for comparison with *P. diadema*. In the Yale specimen of *Daubentonia*, the occiput is damaged.

approximately the same diameter as the orbits of other lemuriforms of comparable body size, the relative inter-orbital distance is appreciably greater in *Daubentonia* (Table 1). The result of these three peculiarities is that, in effect, the medial margin of the orbit of *Daubentonia* is displaced laterally, the anterior margin is displaced still further forward, and the lateral margin is displaced dorsally. The orbits of *Daubentonia* are therefore as convergent as those of other Madagascar lemurs of similar size,[38] but directed much less frontally – that is, they face upwards toward the skull roof, like the orbits of *Nycticebus* or *Perodicticus.*

The basicranium of *Daubentonia* is also aberrant. The sphenoid is greatly inflated, the *sinus sphenoidalis* extending back through the entire basisphenoid.[31] The lateral pterygoid laminae extend from the posterior molar alveoli backward in a virtually straight line to the lateral side of the auditory bulla, where they have a broad articulation with the petrosal,[39] and fuse with the squamosal medial to the glenoid fossa.

Table 1 Relative interorbital breadth in lemuriform prosimians[51]

Species (and sample size)	*Interorbital breadth in orbital diameters*	
	Sample mean	*s.d.*
Lemur catta (6)	0.84	0.08
Lepilemur mustelinus (6)	0.51	0.03
Hapalemur griseus (8)	0.49	0.04
Cheirogaleus major (4)	0.56	0.04
Microcebus murinus (6)	0.31	0.02
Phaner furcifer (2)	0.56*, 0.54*	–
Propithecus verreauxi (6)	0.90	0.03
Avahi laniger (4)	0.62	0.05
Indri indri (6)	0.93	0.11
Daubentonia madagascariensis (3)	1.15*, 1.16*, 1.25*	–

* Values for single specimens.

Table 2 Klinorhynchy in phalangeroids[51]

Species (and sample size)	*Angle between clivus and palate*	
	Sample mean (degrees)	*s.d.*
Eudromicia caudata (6)	165.4	4.4
Trichosurus vulpecula (6)	167.3	1.3
Pseudocheirus lemuroides (7)	165.8	2.5
Phalanger orientalis (6)	163.6	5.4
Phalanger maculatus (6)	162.1	4.2
Phascolarctos cinereus (6)	183.9	4.5
Dactylopsila trivirgata (6)	160.0	4.8
Dactylopsila palpator (6)	154.5	6.0

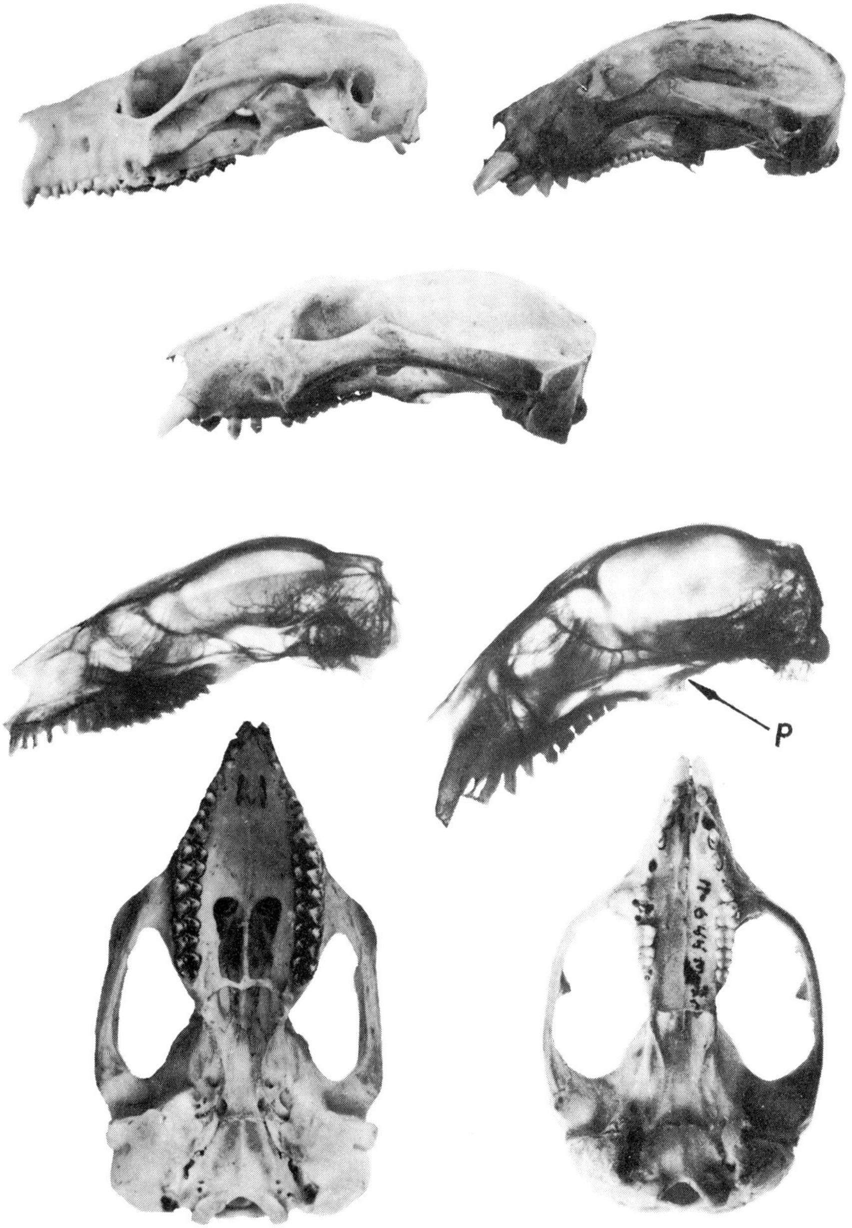

Figure 3. Skulls of phalangeroid marsupials (not to same scale). Top row, *Pseudocheirus* sp. (left), Duke University Dept. of Anatomy collection, and *Dactylopsila palpator* (right), Amer. Mus. Nat. Hist. no. 191043, showing differences in zygomatic arches. Second row, *Dactylopsila trivirgata*, A.M.N.H. no. 143852, a less specialised species of the genus. Third row, lateral radiographs of *Pseudocheirus lemuroides* (left), A.M.N.H. no. 60923, and *D. palpator* (right), A.M.N.H. no. 151964, oriented on clivus; increase in klinorhynchy, in size of olfactory bulbs and frontal cortex, and in size and density of lateral pterygoid processes (p) is evident in *Dactylopsila*. Bottom row, palatal views of same specimens seen in top row; *D. palpator* lacks palatine vacuities.

Most of the peculiarities of cranial morphology which distinguish *Daubentonia* from the other lemuriforms also distinguish *Dactylopsila* from the other phalangeroids, and are especially pronounced in the more specialised species *Dactylopsila palpator.* By comparison with a less specialised phalanger of comparable size (e.g. a small species of *Pseudocheirus*), *Dactylopsila palpator* displays marked klinorhynchy (Table 2), increased relative interorbital breadth (exceeded only by the equally but differently specialised *Phascolarctos* among the species listed in Table 2), and reduction in the length of the snout and augmentation of its vertical height. *Dactylopsila*'s brain is globular and exceptionally large. The dental arcade of *Dactylopsila* is relatively short and narrow. The anterior root of the zygomatic arch extends forward to a point above P^3 in *Dactylopsila*, whereas it lies above the M^{1-2} interspace in most other phalangers; the anterior zygomatic root is also unusually high in *Dactylopsila.* As a consequence, *Dactylopsila*, like *Daubentonia*, has markedly convergent but dorsally-directed orbits. In most phalangeroids, the medial pterygoid laminae are prolonged backward along the cranial base as distinct ridges springing from the basisphenoid; in *Dactylopsila*, the *lateral* pterygoid plates are also prolonged backward as low ridges of dense bone, which extend caudad almost to the glenoid fossa (Fig. 3).

Interpretation

Since *Daubentonia* and *Dactylopsila* have had no common ancestor since at least the mid-Cretaceous, it must be assumed that all the peculiar cranial traits which they share with each other but not with their closest relatives are habitus traits, functionally related to their other adaptations for preying on wood-boring insects. This assumption is further supported by the fact that comparable peculiarities – a shortened face, a foreshortened and rounded braincase, a wide and densely ossified interorbital region, moderate klinorhynchy, and so on – are characteristic of woodpeckers, and are most pronounced in those woodpecker species whose diets contain the largest proportions of wood-boring insects.[40,41] Evidently, all these traits of cranial morphology are primary or secondary adaptations for generating and dissipating the large forces needed to chisel or bite through wood.

When a mammal bites into wood with its incisors, the biting force exerted by the masticatory musculature is transmitted through the wood to the upper incisors. Since the mandible cannot as a rule be protruded far enough to direct the transmitted force along the dorsal surface of the snout (Fig. 4*A*), incisal biting tends to bend the end of the rostrum dorsally, inducing compressive stresses in the roof of the rostrum and tensile stresses in its floor. The amount of force that can be exerted in biting is limited by the capacity of the facial skeleton

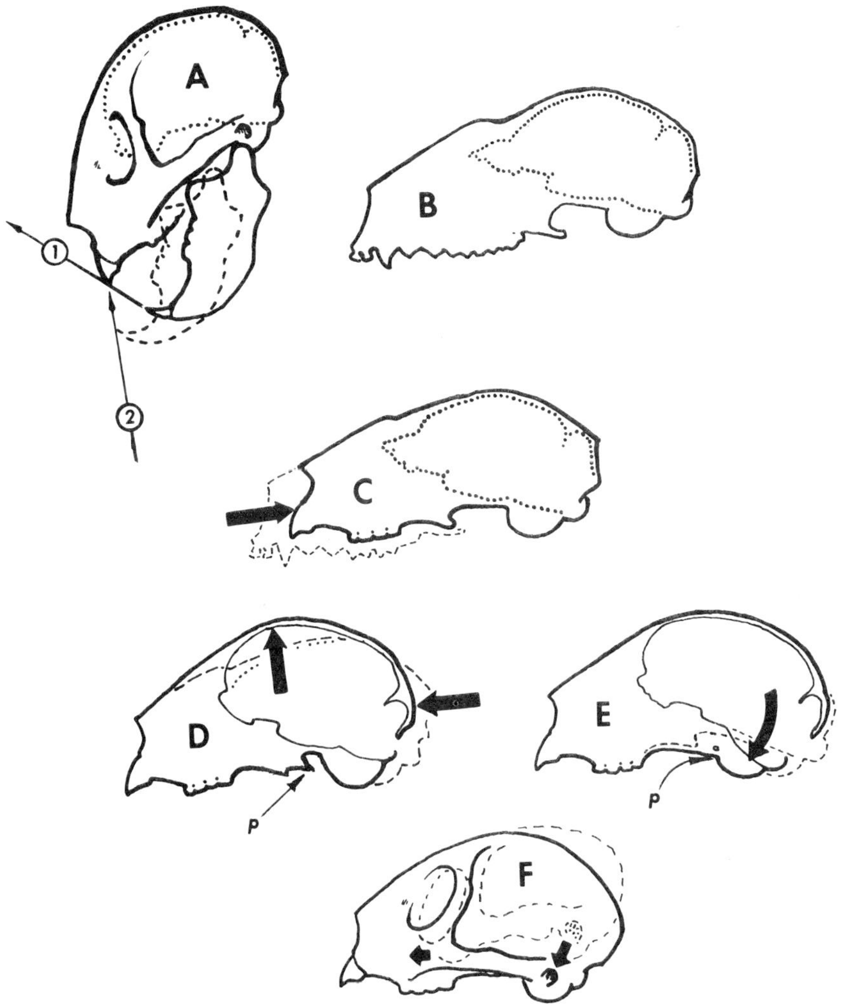

Figure 4. Generation and dissipation of biting stresses in the skull of *Daubentonia*. (*A*) Lateral view of skull. When the jaws are parted in biting into wood, the force transmitted to the rostrum is roughly normal to the rostral axis (arrow 1). To align this force with the rostrum (arrow 2) would require an impossible protrusion of the mandible (dashed line). (*B*) Diagrammatic outline of indriid skull with superimposed silhouette of brain (dotted line). (*C*) Recession of the rostrum (arrow) relative to the proportions seen in other lemurs of similar size (dashed line) reduces the bending moments acting on the rostrum. (*D*) Remodelling of the braincase (arrows) increases the vertical height of the interface between splanchnocranium and neurocranium. (*E*) Klinorhynchy (curved arrow) permits further expansion of the area through which the face is hafted onto the braincase, by presenting the cranial base for attachment of caudad projections from the lateral pterygoid processes (p). (*F*) Klinorhynchy and forward displacement of the anterior zygomatic root (arrows) allow the zygomatic arch to act under tension as a guy wire resisting upward displacement of the end of the rostrum.

to withstand these bending stresses.

Most of the shared cranial peculiarities of *Daubentonia* and *Dactylopsila* have the effect of reducing and countering bending forces in the snout, thus increasing the amount of force that these animals can exert when they tear into infested wood in search of wood-boring insects. Their most obvious adaptation to this end is a reduction in the relative length of the dental arcade. This has the effect of drawing the maxillary incisors back toward the braincase, therefore reducing all the bending moments of the force exerted on the facial skeleton through the maxillary incisors (Fig. 4*C*).

Most of the other cranial specialisations of *Daubentonia* and *Dactylopsila* have the effect of countering bending forces rather than reducing them. In less aberrant lemurs and phalangers, the facial skeleton resembles a long truncated cone affixed to the narrow end of the prolate braincase (Fig. 4*B*). In *Daubentonia,* the height of the braincase is increased, thus, in Radinsky's words, 'strengthening the skull against the biting stresses from the incisors' by increasing the diameter of the facial cone's base in the midsagittal plane (Fig. 4*D*).[33] The same adaption is evident in *Dactylopsila.* In both animals, the braincase is also rotated or tilted back and up with respect to the axis of the snout, bringing the occipital end of the cranial base forward and downward (Fig. 4*E*) and so further increasing the neurocranial surface available for attachment of the facial skeleton. This is accompanied by an extension of the lateral pterygoid laminae back along the ventral surface of the braincase toward the ear region.[42] As the face has rotated downward and shifted back with respect to the cranial base, the cribriform plate has rotated with the olfactory fossa, so that in *Daubentonia* and *Dactylopsila* the olfactory lobes of the brain face more nearly ventrally than in other lemurs or phalangers when the skull is oriented on the clivus. In *Daubentonia*, the vertical height of the facial cone's base is also augmented by pneumatisation of the frontal region and the sphenoid.

Since the biting force tends to bend the end of the rostrum upward, it places the frontal region under compression and the palate under tension. Relative interorbital breadth is increased somewhat in both *Daubentonia* and *Dactylopsila*, enabling this region to withstand more compressive stress. Tensile stresses in the floor of the snout are transmitted backward along the palate and alveolar bone to the pterygoid processes, which in *Daubentonia* and *Dactylopsila* are broadly fused with the cranial base back as far as the glenoid region (Fig. 4*D-E*).

In both *Daubentonia* and *Dactylopsila*, the anterior root of the zygomatic arch arises relatively further rostrad than it does in less specialised lemurs and phalangers. In *Daubentonia*, this has the effect of bringing the post-orbital bar more nearly into alignment with the long axis of the rostrum; and it might be argued that this serves to permit more efficient transmission of compressive stresses from the end of the rostrum backward and upward through the post-orbital bar.[43] But this analysis fails to account for the similar specialisation seen in *Dactylopsila*, which lacks a post-orbital bar. The forward shift of the anterior zygomatic root seen in these animals makes more sense when we note that the associated klinorhynchy involves a depression of the posterior zygomatic root with respect to the long axis of the snout (Fig. 4*F*). These changes in the relative positions of the two ends of the zygomatic arch permit the entire arch to act like a guy wire or tie beam connecting the anterior end of the snout to the depressed glenoid region, thus helping to counteract bending moments acting on the rostrum.

Increase in the vertical height of the anterior zygomatic root serves the same function. Elastic deformation of the rostrum by bending forces transmitted through the maxillary incisors would displace the lacrimal region backward and the molar alveoli forward, and therefore produce torsion in the zygomatic arches. In Fig. 5, two points (x, y) are diagrammed as alternative centres of rotation for this torsion. If the bite force transmitted through the upper incisors produced torsion around x, the moment arm of this force would be p_1. Rotation around x requires anterior displacement of y, and the forces resisting this displacement have a moment arm L equal to the distance from x to y. In effect, this system represents a bent lever. If the anterior zygomatic root is shifted forward, p_1 decreases accordingly; if the vertical height of the anterior zygomatic root is increased, x and y move further apart, and the load arm (L) of the bent lever system increases relative to the power arm p_1. The same analysis applies for a bending force with a moment arm of p_2 acting around y (Fig. 5); or around any point between x and y. The configuration seen in *Daubentonia* and *Dactylopsila* (Fig. 5*B*) therefore permits the arches to resist torsion more efficiently, and so helps to counter bending forces acting on the rostrum.

Several features of the skull of *Dactylopsila* suggest that this marsupial may at times use its rostrum as a chisel, after the fashion of a woodpecker. The enlarged upper incisor in this genus shows pronounced wear on the labial surface; the tip of the tooth is ordinarily worn off flat in a plane roughly perpendicular to the longitudinal axis of the palate. As Thomas pointed out, the lower incisor cannot occlude with this wear facet.[27] If *Dactylopsila* ever thrusts or pries into a piece of wood using its upper incisors alone, as this wear pattern suggests,[44] its entire rostrum will sometimes be placed under compression, and the palate and pterygoid processes must be adapted to carry compressive stresses as well as tensile ones. Two peculiarities of *Dactylopsila* look like adaptations to this end. First, its palate is narrow, dense, and thick, lacking the palatine vacuities seen in all other phalangerid genera; in *D. trivirgata*, the palate is thin and faintly translucent where vacuities might be

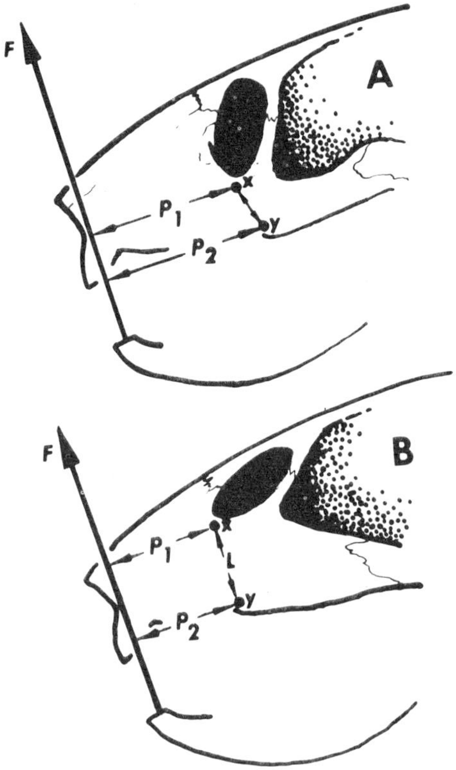

Figure 5. Effect of increase in anterior zygomatic height. (*A*) Unspecialised configuration. (*B*) Configuration seen in *Daubentonia* and *Dactylopsila*. The biting force (F) transmitted to the upper incisors tends to produce torsion in the zygomatic arch; the configuration seen in *Daubentonia* resists this torsion more effectively, and thus reduces strain in the rostrum during gnawing. See Explanation in text.

expected to occur, but in the more specialised *D. palpator* even this vestige of the primitive condition has disappeared (Fig. 3). Secondly, the lateral pterygoid processes or crests which extend backward toward the ear region of *Dactylopsila* are thickened struts which cast a dense x-ray shadow; by contrast, those of *Daubentonia* are relatively delicate bony plates, evidently better suited to carrying loads under tension (through periosteal fibre tracts?) than under compression.

Discussion: gnawing dentitions and skull form

Although a distinguished proponent of the notion that klinorhynchy reflects postural uprightness has been bold enough to assert that *Daubentonia* is 'nearly always bipedal' in its locomotion,[4 5] klinorhynchy clearly has nothing to do with upright posture in either *Daubentonia* or *Dactylopsila*. Radinsky suggests that klinorhynchy in

Daubentonia 'brings the axis of the rostrum more into line with the incisors, so that stresses resulting from biting are transmitted along the long axis of the rostrum'.[33] The difficulty with this interpretation becomes apparent when we translate it into the equivalent suggestion that depression of the occipital end of the braincase (Fig. 4*E*) somehow causes biting stresses to be transmitted along the long axis of the rostrum. In fact, biting stresses in *Daubentonia* cannot be aligned with the long axis of the rostrum, at least when the incisors are parted far enough to bite into wood; the mandible cannot be protruded so far (Fig. 4*A*).

The interpretation advanced here appears preferable: that deflection of the visceral cranium (or retroflexion of the braincase) serves to augment the vertical height of the interface between visceral and neural parts of the skull, thus helping to stabilise the facial skeleton against bending moments in the sagittal plane. How generally applicable is this interpretation? It would be a mistake to attempt to apply it to all instances of pronounced klinorhynchy. In *Tarsius*, klinorhynchy is a secondary effect of the hypertrophy of the eyes,[46] while in *Homo* it probably represents, at least in part, a postural adaptation.[47] But in the interpretation of klinorhynchy in rodents, lagomorphs, and other animals with enlarged anterior teeth used in cutting hard materials, the thesis advanced here should be tested before resorting to explanations involving presumed changes in posture. It seems possible, for instance, that *Lepus* is more klinorhynch than *Ochotona* because it is a larger animal capable of generating larger bending moments in its long snout, not because its facial skeleton has been 'subjected to readjustment by the adoption of upright posture'.[48]

Daubentonia's brain is unexpectedly large for a prosimian of its mass; in 13 of the 14 dimensions of various parts of the brain measured by Stephan et al.,[36] *Daubentonia* exceeds *Indri* in absolute size, though its body weight is less than half that of *Indri*. The suggestion that *Daubentonia* is a dwarf form[49] seems untestable. Moreover, a similar increase in size and height of the braincase is seen in *Dactylopsila*, and is at least as pronounced in *D. palpator* as in the somewhat smaller *D. trivirgata*. Dabelow proposed that central flexion of the cranial base ordinarily results in ('*verursacht*') enlargement of the brain, because the dorsal surface of the braincase must increase in length as the base becomes flexed.[50] In both *Daubentonia* and *Dactylopsila*, klinorhynchy has been accompanied by a slight increase in central flexion between the pre- and post-sellar parts of the cranial floor; more importantly, it has also been accompanied by an increase in the vertical height of the frontal end of the braincase (Fig. 4*D*). It seems possible that the unusual growth of the brain of *Daubentonia* is a secondary result of a change in the pattern of mechanical forces operating on the skull.[51]

Acknowledgments

The data in Tables 1 and 2 are taken from a study funded in part by grants-in-aid from the Wenner-Gren Foundation and from The Society of the Sigma Xi. My attendance at the Research Seminar on Prosimian Biology was made possible by funds provided by the Duke University Department of Anatomy. The skull of *Pseudocheirus* sp. reproduced in Fig. 3 was generously provided by Dr Nancy Bowers of the University of Auckland.

NOTES

1 Mayné, R. (1962), 'Hôtes entomologiques du bois. II: Distribution au Congo, au Rwanda et au Burundi. Observations éthologiques', *Publ. Inst. Nat. Etude Agron. Congo Belge*, sér. sci., 100, 1-514.

2 Graham, S.A. (1952), *Forest Entomology*, New York.

3 Kirkpatrick, T.W. (1957), *Insect Life in the Tropics*, London.

4 Gardner, L.L. (1925), 'The adaptive modifications and the taxonomic value of the tongue in birds', *Proc. U.S. Nat. Mus.* 67 (19), 1-49.

5 Skutch, A.F. (1945), 'Life history of the allied woodhewer', *Condor* 47, 85-94.

6 Hershkovitz, P. (1969), 'The recent mammals of the neotropical region: a zoogeographical and ecological review', *Quart. Rev. Biol.* 44, 1-70.

7 Allen, J.A. (1916), 'Mammals collected on the Roosevelt Brazilian Expedition, with field notes by Leo E. Miller', *Bull. Am. Mus. Nat. Hist.* 35, 559-610.

8 Harrisson, B. (1962), *Orang-Utan*, London.

9 Kaufmann, J.H. (1962), 'Ecology and social behaviour of the coati, *Nasua narica*, on Barro Colorado Island, Panama', *Univ. Calif. Publ. Zool.* 60, 95-222.

10 Peters, J.L. (1948), *Check-List of Birds of the World*, vol. 6, Cambridge, Massachusetts.

11 Short, L.L. (1970), 'Notes on the habits of some Argentine and Peruvian woodpeckers (Aves, Picidae)', *Am. Mus. Novit.* 2413, 1-37.

12 Short, L.L. (1971), 'Systematics and behavior of some North American woodpeckers, genus *Picoides* (Aves)', *Bull. Am. Mus. Nat. Hist.* 145, 1-118.

13 Amadon, D. (1950), 'The Hawaiian honeycreepers (Aves, Drepaniidae)', *Bull. Am. Mus. Nat. Hist.* 95, 151-262.

14 Munro, G.C. (1960), *Birds of Hawaii*, Rutland (Vermont) and Tokyo.

15 Buller, W.L. (1888), *History of the Birds of New Zealand*, 2nd ed., published by the author, London; rev. ed. by Turbott, E.G. Honolulu, 1967.

16 Lack, D. (1947), *Darwin's Finches*, London.

17 Eastman, W.R. Jr., and Hunt, A.C. (1966), *The Parrots of Australia*, Sydney.

18 Frith, H.J. (ed.) (1969), *Birds in the Australian High Country*, Sydney.

19 Perkins, R.C.L. (1903), 'Vertebrata', in Sharp, D. (ed.), *Fauna Hawaiiensis*, London, vol. 1, part 4.

20 Owen, R. (1866), 'On the aye-aye (*Chiromys* Cuvier)', *Trans. Zool. Soc. Lond.* 5, 33-101.

21 Petter, J.-J. and Petter, A. (1967), 'The aye-aye of Madagascar', in S.A. Altmann (ed.), *Social Communication among Primates*, Chicago, 195-205.

22 Le Gros Clark, W.E. (1936), 'The problem of the claw in primates', *Proc. Zool. Soc. Lond.* (1936), 1-24.

23 Petter, J.-J. and Peyrieras, A. (1970), 'Nouvelle contribution à l'étude

d'un lémurien malgache, le Aye-Aye (*Daubentonia madagascariensis* E. Geoffroy)', *Mammalia* 34, 167-93.

24 Cartmill, M. (in press), 'Pads and claws in arboreal locomotion', in Jenkins, F.A. Jr. (ed.,) *Primate Locomotion,* New York, 45-83.

25 Rand, A.L. (1937), 'Results of the Archbold expeditions, no. 17: Some original observations on the habits of *Dactylopsila trivirgata* Gray', *Am. Mus. Novit.* 957, 1-7.

26 Fleay, D. (1941), 'The remarkable striped possum', *Victorian Nat.* 58, 151-5.

27 Thomas, O. (1910), 'A new genus for *Dactylopsila palpator*', *Ann. Mag. Nat. Hist.*, ser. 8, 6, 610.

28 Tate, G.H.H. (1945), 'Results of the Archbold expeditions, no. 55: Notes on the squirrel-like and mouse-like possums (Marsupialia)', *Am. Mus. Novit.* 1305, 1-12.

29 Tate, G.H.H. and Archbold, R. (1937), 'Results of the Archbold expeditions, no. 16: Some marsupials of New Guinea and Celebes', *Bull. Am. Mus. Nat. Hist.* 73, 331-476.

30 Marlow, B. (1965), *Marsupials of Australia*, Brisbane.

31 Starck, D. (1954), 'Morphologische Untersuchungen am Kopf der Säugetiere, besonders der Prosimier, ein Beitrag zum Problem des Formwandels des Säugetierschädels', *Z. wiss. Zool.* 157, 169-219.

32 Hofer, H. (1958), 'Vergleichende Beobachtungen über die kraniocerebrale Topographie von *Daubentonia madagascariensis* (Gmelin, 1788)', *Morph. Jb.* 99, 26-64.

33 Radinsky, L.B. (1968), 'A new approach to mammalian cranial analysis, illustrated with examples of prosimian primates', *J. Morph.* 124, 167-80.

34 Bauchot, R. and Stephan, H. (1966), 'Données nouvelles sur l'encéphalisation des Insectivores et des Prosimiens', *Mammalia* 30, 160-96.

35 Bauchot, R. and Stephen, H. (1969), 'Encéphalisation et niveau évolutif chez les Simiens', *Mammalia* 33, 225-75.

36 Stephan, H., Bauchot, R. and Andy, O.J. (1970), 'Data on size of the brain and of various brain parts in insectivores and primates', in Noback, C.R. and Montagna, W. (eds.), *The Primate Brain*, New York, 289-97.

37 Biegert, J. (1957), 'Der Formwandel des Primatenschädels und seine Beziehungen zur ontogenetischen Entwicklung und den phylogenetischen Spezialisationen der Kopforgane'. *Morph. Jb.*, 98, 77-199.

38 Cartmill, M. (1971) 'Ethmoid component in the orbit of primates'. *Nature,* 232, 566-567.

39 Saban, R. (1963), 'Contribution à l'étude de l'os temporal des Primates'. *Mem. Mus. nat. Hist. nat., Paris,* n.s. 4, sér. A (Zool.), 29, 1-378.

40 Burt, W.H. (1930), 'Adaptive modifications in the woodpeckers'. *Univ. Calif. Publs. Zool.*, 32, 455-524.

41 Hofer, H. (1945), 'Untersuchungen über den Bau des Vogelschädels, besonders über den der Spechte und Steisshühner'. *Zool. Jb.*, 69, 1-158.

42 In this respect, as in several other points of cranial anatomy, *Daubentonia* resembles the indriids.

43 Cf. Tattersall's analysis of *Archaeolemur.* (This volume).

44 I am indebted to Dr. W.L. Hylander for this suggestion.

45 DuBrul, E.L. (1958), *Evolution of the Speech Apparatus*, Springfield (Ill.): Charles C. Thomas.

46 Spatz, W.B. (1968), 'Die Bedeutung der Augen für die sagittale Gestaltung des Schädels von *Tarsius* (Prosimiae, Tarsiiformes)'. *Folia primat.*, 9, 22-40.

47 Weidenreich, F. (1924), 'Die Sonderform des Menschenschädels als Anpassung an den aufrechten Gang'. *Z. Morph. Anthrop.*, 24, 157-189.

48 DuBrul, E.L. (1950), 'Posture, locomotion, and the skull in Lagomorpha', *Am. J. Anat.* 87, 277-313.

49 Stephan, H. (1972), 'Evolution of primate brains: a comparative anatomical investigation', in Tuttle, R.H. (ed.), *The Functional and Evolutionary Biology of Primates*, Chicago, 155-74.

50 Dabelow, A. (1929), 'Uber Korrelationen in der phylogenetischen Entwicklung der Schädelform. I. Die Beziehungen zwischen Rumpf und Schädelform', *Morph. Jb.* 63, 1-49.

51 These data are taken from Cartmill, M. (1970), 'The orbits of arboreal mammals: a reassessment of the arboreal theory of primate evolution', Ph.D. thesis, University of Chicago.

PART II SECTION C
Morphology of the Brain

D. E. HAINES, B. C. ALBRIGHT, G. E. GOODE and H. M. MURRAY

The external morphology of the brain of some Lorisidae

Studies on the evolution of the central nervous system can utilise two tools: palaeoneurology and comparative neurology. The former gives information on the relative size, shape and topography of the endocranium of animals in lineages, while the latter involves comparison of primitive and more advanced extant forms.[1] The members of the family Lorisidae represent, to a limited degree, a series in the development of the central nervous system. The animals used in the present study represent more primitive members of the family (*Galago senegalensis*) and some members that have an advanced degree of CNS development (*Perodicticus potto*). The present report on the external morphology of the brain of some Lorisidae is preparatory to studies in these animals of fibre connections of the cerebellum and red nucleus, ascending projections from the cord, and descending cortical pathways.

Descriptions of the external morphology of the brains of a variety of prosimians are found throughout the literature. Elliot Smith reviewed numerous earlier works.[2] More recently topography of the cortex in several prosimians has been reviewed by Connolly,[3] Friant,[4] and others.[5-8] Le Gros Clark has dealt not only with the cortex, but also with other areas of the CNS of tree shrews,[9-11] mouse lemur,[12] insectivores,[13,14] and tarsier.[15] The anatomy of the brain stem of the lesser galago and slow loris has recently been reported,[16,17] and the cytoarchitectonics of the thalamus have also received some attention.[18] The tarsier has received little attention, even though there is considerable debate as to its phylogenetic significance.[19-21] The phylogenetic changes in the olive, pons and pyramid have been postulated recently in a study that included several lorises and lemurs.[22]

As complete as many of these studies appear to be, many details are still not known. For example, there is disagreement as to the precise exit of cranial nerve rootlets. With the exception of a recent study,[23] the rhomboid fossa of all prosimians has been ignored. The gross relationships of the mesencephalon with the metathalamus also need clarification. These and other areas merit clearer description and illustration.

Materials and methods

The brains of the lesser bushbaby (*Galago senegalensis senegalensis*), greater bushbaby (*Galago crassicaudatus*), slow loris (*Nycticebus coucang*) and potto (*Perodicticus potto*) were used in this study. The animals were anaesthetised with either intraperitoneal or intravenous administration of Nembutal, then perfused with physiological saline followed by 10% formalin. After perfusion, the brain was removed with the dura intact. Special effort was exerted to leave the cranial nerve roots as long as possible. The dura and arachnoid were removed and most observations were made under a dissecting microscope with a zoom lens.

Scale drawings, at 1 mm. = 1 cm., were made and repeatedly checked from all angles for accuracy. A special effort was made when removing the arachnoid not to damage the rootlets of the cranial nerves so that their precise origin from the brainstem could be determined.

Histological sections at 30 to 40 microns were made at some brainstem levels to verify what deep nuclei were responsible for elevations and/or grooves or fissures. These sections were stained either with the luxol fast blue-cresylecht violet method,[24] or with a cresyl acetate violet method.[25]

Medulla oblongata and pons

1. *Ventral and ventro-lateral surfaces*

Owing to variation in the exit of the first cervical nerve, medulla length is measured on mid-sagittal sections. The measurement from the ponto-pyramidal groove to the last visible fibre bundle of the pyramidal decussation corresponds to the external length of the pyramids. *G. crassicaudatus* has the longest pyramidal length (10 mm.) followed by *N. coucang* (9.5 mm.) and *P. potto* (9 mm.) (Figs. 1-3). In *G. senegalensis* the pyramids are 7 mm. in length (Fig. 4). The width of the medulla is considerably greater than its dorso-ventral dimension. In *P. potto* and *G. crassicaudatus*, width measurements were nearly equal (11 mm.) and approximately 1 mm. greater than that of *N. coucang*. The width of the medulla in *G. senegalensis* is 8.3 mm. and of proportionate size to the other specimens.

List of abbreviations

AOT	Posterior accessory optic tract
AS	Aqueduct of Sylvius
A-SMV	Attachment of the superior medullary velum
BCI	Brachium of the inferior colliculus
BCS	Brachium of the superior colliculus

Cranial nerves

I	Olfactory (see O1. B, O1. T)
II	Optic (see ON, OC, OT)
III	Oculomotor
IV	Trochlear
Vm	Trigeminal, motor division
Vs	Trigeminal, sensory division
VI	Abducent
VII	Facial
VIIi	*Nervus intermedius*
VIIIc	Cochlear division
VIIIv	Vestibular division
IX	Glossopharyngeal
X	Vagus
XI	Spinal accessory
(XIb	bulbar rootlets)
(XIs	spinal rootlets)
XII	Hypoglossal

C-1	1st cervical nerve
CC	Corpus callosum
CI	Inferior colliculus
CP	Cerebral peduncle (crus cerebri in text)
CS	Superior colliculus
CT	*Corpus trapezoideum*
CuT	Cuneate tubercle
DCN	Dorsal cochlear nucleus
DCN-CS	Dorsal cochlear nucleus cut surface
FaC	Facial colliculus
FC	*Fasciculus cuneatus*
FG	*Fasciculus gracilis*
GT	Gracile tubercle
HT	Hypoglossal tubercle
ICP	Inferior cerebellar peduncle
LCT	Lateral cuneate tubercle
LF	Longitudinal fissure
LGB	Lateral geniculate body
LL	Lateral lemniscus
MCP	Middle cerebellar peduncle
ME	Median eminence of rhomboid fossa
MGB	Medial geniculate body
MI	*Massa intermedia*
MOR	Maxillo-ophthalmic ramus
O1	Olive
O1B	Olfactory bulb
O1T	Olfactory tract

OC	Optic chiasma
ON	Optic nerve
OT	Optic tract
P	Pituitary gland
PHR	Prehypoglossal region
PM	Ponto-medullary junction (Ponto-pyramidal + Ponto-trapezoidial grooves)
Pn	Pons
Pp	Palaeopallium
PrPn	Pre-trigeminal pons
PtPn	Post-trigeminal pons
Pul	Pulvinar
Py	Pyramids
RP	Reticular prominence
Sulci	
c	Sulcus c
ce	Central sulcus
cg	Cingulate sulcus (intercalary sulcus)
col	Collateral sulcus
cre	Rectus communicating
d	Sulcus d
di	Diagonal sulcus
e	Sulcus e
g	Genual sulcus
ip	Intraparietal sulcus
l	Lunate sulcus
lc	Lateral calcarine
o	Orbital sulcus
oct	Occipito-temporal sulcus
pc	Paracalcarine
r	*Sulcus rectus*
ra	Rhinal fissure anterior
rc	Retrocalcarine
rp	Rhinal fissure posterior
tm	Middle temporal sulcus
ts	Superior temporal sulcus
SCP	Superior cerebellar peduncle
SF	Superior fovea
SL	*Sulcus limitans*
ST-V	Spinal tract of fifth nerve
SyC	Sylvian complex
TG	Trigeminal ganglion
TT	Transpeduncular tract
VA	Vestibular area
VCN	Ventral cochlear nucleus
VT	Vagal tubercle (dorsal motor nucl. of Vagus)

Figure 1. Ventral view of the medulla and pons of *Galago crassicaudatus.* Note the exit of the abducent (VI) nerve, the size of the *corpus trapezoideum* and the distinct ponto-trapezoidial groove, the exit of the rootlets of the hypoglossal (XII) nerve, and the location of the reticular prominence. Scale = 1 mm.

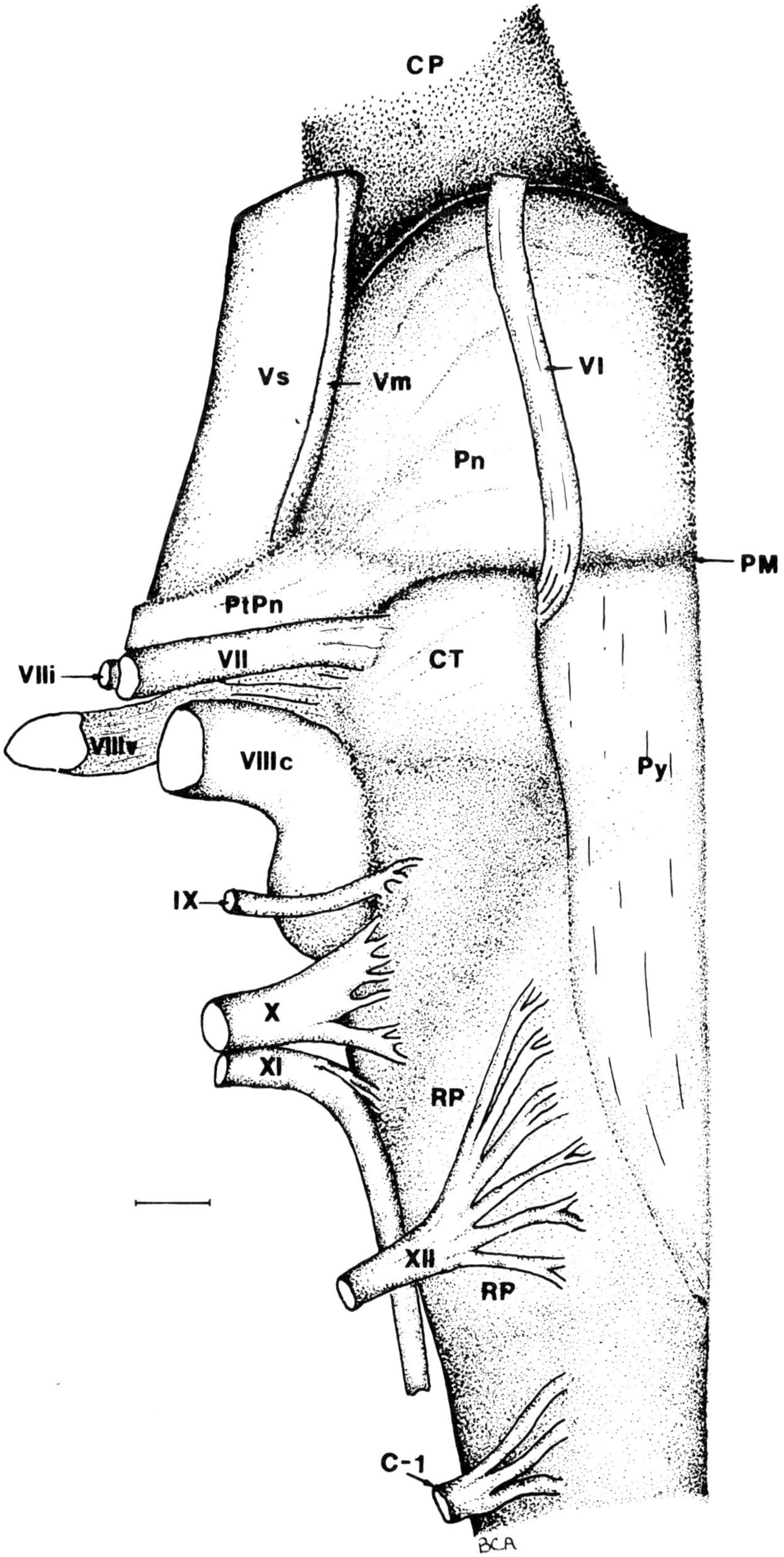

CP
Vs
Vm
VI
Pn
PM
PtPn
VII
CT
VIIi
VIIIv
VIIIc
Py
IX
X
XI
RP
XII
RP
C-1
BCA

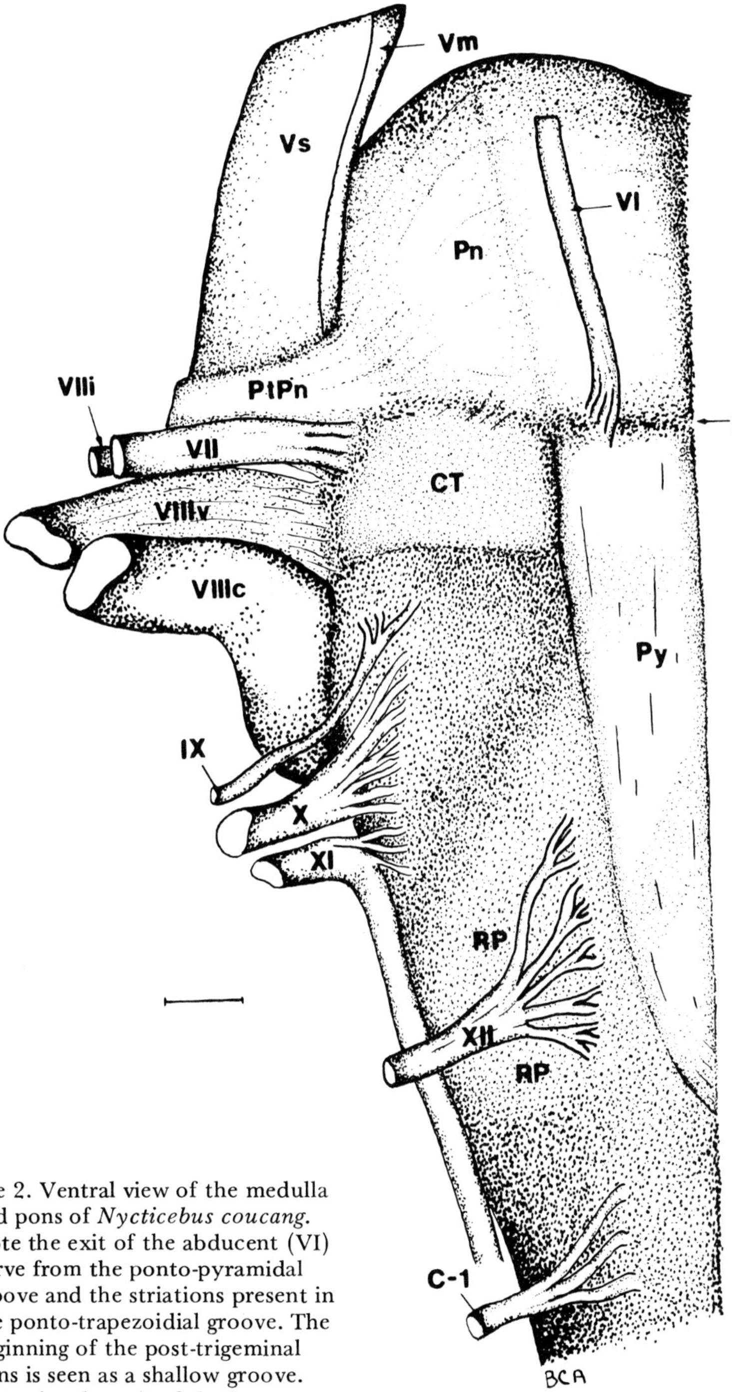

Figure 2. Ventral view of the medulla and pons of *Nycticebus coucang*. Note the exit of the abducent (VI) nerve from the ponto-pyramidal groove and the striations present in the ponto-trapezoidial groove. The beginning of the post-trigeminal pons is seen as a shallow groove. Note also the exit of the hypoglossal (XII) nerve and the location of the reticular prominence.

Figure 2(*a*). The spinal accessory (XI) nerve of *Nycticebus coucang* has distinct bulbar and spinal divisions. Scale = 1 mm.

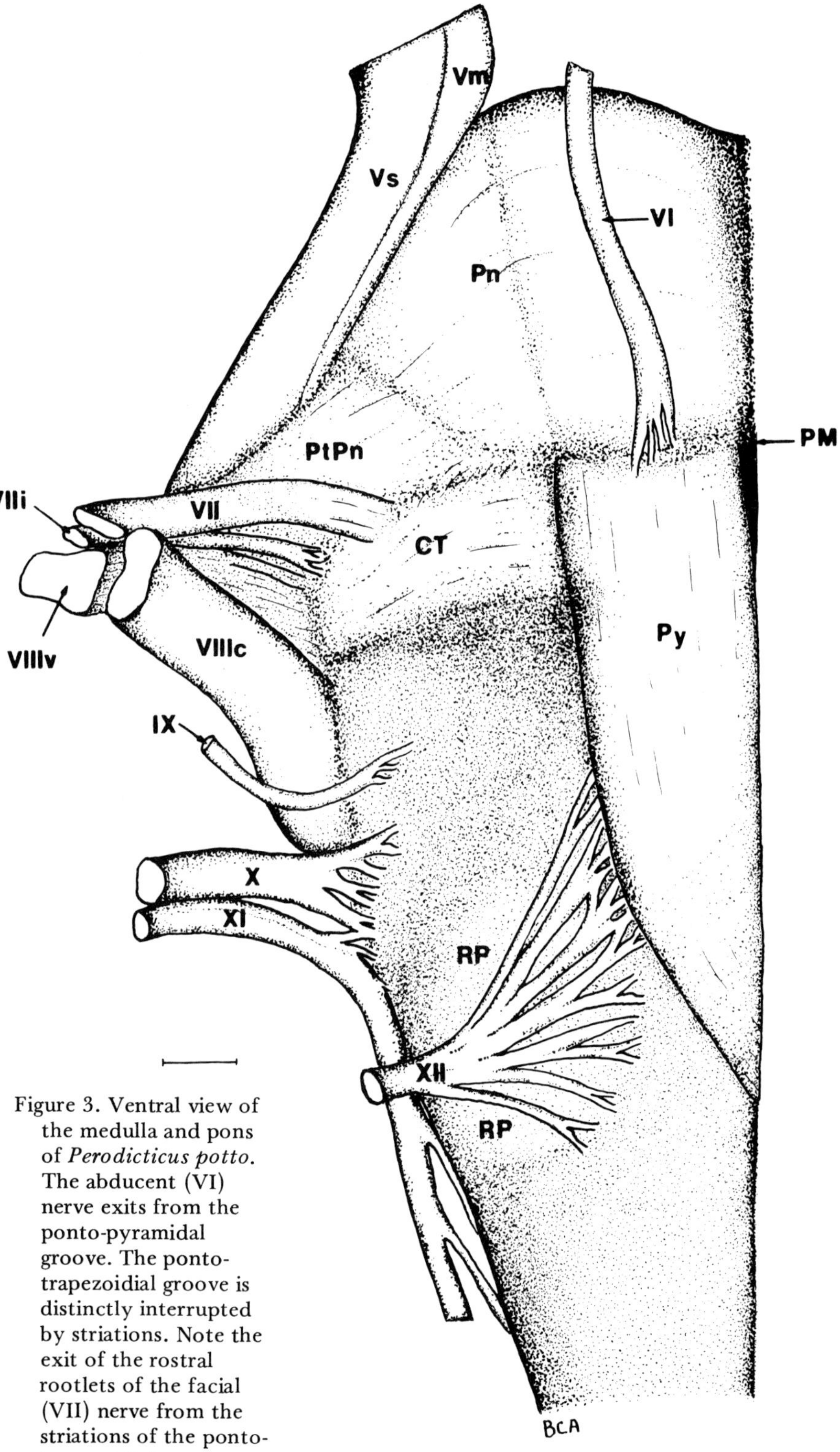

Figure 3. Ventral view of the medulla and pons of *Perodicticus potto*. The abducent (VI) nerve exits from the ponto-pyramidal groove. The ponto-trapezoidial groove is distinctly interrupted by striations. Note the exit of the rostral rootlets of the facial (VII) nerve from the striations of the ponto-trapezoidial groove. Scale = 1 mm.

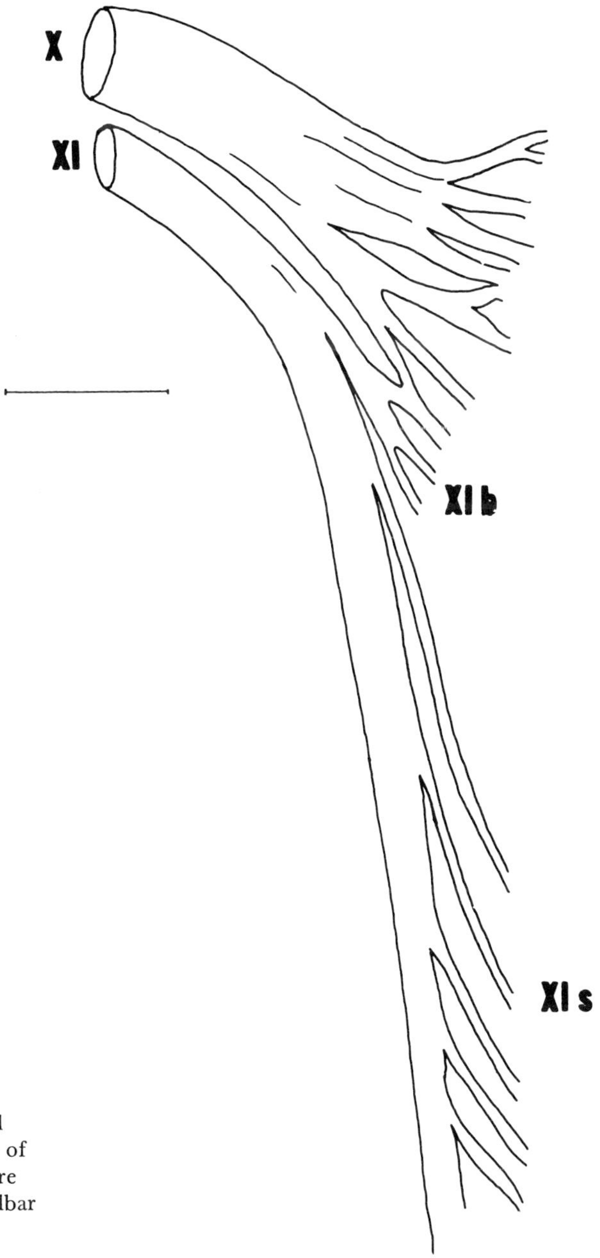

Figure 3(*a*). The spinal rootlets of the spinal accessory (XI) nerve of *Perodicticus potto* are distinct from the bulbar rootlets. Note the association of the bulbar rootlets of XI with the caudal rootlets of the vagus (X) nerve. Scale = 1 mm.

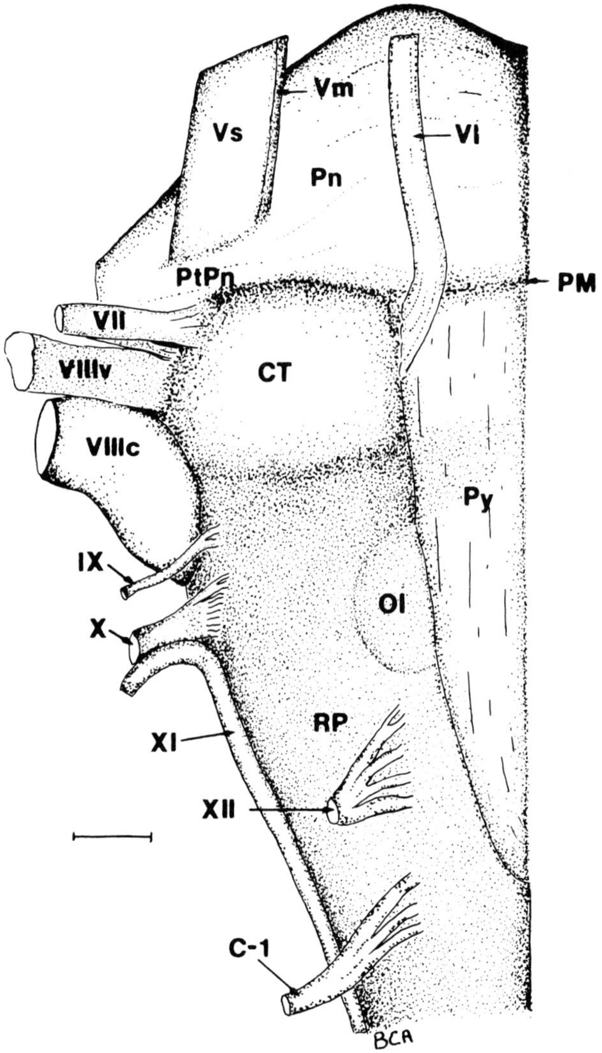

Figure 4. Ventral view of the medulla and pons of *Galago senegalensis.* Note the location of the olive and the size of the *corpus trapezoideum.* The rootlets of the abducent (VI) nerve exit caudal to the ponto-medullary groove and lateral to the pyramids. Scale = 1 mm.

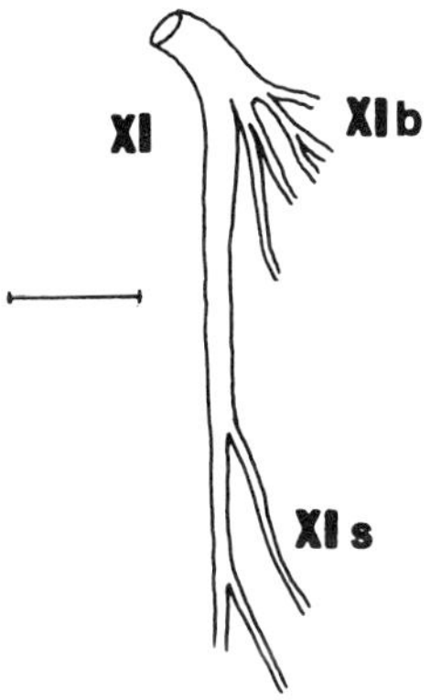

Figure 1(*a*). Rootlets of the spinal accessory (XI) nerve. Note the bulbar (XIb) and spinal (XIs) divisions. Scale = 1 mm.

Excluding *P. potto*, the pyramids are observed to have certain similarities. They are: a gradual uniform tapering, a less distinct lateral border in the region of the olive, and an occasional shallow depression caudal to the *corpus trapezoideum* (Figs. 1, 2, 4). The pyramids of *P. potto* are distinctly different in appearance. In addition to being shorter than *G. crassicaudatus* and *N. coucang*, they appear to be thicker and broader and to taper more abruptly near the region of the hypoglossal (XII) nerve (Fig. 3). The ventral elevation of the pyramids in *P. potto* (0.8 mm.) is 0.2 to 0.3 mm. greater than that measured in *G. crassicaudatus*. The lesser bushbaby and slow loris have similar pyramidal heights of approximately 0.4 mm. The pyramids in slow loris appear flatter and show a less definable ponto-pyramidal groove (Fig. 2). Just caudal to the ponto-pyramidal groove, the pyramids in *P. potto* are 2.5 mm. in width and approximately 0.5 to 0.6 mm. wider than in the greater bushbaby and slow loris respectively. In the lesser bushbaby, pyramidal width is 1.5 to 1.7 mm.

The abducent nerve (VI) is composed of 3-4 rootlets. In both galagos it exits from the lateral border of the pyramids caudal to the ponto-pyramidal groove (Figs. 1, 4). In *P. potto* and *N. coucang*, however, the rootlets exit through the pyramids at the level of the ponto-pyramidal groove (Figs. 2, 3).

The *corpus trapezoideum* is distinct and shows medial to lateral striations; its size and prominence vary considerably. It is similar in the galagos, with well-defined rostral and caudal borders, the rostral forming the ponto-trapezoidial groove. Of the prosimians studied, *G. senegalensis* demonstrates the largest *corpus trapezoideum*, measuring 2.5 mm. in rostrocaudal length (Fig. 4). This measurement is only 0.2 mm. less than that in *G. crassicaudatus* (Fig. 1). In *N. coucang* and *P. potto* it is considerably less prominent, and measures 2 mm. and 1.6 to 1.7 mm., respectively (Figs. 2, 3). In association with the decrease in width, striations appear in the ponto-trapezoidial groove.

The motor root of VII is formed by the union of 3 to 4 short rootlets. Except for *P. potto*, the nerve exits distinctly through the rostral region of the *corpus trapezoideum.* In *P. potto*, the rostral rootlets exit from the striations of the ponto-trapezoidal groove (Fig. 3). The smaller *nervus intermedius*, composed of 2 to 3 fine rootlets, exits through the *corpus trapezoideum* dorsal to the main motor root. The divisions of the vestibulo-cochlear nerve (VIII) are easily distinguished (Figs. 1-4). The vestibular division is striated and exits lateral to the *corpus trapezoideum.* This component is similar in width to that of the cochlear division, yet characteristically flatter. The truncated cochlear division is easily traced toward the cochlear nuclei. It is located caudal to the vestibular division and passes ventral to it. The glossopharyngeal (IX) nerve is small and exits as 2 to 3 fine rootlets ventral to the tuberculum quintae. These rootlets are oriented mediolaterally and in *P. potto* and *N. coucang* appear to be separated into one medial and two lateral rootlets. The vagus (X) nerve exits caudal to the glossopharyngeal nerve and is formed from easily observable rootlets. These rootlets vary from 5 to 6 in *G. senegalensis* (Fig. 4) to approximately 10 in *P. potto* (Figs. 3, 3*a*). In *P. potto* and *N. coucang*, the rootlets have a slightly longer course prior to forming the common trunk. In addition, the rootlets of X in *P. potto* are divisible into dorsal and ventral groups. The ventral roots are shorter and fewer in number. The bulbar and spinal divisions of the accessory nerve (XI) are distinct due to an interval between the two groups of rootlets. The bulbar division is composed of 3 to 6 rootlets, the spinal division 4 to 6 rootlets. In all genera, considerable variation in the number and arrangement of the bulbar rootlets is observed (Figs. 2*a*, 3*a*, 4*a*). These rootlets exit from the postero-lateral surface and are often difficult to observe by direct ventral observation.

In all genera a distinctly raised area, the reticular prominence (see Discussion), is observed on the ventro-lateral medullary surface (Figs. 1-3). In *P. potto*, *N. coucang*, and *G. crassicaudatus* the raised area is approximately 3.8 to 4.0 mm. in length and 2.0 mm. in width. In *G. senegalensis* the area is less definable and accurate measurements are not possible.

In *P. potto* and *G. crassicaudatus* the olives are difficult to identify. They appear to be oriented longitudinally and located adjacent to the pyramids, approximately 1.5 mm. rostrolateral to the pyramidal decussation (Figs. 5*b*, 5*c*). This slightly raised area, approximately 1 mm. in width, is distinguished from the above described reticular prominence by a shallow post-olivary groove. The

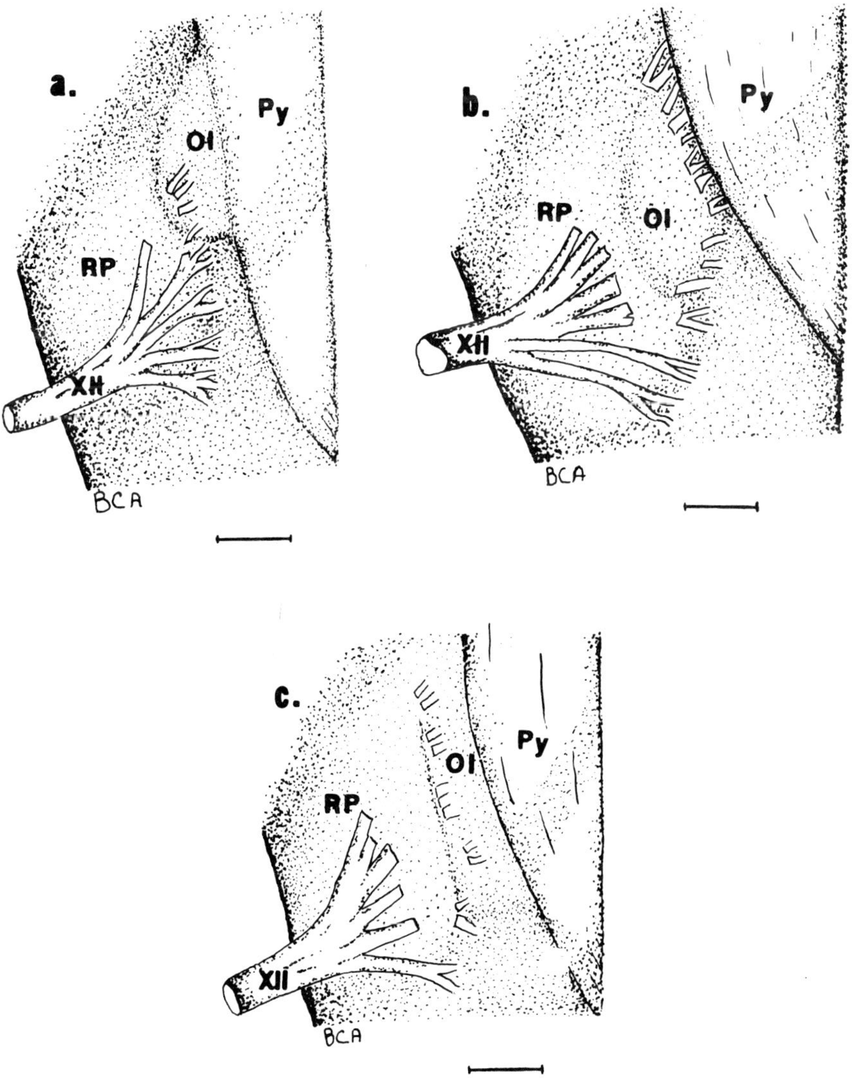

Figure 5. Ventral view of the caudal medulla of *Nycticebus coucang* (*a*), *Perodicticus potto* (*b*), and *Galago crassicaudatus* (*c*). Note the difference in location and shape of the olives. The olive (O1) is distinguished from the reticular prominence (RP) by a shallow postolivary groove. Scale = 1 mm.

olives in *N. coucang* and *G. senegalensis* are similar in appearance (Figs. 4, 5*a*) and are located adjacent to the pyramids, but further rostral than in *P. potto* and *G. crassicaudatus*. Although equal in width (1 mm.), length measurements are 1.5 to 1.7 and 2.5 mm. in *G. senegalensis* and *N. coucang* respectively. The lateral borders of the pyramids are slightly raised and appear to cover the medial aspect of the olive.

P. potto shows an elaborate arrangement of hypoglossal nerve rootlets (Figs. 3, 5*b*). As many as 15 small primary rootlets are observed to join immediately in the formation of 7 to 8 secondary rootlets. The rostral rootlets exit from the lateral border of the pyramids from the preolivary groove. Caudally, the rootlets appear to follow the contour of the reticular prominence. Due to an ill-defined olivary eminence, it is not possible to determine the exact relationship between the exit of XII and the position of the olive. The number, arrangement and location of exit of hypoglossal rootlets in *G. crassicaudatus* and *N. coucang* are similar despite the difference in location of the olive (Figs. 1, 2, 5). In these Lorisidae, 12 to 13 primary rootlets join to form 6 to 7 secondary rootlets. In *G. crassicaudatus* they exit from the mid-longitudinal region of the olive (Fig. 5*c*). In *N. coucang* only the more rostral rootlets exit from the olive, the caudal rootlets follow the reticular prominence (Fig. 5*a*). In *G. senegalensis* the hypoglossal nerve is composed of 4 to 5 rootlets. All the rootlets exit caudal to the olive, from the ventrolateral surface of the medulla (Fig. 4).

Pons

In all genera, the pons appears to be well-developed, having the common features of: a distinct basilar groove, cross-pontine striations, and an obliquely-exiting trigeminal nerve root. The trigeminal nerve divides the lateral pontine surface (middle cerebellar peduncle) into pre- and post-trigeminal bundles (Figs. 1-4, 6). In those animals of comparable brain size, measurements of the rostrocaudal length are equal (5 mm.). In *P. potto* (Fig. 3), however, the pontine width (9 mm.) is approximately 1 mm. greater than that observed in *N. coucang* and *G. crassicaudatus*. In *P. potto* the pons appears thick and its anterior border is abrupt. In the other genera the pons is less pronounced and slopes rostral gradually. The ventrolateral surface of the pons in the galagos is smooth (Figs. 1, 6), except for a shallow groove running caudomedially from the exit of V in *G. crassicaudatus*. This groove, also seen in *N. coucang* and *P. potto*, marks the beginning of the post-trigeminal pons. In *P. potto* a definite ridge is observed on the pontine surface which appears to join the lateral border of the pyramids with that of the cerebral peduncles. There is only a slight indication of this ridge in *N. coucang*. None of the genera exhibited oblique striations that might be indicative of

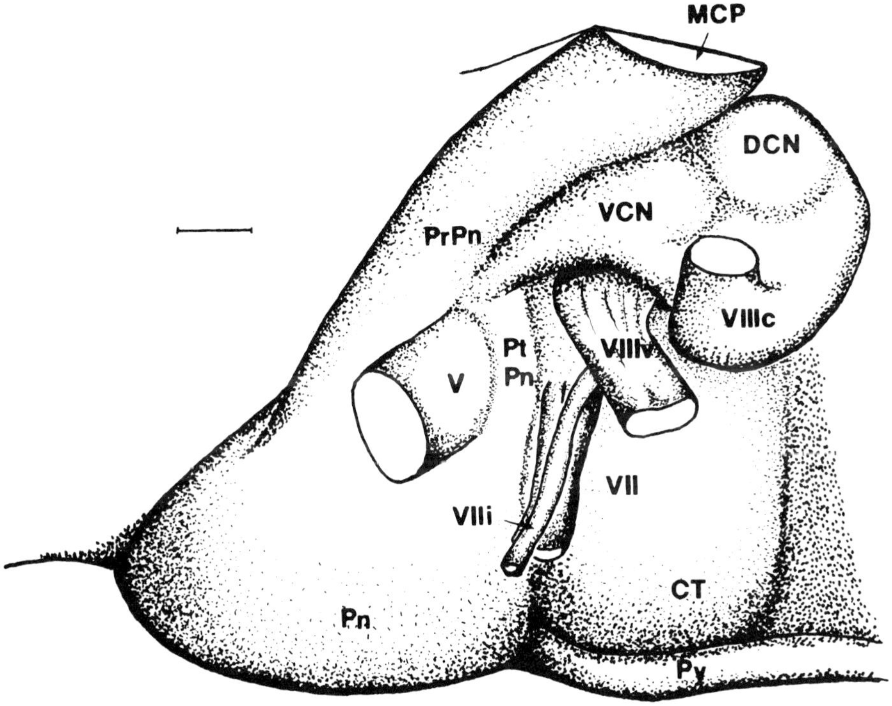

Figure 6. Lateral view of medulla and pons of *Galago senegalensis.* Note the exit of the vestibular division of the vestibulocochlear nerve (VIIIv) and the relative size of the pre-(PrPn) and post-(PtPn) trigeminal pons. Scale = 1 mm.

post-trigeminal pontine fibres originating from rostral pontine areas.

At the dorsal limits of the exit of the fifth nerve, the post-trigeminal pons is approximately half the width of the pre-trigeminal bundle. The post-trigeminal pons in *G. senegalensis* is relatively uniform in width throughout its course (Fig. 6). In contrast, the post-trigeminal pons in *P. potto* shows a greater amount of tapering, which corresponds to the broader width at its ventral origin.

The fifth nerve is relatively large, and both motor and sensory divisions are easily distinguishable. Considerable variation is noted in the number and origin of the rootlets which compose the smaller motor component. The trigeminal ganglion is associated with the ventro-caudal aspect of the pyriform lobe (see later: Figs. 32-4). The motor component of V passes laterally toward the mandibular division. Rostral to the ganglion, the maxillary and ophthalmic divisions occur as a single branch, the maxillo-ophthalmic ramus (Figs. 32-4). This large ramus courses medial to the pole of the pyriform lobe and divides before leaving the skull.

2. Rhomboid fossa

The rhomboid fossa is similar in appearance and proportionately equal in size (Figs. 7-10). The rostral and caudal extensions are separated by a distinct vestibular area. The approximate length and width of .the fossa in *N. coucang, P. potto* or *G. crassicaudatus* is 7.0 to 7.5 mm. by 8.0 mm. In the rostral portion, the facial colliculus and median eminence are separated by a shallow groove which courses from the midline to the superior fovea (Figs. 7, 10). A faint *sulcus limitans* is present medial to the large vestibular area (Figs. 7, 9). The continuation of the sulcus cannot be accurately followed. The inferior fovea can only be identified as a region. The caudal region is characteristically bordered by the obex and two eminences which correspond to those formed by the hypoglossal and external motor nucleus of the vagus (Figs. 8, 10). In *N. coucang* and *P. potto* a pre-hypoglossal region is observed (Figs. 8, 9). The attachments of the superior and inferior medullary vela around the fossa are typical of higher primates.

3. Dorsal medulla

Fasciculi graciles and *cuneati* are easily distinguished. They are separated by the posterior intermediolateral groove (Figs. 7-10). The *fasciculus cuneatus* is bordered laterally by the posterolateral groove, which terminates rostrally at the caudal border of the inferior cerebellar peduncle. In upper cervical and lower medullary regions, the *fasciculus gracilis* is approximately 1/4 the size of *fasciculus cuneatus* (Figs. 7, 8). Both *fasciculi* expand in width as they approach their respective tubercles. This is more evident in the *fasciculus gracilis* (Fig. 8). The posterior intermedio-lateral groove broadens to separate gracilis and cuneate tubercles. This groove is less evident in *N. coucang.* The tubercle *gracilis* in *N. coucang, P. potto,* and *G. senegalensis* is approximately half the size of tubercle *cuneatus* (Figs. 8-10). In *G. crassicaudatus,* the tubercle *gracilis* appears similar in size to that of cuneate tubercle (Fig. 7). Except in *G. senegalensis,* a small eminence is observed rostral to the cuneate tubercle (Figs. 7-9). It may identify the location of the lateral cuneate nucleus. The rostral border of both tubercles serves for the attachment of the inferior medullary velum.

The cochlear nuclei are difficult to distinguish. The dorsal nucleus is usually observed extending to variable distances over the surface of the vestibular area. The inferior cerebellar peduncle courses rostrally from the lateral aspect of the medulla, where it passes deep to the cochlear nuclei (Figs. 8, 10). Apart from being easily traced and serving for the attachment of the superior medullary velum, the superior cerebellar peduncle is separated from the middle cerebellar peduncle by a groove.

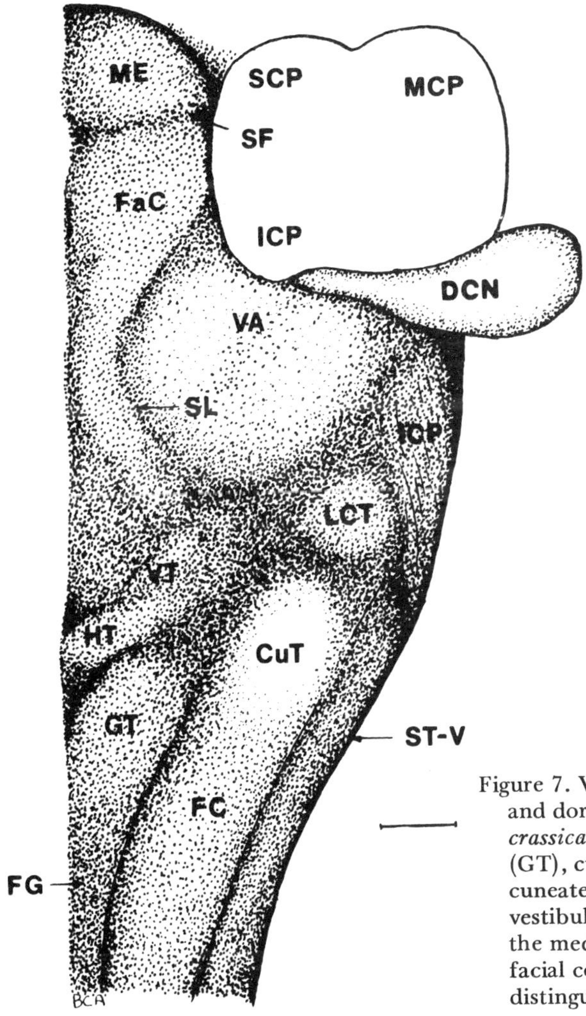

Figure 7. View of rhomboid fossa and dorsal medulla of *Galago crassicaudatus*. Note the gracile (GT), cuneate (CT) and lateral cuneate (LCT) tubercles. The vestibular area (VA) is large and the median eminence (ME) and facial colliculus (FaC) are easily distinguished. Scale = 1 mm.

Mesencephalon and diencephalon

1. Dorsal view

The inferior colliculi appear as dorsally directed structures whose apices are located more laterally than those of the superior colliculi. A series of shallow grooves appear on the caudal face of the inferior colliculus. They are formed by the folia of the anterior lobe of the cerebellum (Figs. 11-14). The trochlear (IV) nerve, after its exit from the medullary velum, traverses the caudolateral surface of the inferior colliculus in a dorsal oblique direction (Figs. 12, 14). Upon

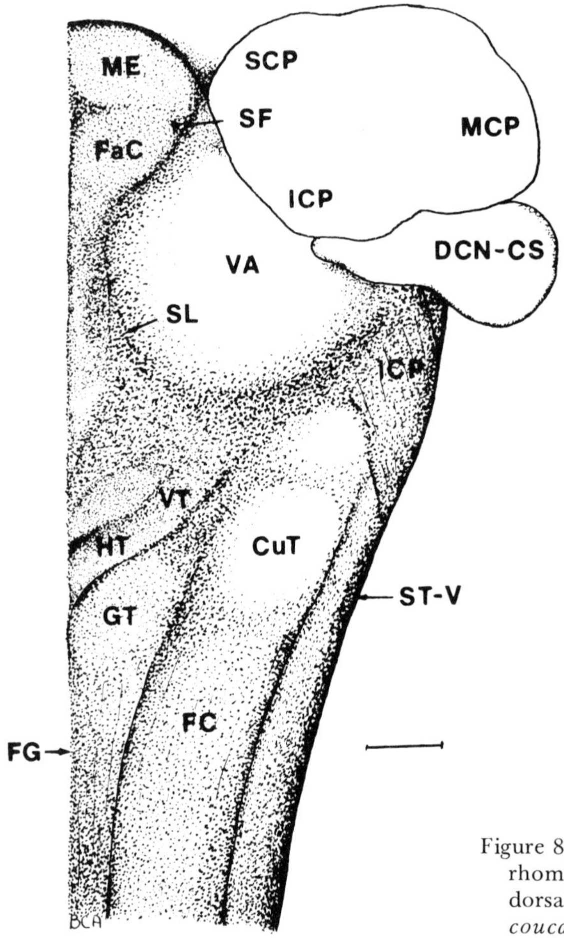

Figure 8. Dorsal view showing rhomboid fossa and tubercles of dorsal medulla in *Nycticebus coucang*. Scale = 1 mm.

reaching the lateral surface of the colliculus, the nerve turns ventrally to travel with the maxillo-ophthalmic ramus of the trigeminal nerve.

The superior colliculi are large, somewhat flattened, oval structures (Figs. 12, 13, 15). The inferior and superior colliculi are separated by the intercollicular groove (sulcus) which is narrow in the lesser galago (Figs. 14, 15),[20] the colliculi abutting along the greater part of their common borders. The groove appears somewhat wider in both the greater galago and slow loris (Figs. 11, 12), and reaches its greatest dimension, 2 to 3 mm., in the potto (Fig. 13).

2. *Lateral view*

The lateral lemniscus is observed to course in a slightly dorsal oblique

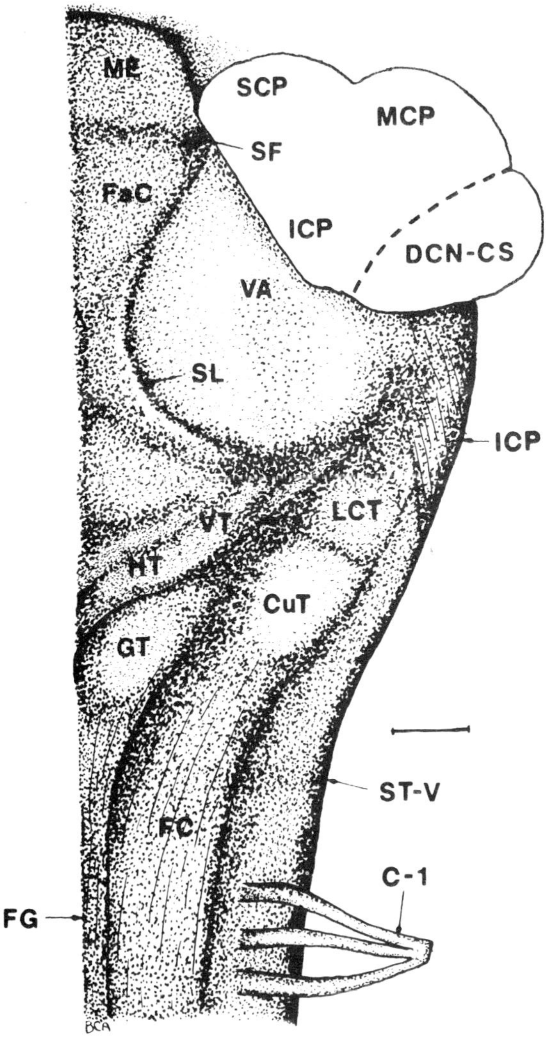

Figure 9. Dorsal view of rhomboid fossa and dorsolateral medulla in *Perodicticus potto*. Scale = 1 mm.

direction across the ventrolateral surface of the inferior colliculus (Figs. 11-14). The brachium of the inferior colliculus leaves the rostrolateral surface and passes obliquely in an anteroventral direction to join the medial geniculate body (Figs. 11, 12, 15).

The medial geniculate body is a distinct rounded tubercle located lateral to the superior colliculus. The medial-lateral diameter is greatest in the greater galago (Fig. 11), (2.8 mm.), followed by the potto, slow loris and lesser galago (2.6, 2.4, 2.3 mm. respectively) (Figs. 13-15). The dorsoventral diameter is greatest in the potto, 3.2 mm., while in the other animals it is 2.5 mm.

The lateral geniculate body appears as a somewhat flattened,

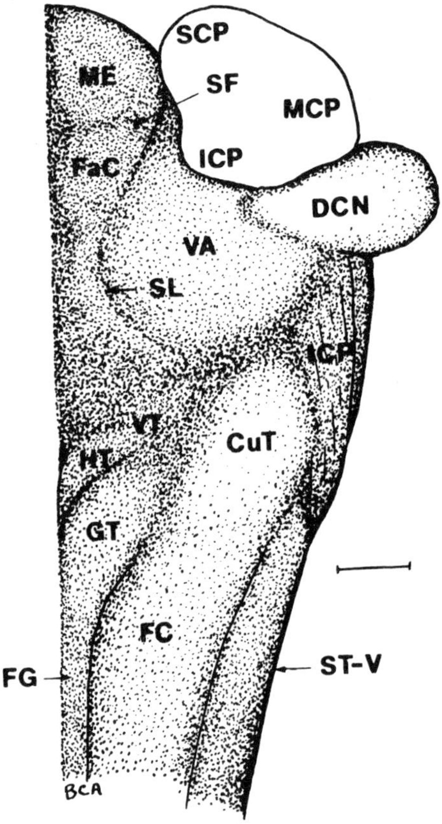

Figure 10. Rhomboid fossa and dorsal medulla in *Galago senegalensis.* Note the size of the vestibular area (VA) and the location of the inferior cerebellar peduncle (ICP). Scale = 1 mm.

crescentic structure located rostrolateral to the medial geniculate body. It attains its greatest medial-lateral width in the potto (Fig. 13), 2.8 mm., while in the greater galago (Fig. 11) it is 2.4 mm.; in the slow loris and lesser galago it is 2.2 mm. at its widest point (Figs. 12, 14, 15). The dorsal-ventral extent is 4.5 mm. in both the potto and lesser galago, while in the greater galago and slow loris it is 4.2 mm. and 3.0 mm. respectively.

Passing laterally from the superior colliculus to join the dorsomedial border of the lateral geniculate body is the striated brachium of the superior colliculus (Figs. 11-14). A deep groove is observed between the ventrolateral aspect of the superior colliculus and the brachium. There is no distinct margin noted between the brachium and the lateral geniculate body, nor at the ventral junction of the lateral geniculate body and optic tract.

3. *Ventral view*

Figs. 16-19 are ventral views of the mesencephalon and diencephalon of the animals used in this study. The interpeduncular fossa at the anterior border of the pons is widest (3.8 to 4.0 mm.) in the lesser galago (Fig. 19), being almost 1½ times to twice as wide as in the other animals (Figs. 16-18). The medial-lateral extent of the crus cerebri at this point is 2.0 to 2.1 mm. in the lesser galago, 3.2 mm. in the potto and slow loris, and 3.0 to 3.1 mm. in the greater galago. (The medial-lateral measurement did not take the curvature of the crura into account.)

The transpeduncular tract is observed as a dorso-ventrally oriented elevation on the crus cerebri. Fibres of the crus appear to pass either over or through the tract. It is 1.0 mm. in width in the greater galago and potto (Figs. 16, 18), and is slightly narrower (0.7 to 0.8 mm.) in the lesser galago and slow loris (Figs. 17, 19). At the dorsal limit of the crus, the tract continues as the posterior accessory optic tract onto the lateral surface of the mesencephalon, coursing caudally around the medial geniculate body (Figs. 16, 17). Using gross morphological techniques, it cannot be determined whether the tract joins the superior colliculus or the brachium of the superior colliculus. A discrete bundle of fibres is observed ventral to the lateral surface of the medial geniculate body. It is probable that this bundle represents portions of the accessory optic system. The optic tract flattens as it courses caudodorsally from the optic chiasma to terminate in the lateral geniculate body.

The 5 to 7 rootlets of the oculomotor (III) nerve exit from the interpeduncular fossa in all the Lorisidae studied except *N. coucang* (Figs. 16, 18, 19). The oculomotor nerve rootlets of individuals of this species exit from the interpeduncular fossa and from the crus cerebri (Fig. 17). At least 2 rootlets are observed to exit bilaterally from the crus in each animal.

4. *Midsagittal view*

The height of the colliculi above the roof of the aqueduct is greatest in *G. crassicaudatus* (inferior colliculus, 3.8 mm.; superior colliculus, 3.6 mm.). In the greater galago, the rostrocaudal length of the inferior colliculus is greatest, while this parameter of the superior colliculus is increased in *P. potto* (3.4 mm.) as compared to the other animals.[26]

The height of the cerebral aqueduct varies among the different Lorisidae examined, being greatest in the greater galago. A shallow groove is noted in the lateral wall of the aqueduct in both galagos.

The oval to circular massa intermedia (or commissure mollis) is large in all animals.[26] Its long axis is 3.7 mm. in *G. senegalensis*, 5.7 mm. in *G. crassicaudatus* and *N. coucang*, and 6.0 mm. in *P. potto*.

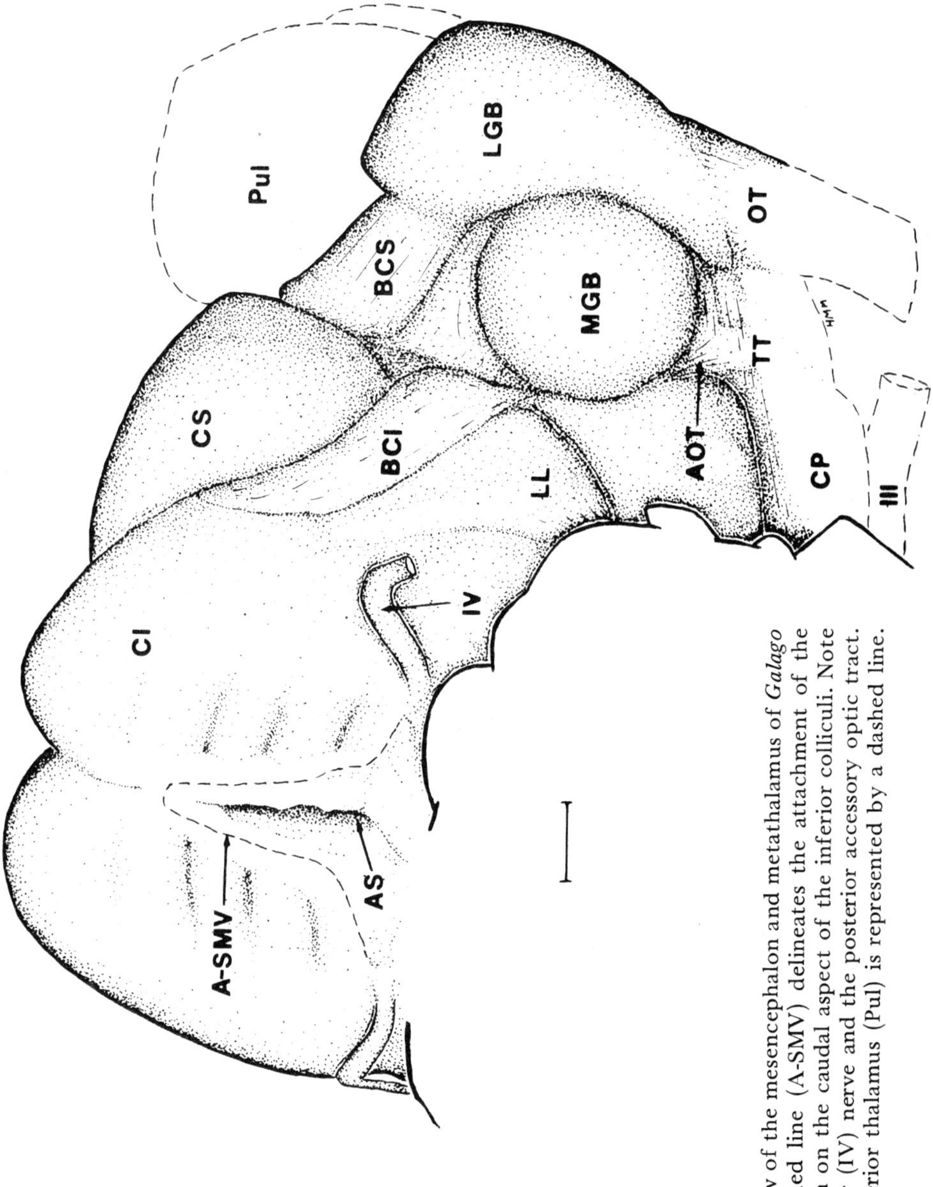

Figure 11. Caudolateral view of the mesencephalon and metathalamus of *Galago crassicaudatus.* The dashed line (A-SMV) delineates the attachment of the superior medullary velum on the caudal aspect of the inferior colliculi. Note the exit of the trochlear (IV) nerve and the posterior accessory optic tract. Cut surface of the posterior thalamus (Pul) is represented by a dashed line. Scale = 1 mm.

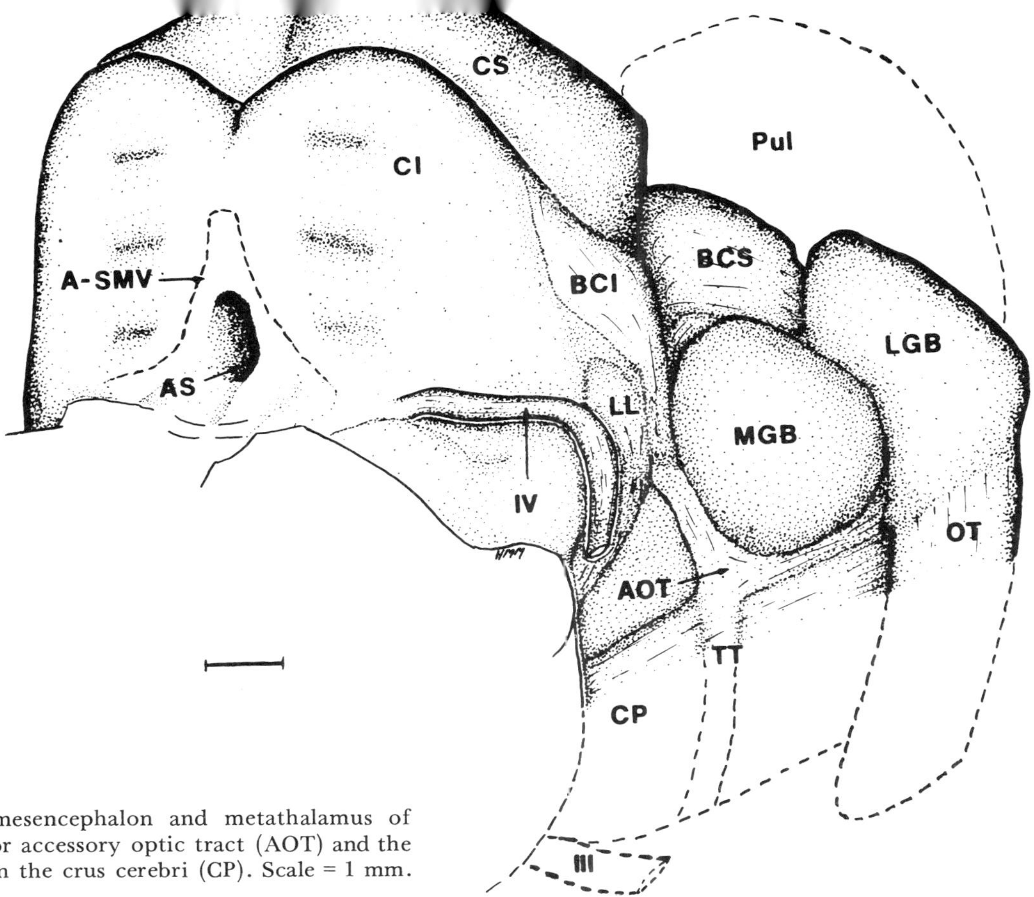

Figure 12. Caudolateral view of the mesencephalon and metathalamus of *Nycticebus coucang.* Note the posterior accessory optic tract (AOT) and the transpeduncular tract (TT) coursing on the crus cerebri (CP). Scale = 1 mm.

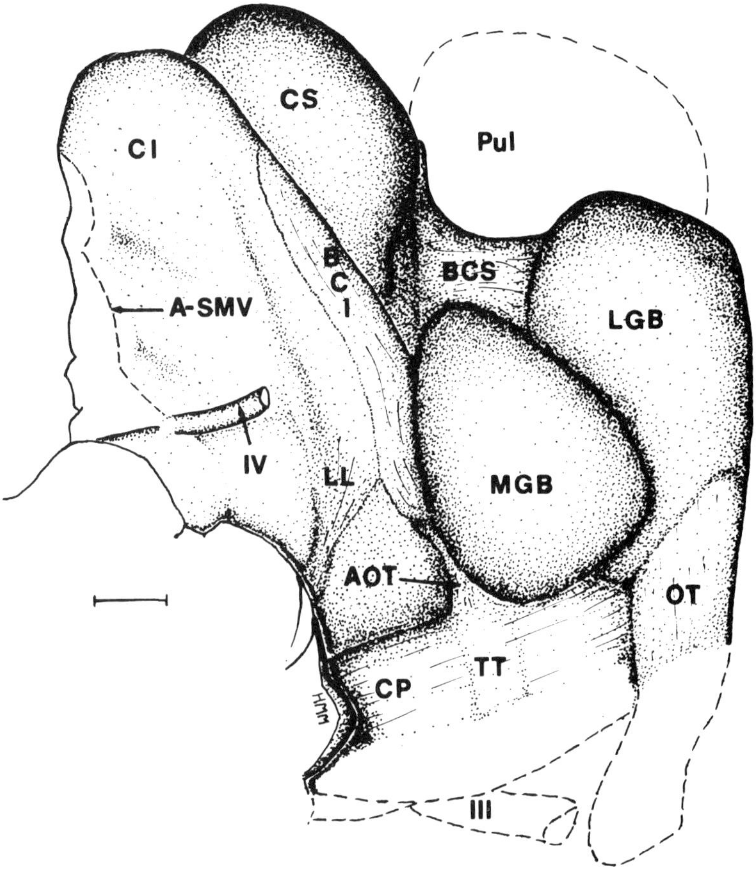

Figure 13. Lateral view of the mesencephalon and metathalamus of *Perodicticus potto* from a slightly caudal aspect. The medial geniculate body (MGB) and lateral geniculate (LGB) are distinct metathalamic structures located on the lateral aspect of the mesencephalon. Scale = 1 mm.

The pituitary gland (see later – Figs. 32, 33) is large in the greater galago, having a rostrocaudal measurement of 5.5 mm. and a dorsoventral measurement of 2.6 mm. The diameter of the gland at its widest point is 4.0 mm. The pituitary of the potto and slow loris is somewhat smaller, while that of the lesser galago is the smallest (rostrocaudally, 3.0 mm.; dorsoventrally, 2.0 mm.; diameter at widest point, 3.2 mm.).

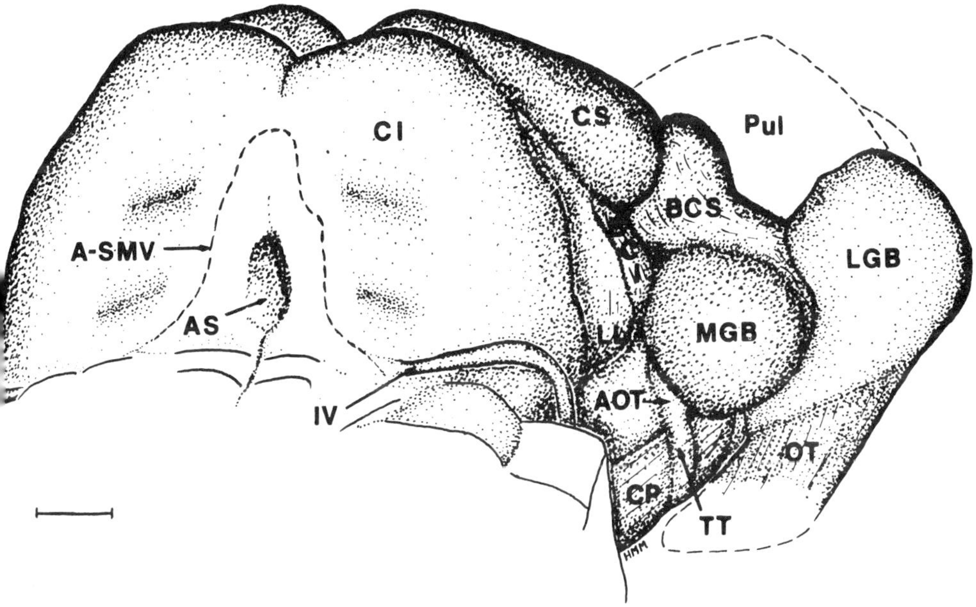

Figure 14. Caudolateral view of the mesencephalon and metathalamus of *Galago senegalensis.* Note the course of the trochlear (IV) nerve, the transpeduncular tract, and the relatively large size of the lateral geniculate body (LGB). Scale = 1 mm.

Telencephalon

1. Dorsolateral surface

The sulcal pattern on the dorsolateral surface of the specimens examined increased in number and complexity in the following order: *G. senegalensis, G. crassicaudatus, N. coucang*, and *P. potto* (Figs. 20-3). These patterns are discussed in relation to the sylvian complex, i.e. sulci that are dorsal, rostral and caudal to the complex.

An arcuate sulcus (intraparietal s.) is evident dorso-mesially from the termination of the sylvian complex. This indentation varies in the lesser galago from a pit to a sulcus not longer than 3 mm. In all specimens except the slow loris the sulcus curves around the termination of the sylvian complex (Figs. 20-3). The sylvian complex in slow loris appears to extend dorso-mesially to within 3 mm. of the longitudinal fissure. This extension is the intraparietal sulcus. Between this arcual sulcus and the midline (Fig. 26), a sagittal sulcus (sulcus *c*) begins approximately 12 mm. from the frontal pole and extends not more than 6 mm. toward the occipital pole. This sulcus is slightly concave toward the midline.

Rostral to the sylvian complex a variety of grooves and sulci are

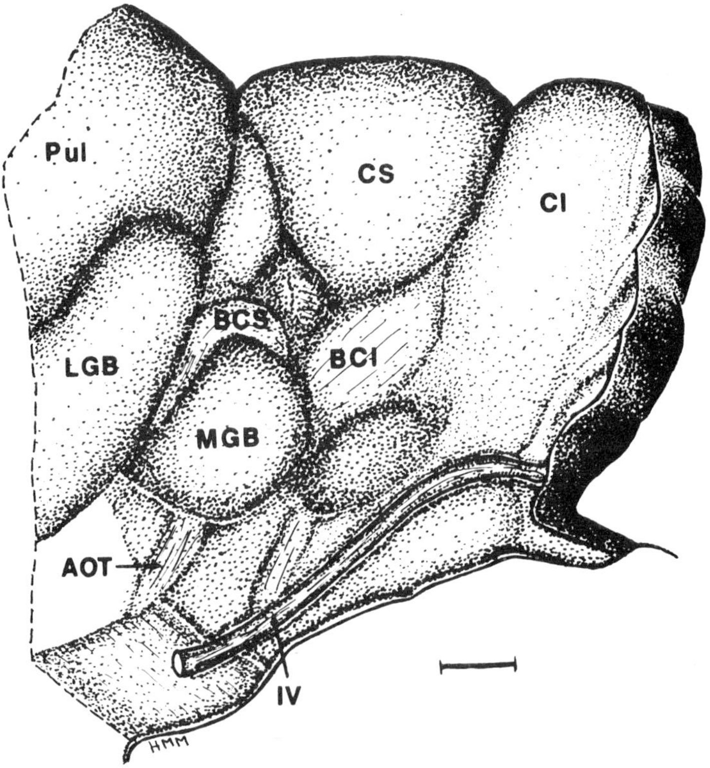

Figure 15. Lateral view of the mesencephalon and metathalamus of *Galago senegalensis.* Note the relative size of the superior colliculi (CS), the lateral geniculate body (LGB), and the indentations present on the caudal face of the inferior colliculi (CI). Dashed line represents cut surface. Scale = 1 mm.

seen. A pit is sometimes seen in the lesser galago, whereas definite sulci are present in the other Lorisidae. A straight sulcus (s. rectus) occurs caudal to the frontal pole. This sulcus is absent in lesser galago (Fig. 24) whereas variations are numerous in the slow loris, the sulcus being either straight, arcuate or triradiate. A sulcus just rostral to the sylvian complex is seen in the slow loris and greater galago. This sulcus (sulcus *e*) is large and arcuate in the slow loris, but small and variable in the greater galago (Figs. 25, 26). In the majority of our specimens of these animals, there is a communicating groove (rectus communicating s.) connecting the straight sulcus (s. rectus) with sulcus *e*. Rarely does the straight sulcus extend inferiorly to the orbital surface of the frontal pole. There is a central sulcus approximately midway between the frontal pole and sylvian complex in the potto (Figs. 23 and 27). It begins approximately 4 mm. from the midline and extends, in the coronal plane, inferolaterally for 11 mm. On the right hemisphere, the central sulcus continues as a

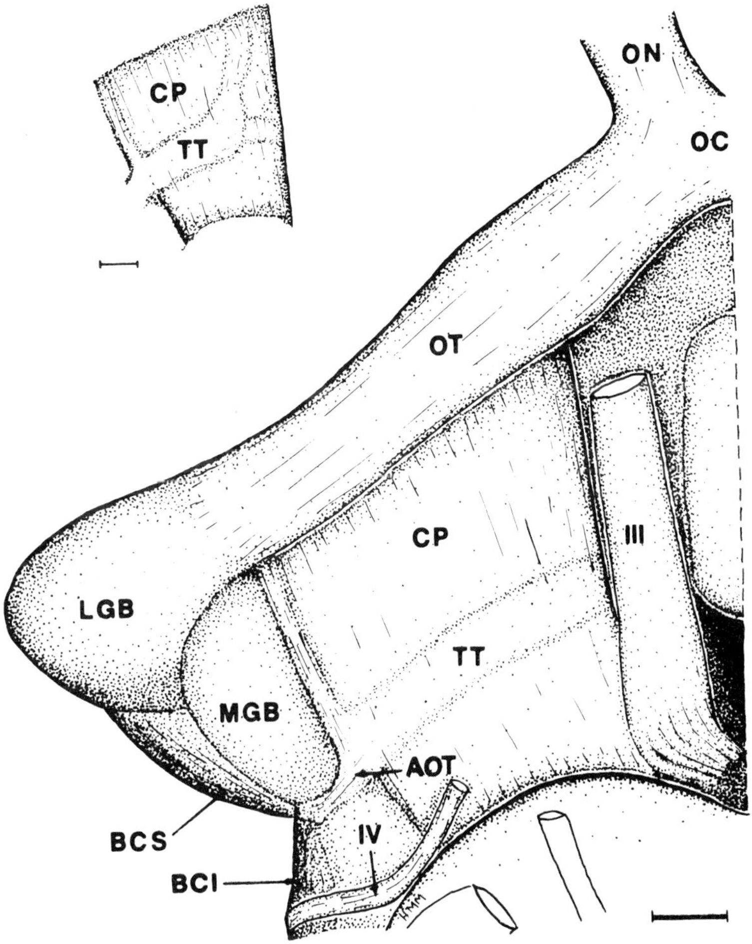

Figure 16. Ventral view of the mesencephalon and diencephalon of *Galago crassicaudatus*. The rootlets of the oculomotor (III) nerve exit through the interpeduncular fossa. Scale = 1 mm.

Inset: Variation in the course of the transpeduncular tract noted in one animal. Scale = 1 mm.

groove to join a caudal pit located approximately 3 mm. from the midline.

Caudal and parallel to the sylvian complex there is a groove or a sulcus (superior temporal s.). The lesser galago consistently shows a gyrus rostral and caudal to this groove, whereas slow loris and potto exhibit a more rounded temporal area (Figs. 20-3). The dorsal termination of the superior temporal sulcus is consistently more complex than its inferior terminus. In the slow loris, the dorsal terminus is a groove (sulcus *d*) which extends toward the sylvian complex (Fig. 26). In the potto and slow loris the area also exhibits a

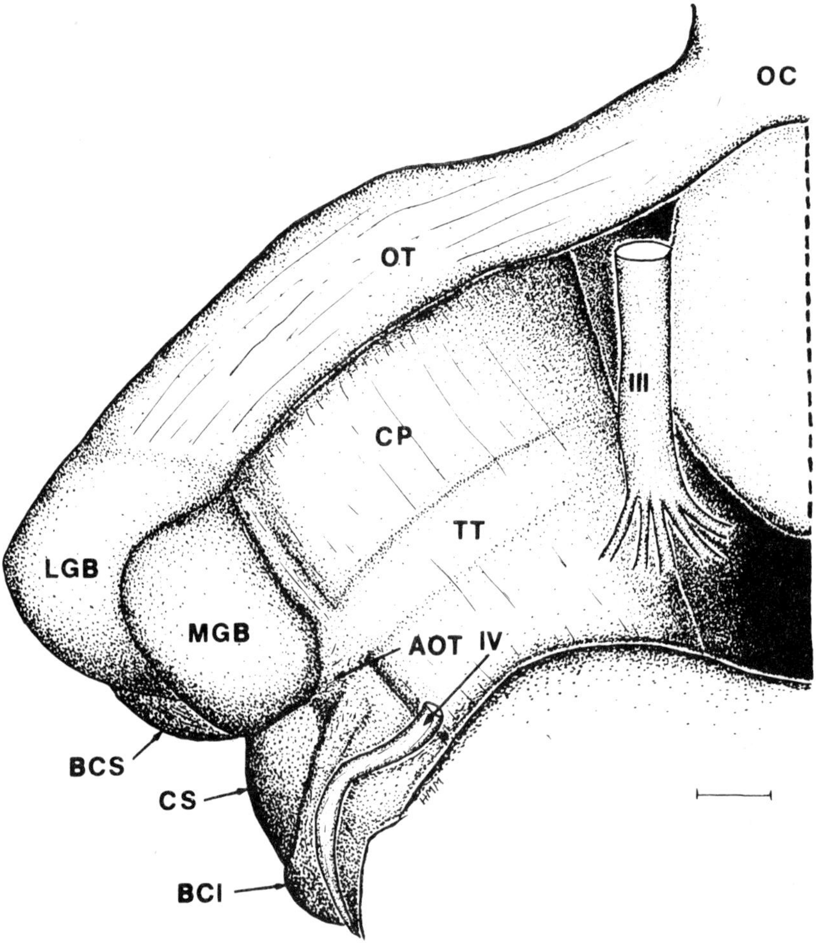

Figure 17. Ventral view of the mesencephalon and diencephalon of *Nycticebus coucang.* Note the rootlets of the oculomotor (III) nerve as they exit from the interpeduncular fossa and through the crus cerebri. Scale = 1 mm.

deep triradiate pit. In the right hemisphere of the potto, a groove arches rostrally to the sylvian complex. Caudally, this groove extends not more than 3 mm. toward the occipital pole (Fig. 23).

Sulci associated with the occipital region are consistently seen only in the slow loris. A sulcus in the coronal plane (lunate s.) indents the superomedial border and extends laterally from 3 to 7 mm. (Fig. 22). A similar groove is evident in both galagos 5 to 6 mm. rostral to the occipital pole (Fig. 32*b*). In one specimen of slow loris, this sulcus was found to be continuous with a groove which extends over the infero-lateral border on to the tentorial surface. A similar pattern is observed in the potto (Fig. 34*c*).

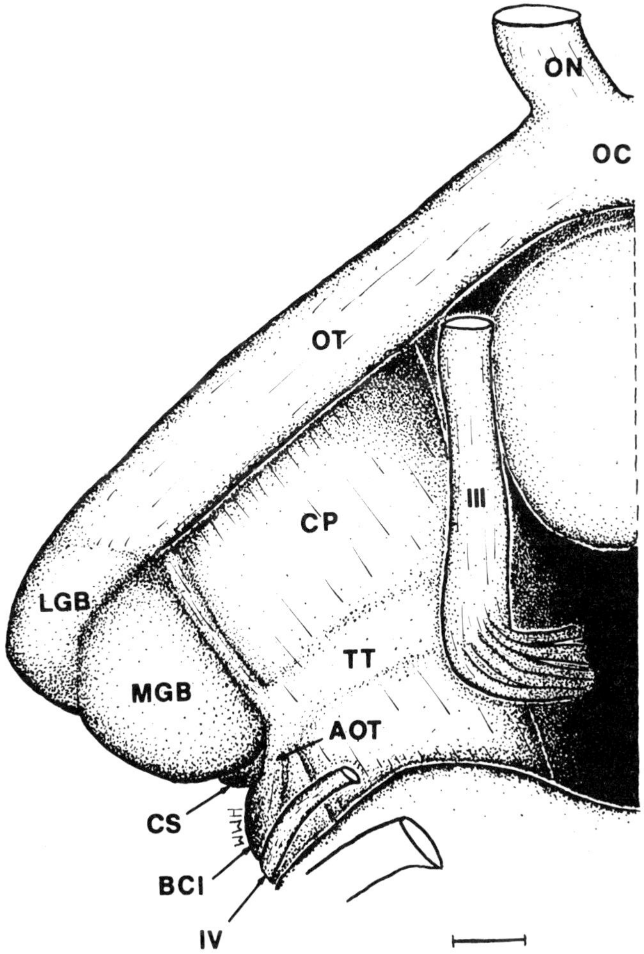

Figure 18. Ventral view of the mesencephalon and diencephalon of *Perodicticus potto*. Scale = 1 mm.

2. *Ventral and mesial surfaces*

The rhinal fissure exhibits an anterior and posterior division in these animals. It is shallow in the lesser galago, but more definite in the larger animals (Figs. 24-7 and 32-4). The posterior division is attenuated caudally. The olfactory bulbs project beyond the frontal pole approximately 3 mm. in the lesser galago and potto, and 5 mm. in the greater galago and slow loris. There is a distinct lateral olfactory tract. The pyriform lobe extends inferiorly and is visible below the posterior rhinal fissure from a lateral view.

On the orbital surface a straight sulcus (orbital s.) is seen (Figs. 32-4). In the lesser galago only a shallow groove is evident. Variations

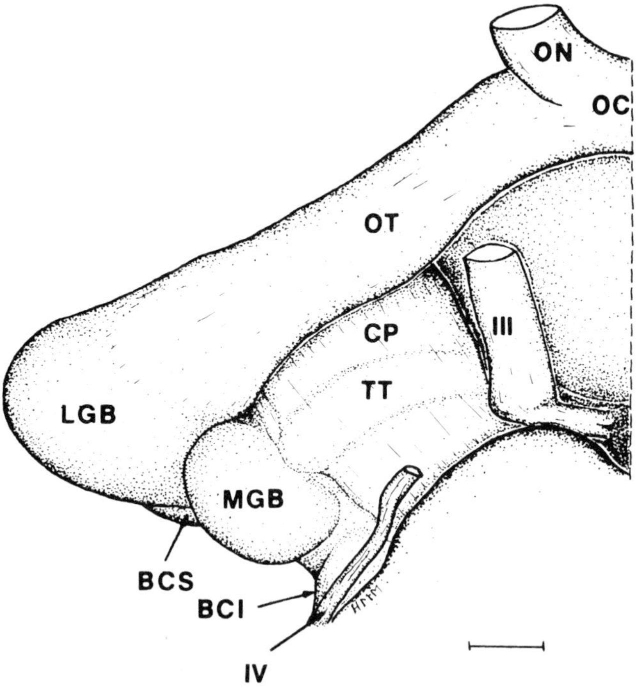

Figure 19. Ventral view of the mesencephalon and diencephalon of *Galago senegalensis.* Scale = 1 mm.

are seen in three of the lorises, the orbital sulcus being continuous with s. rectus (Fig. 33*c*).

Other than the calcarine complex, the sulci of the tentorial surface are consistent only in slow loris. Ventrolaterally, a straight sulcus emerges to indent the inferolateral border. In one specimen, three parallel sulci crossed the surface 7, 10 and 12 mm. from the occipital pole. There is no sulcus coursing parallel to the temporo-occipital polar axis.

On the mesial surface a long cingulate sulcus extends from the frontomesial to the occipitomesial surface. It branches rostrally in the potto; caudally, it continues as a groove to the paracalcarine sulcus in the greater galago and slow loris. In the lesser galago it joins the paracalcarine directly (Figs. 28-31). A short superiorly directed sulcus is interposed between the frontal pole and the origin of the cingulate s. in the greater galago. A similar sulcus is seen in the potto, but not in the slow loris.

The calcarine complex is similar in these animals. The retrocalcarine s. extends toward the occipital pole; the paracalcarine s. is directed superiorly (Figs. 28-31).

The rostrocaudal length of the corpus callosum is 11.5 mm. in the lesser galago, 13.6 mm. in the slow loris, 13.8 mm. in the potto, and

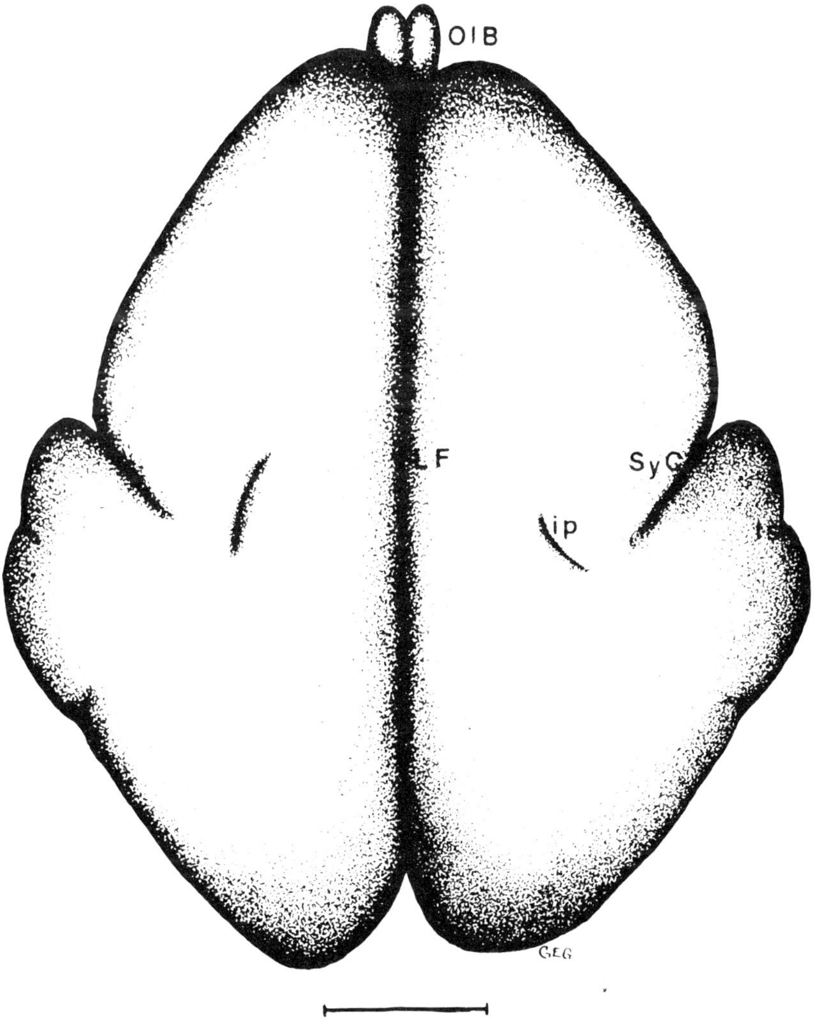

Figure 20. Dorsal view of the telencephalon of *Galago senegalensis.* The Sylvian complex (suprasylvian plus pseudosylvian) separates temporal cortex from frontal cortical areas. Scale = 0.5 cm.

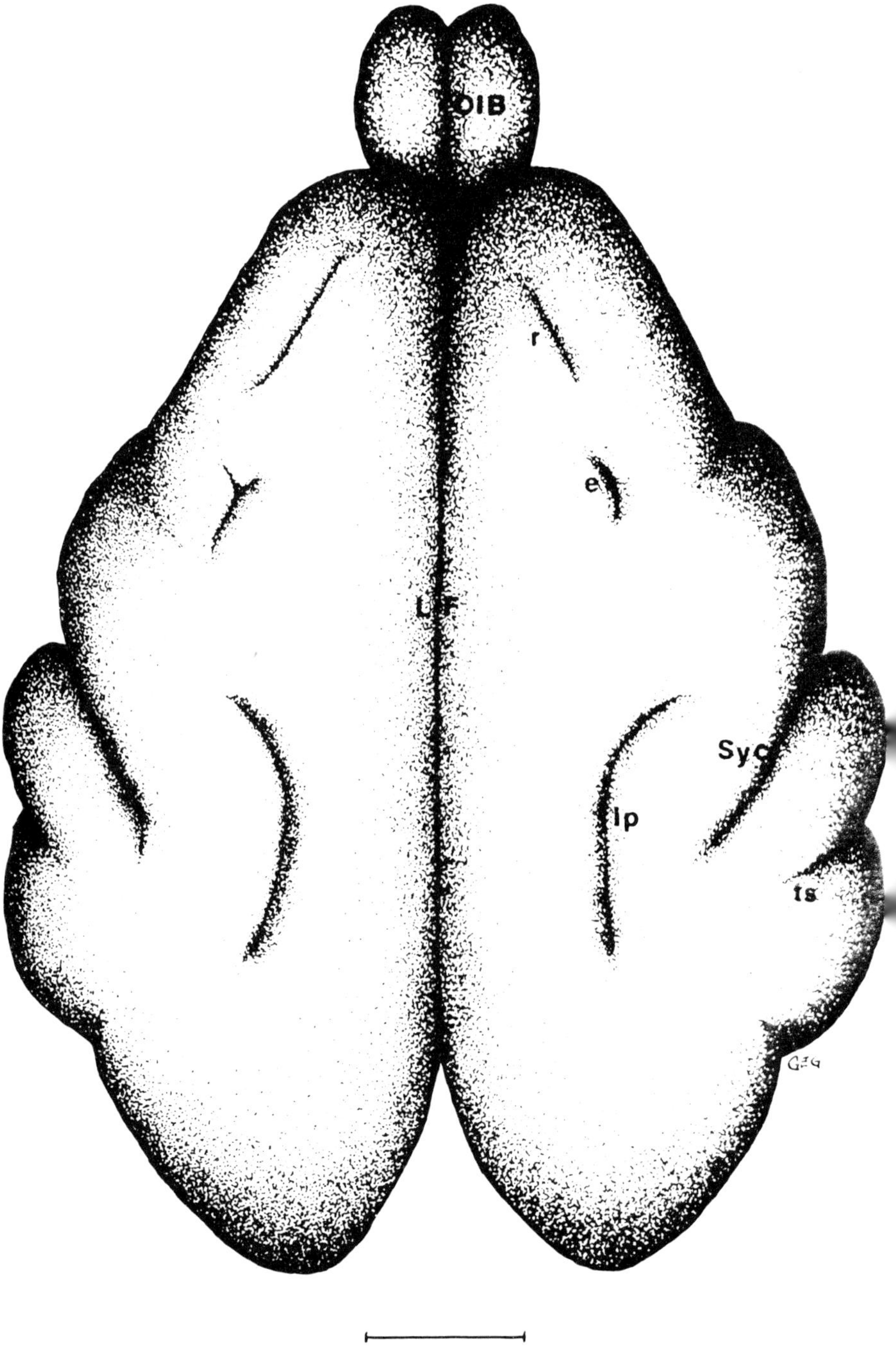

Figure 21. Dorsal telencephalon of *Galago crassicaudatus.* Note the variation of sulcus *e* on the two hemispheres as well as the development of the parieto-temporal cortex. Scale = 0.5 cm.

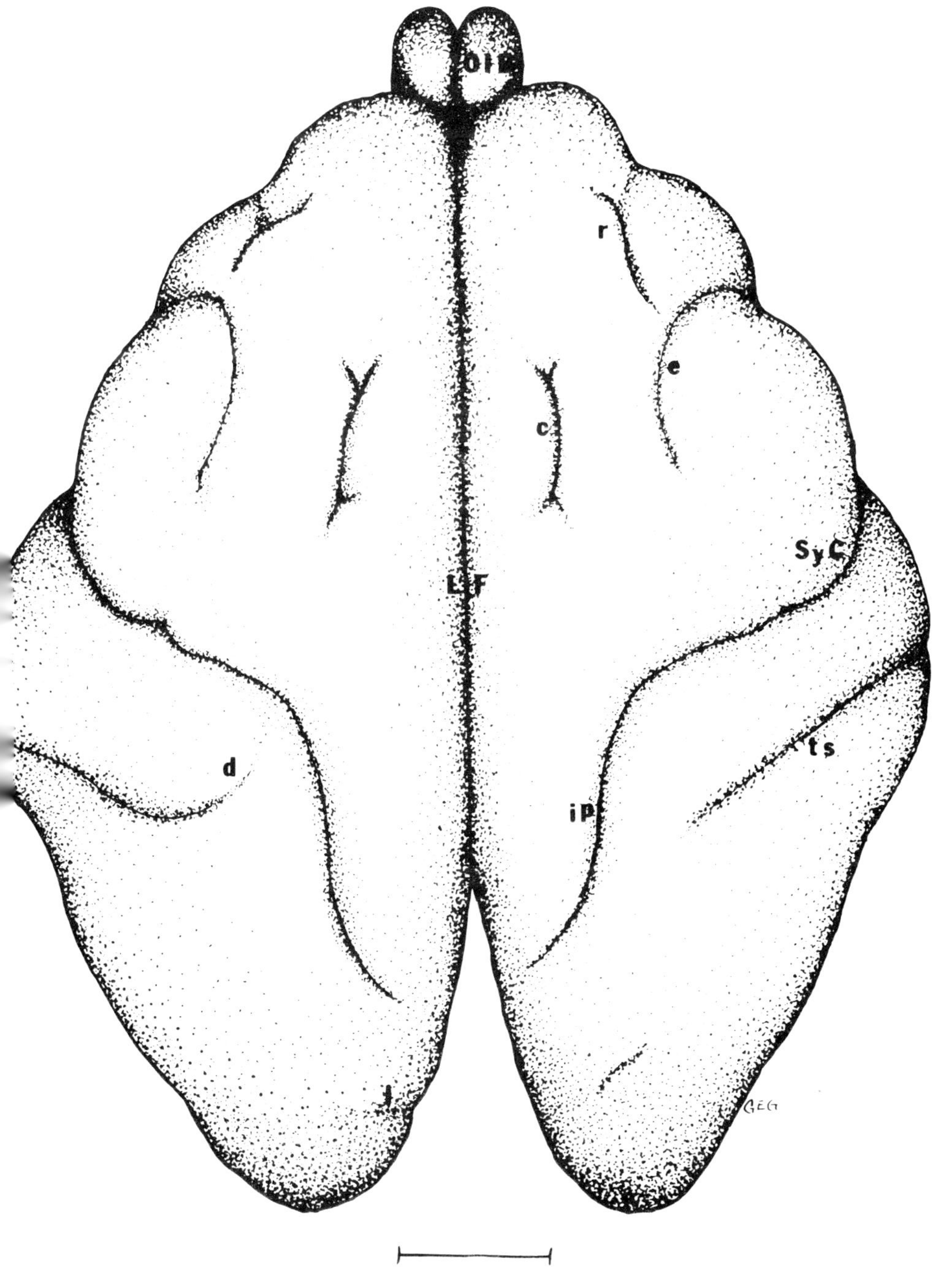

Figure 22. Telencephalon of *Nycticebus coucang.* Notice the indentation of the frontal margin by the sulcus rectus and the position of the lunate sulcus (1). Scale = 0.5 cm.

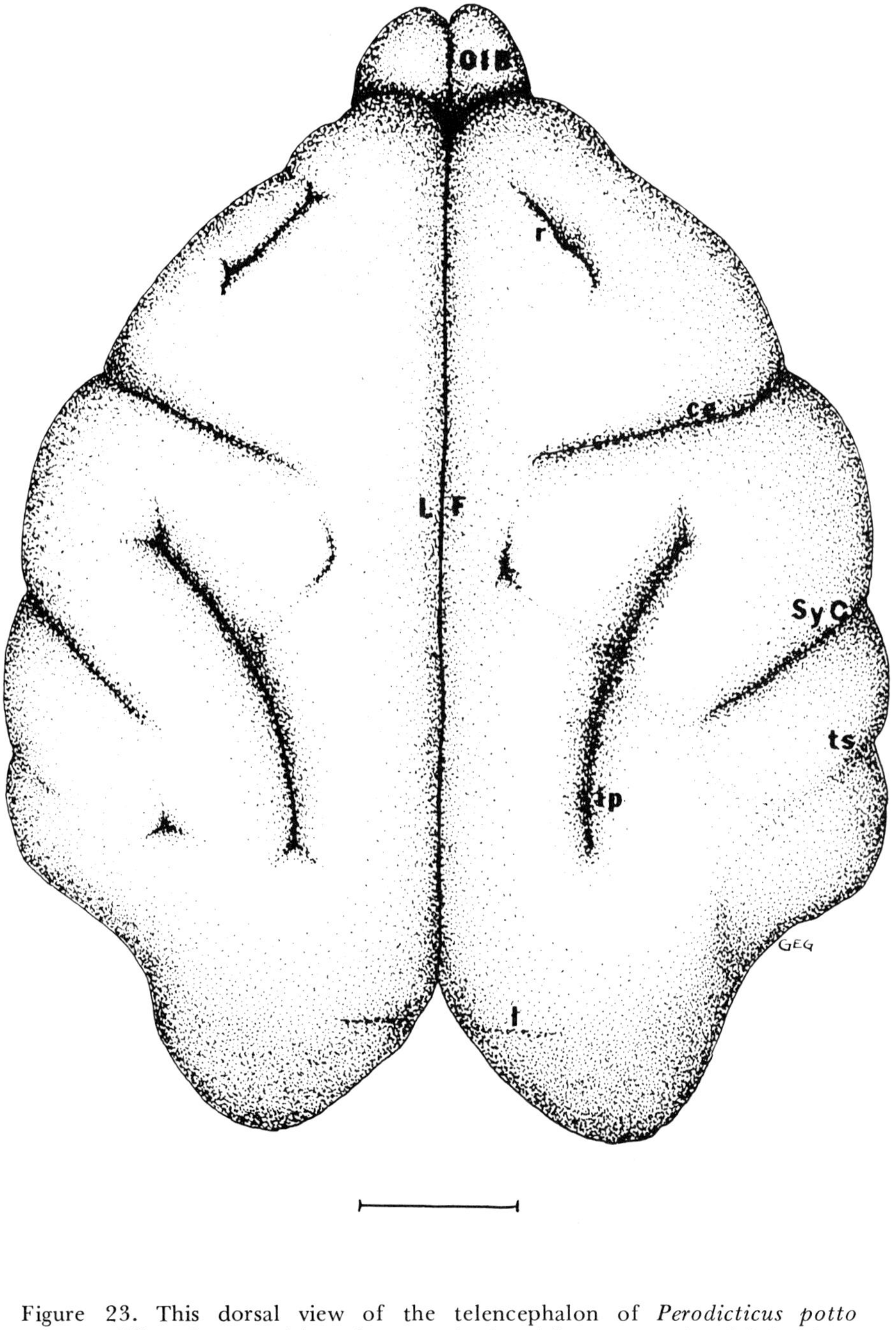

Figure 23. This dorsal view of the telencephalon of *Perodicticus potto* exemplifies the transition from sagittal sulci to coronal sulci in these prosimians. Scale = 0.5 cm.

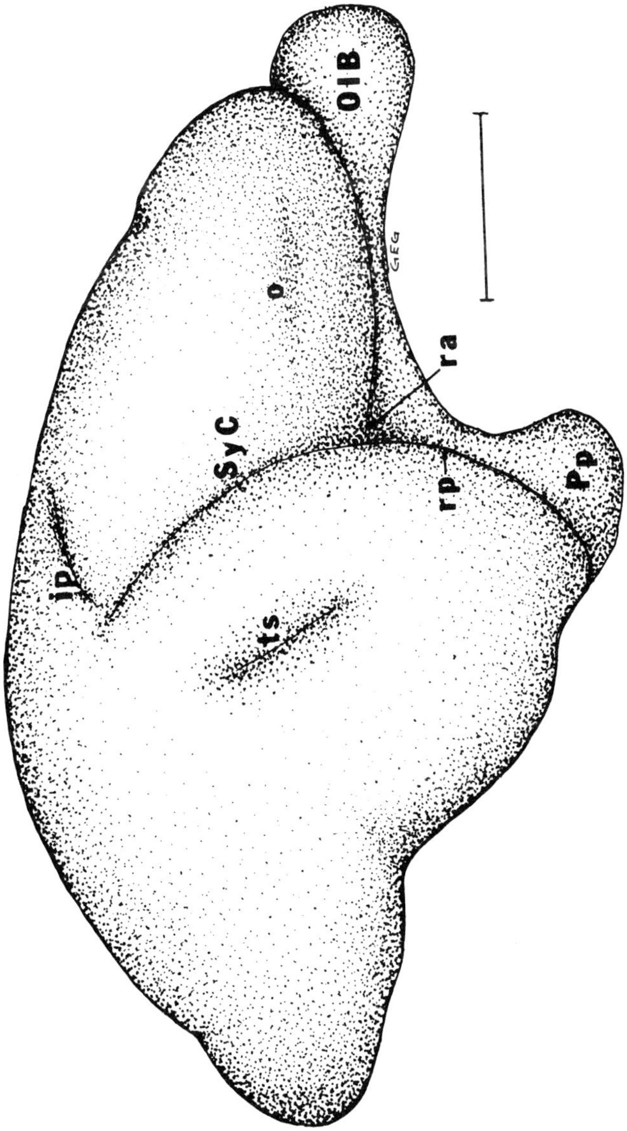

Figure 24. Lateral view of the telencephalon of *Galago senegalensis*. The majority of cortical infoldings are dimples or grooves, except the Sylvian complex (SyC) and intraparietal sulcus (ip). Scale = 0.5 cm.

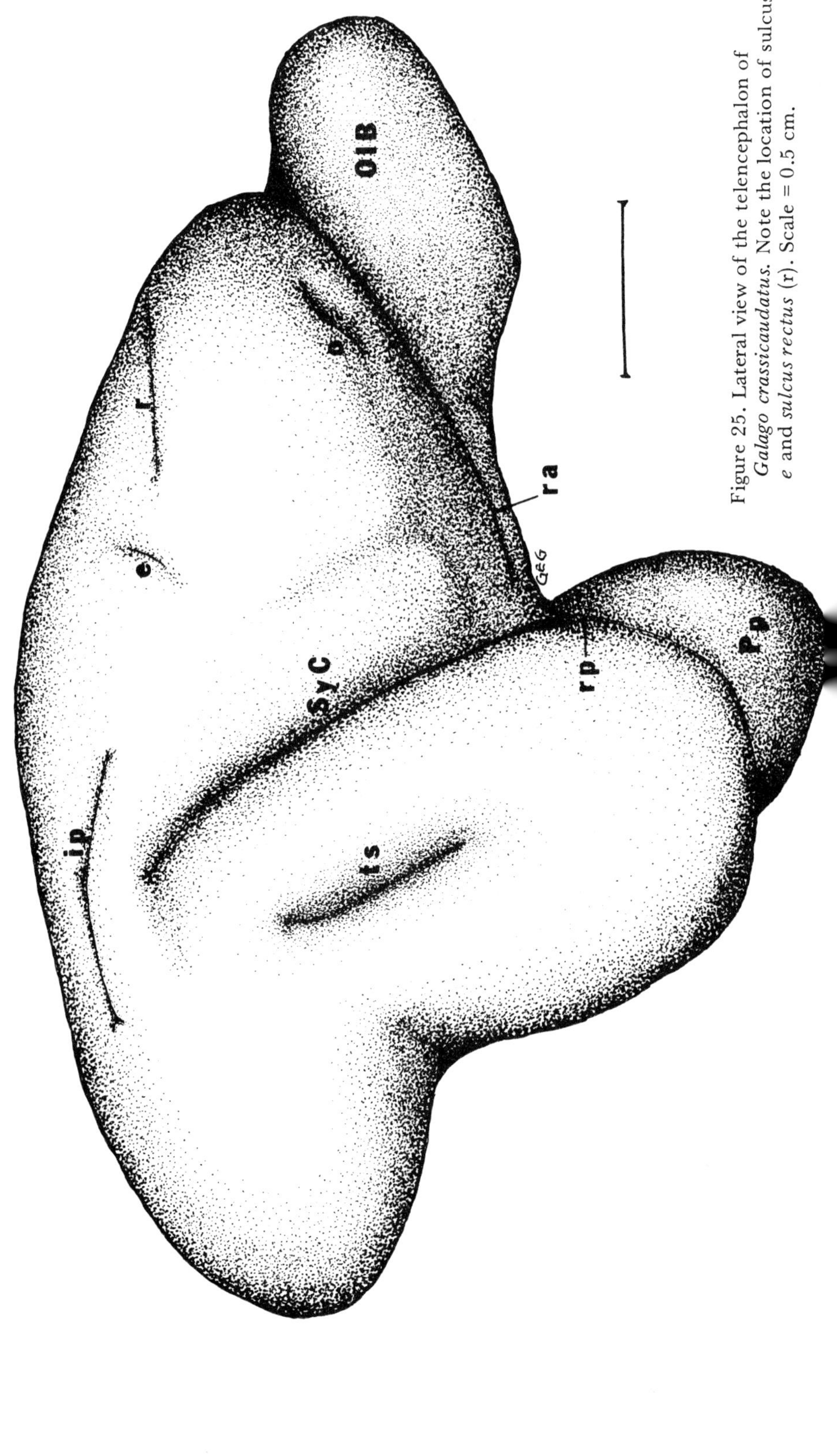

Figure 25. Lateral view of the telencephalon of *Galago crassicaudatus.* Note the location of sulcus *e* and *sulcus rectus* (r). Scale = 0.5 cm.

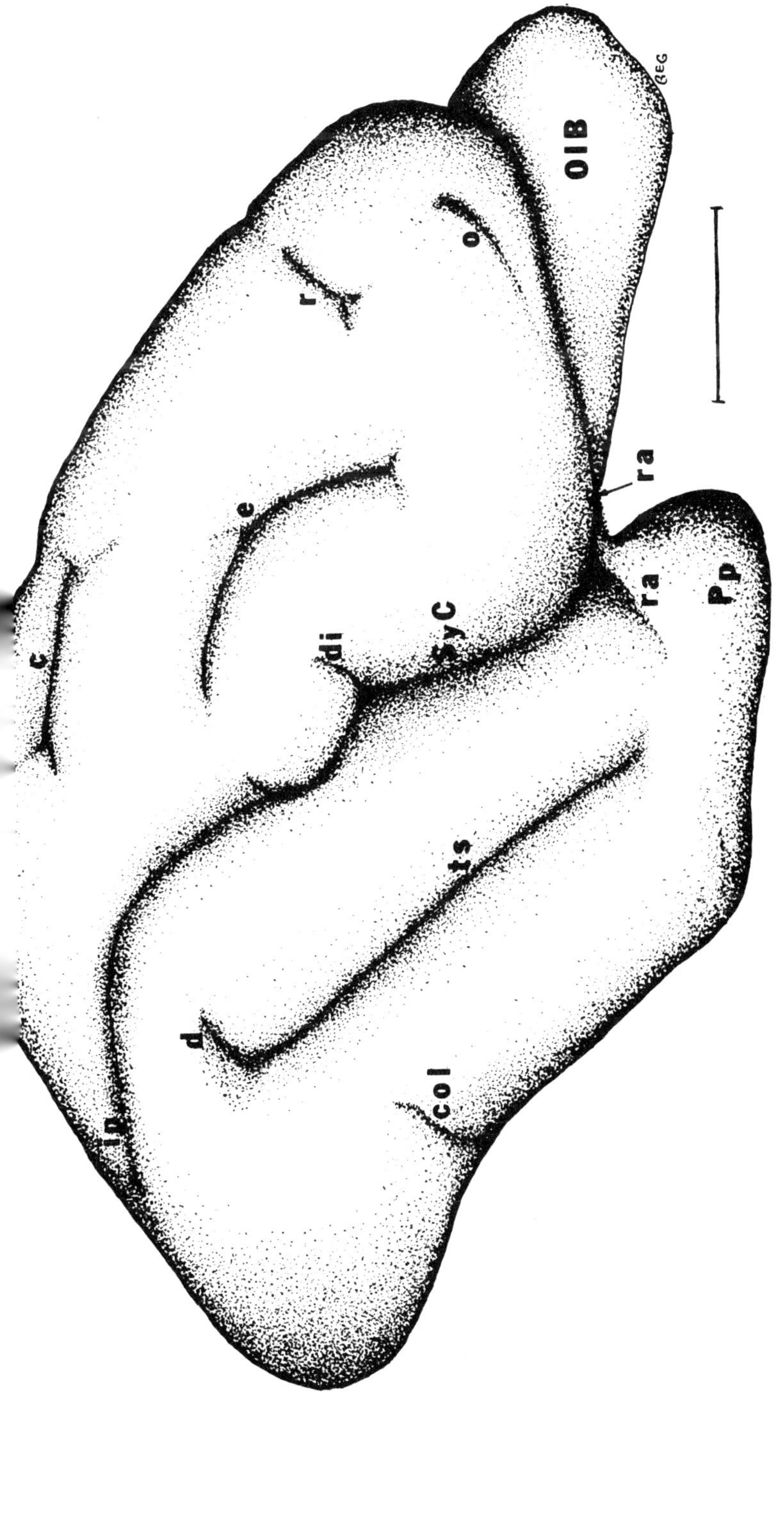

Figure 26. Lateral view of the telencephalon of *Nycticebus coucang*. Note the variations in the sulcal pattern compared to *Galago*, with the addition of sulcus *c*, diagonal sulcus (di) and the continuity of the intraparietal with the Sylvian complex. Scale = 0.5 cm.

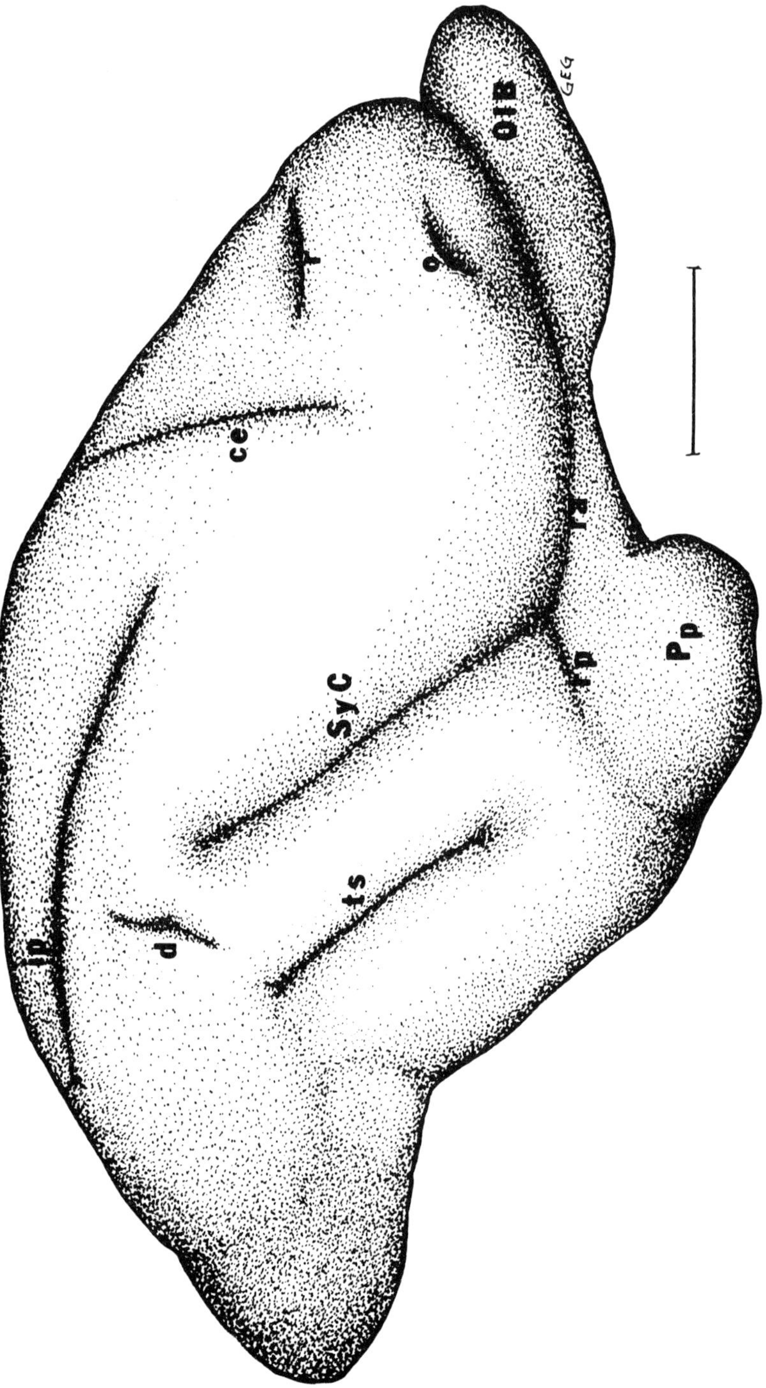

Figure 27. Lateral view of hemisphere of *Perodicticus potto*. The intraparietal sulcus in separate from Sylvian complex. Note also the position of the central sulcus (ce). Scale = 0.5 cm.

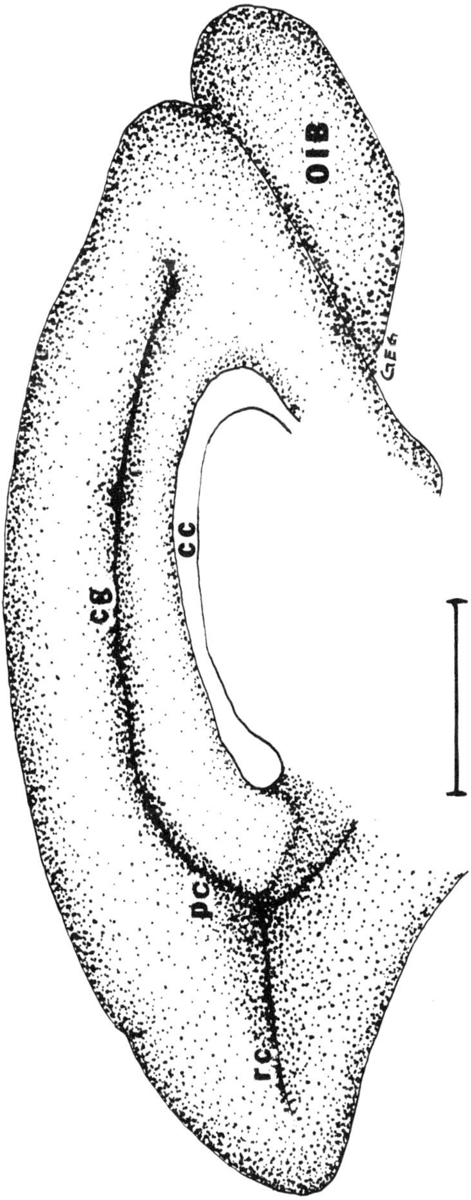

Figure 28. Medial aspect of the hemisphere of *Galago senegalensis.* The cingulate sulcus is continuous with the paracalcarine. Scale = 0.5 cm.

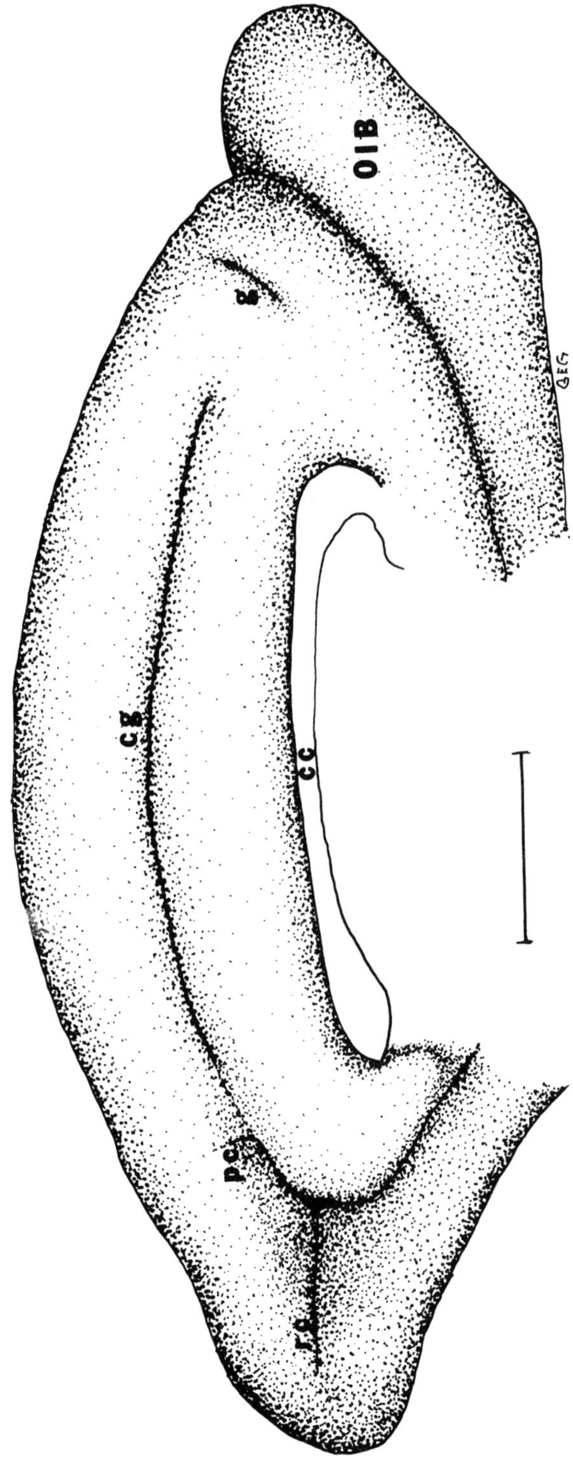

Figure 29. In *Galago crassicaudatus*, the cingulate sulcus is connected to the paracalcarine by a groove. Rostrally a distinct genual sulcus is present. Scale = 0.5 cm.

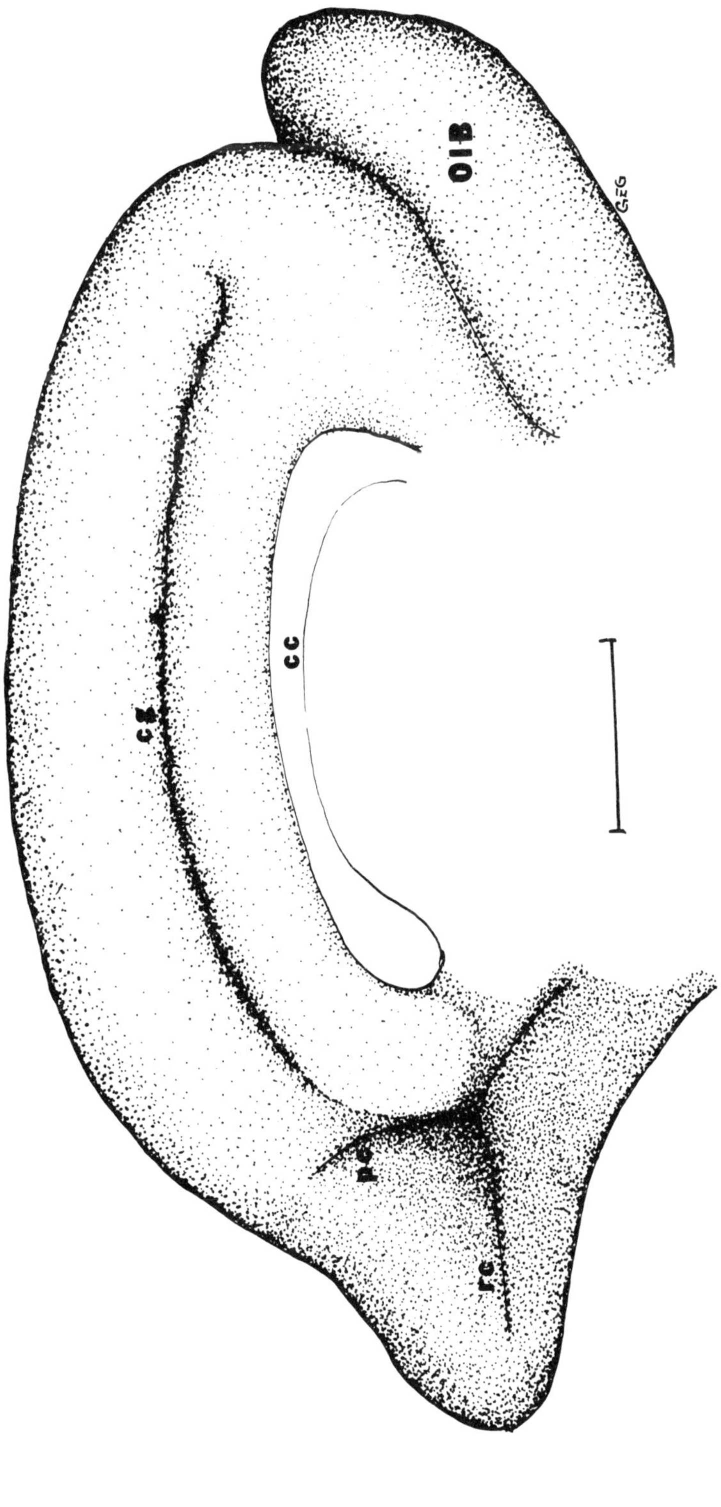

Figure 30. Medial view of the telencephalon of *Nycticebus coucang*. A genual sulcus is not present. The paracalcarine is reflected dorsally and is not associated with the cingulate sulcus as seen in *Galago*. Compare with Fig. 29. Scale = 0.5 cm.

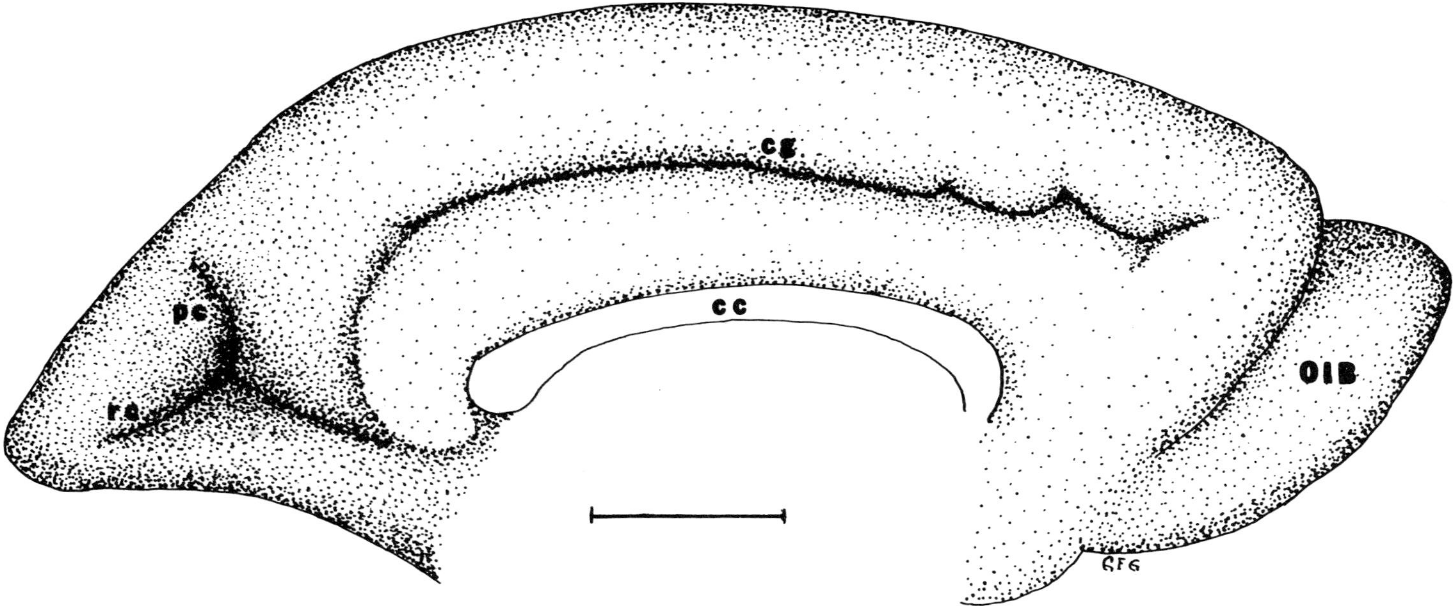

Figure 31. The medial aspect of the hemisphere of *Perodicticus potto* is similar to *N. coucang*, although variations of the rostral terminus of the cingulate sulcus are not seen in the slow loris. Scale = 0.5 cm.

15.0 mm. in the greater galago. The splenium of the corpus callosum overlies the rostral portion of the superior colliculus in all the animals except *G. senegalensis.*[26] The fornix is 0.8 mm. in thickness in the greater galago, while it is 0.5 to 0.6 mm. in the other animals.

Discussion

In Monotremata and Marsupialia, the pons, olive and pyramids are simple in structure and remain relatively unchanged throughout these groups.[22,28] The spiny anteater (*Tachyglossus aculeatus*) has a post-trigeminal pons; the *corpus trapezoideum*, olive and pyramids cannot be seen grossly. In the wallaby, the pons is entirely pre-trigeminal, the pyramids are thin and the olive cannot be seen.[22] The pontine and medullary components in the Insectivora are similar to that observed in the wallaby. In the elephant shrew, the pons and *corpus trapezoideum* are approximately equal in size and are separated by a transverse groove. The fifth nerve exits laterally from the groove.[13]

The pons in *Tupaia minor* measures 2.5 mm. in length and is located rostral to a broad *corpus trapezoideum.*[9] The fifth nerve exits from the lateral pons, and the pyramids are attenuated.[9] We found the pons and medulla of *Tupaia glis* and *Tupaia minor* to be similar in appearance. The pontine surface measures 3 mm. in length and is only 0.5 mm. greater than the *corpus trapezoideum*. The lateral pons shows a very small post-trigeminal bundle. The pyramids are uniform, 6.5 mm. in length and 0.8 mm. in width. The pons in *Tarsius* is notably smaller but more prominent than in *Tupaia.*[23]

In primates the pons, in particular, undergoes considerable increase in size. Phylogenetically, pontine growth occurs in all dimensions, but characteristically more caudally. This caudal increase is believed to be partially responsible for the decrease in the size of the *corpus trapezoideum*;[22] exceptions may be due to specialised sensory mechanisms such as audition. In general, the lower primates have a large *corpus trapezoideum.* In the higher primates (macaque), the *corpus trapezoideum* is covered rostrally by the pons.[22] In chimpanzee and man it is no longer visible.[22,29] It has been proposed that in primates the ponto-trapezoidial groove is obliterated by the caudal growth of the pons.[22]

The pons and medulla of the genera included in this study are typically primate in appearance. The striations observed in the ponto-trapezoidial groove of *N. coucang* and *P. potto* do not necessarily imply a more caudal-pontine growth. This could, in fact, be related to a decrease in the size of the *corpus trapezoideum* and other structures related to audition. Despite the location and shape of the olive, *P. potto* has a well-developed ventral brainstem.[23] This observation is based on the appearance of the pons, the pyramids and

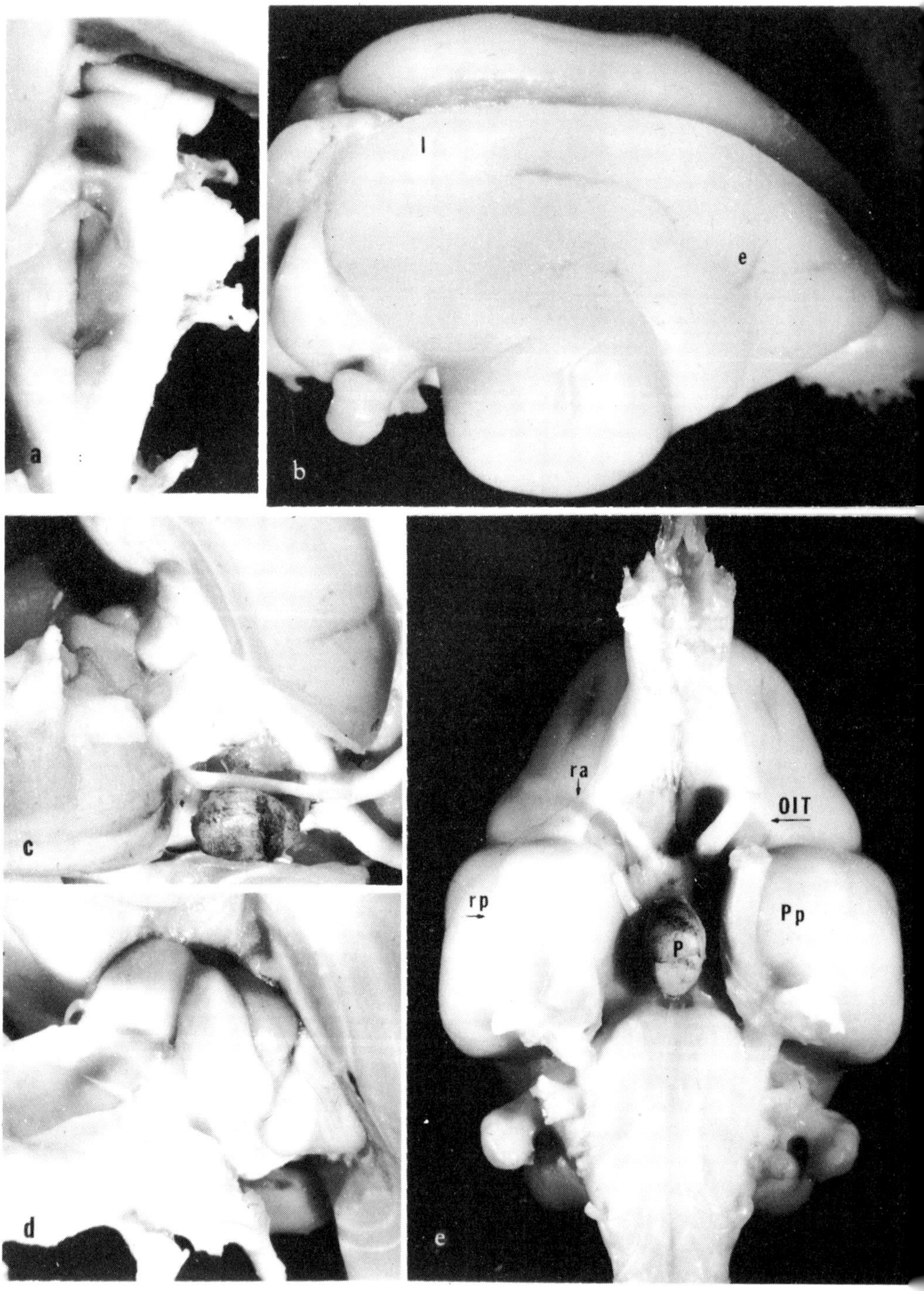

Figure 32. Photographs of the cerebral cortex and brain stem of *Galago crassicaudatus.*

(*a*) Dorsal view of the rhomboid fossa, dorsal and dorsolateral medulla and superior and inferior colliculi. Compare with Fig. 7.

(*b*) Dorsolateral view of the cerebral cortex. Compare with Fig. 25.

(*c*) Ventrolateral view of the myelencephalon and mesencephalon. Note the trapezoid body, pons, crus cerebri and optic tract. Compare with Figs. 1 and 16.

other brainstem structures when compared with the other Lorisidae.

In the lesser bushbaby and slow loris, the *corpus trapezoideum* was described as being located within the caudal portion of the pons.[16,17] Since the *corpus trapezoideum* was considered to be a portion of the pons, some considered the pyramids to be superficial to the caudal pons. However, this caudal area is not pons, but the *corpus trapezoideum*, and should not be included within the borders of the pons in lower mammalian forms. Our observations show that the pons has distinct rostral and caudal borders. The caudal border corresponds to the ponto-medullary groove. The *corpus trapezoideum* is located caudal to the pons and is covered medially by the pyramids. The ponto-medullary groove can therefore be described regionally as the ponto-trapezoidial groove and the ponto-pyramidal groove. The pyramids are located in a superficial position only in the medulla. Rostrally the pyramids pass into the basilar pons.

It is our experience that the small primary rootlets of cranial nerves were extremely difficult to observe without adequate removal of the arachnoid. This may account for the difference in the number of rootlets comprising the vagus and hypoglossal nerves reported in our study and the observations reported for *N. coucang.*[17] We noted in excess of six primary rootlets for the vagus nerve (versus three) and 12 to 13 primary rootlets (versus four) for the hypoglossal nerve.

Histological studies on the potto show a large *nucleus reticularis lateralis.*[23] This nucleus is responsible for the raised area on the ventrolateral surface of the medulla. It is similar in the greater bushbaby, slow loris, and potto, but less evident in the lesser bushbaby. We suggest that this elevation be called the reticular prominence.

The features of the prosimian mesencephalon that call for special note are the *corpora quadrigemina.*[2] The general features of these structures are remarkably constant in all the Metatheria and Eutheria.[2] Wood Jones described the dorsal surface of the midbrain of a generalised marsupial as having well-developed *corpora quadrigemina*, the inferior colliculi being larger and more conspicuous (his Fig. 95).[28]

In *Tupaia* the *corpora quadrigemina* are noteworthy for the large size of the anterior pair, which indent the occipital lobe of the cerebrum and the anterior surface of the cerebellum.[9] Large, anterior colliculi are reminiscent of the optic lobe of the reptilian brain.[10] On

(*d*) Caudolateral view of the mesencephalon. Note the anterior medullary velum, brachium of the inferior colliculus, medial and lateral geniculate bodies. Compare with Fig. 11.

(*e*) Ventral view of the brain and brain stem. Both divisions of the rhinal fissure can be seen as well as the lateral olfactory tract (OlT). Note the size of the pituitary gland.

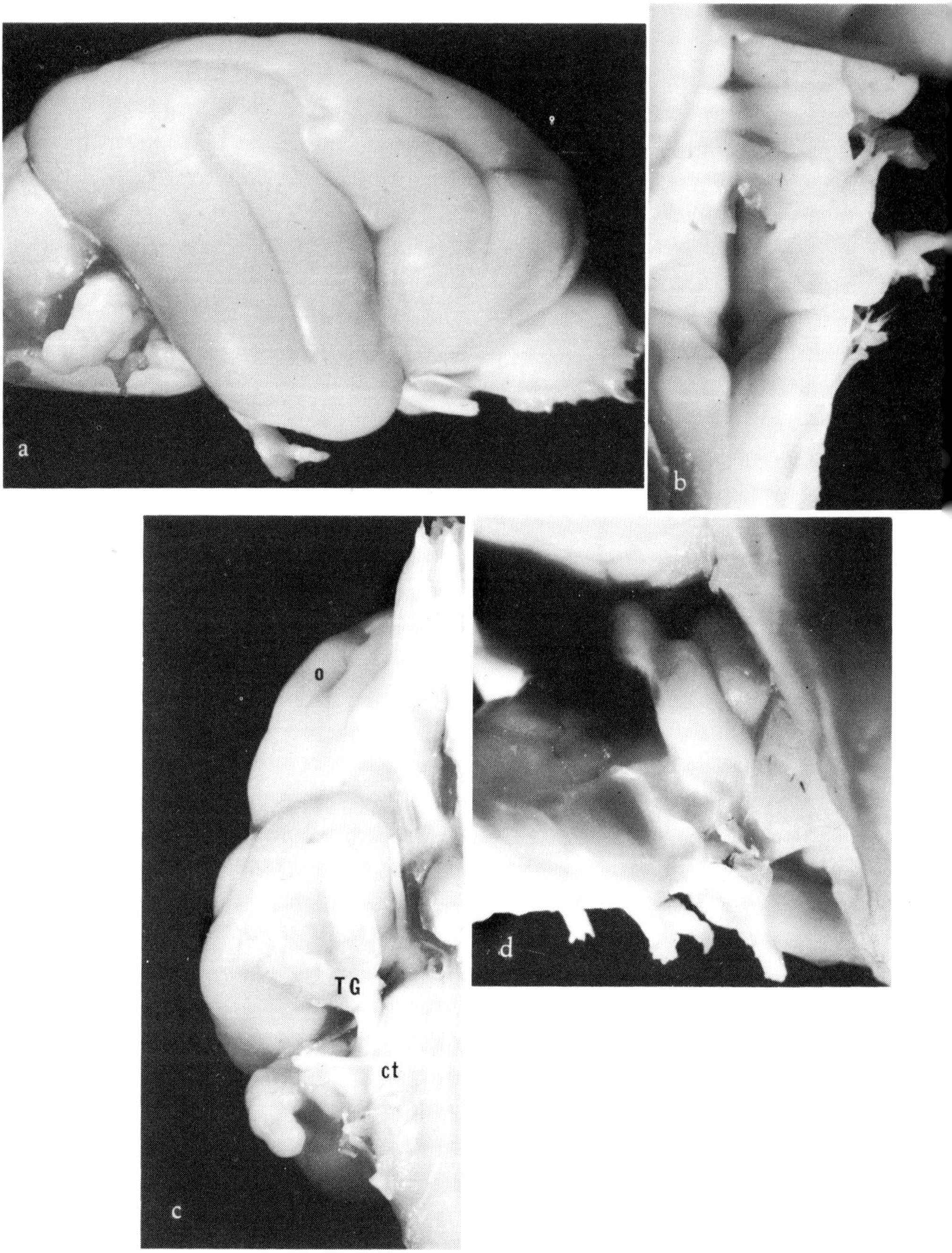

Figure 33. Photographs of the cerebral cortex and brain stem of *Nycticebus coucang*.

(*a*) Lateral view of the cerebral cortex. Compare with Fig. 26.

(*b*) Dorsal view of the myelencephalon and mesencephalon. Compare with Fig. 8.

(*c*) Ventral view of the cerebral cortex and brain stem. Note the trigeminal ganglion (TG) and trapezoid body (CT). Compare with Fig. 2.

(*d*) Caudolateral view of the mesencephalon. Compare with Fig. 12.

the other hand, the posterior colliculi are very inconspicuous in *Tupaia*,[9] forming a transversely elongated strip which lies under the posterior border of the anterior colliculi, again reminiscent of the reptilian condition.[10]

The brain of *Tupaia* exhibits a 'definite hypoplasia' of secondary auditory centres which can be correlated 'with the atrophy of the external ear'.[10] The anterior colliculus of *Ptilocercus* is described as being much smaller relative to the posterior colliculus than is the case in *Tupaia*.[11]

The dorsal surface of the mesencephalon in *Tarsius* exhibits well-developed inferior and superior colliculi.[21] The superior colliculus in this animal attains dimensions that almost warrant the designation of optic lobe.[19,20] While Woollard describes the inferior colliculus and the medial geniculate body as being of normal size,[21] Tilney writes that these two structures are larger and better defined than in all other primates.[20] Both collicular elevations of the mesencephalon are prominent in *Lemur mongoz*.[20]

The anterior quadrigeminal bodies are smaller in *Lemur* than in *Tarsius*.[2] The lateral geniculate body and the superior colliculus in *Galago* and the slow loris are described as distinct features of the brain stem.[16,17] Their prominence has suggested the importance of vision in these nocturnal animals.[16,17]

The inferior and superior colliculi are large elevations on the dorsal surface of the mesencephalon of the Lorisidae. The size of the inferior colliculus may reflect a retention of primordial or rudimentary control over audition.[20] In the present study the inferior colliculus is more prominent in the galagos than in the slow loris and potto. This parameter may be correlated with the size of the external ear. Both galagos have large pinnae, and as Osman Hill has noted, 'the external ear reaches its maximum expression in Galagidae. In addition to its larger size it differs from that of the lorises in its greatly expanded, spoon-shaped, transversely ribbed cymba.'[30] The size of the inferior colliculus and the pinna may also be correlated with that of the *corpus trapezoideum* of the medulla. As was noted above, the *corpus trapezoideum* in the galagos is larger than that in the potto or slow loris.

The superior colliculus is a large, oval somewhat flattened structure in all the animals studied, but in proportion to the size of its brain, the lesser galago has the largest superior colliculus. This correlates with the size of the eyes in proportion to the size of the head in this animal.

The accessory optic system consists of the transpeduncular tract, its nucleus and the accessory optic tract and its nuclei.[31] A distinct transpeduncular tract (*tractus peduncularis transversus*)[32] has been shown in *Tupaia*,[10] *Tarsius* and *Lemur*.[15,2] It was not described in the slow loris or lesser galago.[17,16] In the present study, a well-defined accessory optic system is seen. An account of reduction

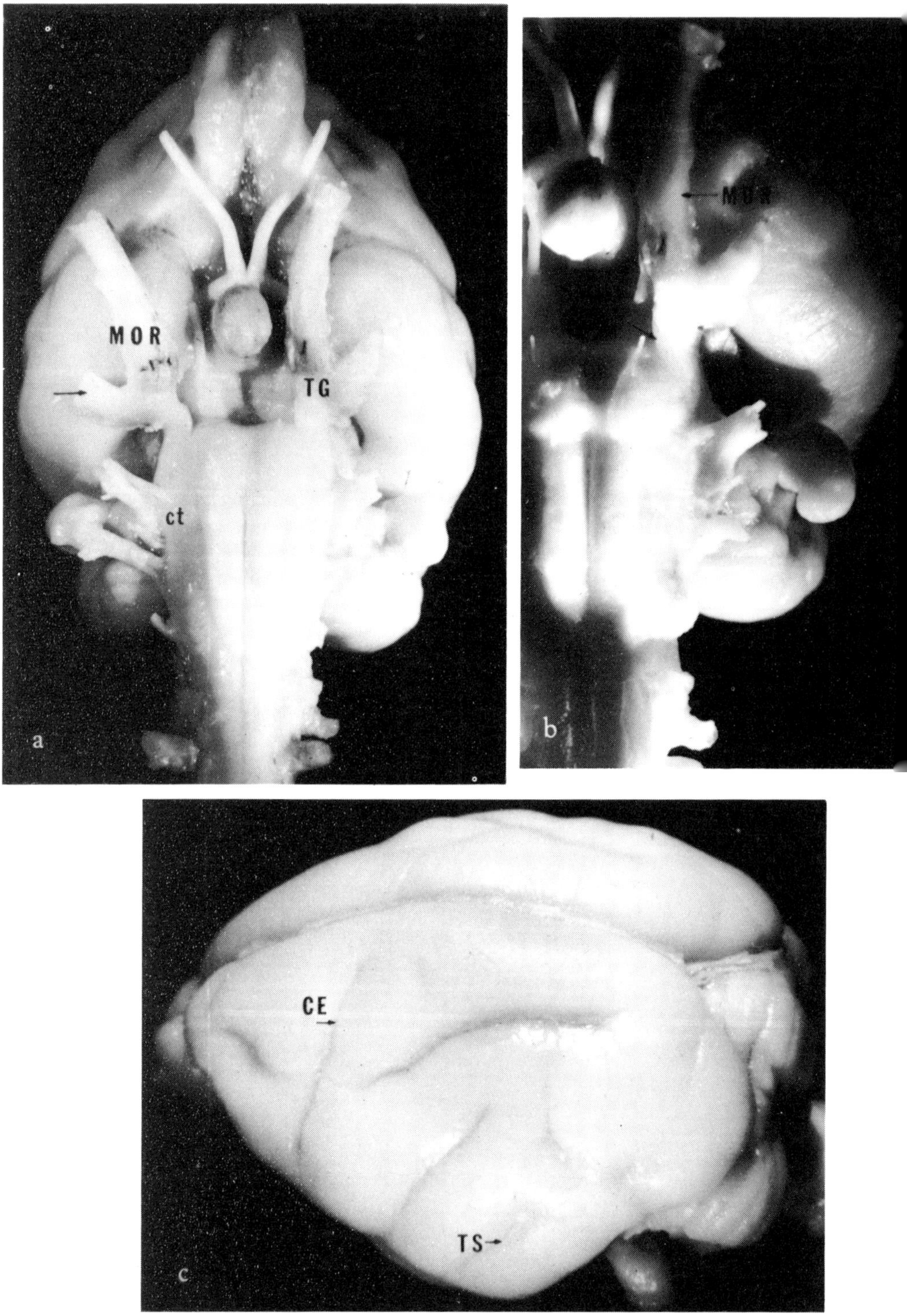

Figure 34. Photographs of the cerebral cortex and brain stem of *Perodicticus potto*.

(*a*) Ventral view of the cerebral cortex and brain stem. Note the maxillo-ophthalmic ramus (MOR) and the mandibular division (arrow) from the trigeminal ganglion (TG). Compare with Fig. 3.

(*b*) Ventral view of the brain stem. Note the small motor (arrow) and larger in the projections of the accessory optic system of some prosimians over that of *Tupaia* was recently published.[33]

It has been reported that in one of eighteen specimens of *N. coucang* the rootlets of the oculomotor nerve left the cerebral peduncle.[17] In all of the individuals of this species in our study, at least two rootlets of the nerve left the crus while the rest left from the interpeduncular fossa. We have seen oculomotor nerve rootlets leave the crus in only one human brainstem. This is considered to be an embryological variant in man, but it may describe the normal condition in the brain of *Nycticebus*.

Le Gros Clark discussed the hypothetical brain of a common progenitor of Primates.[12] This brain would have been lissencephalic except for calcarine, hippocampal and ectorhinal fissures and perhaps an indication of the suprasylvian fissure. Modern tree shrews have advanced little from this basic pattern.[9] *Tarsius spectrum* has only a small suprasylvian element on the orbital surface and a triradiate calcarine complex.[3] The brain of *Microcebus murinus*, although advanced in comparison to *Tupaia*, is still simple when compared to the other lemurs.[4,12] The brain of the lesser galago is advanced in its sulcal pattern when compared to the dwarf and mouse lemurs.[27]

Following the rule of gyrification,[6] the cortex of the animals in our study exhibit an increase in the variety and number of grooves and sulci. The lesser galago exhibits a more complex sulcal pattern than has previously been reported for the dwarf galago,[7,34] or lesser galago.[5,16]

By its sulcal pattern, the lesser galago shows differential cortical growth around the sylvian complex. This is seen in the development of the intraparietal sulcus and a deep superior temporal groove. A triradiate depression is present in the frontal region and a transverse groove noted in the occipital area. With the addition of sulcus *e* in the greater galago the same cortical pattern is seen, but the individual elements of the pattern are more distinct. This gives the impression that the smaller brain has been magnified. The slow loris and potto both have complex sulcal patterns. The majority of sulci are in the sagittal plane, although there are some located coronally (Figs. 32-4).

Compared to the general mammalian cortex, the lower primates may characterise trends of neocortical growth associated with modifications and development of visual and association cortex. These modifications can be correlated with an increase in the number of coronally oriented sulci.[35]

The importance of a definitive description of cortical morphology, including dimples, fissurettes, and subzonal sulci is seen in the studies

---

sensory roots of the trigeminal nerve. The abducent nerves exit the pontopyramidal groove through the pyramids. Compare with Fig. 3.

(*c*) Dorsolateral view of the cerebral cortex. Note the central sulcus and the modifications dorsocaudal to the terminus of the superior temporal sulcus. Compare with Fig. 23.

of Welker and Campos,[36] Sanides and Krishnamurti,[8] and Le Gros Clark.[12] Although the approach of form analogy to suggest neocortical evolution has proved its merit, it relies too heavily on a strict orthogenetic interpretation. The interpretation of neocortical evolution initiated by Abbie,[37,38] and independently proposed by Sanides,[39] reveals a more fundamental correlate between gyrus formation and cortical cytoarchitecture in phylogeny. The expansion of the cortex may also be correlated with brainstem development. This may be associated with the size and complexity of the pons and cerebellum, and the development of the pyramids in these animals. Emphasis can now be placed on the evaluation of gross morphology in the light of combined comparative studies of fibre projections, cytoarchitecture and electrophysiology.

Acknowledgments

The authors wish to express their appreciation to Dr Y. Rumpler for generously supplying two brains of the mouse lemur. This paper was presented at the Research Seminar on Prosimian Biology at The Institute of Archaeology, April 14-17, 1972 by George E. Goode. Partial support for this project was provided by an A.D. Williams Grant (3558-550). The authors are especially indebted to Mrs M.P. Harrell for typing the preliminary and final drafts of this paper.

NOTES

1 Holloway, R.L. (1968), 'The evolution of the primate brain: some aspects of quantitative relations', *Brain Res.* 8, 1-52.

2 Elliott Smith, G. (1903), 'On the morphology of the brain in the Mammalia, with special reference to that of the lemurs recent and extinct', *Trans. Linn. Soc. Lond.* 8, 319-432.

3 Connolly, C.J. (1950), *External Morphology of the Primate Brain*, Springfield, Illinois.

4 Friant, M. (1970), 'Le cerveau des lémuriens de Madagascar', *Acta Neurol. Belg.* 70, 439-470.

5 Kanagasuntheram, R. and Leong, C.H. (1966), 'Observations on some cortical areas of the lesser bushbaby (*Galago senegalensis senegalensis*)', *J. Anat.* 100, 317-33.

6 Ariens Kappers, C.U., Huber, G.C. and Crosby, E.C. (1936), *The Comparative Anatomy of the Nervous System of Vertebrates, Including Man*, New York.

7 Zuckerman, S. and Fulton, J.F. (1941), 'The motor cortex in *Galago* and *Perodicticus*', *J. Anat.* 75, 447-56.

8 Sanides, F. and Krishnamurti, A. (1967), 'Cytoarchitectonic subdivisions of sensorimotor and prefrontal regions and of bordering insular and limbic fields in slow loris (*Nycticebus coucang coucang*)', *J. Hirnforschung* 9, 225-52.

9 Le Gros Clark, W.E. (1924), 'On the brain of the tree shrew (*Tupaia minor*)', *Proc. Zool. Soc. Lond.* 1053-74.

10 Le Gros Clark, W.E. (1929), 'The thalamus of *Tupaia minor*', *J. Anat.* 63, 177-206.

11 Le Gros Clark, W.E. (1926), 'The anatomy of the pen-tailed tree shrew (*Ptilocercus lowii*)', *Proc. Zool. Soc. Lond.* 1179-308.

12 Le Gros Clark, W.E. (1931), 'The brain of *Microcebus murinus*', *Proc. Zool. Soc. Lond.* 463-86.

13 Le Gros Clark, W.E. (1928), 'On the brain of *Macroscelides* and *Elephantulus*', *J. Anat.* 62, 245-75.

14 Le Gros Clark, W.E. (1932), 'The brain of the Insectivora', *Proc. Zool. Soc. Lond.* 975-1013.

15 Le Gros Clark, W.E. (1930), 'The thalamus of *Tarsius*', *J. Anat.* 64, 371-414.

16 Kanagasuntheram, R. and Mahran, Z.Y. (1960), 'Observations on the nervous system of the lesser bushbaby (*Galago senegalensis senegalensis*)', *J. Anat.* 94, 512-27.

17 Krishnamurti, A. (1966), 'The external morphology of the brain of the slow loris (*Nycticebus coucang coucang*)', *Folia primat.* 4, 361-80.

18 Kanagasuntheram, R., Wong, W.C. and Krishnamurti, A. (1968), 'Nuclear configuration of the diencephalon in some lorisoids', *J. Comp. Neurol.* 133, 241-68.

19 Tilney, F. (1927), 'The brain stem of *Tarsius*: A critical comparison with other primates', *J. Comp. Neurol.* 43, 371-432.

20 Tilney, F. (1928), *The Brain From Ape to Man*, New York.

21 Woollard, H.H. (1925), 'The anatomy of *Tarsius spectrum*', *Proc. Zool. Soc. Lond.* 1071-185.

22 Marsden, C.D. and Rowland, R. (1965), 'The mammalian pons, olive and pyramid', *J. Comp. Neurol.* 124, 175-88.

23 Gerhard, L. and Olszewski, J. (1969), 'Medulla oblongata and pons', in Hofer, H., Schultz, A.H. and Starck, D. (eds.) *Primatologia, Handbook of Primatology*, Basel.

24 Klüver, J. and Barrera, E. (1953), 'A method for the combined staining of cell and fibers in the nervous system', *J. Neuropath. Exp. Neurol.* 12, 400-3.

25 Hall, J. (1970), personal communication.

26 For midsagittal views, see Fig. 31 in Haines, D.E., this volume.

27 *Microcebus murinus* has been studied in our laboratory.

Affinities of Cheirogaleinae and Galaginae based on other characters have recently been discussed by Charles-Dominque, P. and Martin, R.D. (1970), 'Evolution of lorises and lemurs', *Nature* 227, 257-60.

28 Wood Jones, F. (1947-1950), 'The study of a generalized marsupial (*Dasycercus cristicauda*) (Krefft)', *Proc. Zool. Soc. Lond.* 26, 493-9.

29 Shantha, T.R. and Manocha, S.L. (1969),s'The brain of the chimpanzee (*Pan troglodytes*)', Basel.

30 Osman Hill, W.C. (1953), *Primates, Comparative Anatomy and Taxonomy*, I: *Strepsirhini*, Edinburgh.

31 Giolli, R.A. (1963), 'An experimental study of the accessory optic system in the cynomolgus monkey', *J. Comp. Neurol.* 121, 89-107.

32 Marburg, O. (1903), 'Basale Opticuswürzel und Tractus peduncularis transversus', *Arb. a.d. Neurol. Inst. a.d. Wien. Univ.* 10, 66.

33 Giolli, R.A. and Tigges, J. (1970), 'The primary optic pathways and nuclei of primates', in Noback, C.R. and Montagna, W. (eds.), *The Primate Brain*, New York.

34 von Bonin, G. (1945), *The Cortex of Galago*, Urbana, Illinois.

35 Le Gros Clark, W.E. (1947), 'Deformation patterns in the cerebral cortex', in *Essays on Growth and Form Presented to d'Arcy Wentworth Thompson*, Oxford.

36 Welker, W.I. and Campos, G.B. (1963), 'Physiological significance of sulci in somatic sensory cerebral cortex in mammals of the family Procyonidae', *J. Comp. Neurol.* 120, 19-36.

37 Abbie, A.A. (1940), 'Cortical lamination in the Monotremata', *J. Comp. Neurol.* 72, 428-67.

38 Abbie, A.A. (1942), 'Cortical lamination in a polyprodont marsupial, *Perameles natusa*', *J. Comp. Neurol.* 76, 509-36.

39 Sanides, F. (1942), 'Die Architektonik des menschlichen Stirnhirns', *Mon. Neurol. Psych.*, Berlin-Göttingen-Heidelberg, 98.

D. E. HAINES

The cerebellum of some Lorisidae

Introduction

Few portions of the mammalian central nervous system have undergone as obvious a development in the course of phylogeny as the cerebellum. Its enlargement in the higher mammals, especially the primates, is related to subsequent development of many important brain stem areas, for example the pons, ventrolateral nucleus of the thalamus, reticular formation, inferior olivary nucleus, red nucleus, and others.

Elliot Smith studied the cerebelli of *Lemur, Tarsius* and several other forms.[1] He proposed that the cerebellum be divided in the following manner: 'The mesial part of the cerebellum is divided into three regions by means of the deep *fissura prima* . . . and by the shallower *fissura secunda.* The region in front of the *fissura prima* (which is homologous with the so-called "preclival" fissure)[2] may be called the anterior lobe, that between the *fissura prima* and *fissura secunda* the central, or, better middle lobe; the third region being the posterior lobe.' Bolk studied the cerebelli of *Lemur, Nycticebus* and *Perodicticus*, and proposed a schema considerably different from that of Elliott Smith.[3] Subsequent investigators used the terminology of Elliott Smith.[1] Woollard and Tilney briefly discussed the cerebellum of *Tarsius*;[4,5] Le Gros Clark mentioned the cerebellum of *Tupaia, Ptilocercus* and *Microcebus*;[6-8] Kanagasuntheram and Mahran commented on the cerebellum of *Galago senegalensis*,[9] and Krishnamurti briefly mentioned the cerebellum of *Nycticebus coucang.*[10] With the exception of the work of Bolk,[3] the above-mentioned studies did not deal primarily with the cerebellum; consequently, this structure received little or no detailed attention. The above-mentioned papers have been reviewed elsewhere.[11]

Bradley, although he did not deal with the prosimian cerebellum, proposed a different scheme. He stated:[12,13]

> The simplest workable type of cerebellum was found to be one in which a mesial sagittal section shows five lobes separated from each other by four fissures. In order to avoid confusion, no pre-existing names were applied to these lobes and fissures, but the simplest method of designating them – that of letters and figures – was employed. The five lobes were called A, B, C, D, and E. . . . The fissures were similarly designated as I, II [*fissura prima*],[14] III, and IV.

This method did not receive wide support and was not adopted, to any extent, by subsequent investigators.

Over a period of years Larsell developed the method of dividing the mammalian cerebellum into three main lobes and ten lobules.[15,16] His technique was developed by studying the sequence of development of lobules and fissures in the embryo. The postlateral fissure (fissure IV of Bradley) is the first to appear in the embryo and it divides the developing cerebellum into the flocculonodular lobe and the *corpus cerebelli.* The primary fissure (*fissura prima* of Elliot Smith; fissure II of Bradley) is the second to develop and it divides the *corpus cerebelli* into anterior and posterior lobes. The anterior lobe is composed, in the adult, of lobules I to V; the posterior lobe of lobules VI to IX; the flocculonodular lobe is lobule X. A more detailed discussion of this method is given elsewhere.[11,15,16]

The method of Larsell has been applied to the cerebelli of early postnatal and adult specimens of *Galago senegalensis* and *Tupaia glis.*[11,17] In these studies it was concluded that the cerebellum of *G. senegalensis* has all the characters of higher primates, but at a basic level of organisation. The cerebellum of *Tupaia* has morphological characteristics which are seen both in non-primate mammals and in some basic primate forms.

The morphology of the deep cerebellar nuclei of prosimians has been virtually ignored. Tilney stated that the cerebellar nuclei of *Tarsius* 'emphasise their general resemblance to lower mammalian forms'.[5] The cerebellar nuclei of *Lemur catta, Avahi laniger, Loris tardigradus* and *Tarsius spectrum* have been divided into a *nucleus lateralis* (NL), *nucleus interpositus anterior* (NIA), *nucleus interpositus posterior* (NIP), and *nucleus medialis* (NM).[18] In *G. senegalensis* and *T. glis* the cerebellar nuclei are also divided into NL, NIA, NIP, and NM.[19] In *Galago* the nuclei are relatively easy to distinguish as separate cell masses, and the lateral nucleus (dentate) appears concave medially and shows a primitive hilus. However, in *Tupaia* the NL, NIA, and NIP are joined in caudal and ventral areas, and the lateral nucleus is merely an irregular mass of cells with no distinct hilus. In this respect, the morphology of the deep cerebellar nuclei of *Tupaia* closely resemble the basic pattern seen in the hedgehog and the mole.[18,19] The characteristic undulatory appearance of the lateral cerebellar nucleus of higher primates is exemplified at a basic level of organisation in the lesser *Galago.*

The cerebellar corticonuclear projections and cerebellar efferents of prosimians have not been investigated. The concept of medial, paravermal, and lateral functional zones, as originally proposed using the Marchii technique, has undergone some modification in deference to recent results of cerebellar corticonuclear studies using the newer silver degeneration techniques (see Discussion). The present paper will include a discussion of corticonuclear projections in the lesser *Galago* following lesions of the so-called lateral and medial functional zones.

Rationale

The cerebellum functions in the area of coordination and integration of unconscious proprioception with motor function. The animals most often used in a comparative study of cerebellar morphology or cerebellar corticonuclear connections represent divergent (and to a certain extent specialised) evolutionary forms such as the cat, rat and monkey. Not only are these animals divergent forms, but they have developed distinctly different locomotor patterns. There are undoubtedly neurological specialisations, coupled with osteological and myological specialisations, which are related to the evolution of the locomotor characteristics of each group.

The neurological basis for the evolution of bipedal locomotion has not been adequately explored. The family Lorisidae has members which illustrate different locomotor patterns, coupled with varying degrees of brain development. A study of the cerebellum as one centre related to postural orientation may provide clues relative to the neurological specialisations which enhanced the evolutionary development of bipedal locomotion. Some results of work on the cerebellum of the tree shrew (*Tupaia*) are included, since this animal may represent a living example of the primitive stock from which primates, and possibly all placental mammals, arose.

Materials and methods

The cerebelli of adult specimens of the greater bushbaby (*Galago crassicaudatus*), the lesser bushbaby (*Galago senegalensis senegalensis, G. s. moholi*), the slow loris (*Nycticebus coucang*), and the potto (*Perodicticus potto*) were used in this study. Several specimens of the tree shrew (*Tupaia glis*) were also used.[20] Gross examinations of the cerebelli were made under a dissecting microscope with a zoom lens. The gross drawings were made using calipers and metric rules, and repeatedly checked from all angles for accuracy.

The deep cerebellar nuclei of *G. senegalensis, G. crassicaudatus, N.*

coucang and *P. potto* were studied in transverse and horizontal serial sections. The method used in our laboratory for serial sectioning allows exact determination of all dimensions of the nuclei under consideration. The Klüver and Barrera method for combination staining,[21] and a cresyl acetate violet Nissl stain, were used to stain the deep nuclei and surrounding fibre bundles. All sections used in the present study were cut at 30 or 40 microns on a sliding microtome modified for frozen sections.

Lesions were placed in the cerebellar cortex of nine lesser bushbabies (*G. s. senegalensis, G. s. moholi*) and two tree shrews (*Tupaia glis*). The animals were anaesthetised intravenously with sodium pentobarbital. The nuchal musculature was retracted, a small area of occipital bone removed, the dura incised, and a small portion of cortex was aspirated. The animals were allowed to survive 3 to 5 days, then sacrificed by perfusion with saline followed with 10% formalin. The brain and spinal cord were removed *in toto.* The entire brainstem was serially sectioned and stained with the Fink and Heimer method for degenerating fibres.[22] Alternating sections were stained with a cresyl acetate violet method for cell bodies.[23] Projected drawings of the deep cerebellar nuclei of normal and lesioned animals were made on a Leitz Prado projector.

In the present study, horizontal sections are defined as those cut on a plane parallel to the floor of the fourth ventricle. A transverse section is cut on a plane perpendicular to the horizontal plane.

Results

1. Nomenclature

The nomenclature of the gross morphology of the cerebellum in this study is that of Larsell. The terminology of the cerebellar nuclei and their homologisation with the nuclei of other animals has been established by Rüdeberg.

2. Anterior lobe of the corpus cerebelli

The anterior lobe of the *corpus cerebelli* is composed of all portions rostral to the primary fissure. It can be divided into either four or five lobules; the merits of both approaches will be discussed below.

The anterior lobe of the cerebellum is well developed in all the Lorisidae examined in the present study. The anterior vermis is relatively large and, with the exception of a modest lobule I (lingula), the hemispheric portions of the lobules II through V are well developed.

(*a*) *Lobule I.* Vermian lobule I, the lingula, is small and composed of two or three folia. The lingula is larger in *G.*

crassicaudatus than in *P. potto, G. senegalensis*, and *N. coucang* (Fig. 1). In the latter three, it is composed of either a single lobule, or two sublobules. The principal pattern appears to be that of two sublobules. In the greater galago, the lingula is large and occasionally shows three sublobules. In greater and lesser galago, the lingula has no hemispheric extension (see later, Fig. 18).[11] However, the potto and slow loris have what appears to be a laterally attenuated portion of the lingula that may represent the hemisphere (HI) of the lingula in these animals. The precentral fissure separates the lingula from the ventral central lobule.

(*b*) *Lobules II and III.* Lobules II and III of the vermis represent the ventral and dorsal central lobules respectively. The ventral central lobule II in all specimens examined, with the exception of the greater galago, consists of a single folium (Figs. 3, 4; Haines[11] Fig. 1). In the greater galago, lobule II has two sublobules (II*a*, II*b*) with sublobule II*a* being small in most animals. The shallow fissure separating sublobules II*a* and II*b* does not extend to the cerebellar margin (see also Fig. 18). The predominant pattern appears to be that of a single folium for lobule II in the Lorisidae.

(*c*) *Lobule III*, the dorsal central lobule, is composed of sublobules III*a* and III*b* in slow loris, potto, and lesser bushbaby (Figs. 3, 4; Fig. 1 of Haines 1969).[11] In the greater galago, two or three sublobules may be present (Fig. 2). The intracentral fissure extends to the lateral cerebellar margin only in *G. crassicaudatus* (Fig. 2), while in the lesser galago, slow loris and potto it terminates before reaching the margin (Figs. 3, 4). The hemispheric portions of lobule II (HII) and III (HIII) are separated from each other by the intera-central fissure in greater galago. Since the intracentral fissure terminates prior to reaching the cerebellar margin, the hemisphere portions HII and HIII in potto, slow loris, and lesser bushbaby share a common area (HII + HIII).[24] The preculminate fissure extends to the lateral cerebellar margin separating the dorsal central lobule from the ventral culminate lobule.

(*d*) *Lobules IV and V.* The ventral and dorsal culminate lobules (IV and V respectively) are located between the preculminate and primary fissures and constitute the larger portions of the anterior lobe. In all genera studied, lobule IV consists of two sublobules and apposes the flocculus and/or middle cerebellar peduncle laterally (Figs. 2-4; Figs. 1, 2 of Haines).[11] The hemisphere area, HIV, is a single folium in all animals. Vermian lobule V is composed of two or three main sublobules in the slow loris, and in greater and lesser bushbabies (Figs. 2, 3). In the potto, lobule V had four sublobules and appears somewhat enlarged over the same area in the other genera. The intraculminate fissure extends to the lateral cerebellar margin, dividing a rather small HIV from the relatively well developed HV region. It is interesting to note that the hemispheric portions of the anterior lobe (HII-HV) of the Lorisidae of the present study are well developed, while in the tree shrew (Fig. 11 in Haines 1969)[11] and other basic forms this region is usually poorly developed.

List of abbreviations

Roman numerals associated with lower case letters, either in the text or in figures, indicate sublobules. Roman numerals in the discussion or in figures prefixed by an upper case H indicate the hemispheric portion of that specific vermian lobule.

Ac	anterior commissure
AMV	anterior medullary velum
ANS	ansiform lobule
APML	anterior paramedian lobule
CP	*copula pyramidis*
Cr I	crus I of the ansiform lobule
Cr II	crus II of the ansiform lobule
DPF	dorsal paraflocculus
F	flocculus
For	fornix
FB	fibre bundle (position of efferents of the lateral cerebellar nucleus)
FAP	ansoparamedian fissure
FIC	intracentral fissure
FICL	intraculminate fissure
FN-L	flocculonodular lobe
FNP	flocculonodular peduncle
FP	primary fissure
FPC	preculminate fissure
FPn	prenodular fissure (posterolateral fissure)
FPo	postpyramidal fissure
FPr	prepyramidal fissure
FPRC	precentral fissure
FPS	posterior superior fissure
h	hilus of the lateral cerebellar nucleus
Hab	habenula
ICP	inferior cerebellar peduncle
ICS	intercrural sulcus
LS	simple lobule (*lobulus simplex*)
MB	mammillary body
MCP	middle cerebellar peduncle
NIA	anterior interposed nucleus (*nucleus interpositus anterior*)
NIP	posterior interposed nucleus (*nucleus interpositus posterior*)
NL	lateral cerebellar nucleus (*nucleus lateralis*)
NM	medial cerebellar nucleus (*nucleus medialis*)
p	position of the parafloccular stalk
Pc	posterior commissure

Pi	pineal body
PL-CC	posterior lobe of the corpus cerebelli
PPML	posterior paramedian lobule
PS	parafloccular stalk
SCC	splenium of corpus callosum
SCP	superior cerebellar peduncle
VL	lateral vestibular nucleus
VM	medial vestibular nucleus
VPF	ventral paraflocculus
VS	superior vestibular nucleus
4th	fourth ventricle
5th	trigeminal nerve root

I	lingula
II	ventral central lobule
III	dorsal central lobule
IV	ventral culmen
V	dorsal culmen
VI	declive (*lobulus simplex*)
VII	folium tuber
VIII	pyramis
IX	uvula
X	nodulus

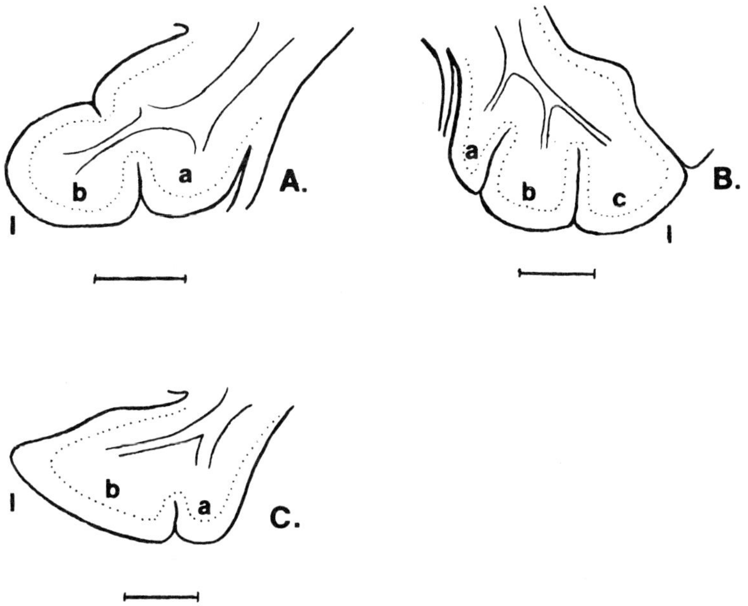

Figure 1. Mid-sagittal view of the lingula (lobule 1) of *P. potto* (*A*), *G. crassicaudatus* (*B*), and *N. coucang* (*C*). Scale = 1 mm.

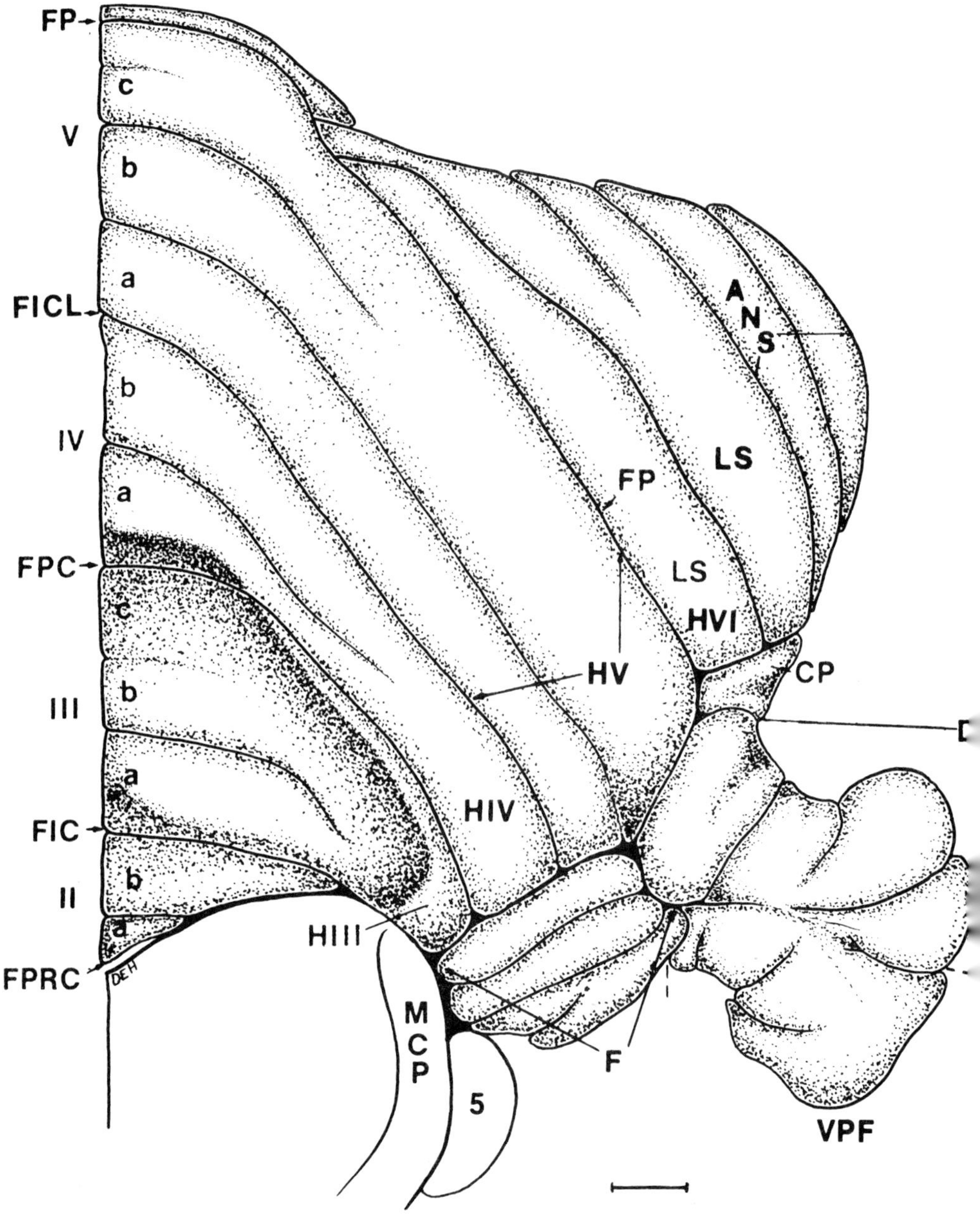

Figure 2. Rostral and slightly dorsal view of the anterior aspect of the cerebellum of *Galago crassicaudatus*. Note the indentation of the anterior lobe formed by the inferior colliculus. Scale = 1 mm.

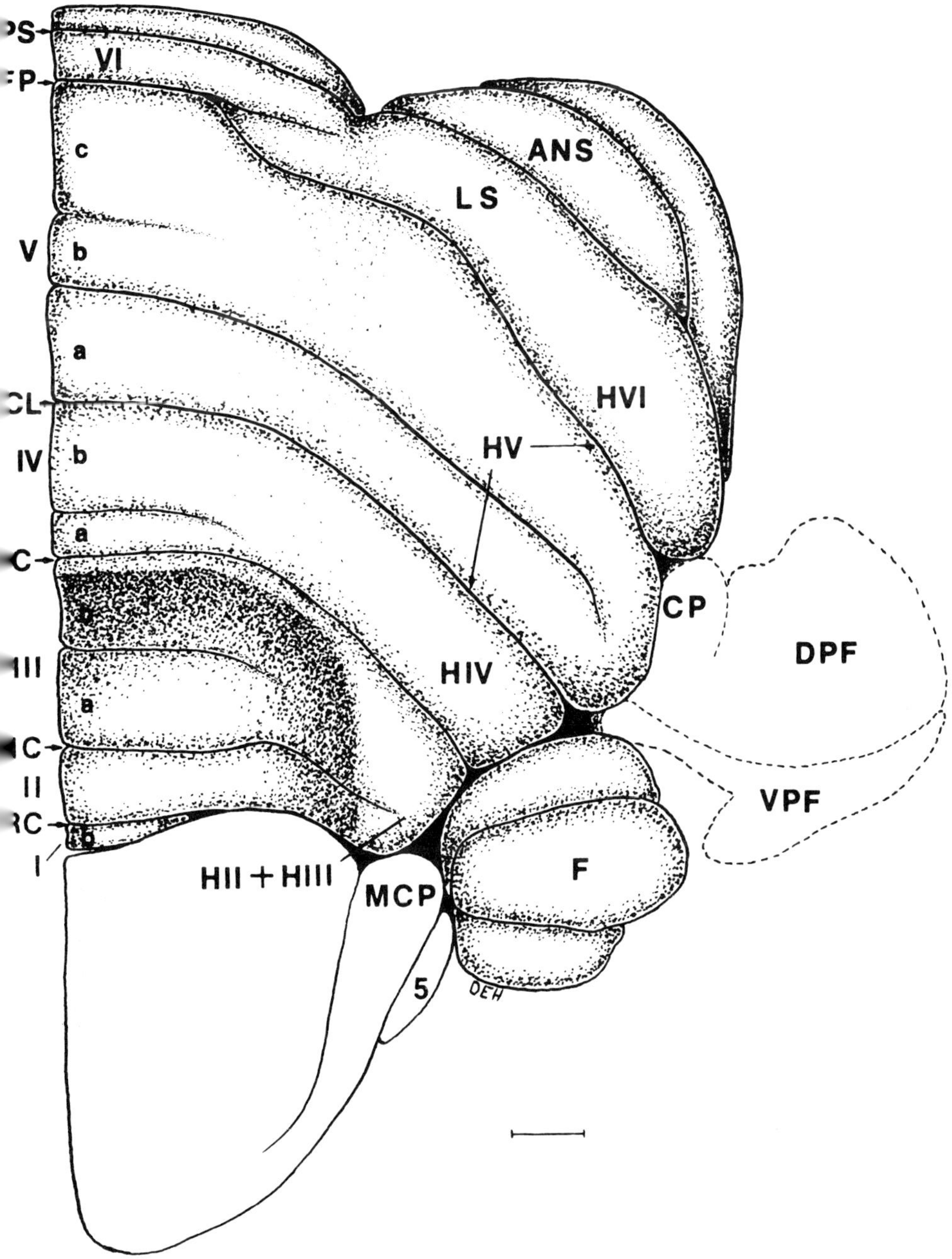

Figure 3. Rostral and slightly dorsal view of the anterior aspect of the cerebellum of *Nycticebus coucang*. Scale = 1 mm.

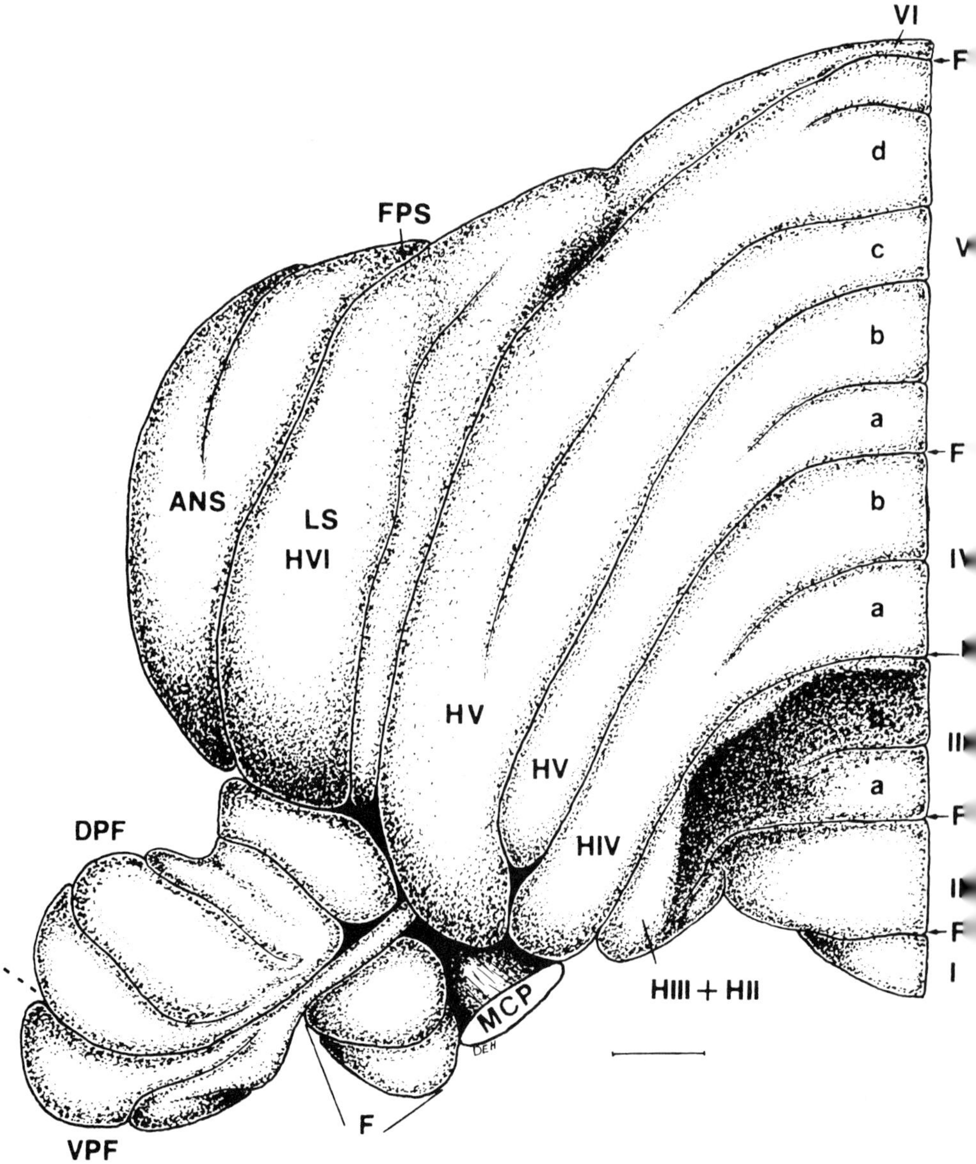

Figure 4. Rostral and slightly dorsal view of the anterior aspect of the cerebellum of *Perodicticus potto.* Scale = 1 mm.

3. *Posterior lobe of the corpus cerebelli*

The posterior lobe of the *corpus cerebelli* is located between the primary fissure and posterolateral fissure and it is the largest of the three principal lobes. The posterior lobe is divisible into a simplex lobule (VI), a folium-tuber region (VII), a pyramidal lobule (VIII) and the uvular lobule (IX).

(*a*) *Lobule VI.* The lobulus simplex in the vermis is a single folium, and its continuation with the hemisphere is distinct (Figs. 6, 8). Laterally, the HVI portion of the lobule is in contact with portions of the dorsal paraflocculus and *copula pyramidis.* The posterior superior fissure separates lobules VI and VII, and extends, almost uninterrupted, from the vermis to the lateral cerebellar margin. The hemisphere portion (HVI) of the lobulus simplex is usually a single folium of variable width in slow loris (Fig. 5), while in greater and lesser galago and potto this lobule is composed of two folia (Figs. 4, 6, 7). When two folia are present, the rostral folium is usually the narrowest (Figs. 4, 7).

Even though the vermis portion of the *lobulus simplex* appears relatively small in size, the HVI portion of this lobule in all genera is more extensive. In the slow loris, lobule HVI is not foliated on the surface; but several folia are present on the anterior aspect of the lobule (HVI) deep within the primary fissure (Fig. 9). A similar pattern to that shown in the loris (Fig. 9) is also observed on HVI within the primary fissure of the other genera studied.

(*b*) *Lobule VII.* Vermian lobule VII appears to be subdivided into two primary sublobules. These are divided by the ansoparamedian fissure into VIIA and VIIB. In its course from the vermis to the lateral cerebellar margin, the ansoparamedian fissure is discontinuous in the paravermal area in both greater and lesser bushbaby and potto (Figs. 6, 8, 12). In the slow loris it is a continuous fissure (Fig. 10). The hemisphere portions of VIIA and VIIB are the ansiform lobule and anterior paramedian lobule respectively. In potto and greater galago the ansiform lobule is made up of three folia, the rostral two folia interpreted as crus I and the caudal folium as crus II (Figs. 4, 6, 8). In the slow loris and lesser galago, the anisform lobule (HVII) is usually composed of two large folia, each interpreted as a crus (Figs. 7, 10, 11). In some specimens of greater galago, crus II of the ansiform lobule (HVII) is located on the caudal aspects of crus I deep within the ansoparamedian fissure (see inset, Fig. 6). Variations of the ansiform lobule in the lesser galago have been reported previously.[11]

The anterior paramedian lobule is a lateral continuation of vermian lobule VIIB. In the potto it appears to be composed of a wide folium subdivided by a shallow intrinsic sulcus. The continuity between vermian lobule VIIB and the anterior paramedian lobule is obvious (Fig. 8). The anterior paramedian lobule of the slow loris is slightly larger than that of the potto, and the continuation between this lobule and vermian lobule VIIB is relatively distinct (Fig. 10). The anterior paramedian lobule of the lesser galago is evaluated anew in light of the comparative data obtained in the present report on other Lorisidae. The ansoparamedian fissure extends to the cerebellar margin caudal to crus II of the ansiform lobule, and the prepyramidal fissure passes laterally around the hemisphere to join the former fissure. The

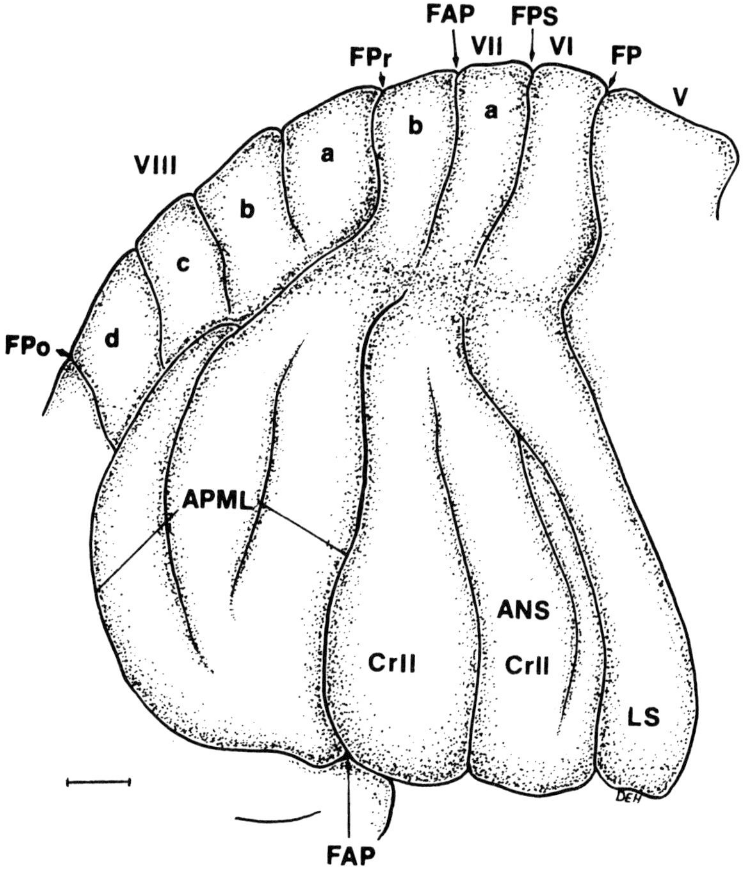

Figure 5. Dorso-lateral view of the hemisphere of the posterior lobe of the cerebellum of *Nycticebus coucang.* Note that some fissures (FAP and FPS) are discontinuous in the paravermal region. Scale = 1 mm.

anterior paramedian lobule is interpreted as the three folia between these fissures (Fig. 7). This would suggest that this lobule is slightly larger than previously interpreted.[11] In the greater galago the anterior paramedian lobule consists of four

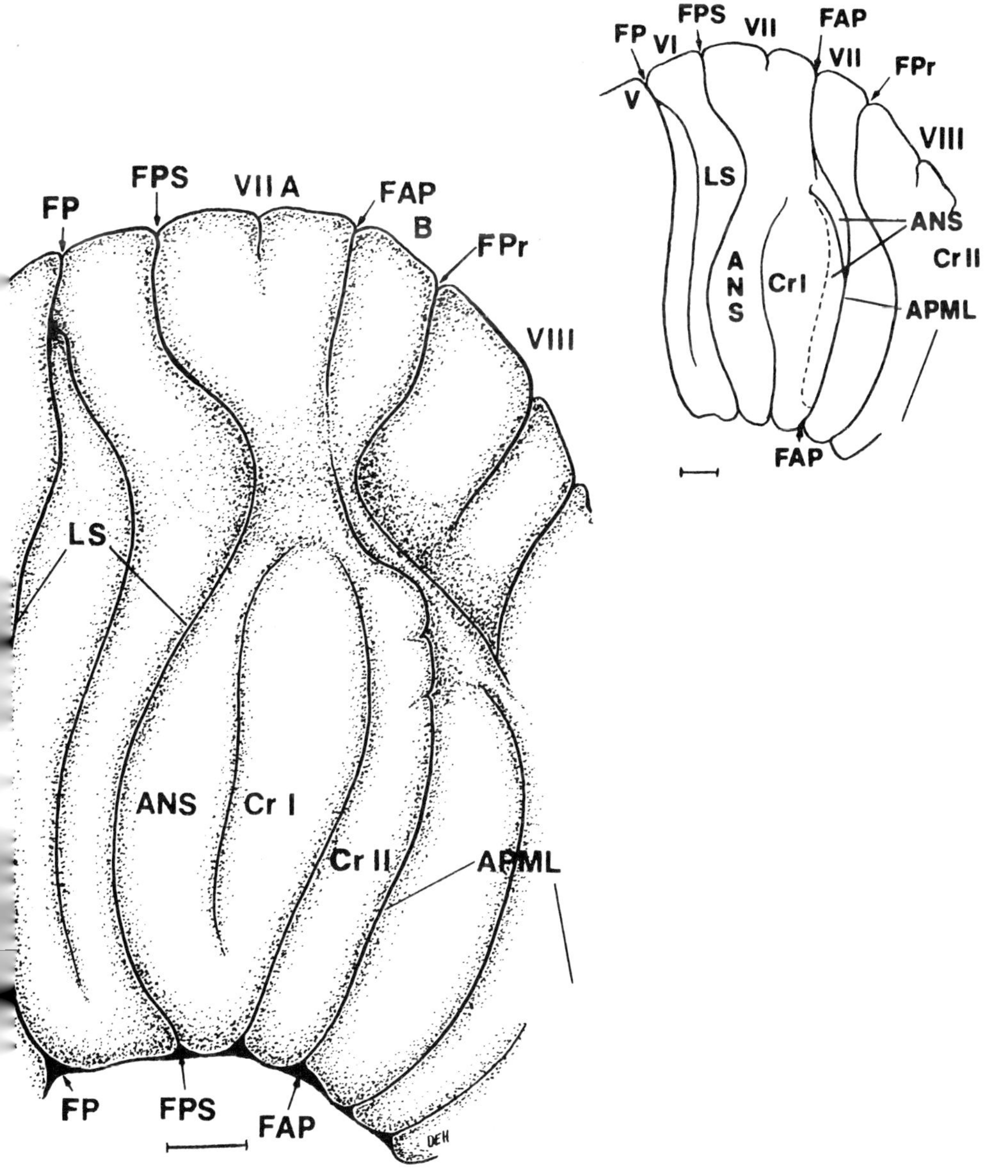

Figure 6. Dorso-lateral view of the hemisphere of the posterior lobe of the cerebellum of *Galago crassicaudatus.* In the inset, note that ansiform lobule crus II is located on the caudal aspect of crus I within the ansoparamedian fissure. Scale = 1 mm.

principal folia covering a large portion of the hemisphere of the posterior lobe (Figs. 12, 13). This conclusion is based on the fact that, when vermian lobule VIII is carefully dissected, no fibres from the pyramidal lobule seem to enter these folia. The observation is in accord with the embryological development of this area in other mammals.

(*c*) *Lobule VIII.* The prepyramidal fissure separates the *folium tuber* region from the pyramidal lobule, vermian lobule VIII. In the foetus, this fissure passes laterally and terminates by dividing the area of the future posterior paramedian lobule from the more rostrally located ansoparamedian lobule. The latter will subdivide into the ansiform and anterior paramedian lobule of the adult. Caudal to the prepyramidal

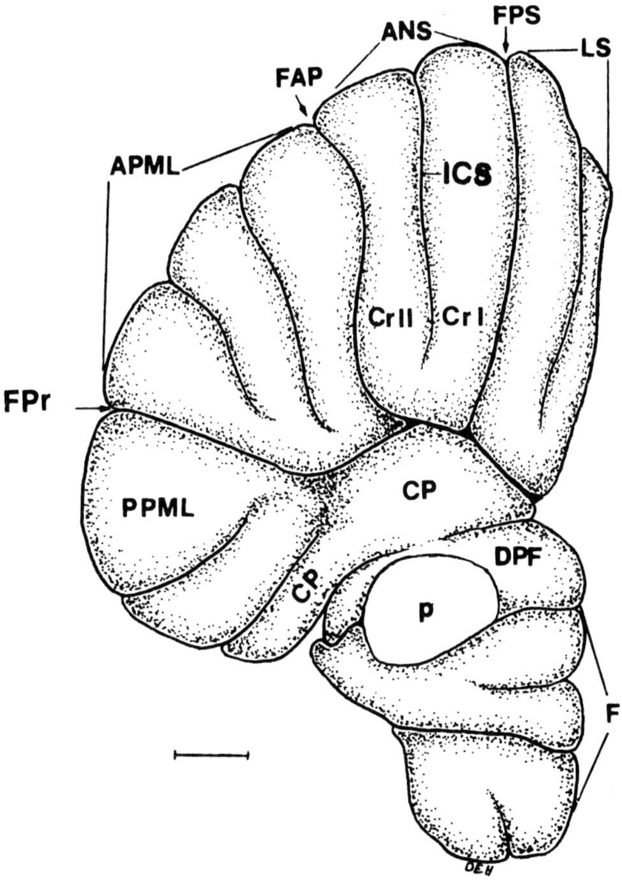

Figure 7. Lateral view of the cerebellar hemisphere of *Galago senegalensis.* The paraflocculus (p) has been removed. Note the anterior and posterior paramedian lobules and the continuation of the copula pyramidis with the dorsal paraflocculus. Scale = 1 mm.

fissure, in the foetus and adult, are the hemisphere portions (HVIII) of the pyramidal lobule. These are the posterior paramedian lobule, *copula pyramidis*, and the dorsal paraflocculus.

The prepyramidal fissure in the Lorisidae of the present

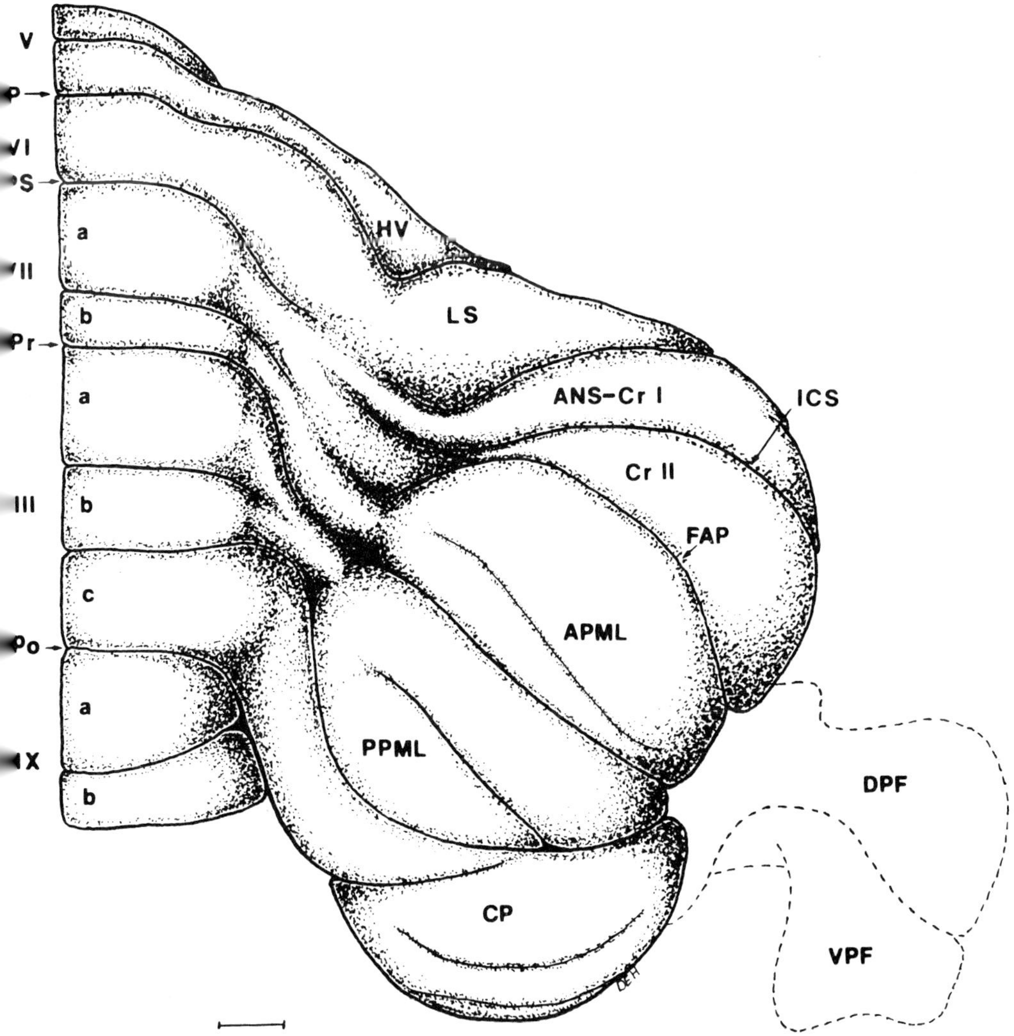

Figure 8. Caudal and slightly dorsal view of the posterior lobe of the cerebellum of *Perodicticus potto*. Note the obvious continuations of vermian lobules, V, VI, VII, and VIII with their respective portions of the hemisphere. The paraflocculus is outlined to scale. Scale = 1 mm.

study extends from the vermis to the lateral cerebellar margin. In the area of the paramedian fissure, it is somewhat obliterated since it travels from the vermis into the paramedian fissure, where it turns abruptly caudal before passing laterally between the anterior and posterior portions of the paramedian lobule.

The posterior paramedian lobule in the slow loris and potto is made up of two folia; in the lesser galago it may be one or two folia, but in the greater galago three folia are usually present (Figs. 7, 8, 10, 13). The posterior paramedian lobule is medially continuous with the base of vermian lobule VIII. The remaining portions of HVIII are the *copula pyramidis* and the dorsal paraflocculus. In the potto, the lower portion of lobule VIII is continuous into the lower to

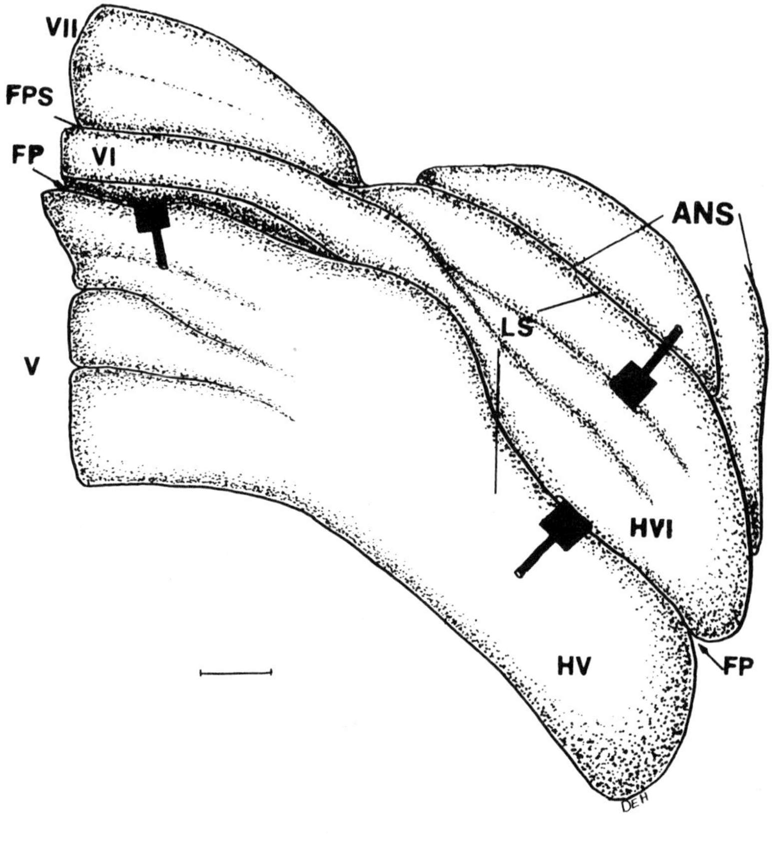

Figure 9. Anterior view of lobules V, VI, and VII of *Nycticebus coucang*. The caudal lip of lobule V is retracted so that the primary fissure is opened. Note the fissuration of the *lobulus simplex* within the fissure. See Results. Scale = 1 mm.

three folia which pass laterally toward the stalk of the dorsal paraflocculus (Figs. 8, 15). Because of its obvious continuation with the pyramidal lobule this area is interpreted as *copula pyramidis* in the potto (Fig. 8). The cortex portion representing the *copula pyramidis* in the slow loris and greater galago is somewhat reduced in size and located ventral to the posterior paramedian lobule (Figs. 10, 13). In both greater galago and slow loris, this cortical area consists of two

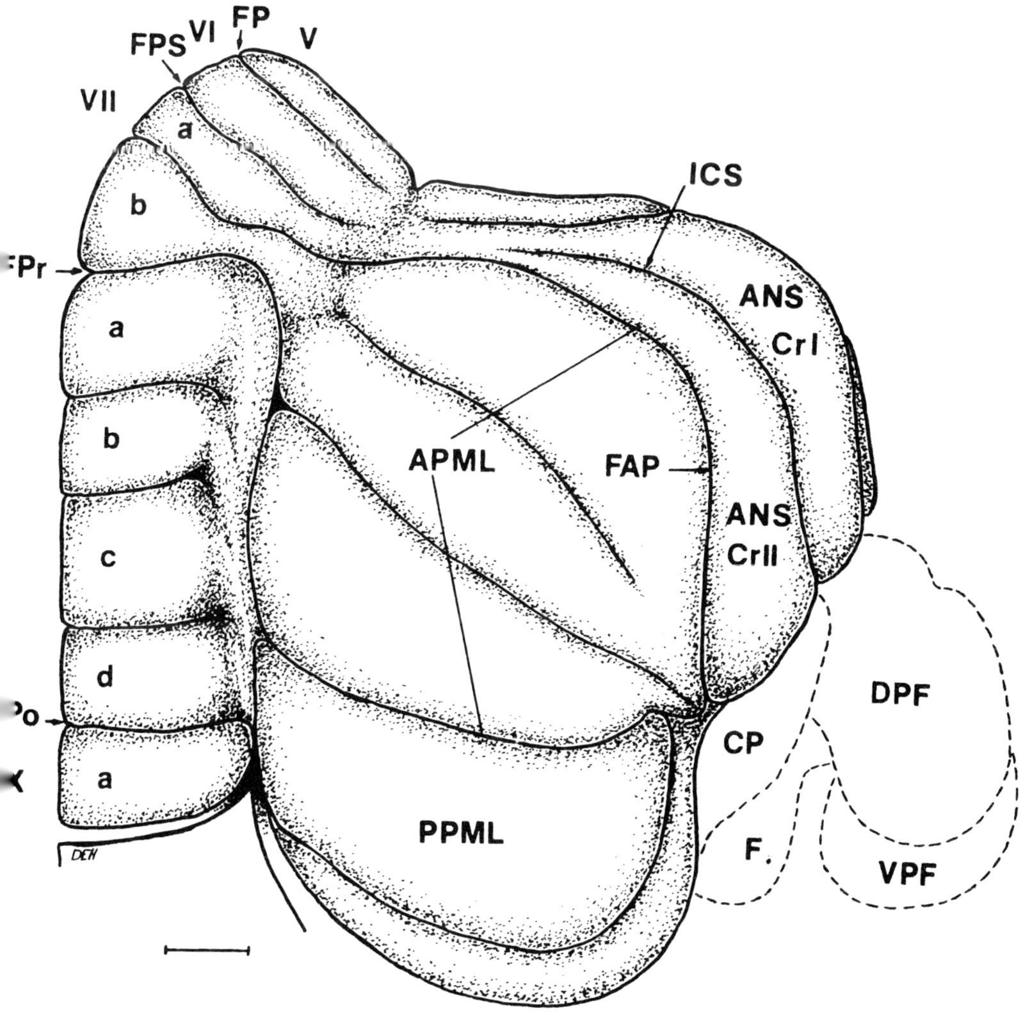

Figure 10. Caudal and slightly dorsal view of the posterior lobe of the cerebellum of *Nycticebus coucang*. Note the distinct continuation of vermian lobules VI and VII with their respective portions of the hemisphere. Scale = 1 mm.

folia (Figs. 11, 13, 18, 19). In the slow loris, the division of the copula is ventral with the lateral-most folium passing laterally to join with the dorsal paraflocculus (Figs. 16, 19). In the greater galago, a moderately-sized folium passes from ventral to lateral and a second flattened folium extends onto the parafloccular stalk in close association with the dorsal paraflocculus (Figs. 13, 18). In general, the appearance of the copula of greater galago and slow loris is similar, while the same area in potto is tentatively interpreted as being somewhat larger (Figs. 8, 15). These observations lead to a reconsideration of the same area in the lesser galago. The *copula pyramidis* had been interpreted as the large area on the ventral aspect of the posterior lobe hemisphere.[11] When the limits of the anterior paramedian lobule were reconsidered and subsequently modified, the area of the copula was also reconsidered. In view of the consistent appearance of this area in the greater galago and slow loris, it was concluded that the copula in the lesser galago may possibly be smaller. The pyramidal lobule and surrounding structures were dissected and fibre bundles followed into hemisphere folia. The posterior paramedian lobule is the lower two folia of the hemisphere in the lesser galago and the copula is the

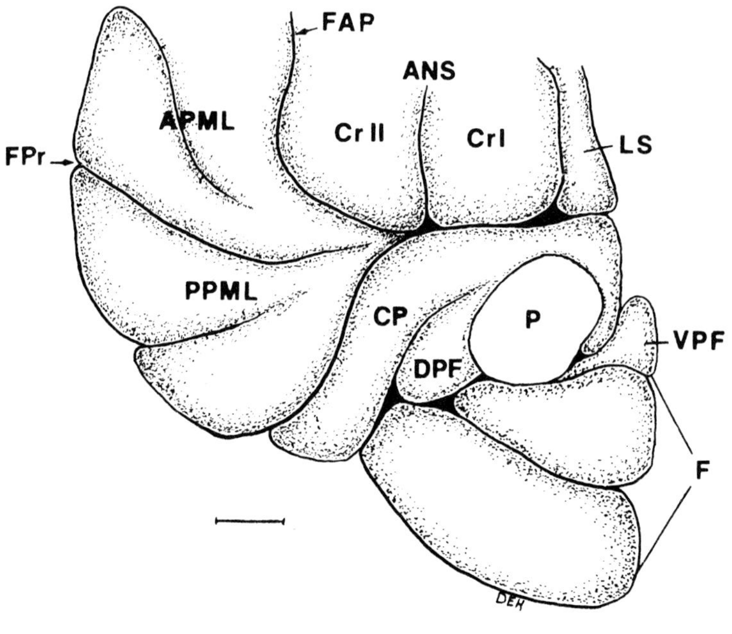

Figure 11. Lateral view of the ventrolateral aspect of the posterior lobe of the cerebellum of *Nycticebus coucang.* Note the continuation of the *copula pyramidis* onto the dorsal paraflocculus. The paraflocculus has been removed. Scale = 1 mm.

lowermost folium on the ventral aspect (Figs. 7, 21). This relatively large folium passes laterally into the caudodorsal aspect of the dorsal paraflocculus where it is broad and flattened (Fig. 7). In general, the configurations of the greater and lesser galagos resemble each other on the

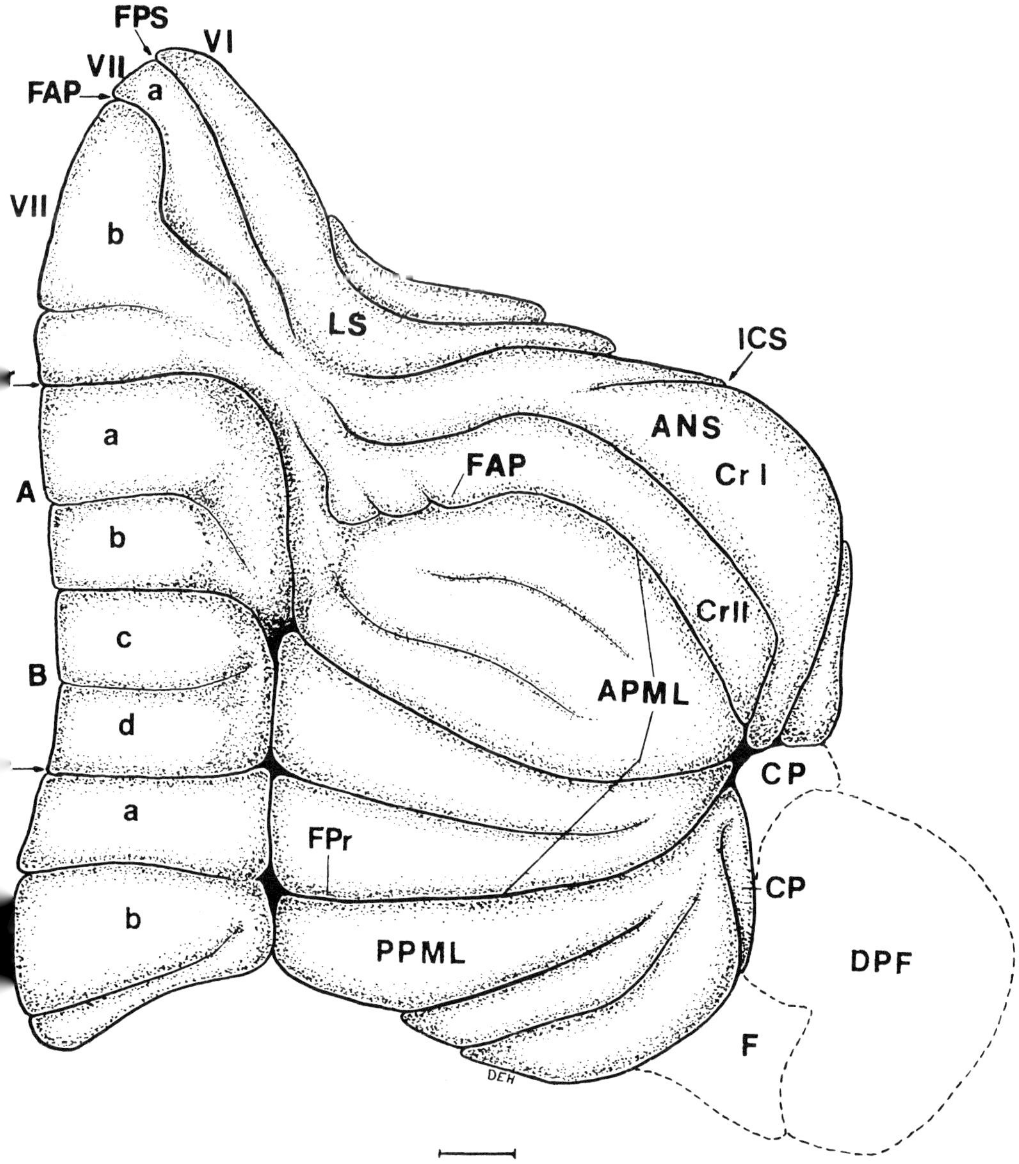

Figure 12. Caudal and slightly dorsal view of the posterial lobe of the cerebellum of *Galago crassicaudatus.* The continuations of vermian lobules V, VI, and VII with their respective areas of the hemisphere is distinct. Note the large anterior paramedian lobule. Scale = 1 mm.

ventrolateral aspect of the posterior lobe of the hemisphere (Figs. 7, 13).

The most lateral portion of HVIII is the dorsal paraflocculus. In all genera studied it is only modestly developed, being composed of two to three principal folia (Figs. 14, 16). The dorsal paraflocculus is slightly larger in the potto than in the other genera (Fig. 15). The division between dorsal and ventral portions is distinct on the rostral surface, but somewhat obliterated on the caudoventral surface. In the slow lorises of the present study a slight variation in the size and shape of the paraflocculi is noted (Fig. 17), although they did show a distinct general pattern.

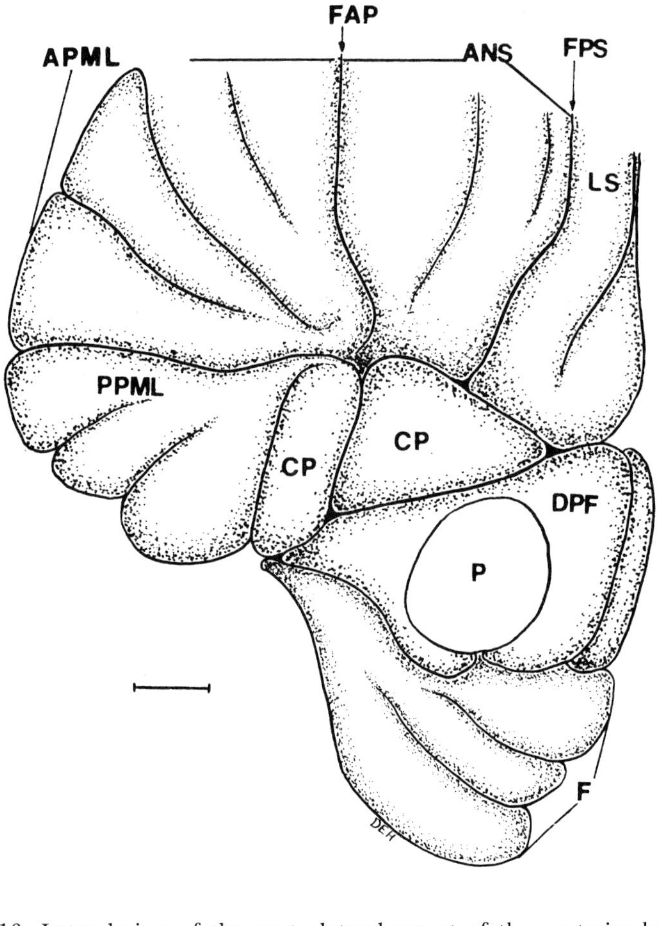

Figure 13. Lateral view of the ventrolateral aspect of the posterior lobe of the cerebellum of *Galago crassicaudatus.* Note the large anterior paramedian lobule, and the arrangement of the *copula pyramidis.* The paraflocculus has been removed. Scale = 1 mm.

(*d*) *Lobule IX.* The postpyramidal fissure extends laterally from the vermis to terminate by dividing the paraflocculus into dorsal and ventral portions. The fissure is continuous from the midline into the hemisphere, but its extreme lateral course is somewhat obliterated by the relation of the paraflocculus and development of the posterior lobe cortex during morphogenesis. Vermian lobule IX, the uvula, is

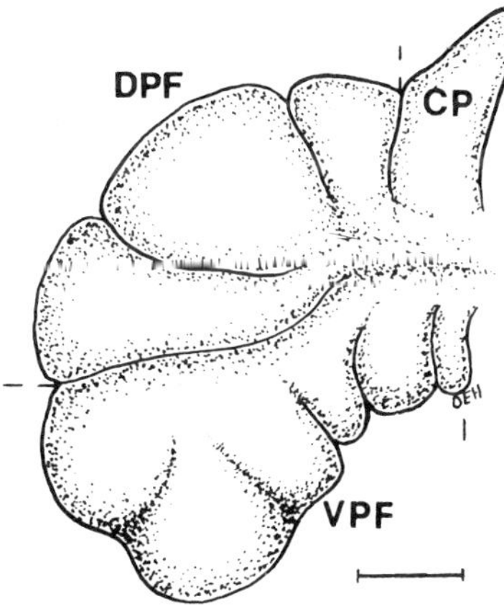

Figure 14. Dorsal view of the paraflocculus of *Galago crassicaudatus.* Note the dorsal and ventral portions, and the intimate association of the *copula pyramidis* with the dorsal paraflocculus. Scale = 1 mm.

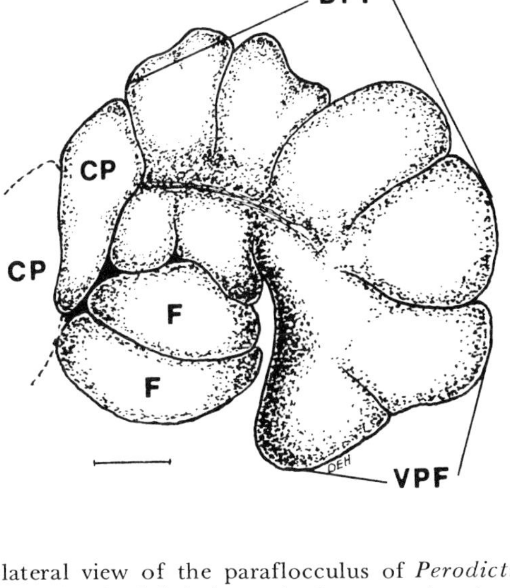

Figure 15. Dorsolateral view of the paraflocculus of *Perodicticus potto.* Note that the dorsal division is slightly larger in the potto than in the other animals of the present study. Scale = 1 mm.

composed of three main sublobules in all the Lorisidae studied, with the exception of the potto. In the latter, the uvula consists of two elongated sublobules.

The only hemispheric representation of the uvula is the ventral paraflocculus, lobule HIX, and in the adult specimens, the continuation between these two structures is reduced to a bundle of fibres located deep with respect to the flocculonodular bundle (Figs. 6, 14 in Haines).[11] The ventral paraflocculus is also only moderately developed. In the potto it consists of four folia (Fig. 15), while in the greater and lesser galago and slow loris it is somewhat smaller, being composed of two to three main folia (Figs. 14, 16, 17; Fig. 7 in Haines).[11]

4. *Flocculonodular lobe*

The prenodular or posterolateral fissure (of Larsell) separates the uvula from the nodulus in the vermis and laterally separates the flocculus from the ventral paraflocculus. The flocculus in the slow loris, potto, and greater galago is composed of two to four rostro-caudally elongated folia (Figs. 4, 11, 13, 18, 20). In the lesser bushbaby the folia are somewhat shortened but four are usually present (Fig. 7). The nodulus is relatively simple in all specimens. It appears to have two main sublobules in greater and lesser bushbabies and slow loris. In the potto it is slightly larger and has three sublobules.

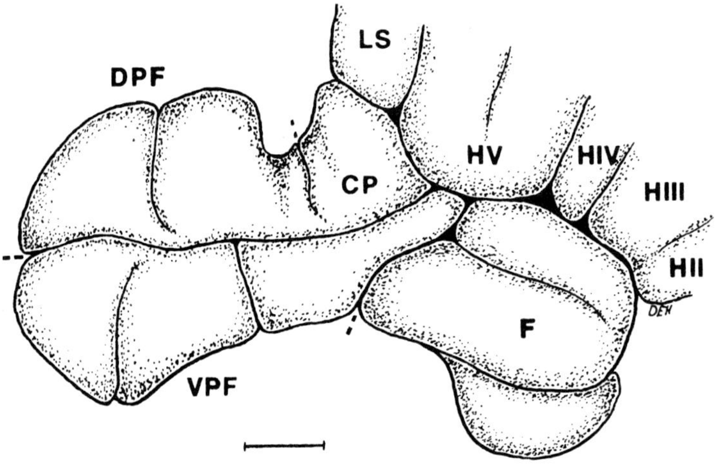

Figure 16. Rostral view of the paraflocculus, flocculus, and ventrolateral portions of the hemisphere of *Nycticebus coucang*. The separation between dorsal and ventral divisions of the paraflocculus is distinct. Note the continuation of the *copula pyramidis* onto the dorsal paraflocculus. Scale = 1 mm.

The flocculus is the hemispheric portion of lobule X, (HX), and it is connected with the nodulus by a medullated bundle of fibres, the flocculonodular peduncle (floccular peduncle of Larsell). This connection is a low ridge of tissue in the slow loris and greater galago, in which delicate striations are visible (Figs. 18, 19). In the potto, however, only a low ridge is evident and no striations can be seen (Fig. 20). In the lesser bushbaby the connections are similar to those seen in the greater galago and slow loris (Fig. 21). This striated ridge passes directly between the most ventral portions of the flocculus and the nodulus of the vermis. The gross separation of the fibres of the flocculonodular peduncle and the fibres passing from the uvula (IX) to the ventral paraflocculus (HIX) is obliterated in the adult.[11,17]

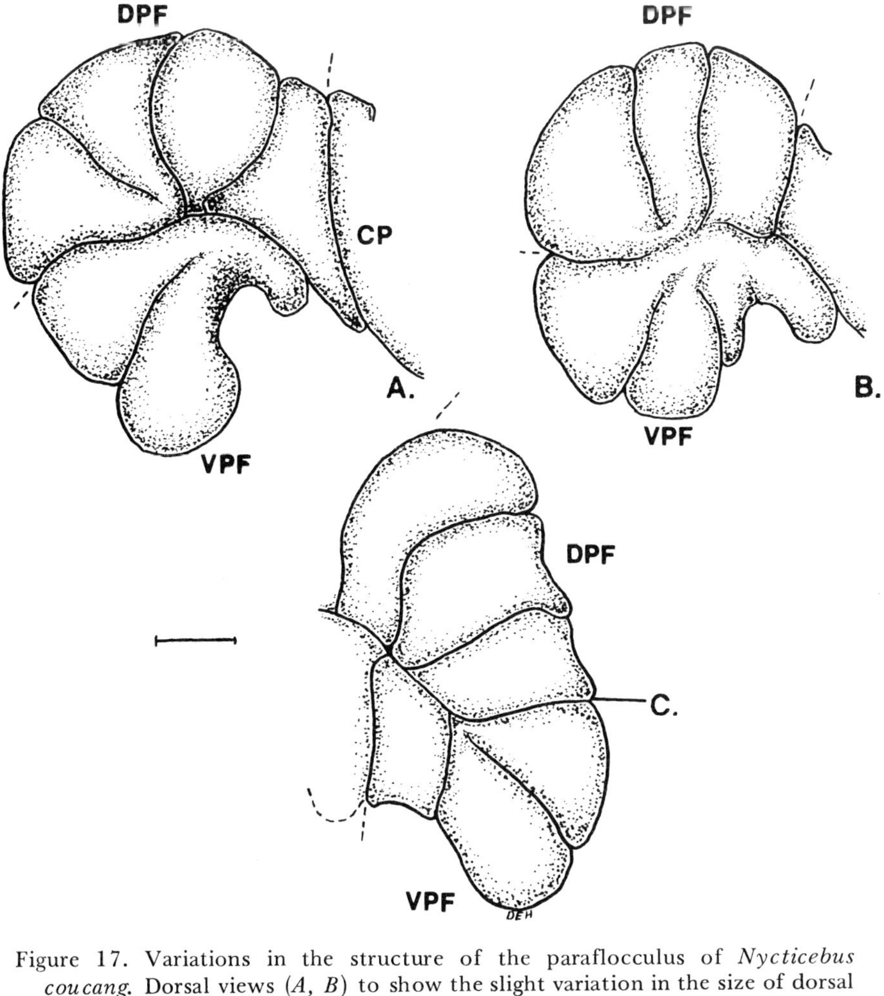

Figure 17. Variations in the structure of the paraflocculus of *Nycticebus coucang*. Dorsal views (*A*, *B*) to show the slight variation in the size of dorsal and ventral divisions. Ventral and slightly caudal view (*C*) of the paraflocculus. Scale = 1 mm.

5. Cerebellar nuclei of Lorisidae

The deep cerebellar nuclei are the principal centres for the origin of efferent information passing out of the cerebellum. In many mammalian forms, the deep nuclei are divisible either partially or totally into a *nucleus lateralis* (NL), a *nucleus interpositus anterior*

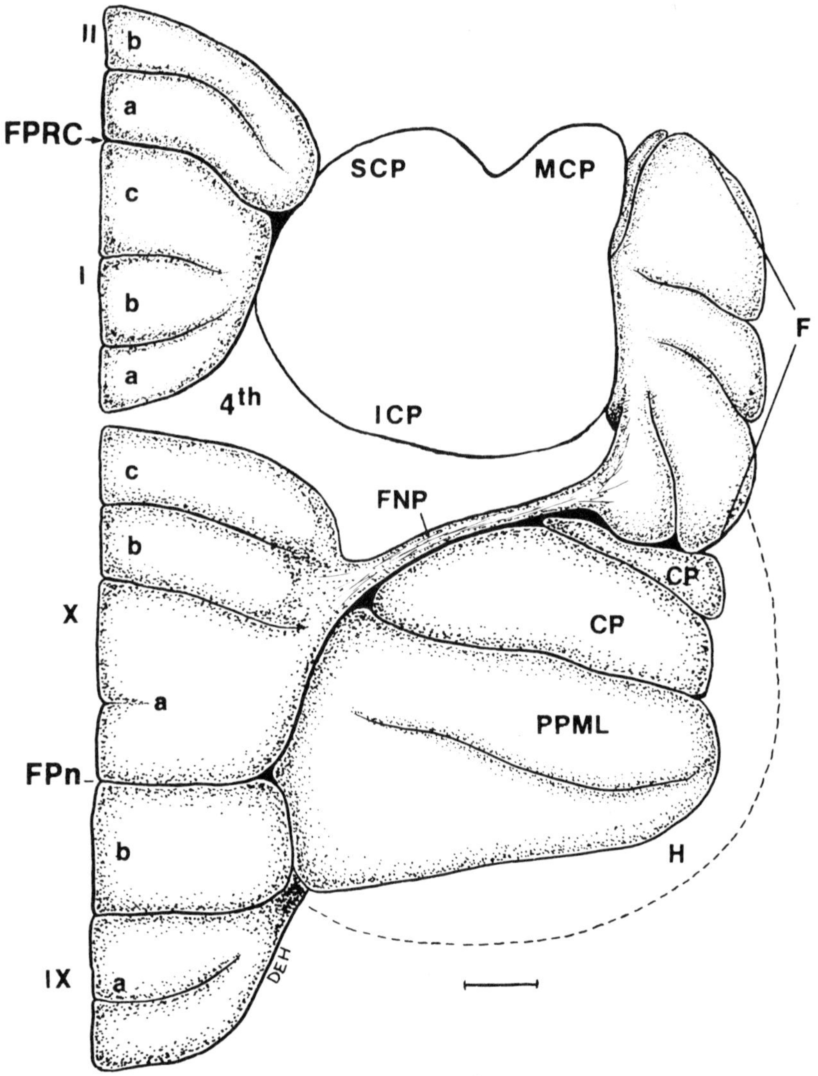

Figure 18. Ventral view of the cerebellum of *Galago crassicaudatus*. The cerebellar peduncles (SCP, MCP, ICP) have been cut. Note the nodulus (X), the flocculus (IX) and the flocculonodular peduncle. The lingula shows no hemisphere, and the hemisphere area of the ventral central lobule is limited. The posterior paramedian lobule and two divisions of the *copula pyramidis* are clearly evident. Scale = 1 mm.

(NIA), a *nucleus interpositus posterior* (NIP) and a medial cerebellar nucleus (NM), on each side.

(*a*) *Nucleus lateralis.* The lateral cerebellar nucleus of slow loris and potto closely resemble each other (Figs. 22, 23). The NL of the slow loris measures approximately 2500 by 1900 microns at its greatest diameter and about 2400 microns at

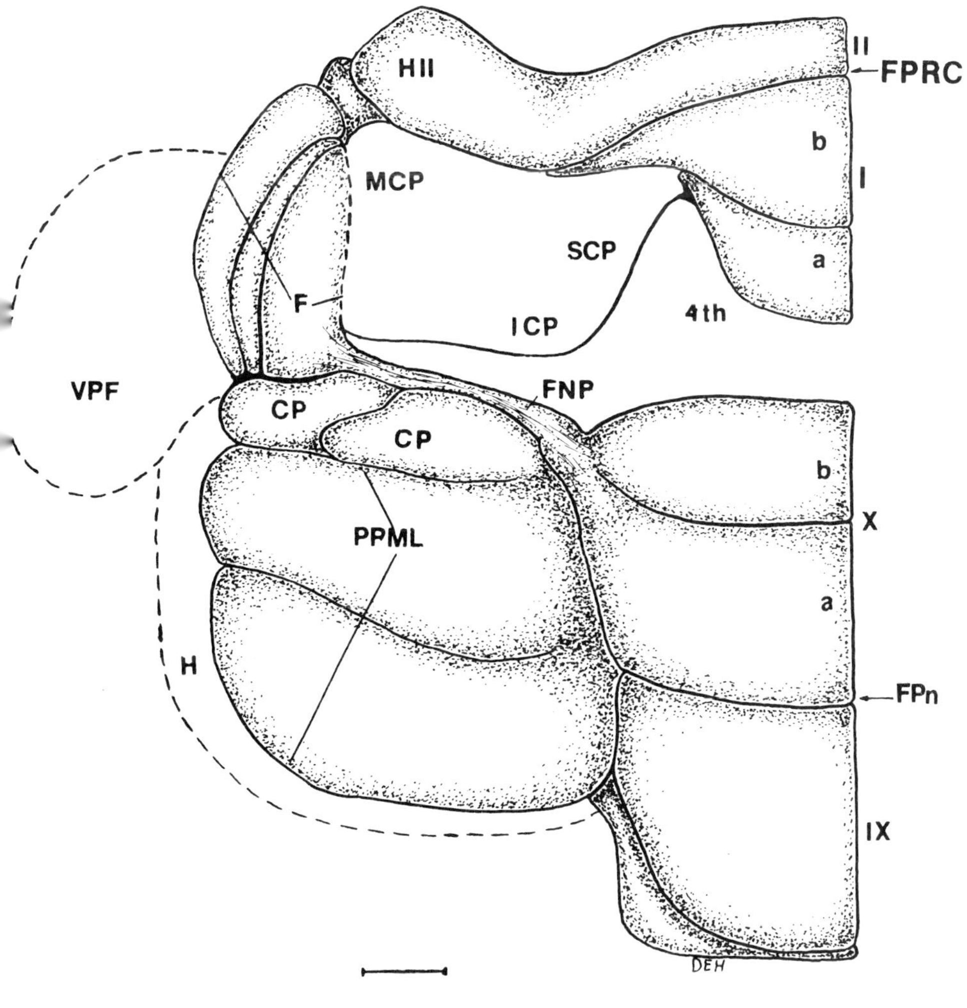

Figure 19. Ventral view of the cerebellum of *Nycticebus coucang*. The cerebellar peduncles (SCP, MCP, ICP) have been cut, exposing the entire ventral area. Note the orientation of the folia of the flocculus, the size of the nodulus, and the striations of the flocculonodular peduncle. The posterior paramedian lobule is large and the two divisions of the *copula pyramidis* are distinct. There is a well developed HII area, and lobule I has a small lateral projection (see Discussion). Scale = 1 mm.

its greatest rostro-caudal length. In the potto the NL is of similar size, measuring about 2600 by 1800 microns at its greatest diameter and approximately 2200 microns at its greatest rostro-caudal length. Throughout their dorsal area, the NL and NIA are joined so closely that it is not possible to define the borders of each nucleus exactly. At approximately mid-horizontal level, the NL and NIA are separated from each other by a delicate lamina of fibres in their middle third (Fig. 23). This separation becomes more evident toward the

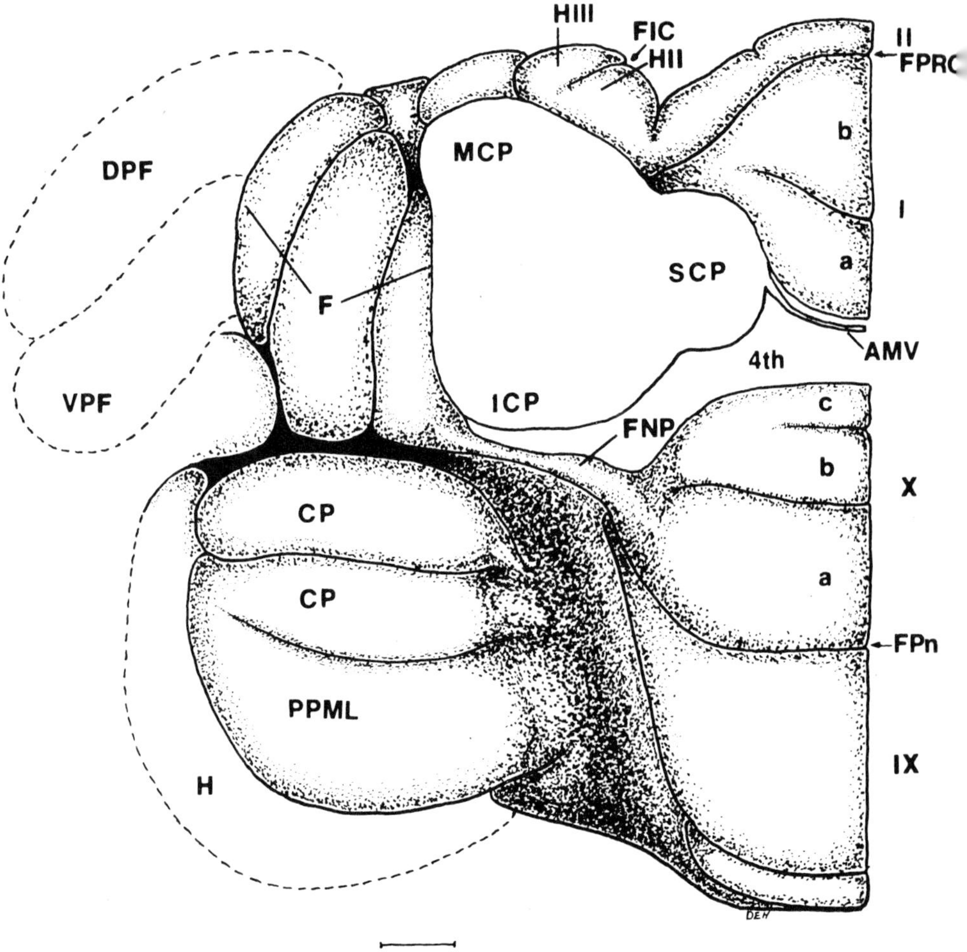

Figure 20. Ventral view of the cerebellum of *Perodicticus potto*. The cerebellar peduncles (SCP, MCP, ICP) have been cut, exposing the entire ventral surface. The nodulus is only moderate in size, the flocculus is distinct, and the flocculonodular peduncle is non-striated. The copula is well developed. Note the lateral extension of lobule I, and the moderate development of lobules HII and HIII. Scale = 1 mm.

rostral pole of the NL in the ventral area. The caudomedially directed pole of the NL is continuous with the NIA throughout the dorsoventral extent (Figs. 22, 23). This is especially evident in potto (Fig. 23*B-E*). In the slow loris and potto, clusters of cells are present between the NL and NIA in the ventro-caudal region (Figs. 22*F*, 23*E-F*).

In overall observation, the *nucleus lateralis* of the potto and slow loris is somewhat semilunar in shape, with a distinct ventromedially directed hilus (Fig. 32*c*, *d*). It is significant to note, however, that the NL and NIA are difficult to divide morphologically into distinct separate cell masses.

Identification of the hilus area is facilitated by the presence of fibre bundles presumed to constitute efferents of the NL. The cell population of the NL in potto and slow loris consists of small and medium-sized fusiform and stellate shaped cells (14-23 microns) dispersed throughout a population of medium and large fusiform and stellate shaped cells (24-35 microns). In the ventrocaudal region of the NL of the slow loris there appears to be a higher population of smaller cell bodies (14-21 microns); however, larger cells (22-28 microns) are also seen in this area. In the potto the range of cell size is essentially the same as in the slow loris, although

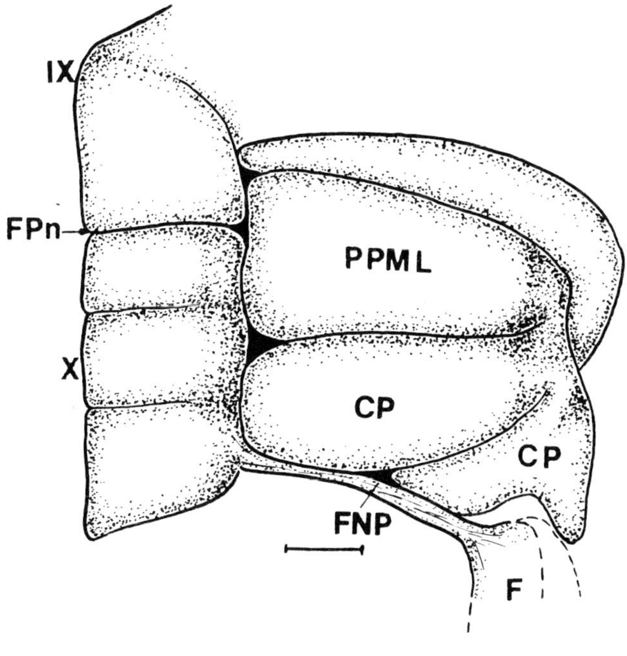

Figure 21. Ventral view of caudal area of the cerebellum of *Galago senegalensis*. Note the nodulus, flocculus, and the delicate flocculonodular peduncle. The *copula pyramidis* is large. Scale = 1 mm.

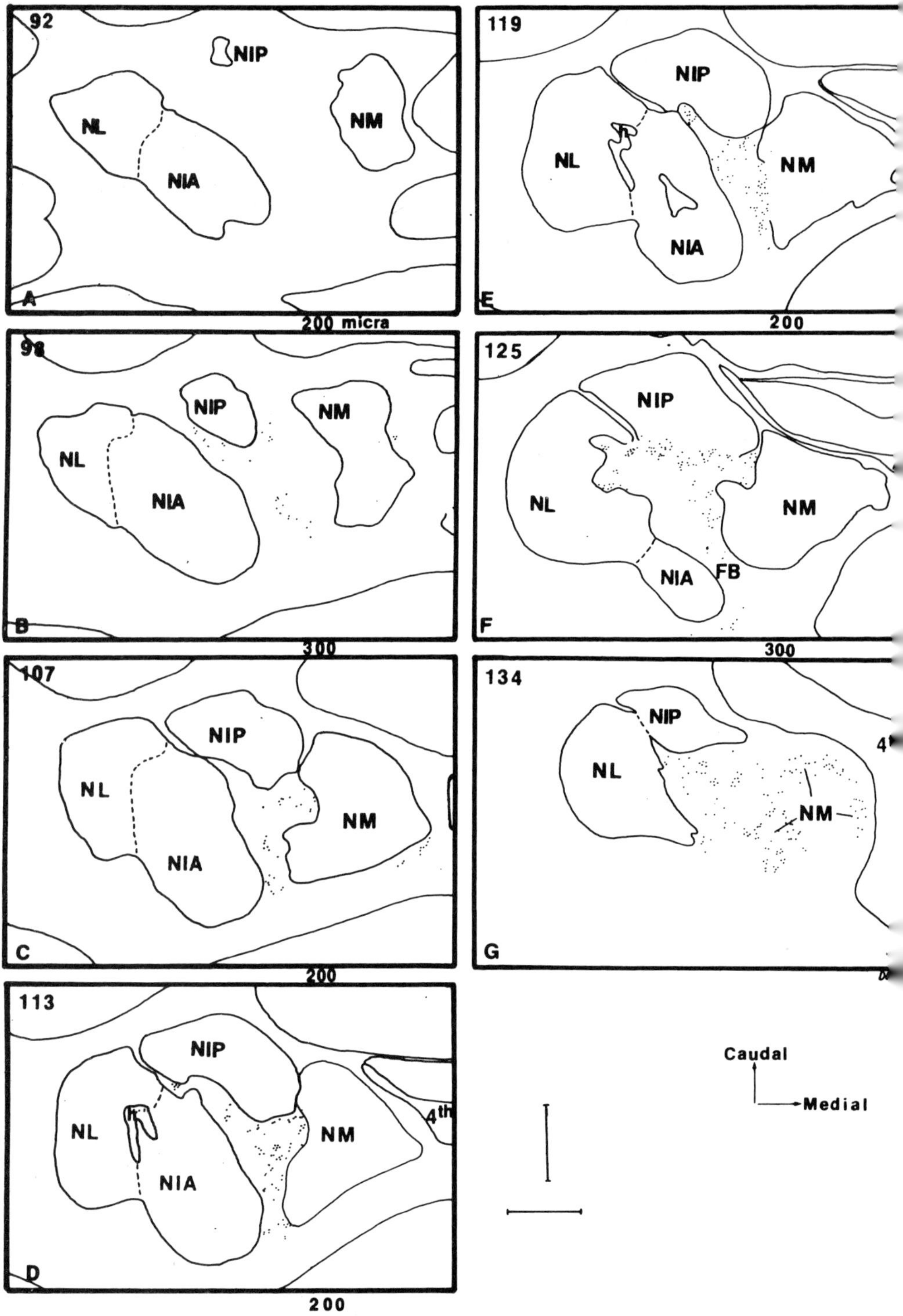

Figure 22. Tracings of the cerebellar nuclei of *Nycticebus coucang* in horizontal section. The upper left number in each tracing is the section number of that specific tracing, and the numbers between each tracing represent the depth in microns separating one from the next. The individual dots represent the positions of cell bodies seen in that particular 40 micron section. These cells did not appear to be located within the boundary of any individual nucleus, nor did they form a subnucleus. These tracings are from dorsal (*A*) to ventral (*G*). Scale = 1 mm.

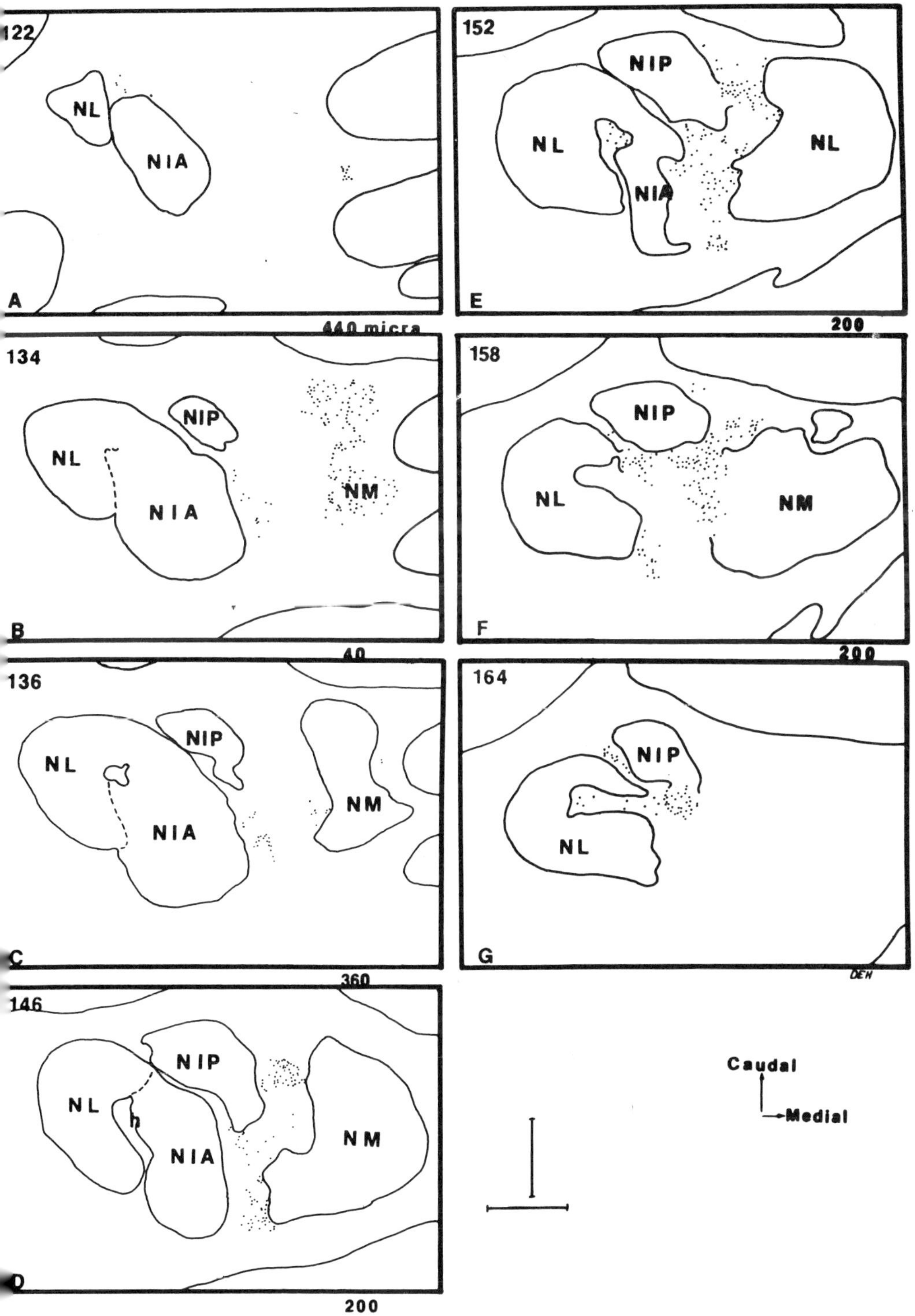

Figure 23. Tracings of the cerebellar nuclei of *Perodicticus potto* in horizontal section. The broken line in some drawings represents an arbitary division of nuclei. For additional explanation see Figure 22. Tracings are from dorsal (*A*) to ventral (*G*). Scale = 1 mm.

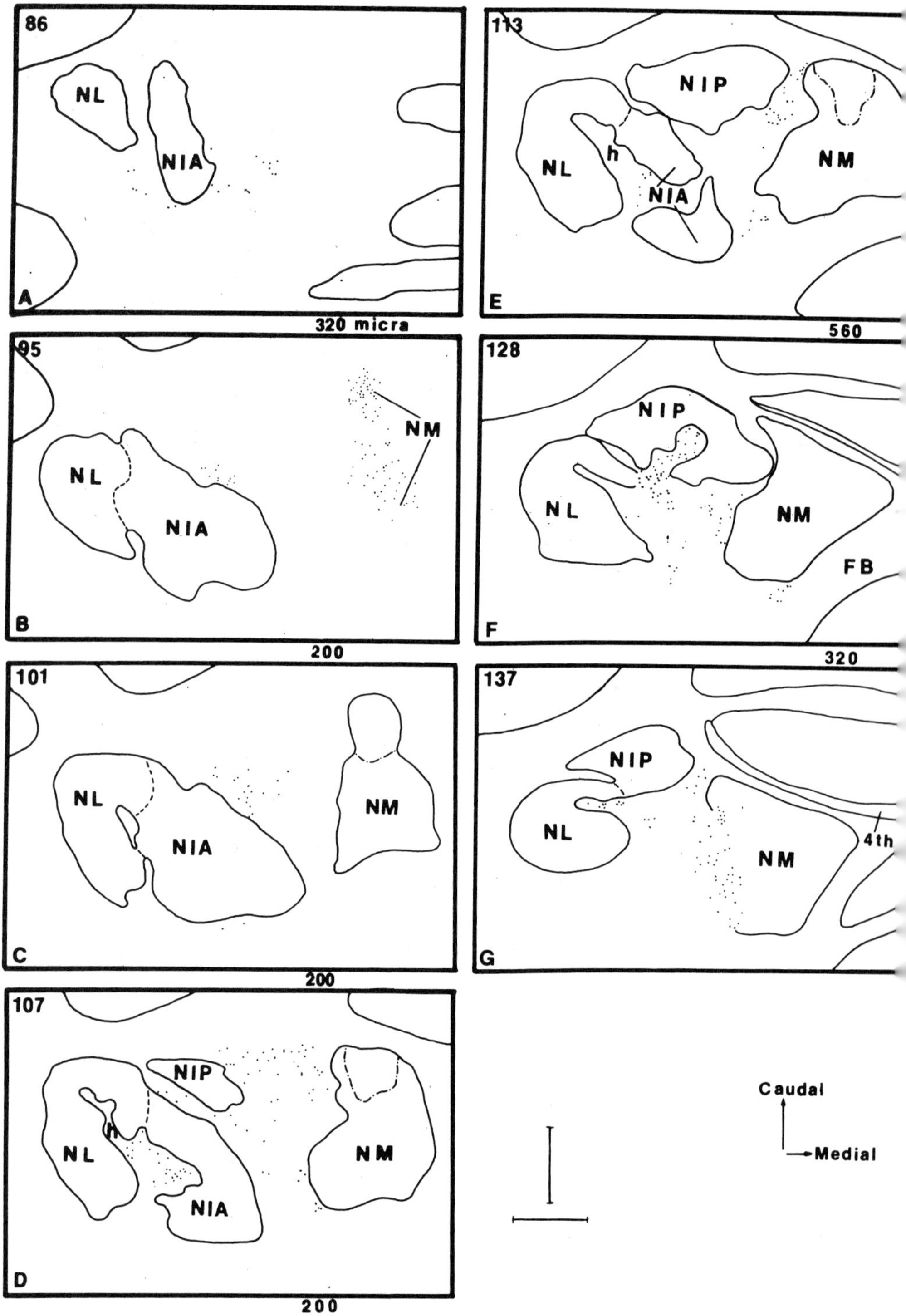

Figure 24. Tracings of the cerebellar nuclei of *Galago crassicaudatus* in horizontal section. Note the compact appearance of cells in the caudal area of NM (*B*). The area of compact cells is indicated by the dot-dash line (*C-E*). For additional explanation see Fig. 22. Tracings are from dorsal (*A*) to ventral (*G*). Scale = 1 mm.

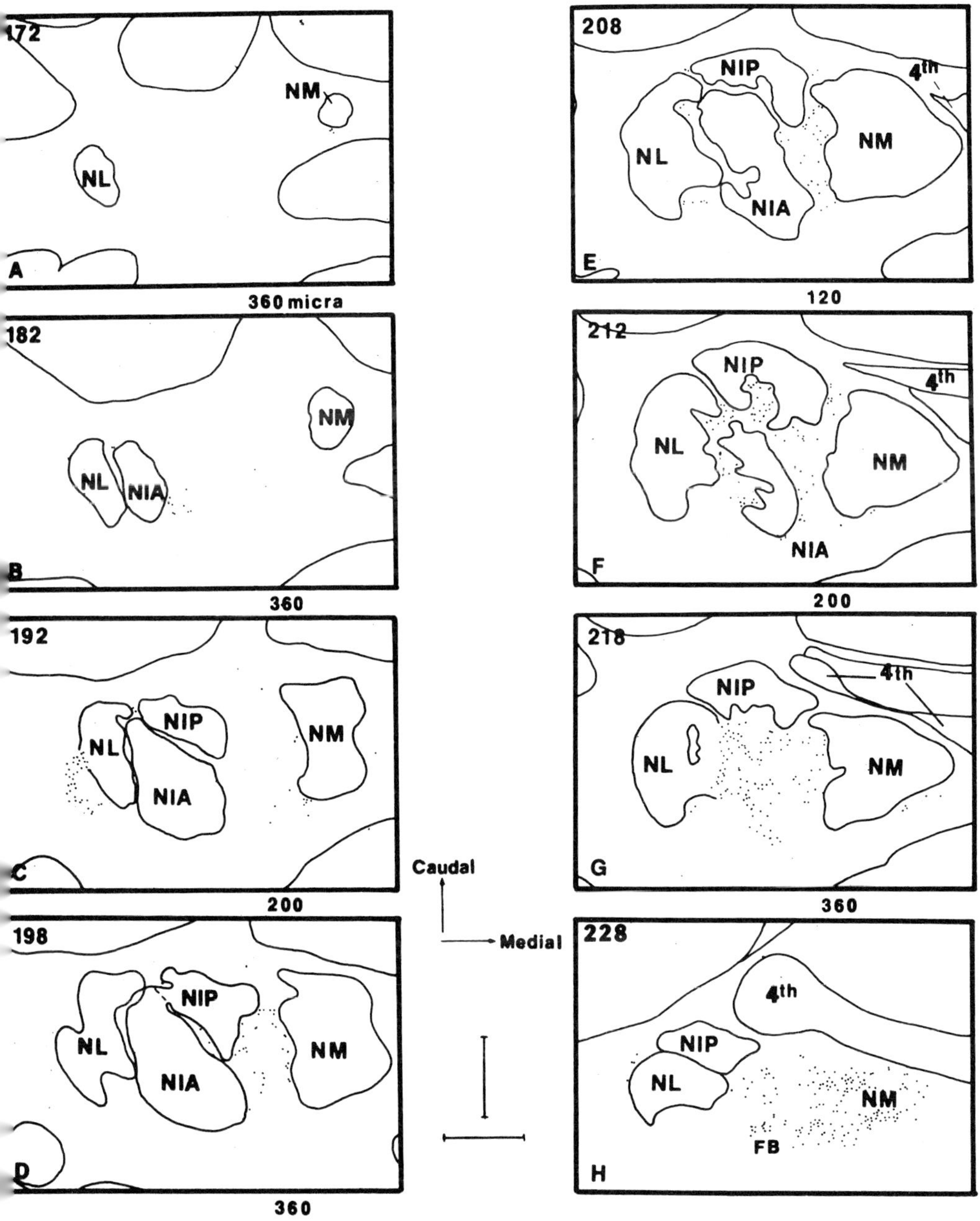

Figure 25. Tracings of the cerebellar nuclei of *Galago senegalensis* in horizontal section. Tracings are from dorsal (*A*) to ventral (*H*). For additional explanation see Fig. 22. Scale = 1 mm.

there is a predominant population of oval-shaped cell bodies (Fig. 33*e*). There is no regional concentration of large or small cell bodies, therefore the NL of the potto and slow loris does not appear to be subdivisible into subnuclei.

In the greater and lesser galago the NL is separated from the NIA throughout by a lamina of delicate medullated fibres. The NL of the greater bushbaby is approximately 2800 by 1700 microns at its greatest diameter and about 2300 microns at its greatest rostrocaudal length. The approximate size of the NL in the lesser bushbaby has been reported elsewhere.[19] The only point at which it is difficult to define the boundary of NL and NIA is at the caudomedially directed pole of the NL in the greater galago (Fig. 24*D, E*). The NL of both galagos is also in intimate contact with a small portion of the NIP in the most caudal and ventral area (Figs. 24*G*, 25*H*). In both the bushbabies the NL is medially concave, being more so in the greater than in the lesser bushbaby (Fig. 32*a*). A medially oriented hilus is apparent in the lesser galago (Fig. 25*D-F*). In the greater galago the NL is somewhat larger and the hilus is more distinct and oriented ventromedially (Fig. 24*D-F*). No large- or small-celled areas are identifiable; therefore, no subnuclei are definable. The size and shape of the cell bodies seen in the NL of the greater and lesser galago are within the ranges mentioned above for the slow loris.

(*b*) *Interposed nuclei.* The interposed nuclei in the Lorisidae studied can be divided into a *nucleus interpositus anterior* (NIA) and a *nucleus interpositus posterior* (NIP). The NIA extends most rostrally and the NIP extends most caudally in all genera studied (Figs. 21-25). As previously noted, the NIA is closely associated with the NL throughout the dorsal area in the slow loris and potto, but separated from NL, in dorsal and rostral areas, by a delicate lamina of fibres in both galagos. The NIA is strikingly similar in all genera studied; it has a rounded bulbous anterior pole, has its most intimate contact with the NL, disappears first in serial section passing dorso-ventrally, and in traverse section the NIA terminates caudolaterally in close association with the NL and NIP (Figs. 22-25, 32*d*). The NIA and NIP in the slow loris form a group of cells that measures approximately 2200 by 2100 microns at its greatest diameter and about 3400 microns at its greatest rostro-caudal length. These nuclei in the potto, also taken collectively, form a group of cells about 2000 by 1900 microns at its greatest diameter and about 2900 microns at its greatest rostro-caudal length.

The NIP also presents a rather consistent pattern in the greater and lesser bushbabies, slow loris and potto (Figs.

22-25). It appears in horizontal section (passing dorsal to ventral) at about the same time as the *nucleus medialis* (NM). The NIA and NIP collectively form a group of cells in the greater bushbaby about 2000 by 2200 microns at its greatest diameter and approximately 3000 microns at its greatest rostro-caudal length. These dimensions are somewhat smaller in the lesser bushbaby.[19] The NIP is located caudomedial in relation to the NIA, and in all specimens has a lateral portion extending caudal to the NL. The NIP is largest in the potto and slow loris, and it comes in intimate contact with the NM in caudal areas (Figs. 22, 23). In the greater and lesser galago, the NIP is slightly smaller (Figs. 24, 25). In all genera, the NIP is separated from the NL and NIA by a relatively consistent lamina of fibres. There appear to be no cyto-architectural subdivisions of either the NIA or NIP, and the predominent population is of medium (16-24 microns) and large (25-36 microns) fusiform and stellate-shaped cell bodies (Fig. 32*b*, *d*).

(*c*) *Medial nucleus.* The *nucleus medialis* (NM), or tectal nucleus, is large in all genera and located just lateral to the midline and dorsolateral to the fourth ventricle. The NM of the slow loris is an irregular mass of cells, somewhat flattened dorsoventrally, measuring about 2400 by 2000 microns at its greatest diameter and about 2200 microns at its greatest rostro-caudal length. The NM in potto is slightly larger, measuring about 2600 by 2100 microns at its greatest diameter and 2500 microns at its greatest rostro-caudal length. The NM in the greater bushbaby is an irregular mass of cells approximately 2200 by 2200 microns at its greatest diameter and about 2350 microns at its greatest rostro-caudal length. In the lesser bushbaby, the NM is considerably smaller.[19] The NM of all genera is relatively large and somewhat irregular in shape (Figs. 22-25). With the exception of the slow loris, the NM is separated from the NIP and NIA by small bundles of fibres. In the slow loris the NM and NIP are in close contact in the middle third at their caudal aspect (Fig. 22*C-F*). Compared in size with the lateral three nuclei, the NM is proportionately the same size in greater and lesser galagos and slow loris (Figs. 22, 24, 25). In the potto the NM is distinctly larger in comparison with the lateral three nuclei (Fig. 23). The general cell population is of small (14-18 microns) and medium-sized (19-23 microns) oval, fusiform, and stellate-shaped cell bodies. Intermixed with these small cells is a general population of medium and large (24-37 microns) oval, a few fusiform elongate, and stellate-shaped cell bodies. There is no organisation of cell bodies in a manner that would suggest the existence of a subnucleus of the NM in any of the Lorisidae studied. However, in the greater galago the cell bodies in the caudal third of the NM are distinctly compact in their organisation (Figs. 24*B*, 32*b*). This area is composed of small and medium-sized cell bodies with some large cells scattered throughout. Therefore it could not be considered as a subnucleus based strictly on cell size. The caudal portion of the NM of the potto also appears to have a more compact cell population than the rostral areas. However, this condition is not as accentuated as in the greater galago.

6. *Cerebellar corticonuclear projections*

The cerebellar cortex projects onto the deep cerebellar nuclei in a highly organised manner. The zonal concept (see Discussion) proposed that the cortex of the vermis projected onto the medial cerebellar nucleus, the paravermal cortex projected onto the interposed nucleus (NIA and NIP), and the lateral functional zone projected onto the lateral cerebellar nucleus. Recent observations on corticonuclear projections of the cat, a monkey and the rat (see Discussion) has shown that these projections are not necessarily limited to a single specific deep nucleus.

In the present study, lesions were placed in the cortex of the lateral and medial functional zones of the lesser galago (*G. s. senegalensis* and *G. s. moholi*) and the tree shrew (*Tupaia glis*). The animals used for this portion of the present study therefore represent a prosimian with a specialised locomotor pattern and an animal which may or may not be a primate.[59] It has been suggested by several investigators that the tree shrews represent either a living example of the stock from which modern primates evolved or the ancestral stock of all placental mammals. These considerations, coupled with the quadripedal locomotor pattern of tree shrews, have led to its use, on a comparative basis, in a long-term study of the phylogeny of cerebellar function as related to the evolutionary development of upright patterns of locomotion.

(*a*) *Lesions of the lateral functional zone.* The degeneration pattern following lesions of the lateral functional zone is reported (Figs. 26-9). A lesion placed in the lateral functional zone in the cortex of crus II of the ansiform lobule and the rostral folium of the anterior paramedian lobule shows degeneration in the lateral cerebellar nucleus (NL), and in both the anterior and posterior portions (NIA, NIP) of the interposed nuclei (Fig. 26). Fibres of passage enter the NL and the NIA from the caudal and caudolateral aspect (Fig. 26*A*), and fibres of passage extend into the lamina separating the NL and NIA (Fig. 26*A*, *B*). Most of these fibres appear to enter the extreme lateral portion of the NIA. There is terminal degeneration in the caudal half of the NL and in the lateral portion of NIA (Fig. 26*B-D*). A small amount of terminal degeneration is seen in the rostromedial areas of the NIA in dorsal levels (Fig. 26*A, B*) and in the rostral area of the NL in mid-horizontal levels (Fig. 26*C, D*). In more ventral areas, terminal degeneration in the NIA is extremely sparse (Fig. 26*E*). Fibres of passage are noted to course rostromedially through the anterior area of the NL. Some of these fibres terminate on cell bodies in the medial area of the NL and others appear to extend into the most lateral area of the NIA (Fig. 26*D, E*). A small bundle of fibres is noted passing towards, and entering, the caudolateral area of the NIP (Fig. 26*A-D*). This small discrete fasciculus of degenerating fibres terminates on cell bodies in the caudolateral area of the NIP (Figs. 26*D-F*, 35*c*, *e*).

A smaller lesion of crus I of the ansiform lobule in the cortex of the lateral functional zone shows a pattern of degeneration similar to that seen above (Fig. 27). Fibres of passage enter the caudolateral aspect of the NL and NIP, and a small bundle is noted entering the caudolateral area of the NIP (Fig. 27*D–F*). Terminal degeneration is dense in the lateral area of the NL (Fig. 34*a*) and sparse, but consistent in the lateral areas of the NIA and NIP (Fig. 27*B-F*, Fig. 35*d*,*f*). In horizontal and transverse section, the degeneration in the lateral nucleus is most dense in the dorsal 4/5, leaving the ventral area relatively free of terminal debris (Figs. 26, 27). In both animals illustrated (*G. senegalensis senegalensis*, Br. Ser. 6, and *G. senegalensis moholi*, Br. Ser. 12) it is also noted that the degeneration in the NL is most dense in the caudo-lateral portion of the nucleus (cf. Figs. 26 and 27).

As noted above, the dorsal and ventral paraflocculi are hemisphere areas HVIII and HIX, respectively. Due to their embrylogical development and anatomical position in the adult, they represent the most ventrolateral area of the cortex of the lateral functional zone of the posterior lobe. A lesion of the total dorsal and ventral paraflocculus results in dense degeneration in the ventrolateral caudal half of the NL (Figs. 28*A-E*, 34*a*). There is also a heavy amount of terminal degeneration in the medial area of the NL (Fig. 28*C*), and an extremely light amount of degeneration in the caudolateral area of the NIA (Figs. 28*C*, 34*d*, *e*). A small bundle of degenerated fibres extends caudorostrally ventral to the NIP and NIA (Figs. 27*B*, 34*C*). This bundle appears to give off a few fibres which terminate on cell bodies located slightly ventral to the NIA and NIP proper (Fig. 28*B, C*); however, its primary contibution appears to be to a small concentration

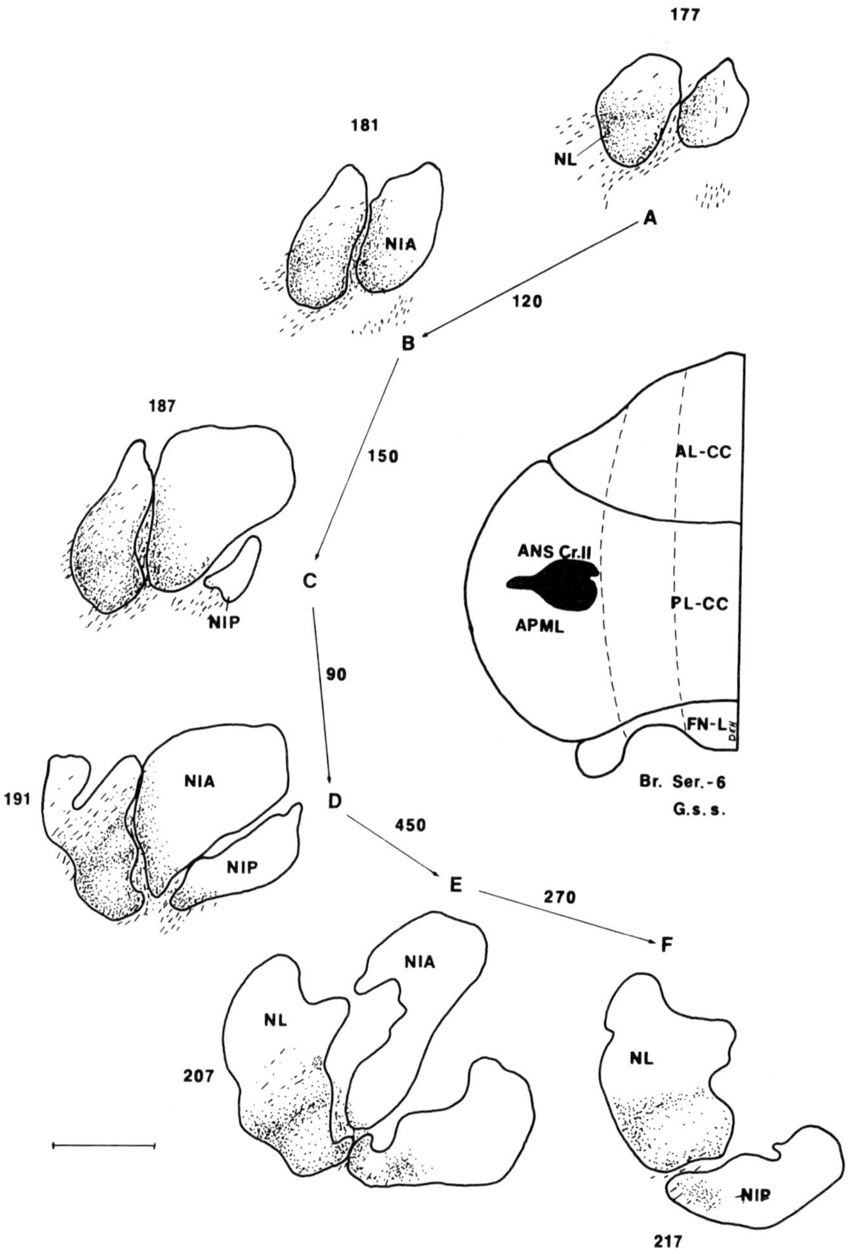

Figure 26. Pattern of degeneration following lesion of the lateral functional zones, which include crus II of the ansiform lobule and the anterior portion of the anterior paramedian lobule. Lesion of an adult *Galago senegalensis senegalensis.* The sections are on the horizontal plane and cut from dorsal (*A*) to ventral (*F*). The numbers between each section represent the distance in microns separating each drawing. The short lines represent fibres of passage and the dots represent areas of terminal degeneration. Note that the terminal degeneration is located in the lateral area of the NL, NIA, and NIP. Scale = 1 mm., in relation to the tracings of the nuclei.

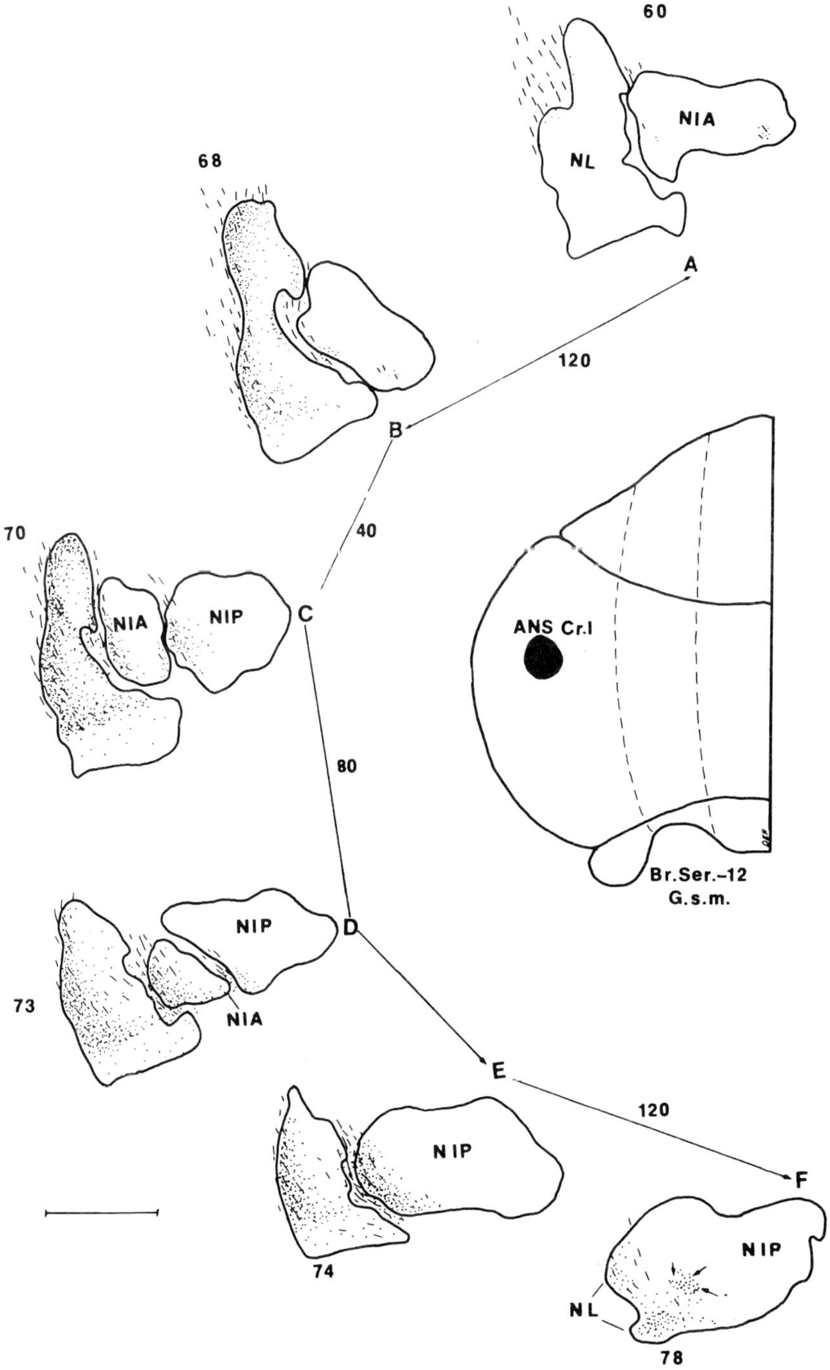

Figure 27. Pattern of degeneration following a lesion of the lateral functional zone, which was located in crus I of the ansiform lobule. Lesion of an adult *Galago senegalensis moholi*. The sections are cut on the transverse plane from rostral (*A*) to caudal (*F*). Note the pattern of degeneration and compare with Fig. 26. Terminal degeneration is seen in the lateral areas of the NL, NIA, and NIP. The ventral area of the NL shows only sparse degeneration. Scale = 1 mm., in relation to the tracings of the nuclei.

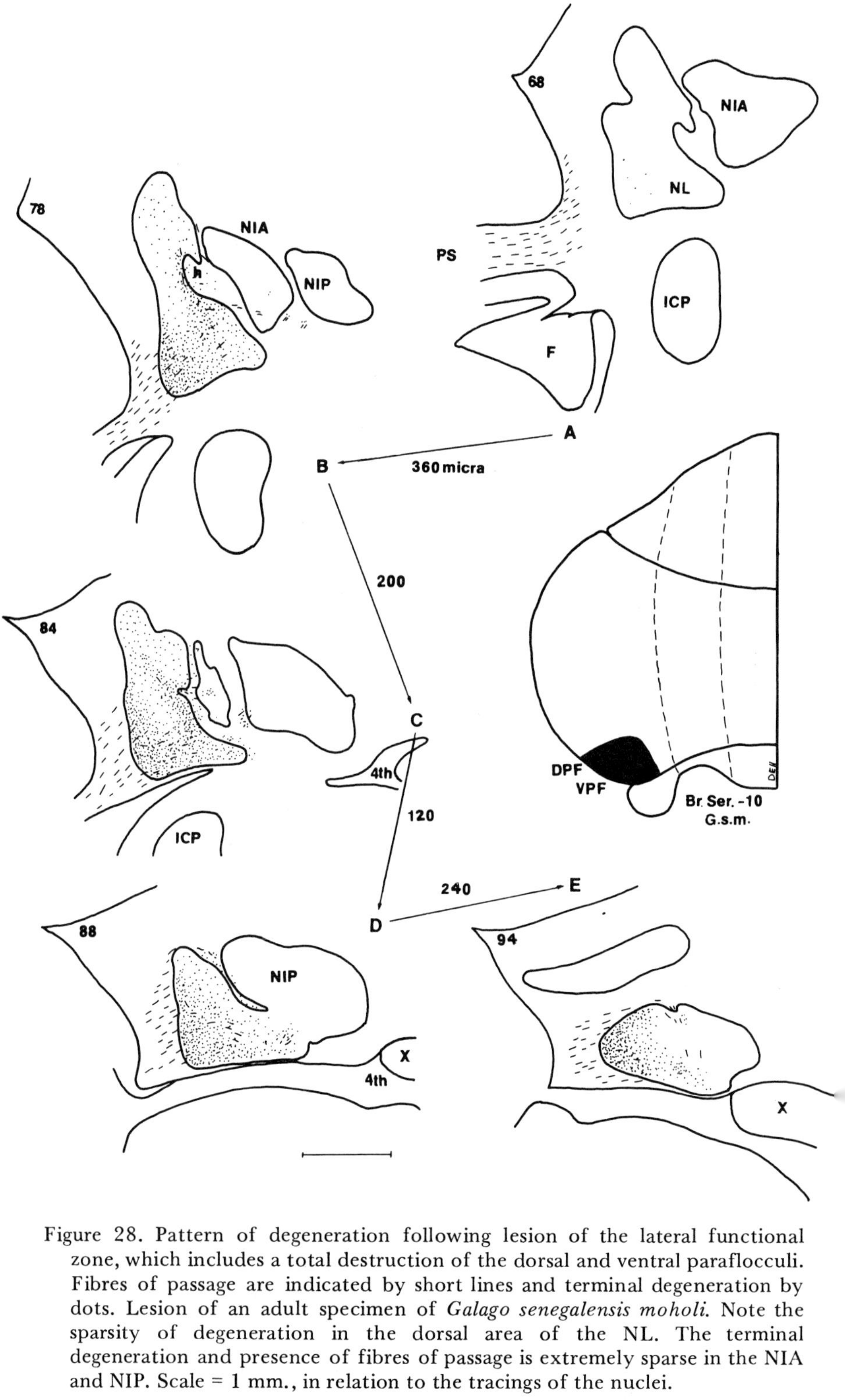

Figure 28. Pattern of degeneration following lesion of the lateral functional zone, which includes a total destruction of the dorsal and ventral paraflocculi. Fibres of passage are indicated by short lines and terminal degeneration by dots. Lesion of an adult specimen of *Galago senegalensis moholi*. Note the sparsity of degeneration in the dorsal area of the NL. The terminal degeneration and presence of fibres of passage is extremely sparse in the NIA and NIP. Scale = 1 mm., in relation to the tracings of the nuclei.

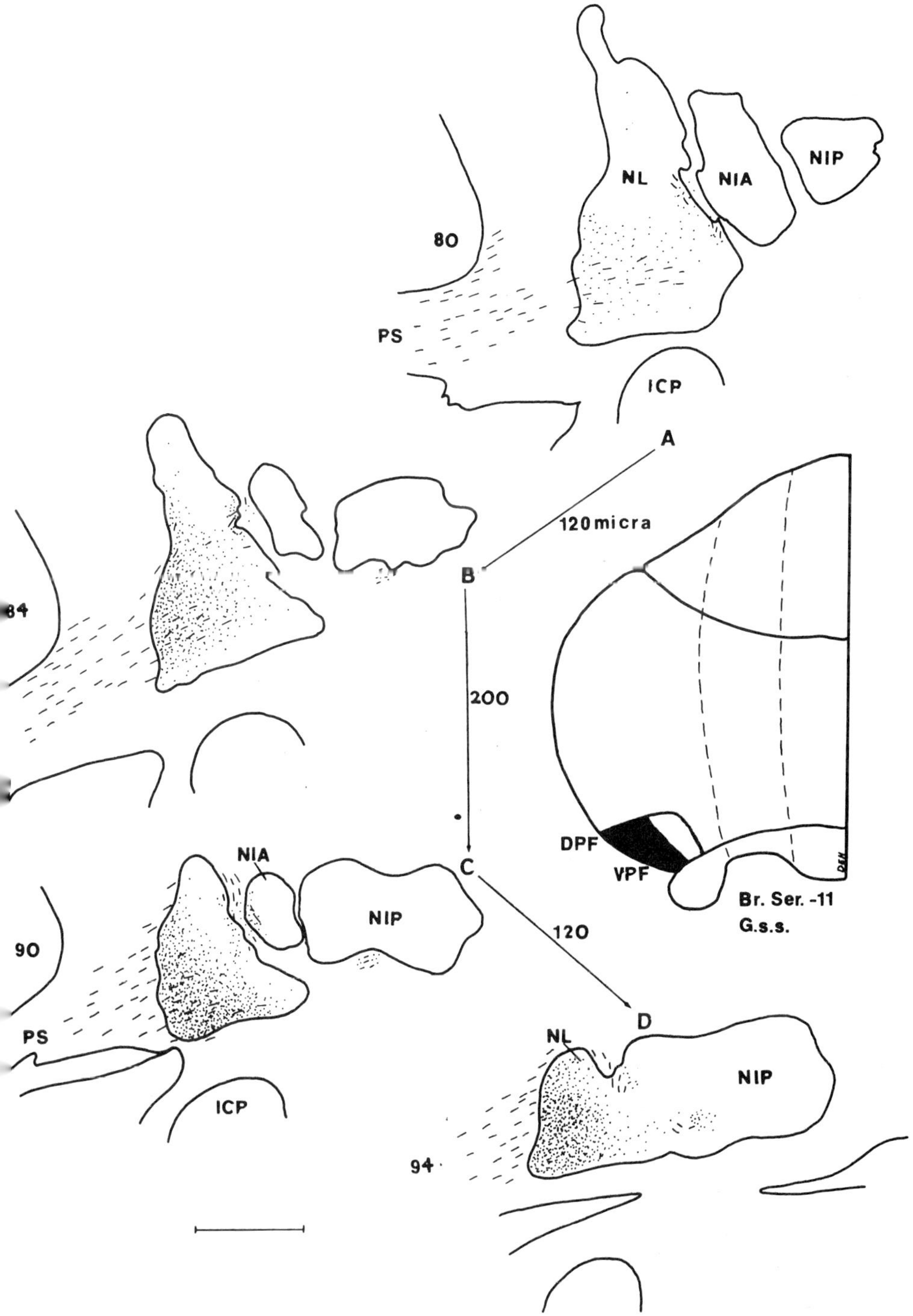

Figure 29. Pattern of degeneration following lesion of the lateral functional zone, which includes total destruction of the lateral two thirds of the dorsal and ventral paraflocculi. Lesion of an adult *Galago senegalensis senegalensis.* Note the sparse degeneration in the medial areas of the NL as compared with Fig. 28. Small areas of degeneration in the NIA and NIP are noted. Terminal degeneration is most dense in the lateral and ventral regions of the NL. (Compare with Fig. 28.) Scale = 1 mm., in relation to the tracings of the nuclei.

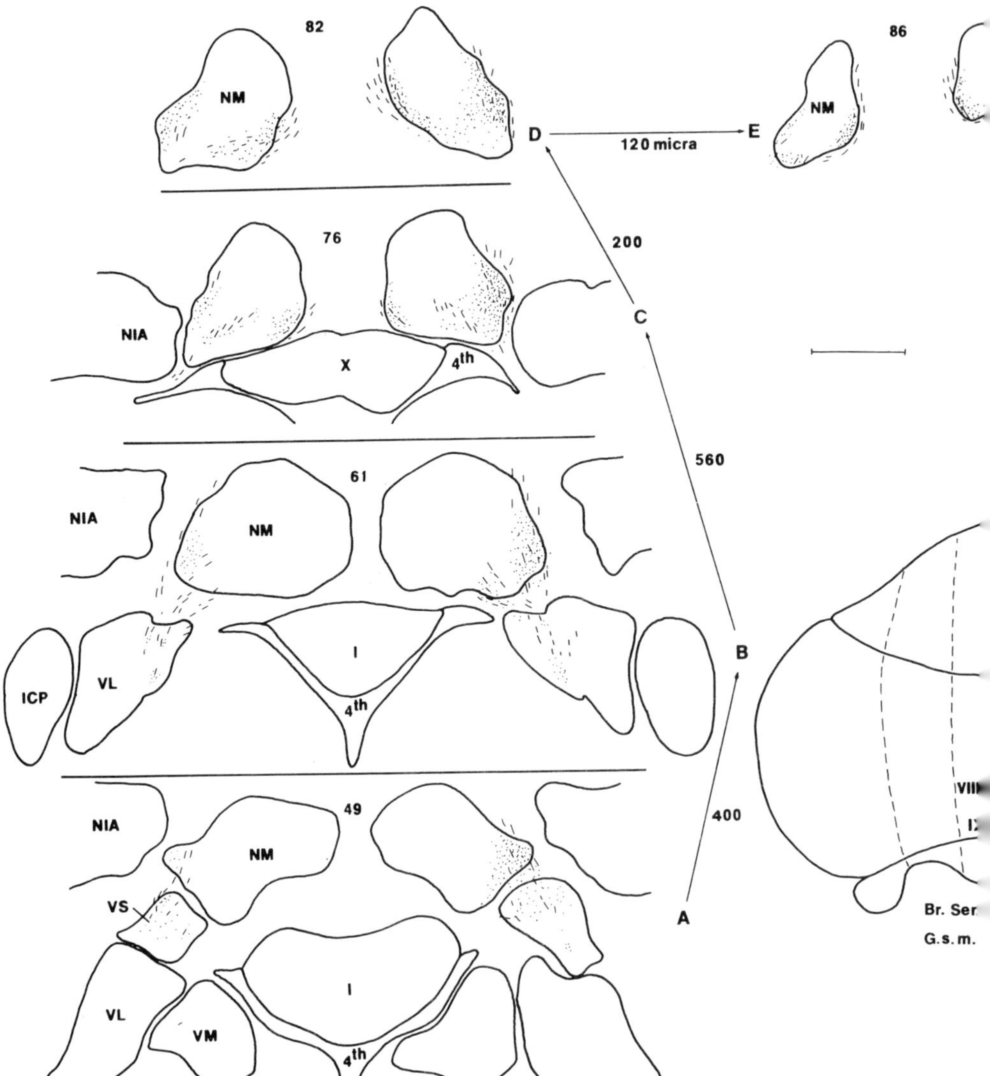

Figure 30. Pattern of degeneration following lesion of the medial functional zone, which includes lobule VIIIB and the upper lip of lobule IX. Lesion of an adult *Galago senegalensis moholi.* The projections are bilateral and pass to the NM, the lateral vestibular nucleus and the superior vestibular nucleus. The heavier degeneration to the nuclei on the right is due to the fact that the lesion is larger on the right side of the midline. Right and left sides are indicated by L. and R. Scale = 1 mm., in relation to the tracings of the nuclei.

of cells in the caudolateral area of the NIP (Fig. 28*D*).

Further substantiation of the sparse fibres to the NIA and NIP following parafloccular lesions was considered mandatory. Therefore, the lateral 2/3rds of both the dorsal and ventral paraflocculi was removed in *G. senegalensis senegalensis*, Br. Ser. 11 (Fig. 29). Fibres of passage entered the lateral aspect of the NL in the caudal half of its length, essentially the same as that noted in a lesion of the entire paraflocculus (Fig. 28). Heavy terminal degeneration is present in the lateral and ventral areas of the NL in its caudal half (Fig. 29*B*-*D*). Terminal degeneration is sparse in the dorsal area of the NL with complete removal of the paraflocculus (Fig. 28), or almost completely absent when only the lateral 2/3 of the paraflocculus is removed (Fig. 29). In both cases illustrated, there is a persistent amount of extremely sparse fibres of passage and terminal degeneration in caudal and lateral areas of the NIA and NIP (Figs. 28*B*-*E*, 29*B*-*D*, 34*c*, *e*). In these descriptions it is *stressed* that these projections to NIA and NIP from the cortex of the paraflocculus are *extremely* sparse.

(*b*) *Lesions of the medial functional zone.* The fibre projections from lesions of the cortex of the medial functional zone are shown in Figs. 30 and 36. In one animal, a lesion was placed in the cortex of lobules VIIIB and IX on the midline with slightly more of the lesion on the right hand side of the vermis (Fig. 30). Corticonuclear fibres pass rostrally toward the caudal pole of the NM. Fibres of passage are seen coursing bilaterally in a caudorostral direction on the medial and lateral sides of the NM (Fig. 30*e*). There is a small amount of terminal degeneration in the medial aspects of the NM and a moderate amount in the ventrolateral areas (Fig. 30*B*-*D*). The amount of terminal degeneration and fibres of passage is slightly greater in the NM on the right. This is due to the fact that the lesion involved more of the cortex on the right side of the midline than on the left.

Fibres of passage course rostrally on the lateral and medial aspects of the NM. Some of the fibres located medial to the NM terminate on small- and medium-sized cell bodies in the ventromedial area of the nucleus. Other fibres travelling in this area course rostrally, then turn laterally to travel through the ventral area of the NM (Figs. 30*C*-*D*, 35*a*, *b*). These fibres show a small amount of terminal degeneration throughout their course, although the majority of the fibres appear to pass out of the NM at its ventrolateral aspect. These fibres then enter, and terminate on cell bodies in, the medial portion of the lateral vestibular nucleus (Figs. 30*B*, 36*b*, *e*). A second distinct fasciculus of degenerating fibres courses rostrally through the lateral portion of the NM. Some fibres enter the centrolateral area of the NM to terminate, while others extend rostral to enter the superior vestibular nucleus. The fibres passing into and terminating on the superior vestibular nucleus are thin fibres and the pattern of terminal degeneration is sparse (Fig. 36*a*, *c*, *b*), and located in the caudodorsal area of the nucleus (Fig. 30*A*).

Lesions of the medial functional zone cortex of the posterior lobe which do not extend over the midline (not illustrated) show the same pattern of projection to the NM, the lateral and superior vestibular nuclei. The projections in cases of lesions restricted to one side of the midline are almost exclusively into the nuclei of the ipsilateral side. In the present study, no degenerated fibres were seen to cross the midline between the medial nuclei.

Discussion

The cerebelli of the Lorisidae considered in this study show many similarities, yet at the same time some interesting differences are noted. The anterior lobe of the *corpus cerebelli* is divided into the five lobules characteristic of many mammalian forms. The division of the anterior lobe into five lobules has been accomplished in a wide variety of mammals ranging from the rat to the higher primates.[16] The anterior lobe of the avian cerebellum also has five characteristic subdivisions which are indicated as folia instead of lobules.[16,25] It has been recently suggested that the anterior lobe may consist of four lobules.[26] The criterion for this is: 'In order to be considered a lobule, the morphological feature in question must be separated from its neighbours by deep fissures that, as seen in mid-saggital section, must approach that fastigial nucleus closely, and must not conjoin them laterally in surface aspect.' This technique, in the form stated,[26] can only be applied to the animals of the present study in a modified form. In the greater galago and slow loris, lobules IV and V of Larsell could correspond to lobule IV of Dillon and Brauer. However, if the above definition is applied to the potto and the lesser galago, lobules II and III of Larsell would correspond to lobule II of Dillon and Brauer.[16,26] Considerably more work must be carried out on other prosimian forms to evaluate which of these methods are applicable to the suborder as a whole. The method of Larsell is utilised in the present investigation since it is well established in a variety of vertebrates.[15,16]

Figure 31. Midsagittal views of the cerebellum and diencephalon of *Nycticebus coucang* (*a*), *Galago crassicaudatus* (*b*), and *Perodicticus potto* (*c*). For abbreviations not listed on p. 730, see Notes.[33]

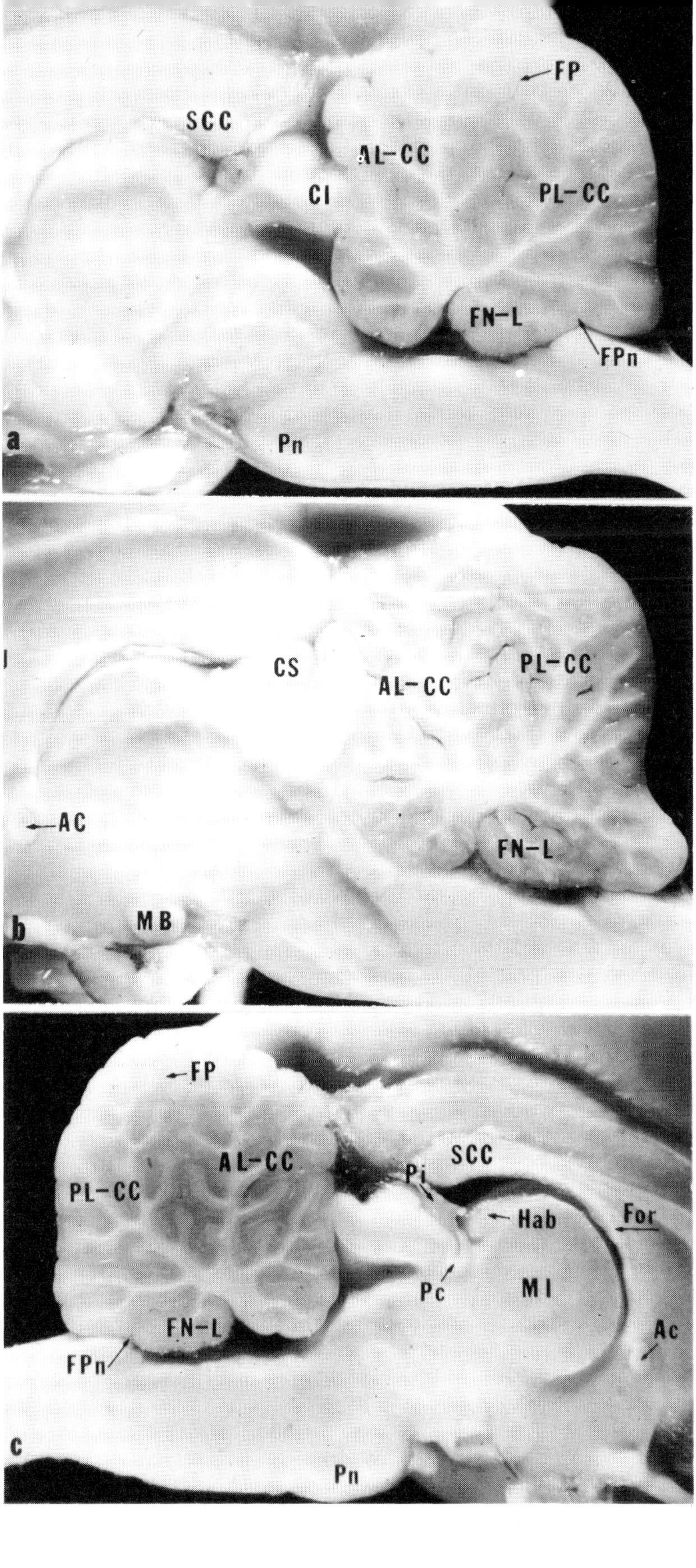
FP
SCC
AL-CC
Cl
PL-CC
FN-L
FPn
a
Pn
CS
AL-CC
PL-CC
AC
FN-L
b
MB
FP
SCC
AL-CC
PL-CC
Pi
Hab
For
Pc
MI
FN-L
Ac
FPn
c
Pn

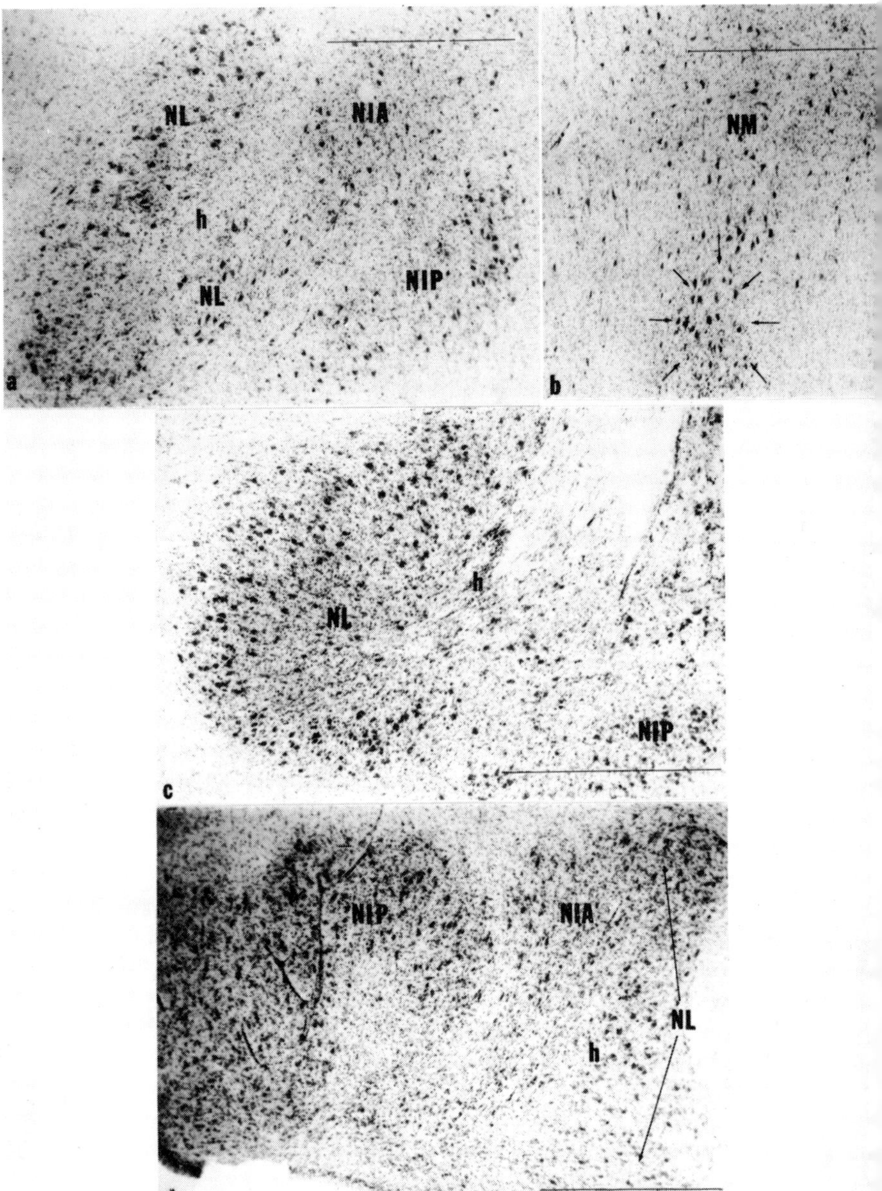

Figure 32. Photomicrograph of the deep cerebellar nuclei of *Galago crassicaudatus* (*a,b*), *Perodicticus potto* (*c*), and *Nycticebus coucang* (*d*). Scale = 1 mm.

(*a*) The NL, NIA, and NIP of *G. crassicaudatus* are sectioned in horizontal section. Note the undulatory appearance of the NL. This photomicrograph is approximately at the level of Figure 24*E-F*.

(*b*) The NM of *G. crassicaudatus*. Note the compact appearance of the cells in the caudal area, versus the reticulated appearance of the cells in the rostral areas. This photomicrograph is approximately at the level of Figure 24*B*.

(*c*) The NL, and NIP of *P. potto*. The NL lacks the undulatory appearance characteristic of the greater galago. Note the hilus. This photomicrograph is approximately at the level of Figure 23*F*.

(*d*) The NL, NIP, and the caudolateral portion of the NIA of *Nycticebus coucang* as seen in transverse section.

Figure 33. Photomicrographs of the typical appearance of cell bodies of the NL (*a*), the NIA (*b*), the NM (*c*), and the NIP (*d*). The more oval shaped cell bodies of the potto are exemplified by the photomicrograph of cells from the NL (*e*). Scale = 50 microns.

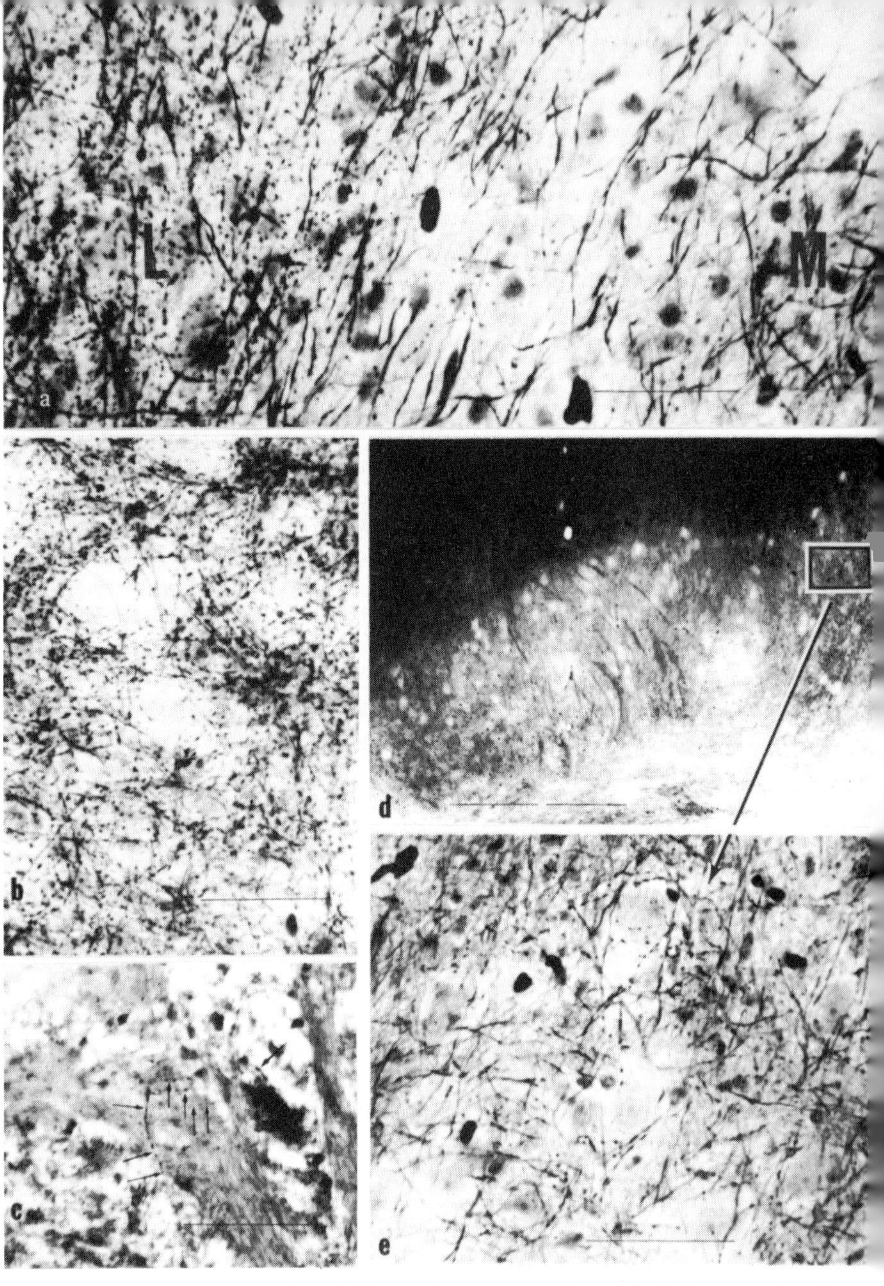

Figure 34. Photomicrographs of fibres of passage and terminal degeneration in the deep cerebellar nuclei.

(*a*) Photomicrograph of heavy degeneration in the lateral area of the NL (L) and sparse degeneration in the medial area of the NL (M). From *G.s. moholi* No.12. Scale = 50 microns.

(*b*) Photomicrograph of dense degeneration in the ventral area of the NL. From *G. s. senegalensis* No.10. Scale = 50 microns.

(*c*) Photomicrograph of degeneration of fibres of passage in a small bundle located ventral to the NIP. This bundle courses in a caudorostral direction. *G. s. senegalensis* No.10. Scale = 50 microns.

(*d*) Low power photomicrograph of the NIP, NIA, and NL. Taken from Fink-Heimer stain. From *G. s. senegalensis* No.10. Scale = 1 mm.

(*e*) Photomicrograph of the light degeneration in the caudolateral area of the NIA. This is high power of the boxed area of Figure 34*d*. Note the fibres of passage and sparse amount of terminal degeneration. *G. s. senegalensis* No.10. Scale = 50 microns.

The anterior lobe of the cerebellum is poorly developed in basic subprimate mammalian forms. In the insectivores and the elephant shrews, it is hidden by forward expansion of the posterior lobe vermis (middle lobe of Elliot Smith in the older literature), and has poorly developed hemisphere areas.[27-9] In *Ptilocercus lowii* the cerebellum is simpler than that of *Tupaia.*[7] In both of these animals, the anterior lobe is small with poorly developed hemispheres.[7,11,17] In a variety of basic forms, the anterior lobe of the *corpus cerebelli* shows a characteristically poor level of development.[15,16,30] It is equally significant to note that in rodents, carnivores, lagomorphs and other mammalian forms the anterior lobe is poorly developed, even though the total size of the cerebellum may equal or exceed that of some primates.[16] In the Lorisidae of the present study the hemisphere of the anterior lobe (HII-HV) is well developed. In a variety of Old and New World primates, regardless of the total size of the cerebellum, the hemisphere of the anterior lobe is well developed.[16,31] It appears that well developed hemispheres of the anterior lobe represent characteristics shared by all primates.

The lingula of the greater and lesser bushbabies was simple and restricted to the vermis. In the slow loris and potto the lingula appeared to have a lateral portion which may represent some development of a very primitive hemisphere. It has been suggested that the hemisphere of the lingula, when present, is related to the development of a prehensile tail.[16] Since the potto and slow loris obviously lack prehensile tails, the development of the lingula in these animals awaits another answer. It has recently been shown that lesions of the lingula in a monkey cause the animal to rotate in a clockwise direction when aroused.[32] This observation would support the concept of vestibulo-cerebellar fibres projecting to the lingula. Since the potto and slow loris do not have a tail that could assist them in maintaining balance, a modest development of the lingula may provide a closer functional relationship between the cerebellum and vestibular centres for maintainence of balance in an arboreal habitat. The greater and lesser bushbabies have long tails to assist them in maintenance of balance, therefore some adaptive specialisation of the nervous system would be unnecessary.

The posterior lobe of the corpus cerebelli is well developed in all genera of the present study. Even though the size and shape of the cerebelli, pons and cerebral cortices vary somewhat,[33] the relative complexity of the posterior lobe is similar in the slow loris, lesser galago and potto. A notable exception is the greater galago. The cerebellum of this animal is consistently more complex than that of the other three genera. The anterior and posterior paramedian lobules are most complex in this genus. The ansiform lobule, size and shape of the dorsal paraflocculus, and limitations of the *copula pyramidis* (the potto excepted) are similar in all groups. Enlargement of the posterior lobe, especially the hemispheres, is characteristic of

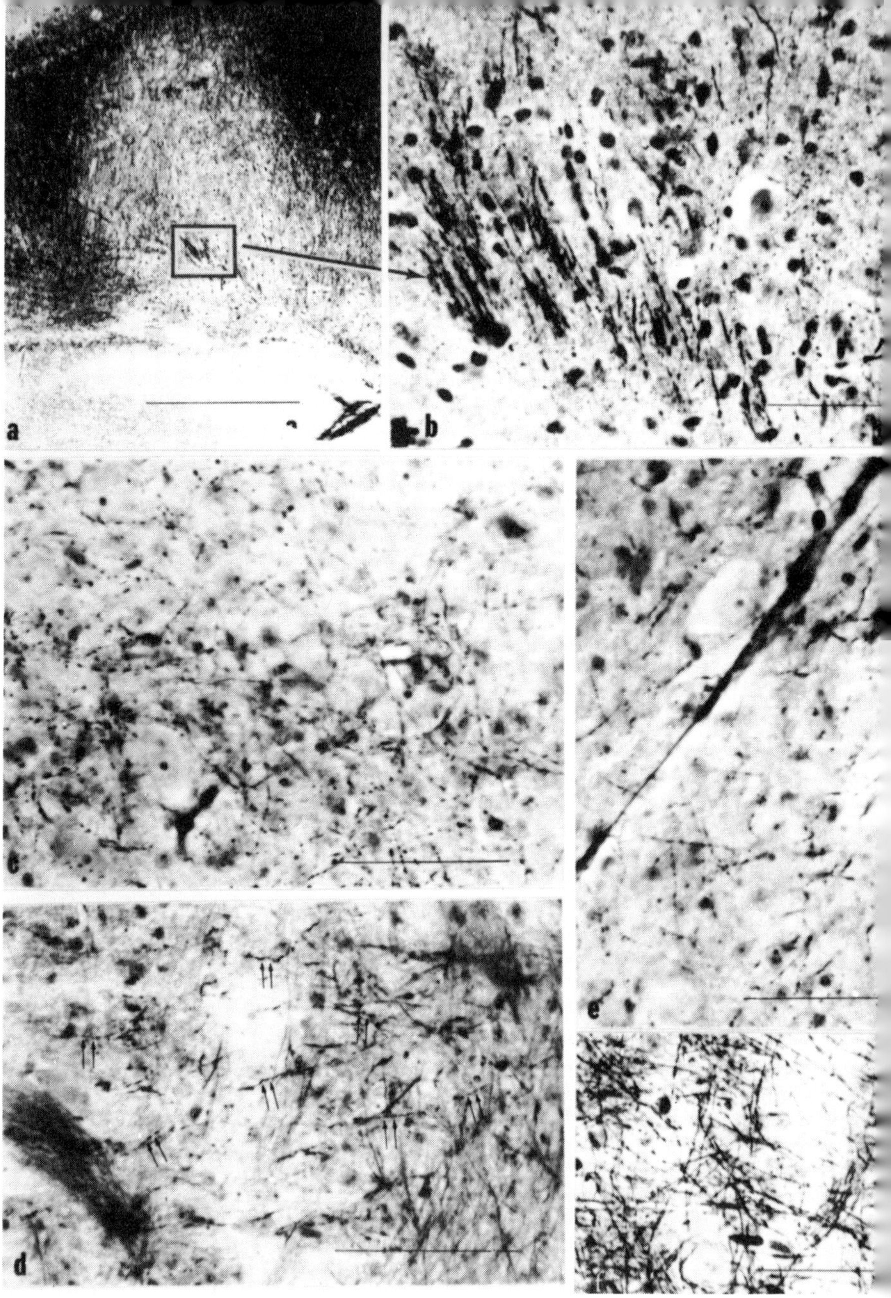

Figure 35. Photomicrographs of fibres of passage and terminal degeneration in deep cerebellar nuclei.

(*a*) Low power photomicrograph of the NM stained with the Fink-Heimer stain. From *G. s. moholi* No.9. Scale = 1 mm.

(*b*) Photomicrograph of the bundle of fibres coursing through the ventral area of the NM. This is a high power view of the approximate area indicated by the box in Figure 35*a*. Note the fibres of passage in the bundle and the terminal degeneration in the area lateral to it (upper right corner of photo). From *G. s. moholi* No.9. Scale = 50 microns.

(*c*) Photomicrograph of fibres of passage and terminal degeneration in the lateral area of the NIP. Note the fibres of passage and the moderate terminal degeneration. From *G. s. senegalensis.* No.6. Scale = 50 microns.

(*d*) Photomicrograph of degeneration of fibres of passage in the small

all primates.[16,34,35] The cerebelli of the Lorisidae are more complex than the same structure in many new world monkeys.[16,31]

The deep cerebellar nuclei of the Lorisidae of the present study can be divided into four cell masses on either side: *nucleus medialis, nucleus interpositus anterior, nucleus interpositus posterior*, and *nucleus lateralis.* The cerebellar nuclei of primates have been reported as four separate cell masses for a variety of different genera.[18,19,36,37] The most obvious change in the morphology of the cerebellar nuclei is the increasingly complex undulatory appearance of the lateral cerebellar nucleus as one ascends the primate scale. The greater and lesser galago show the most advanced degree of differentiation of the cerebellar nuclei in the present study. The potto and slow loris show a lesser degree of differentiation, as indicated by the intimate association of the NL with the NIA and NIP. The NIA and NIP are relatively large in all genera of the present study. In the potto, the NIP is slightly larger than in the other genera, but the NIA is essentially similar. The NIA is a slightly elongated cell mass that expands further rostral than any other cell group. In view of recent evidence of somatotopical projections from the NIA to the red nucleus,[38] the pathways between NIA and the mesencephalic red nucleus may be well developed in the Lorisidae. It has been suggested that the fusion of the cerebellar nuclei represents a basic level of organisation from which higher levels of differentiation arose.[39]

The cerebellar nuclei of a variety of sub-primate mammals have also been divided into four cell masses on each side.[39-42] It should be noted, however, that these animals are specialised mammals that have been evolving divergently from the primates for considerable time. If basic forms which may have a more direct relationship with extant primates are considered, it is noted that there is a progressive degree of phylogenetic development of the cerebellar nuclei from a single cell mass to two bilateral cell masses and eventually four bilateral cell masses.[19,30,43,44] The enlargement of the lateral cerebellar nucleus appears to be correlated with increase in size of the hemisphere of the anterior and posterior lobe. In the present study, the degree of differentiation of the cerebellar nuclei in ascending order of complexity would be: slow loris, potto, greater galago, lesser galago. The development of the NL is greater in *Galago*, especially *G. senegalensis*, and appears to be partially related to the

---

bundle of fibres travelling in a caudorostral direction in the ventral area of the NIA. From *G. s. moholi* No.12. Scale = 50 microns.

(*e*) Photomicrograph of fibres of passage and terminal degeneration in the lateral portion of the NIP. From *G. s. senegalensis* No.6. Scale = 50 microns.

(*f*) Photomicrograph of sparse fibres of passage and terminal degeneration in the ventromedial area of the NIP. From *G. s. moholi* No.12. Scale = 50 microns.

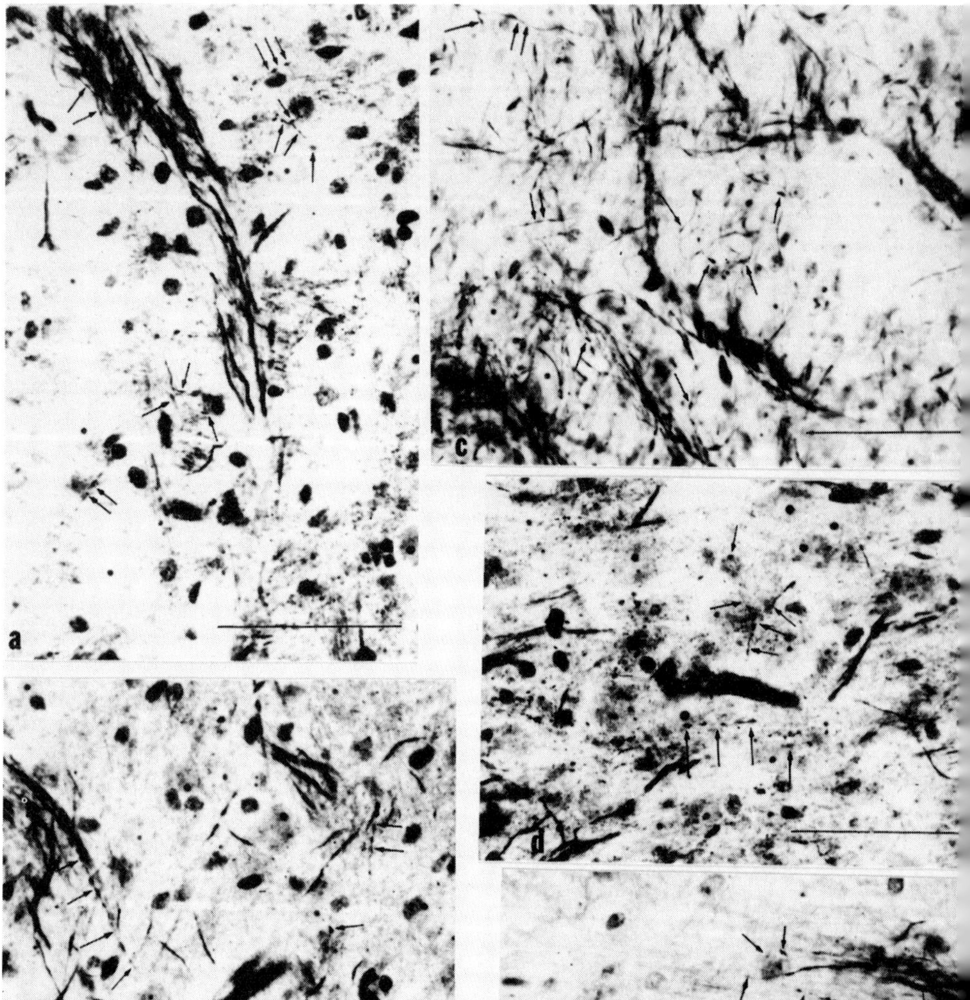

Figure 36. Photomicrograph of fibres of passage (double arrows) and terminal degeneration (single arrows) in the superior vestibular nucleus (*a*,*c*,*d*), and in the lateral vestibular nucleus (*b*,*e*). Note that the projection from the cortex of the medial functional zone passes to two of the vestibular nuclei. These are sparse yet consistent projections, and can be termed cerebellovestibular in nature. *G. s. moholi* No.9. Scale = 50 microns.

saltatory locomotor pattern of these animals. Results of recent work on the *nucleus dorsalis* of Clarke (spinal cord afferents to the cerebellum) further support the concept of neurological specialisations which may be related to locomotion.[45] Rüdeberg has established the homologous characteristics of the deep cerebellar nuclei in a wide variety of vertebrates including man, and his terminology is used in this report.[46]

The zonal concept of cerebellar corticonuclear projections suggested that the vermal cortex (medial functional zone) projects onto the NM, the paravermal cortex (paravermal functional zone) onto the NIA and NIP, and the lateral hemisphere (lateral functional zone) projecting on to the NL.[47] Recent evidence using newer degeneration techniques has not totally supported this concept.[48–53] The results of the present study also show that corticonuclear fibres from the cerebellar cortex project, in overlapping fields, on to the deep nuceli. In the present study, the lateral functional zone of the posterior lobe projects to the NL, NIA, and NIP. The heaviest projection is to the dorso-lateral half of the NL. Fibres passing to the NIA and NIP are, at best, moderate to sparse. These observations are in general agreement with other recent studies that have shown projections from the lateral functional zone on to portions of several of the deep cerebellar nuclei.[47–9,51] If the results of the present study are correlated with previous reports, it appears that the lateral functional zone projects onto the lateral portions of the interposed and lateral cerebellar nuclei. Lesions of the parafloccular cortex in the *Galago* projected to ventrolateral and caudal area of NL, and to extremely small areas of the NIA and NIP. It has been previously suggested, for the rat, that the parafloccular cortex projects to a single cerebellar nucleus.[48] All corticonuclear projections from the cortex of the posterior lobe passed into the caudal one-half to two-thirds of the deep nuclei, leaving rostral areas free of degeneration.

Corticonuclear projections from the vermis of the lesser *Galago* pass into the ventromedial and ventrolateral areas of the NM when the lesion is on the midline. If the lesion is restricted to one side of the midline, the fibres appear to pass principally to the ventrolateral areas of the NM. Vermian lobules VIII and IX in lesser *Galago* give cerebello-vestibular fibres which pass to the lateral and superior vestibular nuclei. In *Tupaia*, it tentatively appears that the vermis of the posterior lobe projects to the NM and lateral vestibular nucleus only.[53] Cerebello-vestibular fibres have been reported in the cat, but are apparently not present in the rat or in *Macaca*.[48,50] In *Galago*, the fibres passing to the superior and lateral vestibular nuclei are sparse. It is not surprising that the fibre projections from the cerebellar cortex are slightly different from those previously reported for other animals.[48–50] It has been acknowledged by previous investigators that true species differences do occur.[48] It is also suspected that the evolutionary development of the locomotor pattern of the particular animal under consideration will be reflected by specialisations within the nervous system.[45]

The question of the phylogeny and taxonomy of the tree shrews has been the subject of considerable interest. To some, they represent primates,[54–5] but to others, they have little or no primate affinities.[56–7] A recent review of neuroanatomical evidence would tend to support the concept that the tree shrews should more correctly be considered as a separate order.[58–9] Romer has suggested that tree shrews may represent the primitive stock from which all placental mammals arose, not just extant primates.[60] The tentative results of the present investigation, and some previous studies on the tree shrews,[58,61] indicate that valuable information can be obtained by comparative studies on these animals. If *Tupaia* is further compared with mammals to which it may be related phylogenetically (for example, the elephant shrews and various prosimian forms), new information relative to its position should be realised.[62]

Acknowledgments

The author wishes to express his sincere appreciation to David Cohen and Donald Wright, and to Betsy Harmeling and Jo Bishop for their technical help. The expert assistance of Janet Hall of Duke University is appreciated. The Department of Anatomy and School of Basic Sciences and Graduate Studies has enthusiastically supported this project, and grants from the MCV Foundation and the A.D. Williams Committee (3558-550) partially supported this research. The author is especially indebted to C.J. Gray for typing the preliminary draft of this paper and to M.P. Harrell for typing the final draft.

NOTES

1 Elliot Smith, G. (1903), 'On the morphology of the brain in the mammals, with special reference to that of the lemurs recent and extinct', *Trans. Linn. Soc. Lond.* 8, 319-432.

2 Parentheses are those of Elliot Smith (1903).

3 Bolk, L. (1906), *Das Cerebellum der Säugetiere: Eine vergleichende anatomische Untersuchung*, Haarlem.

4 Woollard, H.H. (1925), 'The anatomy of *Tarsius spectrum*', *Proc. Zool. Soc. Lond.* 1071-184.

5 Tilney, F. (1925), 'The brain stem of *Tarsius*: A critical comparison with other primates', *J. Comp. Neurol.* 43, 371-432.

6 Le Gros Clark, W.E. (1924), 'On the brain of the tree shrew (*Tupaia minor*)', *Proc. Zool. Soc. Lond.* 1053-74.

7 Le Gros Clark, W.E. (1926), 'The anatomy of the pen-tailed tree shrew (*Ptilocercus lowii*)', *Proc. Zool. Soc. Lond.* 1190-307.

8 Le Gros Clark, W.E. (1931), 'The brain of *Microcebus murinus*', *Proc. Zool. Soc. Lond.* 463-86.

9 Kanagasuntheram, R. and Mahran, Z.Y. (1960), 'Observations on the nervous system of the lesser bushbaby (*Galago senegalensis senegalensis*)', *J. Anat.* 94, 512-27.

10 Krishnamurti, A. (1966), 'The external morphology of the brain of the slow loris *Nycticebus coucang coucang*'. *Folia primat.* 4, 361-80.

11 Haines, D.E. (1969), 'The cerebellum of *Galago* and *Tupaia*. I: Corpus cerebelli and flocculonodular lobe', *Brain Behav. Evol.* 2, 377-414.

12 Bradley, O.C. (1904), 'The mammalian cerebellum: Its lobes and fissures', Part I', *J. Anat.* 38, 448-75.

13 Bradley, O.C. (1904-1905), 'The mammalian cerebellum: Its lobes and fissures, Part II', *J. Anat.* 39, 99-117.

14 Author's parentheses.

15 Larsell, O. and Jansen, J. (1967), *The Comparative Anatomy and Histology of the Cerebellum from Myxinoids through Birds*, Minnesota.

16 Larsell, O. and Jansen, J. (1970), *The Comparative Anatomy and Histology of the Cerebellum from Monotremes through Apes*, Minnesota. (All the earlier works of Larsell and others are reviewed in this volume.)

17 Haines, D.E. (1971), 'The cerebellum of *Galago* and *Tupaia*. II: The early postnatal development', *Brain Behav. Evol.* 4, 97-113.

18 Okawa, K. (1957), 'Comparative anatomical studies of cerebellar nuclei of mammals', *Arch. Hist. Jap.* 13, 21-58.

19 Haines, D.E. (1971), 'The morphology of the cerebellar nuclei of *Galago* and *Tupaia*', *Am. J. Phys. Anthrop.* 35, 27-42.

20 The author is fully aware that *Tupaia* is not a member of the Lorisidae, and is even excluded from the suborder Prosimii by some authorities.

21 Klüver, J. and Barrera, E. (1953), 'A method for the combined staining of cells and fibers in the nervous system', *J. Neuropath. Exp. Neurol.* 12, 400-3.

22 Fink, R.P. and Heimer, L. (1967), 'Two methods for selective silver impregnation of degenerating axons and their synaptic endings in the central nervous system', *Brain Res.* 4, 369-47.

23 The author expresses his sincere thanks to I.T. Diamond for allowing Janet Hall of his laboratory to coach the author in the proper methods for using the Fink and Heimer stain on *Galago* and *Tupaia*. Her expertise has been an invaluable aid.

24 The hemispheric portion of a vermian lobule is indicated by the prefix H. For example, the hemisphere of lobule II is indicated by HII.

25 Larsell, O. and Whitlock, D.G. (1952), 'Further observations on the cerebellum of birds', *J. Comp. Neurol.* 97, 545-66.

26 Dillon, L.S. and Brauer, K. (1970), 'A proposed method for establishing homologies among the lobules of the anterior lobe in the mammalian cerebellum', *J.f. Hirnforsch.* 21, 217-32.

27 Elliott Smith, G. (1902), 'Notes on the brain of *Macroscelides* and other Insectivora', *Linn. Soc. J. Zool.* 28, 443-8.

28 Le Gros Clark, W.E. (1928), 'On the brain of the Macroscelididae (*Macroscelides* and *Elephantulus*)', *J. Anat.* 63, 245-75.

29 Le Gros Clark, W.E. (1932), 'The brain of Insectivora', *Proc. Zool. Soc. Lond.* 975-1013.

30 Obenchain, J.B. (1925), 'The brains of the South American marsupials *Caenolestes* and *Orolestes*', *Chicago Field. Mus. Nat. Hist., Zool. Ser.* 14, 175-232.

31 Haines, D.E. and Leichnetz, G.R. (1970), 'The gross morphology of the cerebellum of the adult marmoset', *Anat. Rec.* 169, 332.

32 Crosby, E.C., Taren, A.J. and Davis, R. (1969), 'The anterior lobe and lingula of the cerebellum in monkeys and man', *Top. Probl. Pyschiat. Neurol.* 10, 22-39.

33 Haines, D.E., Albright, B.C., Goode, G.E. and Murray, H.M., this volume.

34 Larsell, O. (1953), 'The cerebellum of the cat and the monkey', *J. Comp. Neurol.* 99, 135-99.

35 Shantha, T.R. and Manocha, S.L. (1969), 'The brain of chimpanzee (*Pan troglodytes*). I: External morphology', *The Chimpanzee* I, 187-237.

36 Courville, J. and Cooper, C.W. (1970), 'The cerebellar nuclei of *Macaca mulatta*: A morphological study', *J. Comp. Neurol.* 140, 241-54.

37 Tamagaki, R. (1959), 'Comparative anatomical studies on the cerebellar nuclei of the primates', *Bull. Kob. Med. Col.* 17, 387-421.

38 Courville, J. (1966), 'Somatotopical organisation of the projection from the nucleus anterior of the cerebellum to the red nucleus: An experimental study in the cat with silver impregnation methods', *Exp. Brain Res.* 2, 191-215.

39 Ohkawa, K. (1957), 'Comparative anatomical studies of cerebellar nuclei of mammals', *Arch. Hist. Jap.* 13, 21-58.

40 Flood, S. and Jansen, J. (1961), 'On the cerebellar nuclei in the cat', *Acta Anat.* 46, 52-72.

41 Korneliussen, H.K. (1966), 'On the morphology and subdivision of the cerebellar nuclei of the rat', *J. f. Hirnforsch.* 10, 109-22.

42 Snider, R.S. (1940), 'Morphology of the cerebellar nuclei in the rabbit and cat', *J. Comp. Neurol.* 72, 399-415.

43 Hines, M. (1929), 'The brain of *Ornithorhynchus anatinus*', *Phil. Trans. Roy. Soc. Lond.* (*B*) 217, 155-287.

44 Abbie, A.A. (1934), 'The brain stem and cerebellum of *Echidna aculeata*', *Phil. Trans. Roy. Soc. Lond.* (*B*) 244, 1-74.

45 Albright, B.C. and Haines, D.E. (1972), 'Morphology of Clarke's column in *Galago senegalensis*', *Anat. Rec.* 172, 447.

46 Rüdeberg, S-I. (1961), 'Morphogenetic studies on the cerebellar nuclei and their homologization in different vertebrates including man', Dissertation, University of Lund.

47 Jansen, J. and Brodal, A. (1940), 'Experimental studies on the intrinsic fibers of the cerebellum. II: The corticonuclear projection', *J. Comp. Neurol.* 73, 267-321.

48 Goodman, D.C., Hallett, R.C. and Welch, R.B. (1963), 'Patterns of localization in the cerebellar cortico-nuclear projections of the albino rat', *J. Comp. Neurol.* 121, 51-67.

49 Eager, R.P. (1963), 'Efferent cortico-nuclear pathways in the cerebellum of the cat', *J. Comp. Neurol.* 120, 81-104.

50 Eager, R.P. (1966), 'Patterns and mode of termination of cerebellar cortico-nuclear pathways in the monkey (*Macaca mulatta*)', *J. Comp. Neurol.* 126, 551-66.

51 Dow, R.S. (1936), 'The fiber connections of the posterior parts of the cerebellum in the rat and cat', *J. Comp. Neurol.* 63, 527-48.

52 Walberg, F. and Jansen, J. (1964), 'Cerebellar corticonuclear projection studied experimentally with silver impregnation methods', *J.f. Hirnforsch.* 6, 338-54.

53 Unpublished observations on the tree shrew (*Tupaia*).

54 Sorenson, M.W. (1970), 'Behavior of Tree Shrews', in Rosenblum, L.A. (ed.), *Primate Behavior: Developments in Field and Laboratory Research*, New York.

55 Le Gros Clark, W.E. (1959), *The Antecedents of Man*, Edinburgh.

56 Campbell, C.B.G. (1966), 'The relationships of the tree shrews: The evidence of the nervous system', *Evolution* 20, 276-81.

57 van Valen, L. (1965), 'Tree shrews, primates and fossils', *Evolution* 19, 135-51.

58 Haines, D.E. and Swindler, D.R. (1972), 'Comparative neuroanatomical evidence and the taxonomy of the tree shrews (*Tupaia*)', *J. Human Evol.*, in press.

59 Straus, W.L. (1949), 'The riddle of man's ancestry', *Quart. Rev. Biol.* 24, 200-23.

60 Romer, A.S. (1969), 'Vertebrate history, with special reference to factors related to cerebellar evolution', in Llinas, L.L. (ed.), *Neurobiology of Cerebellar Evolution and Development*, AMA Education and Research Foundation.

61 Martin, R.D. (1968), 'Reproduction and ontogeny in tree-shrews (*Tupaia belangeri*), with reference to their general behaviour and taxonomic relationships', *Z.f. Tierpsychol.* 25, 409-95 and 505-32.

62 Since submission of this manuscript the author has become aware of additional papers by Achenbach and Goodman (*Brain Behav. Evol.* 1,43-57, 1968) and Eager (Symp. on the Role of Vestibular Organs in Space Exploration, Pensacola, Fla., 1967) which report cerebellar corticovestibular fibres in the rat and *Macaca.*

L. RADINSKY

Prosimian brain morphology: functional and phylogenetic implications

Introduction

Two sources of information are available for studies of prosimian brain evolution: the brains of modern prosimians and endocranial casts of fossils. The scanty fossil record of prosimian brains has been reviewed recently,[1] and not enough new information has become available since then to warrant another review. Studies of modern prosimian brains offer access to much more information from which evolution may be interpreted, but few broad comparative surveys have been carried out. A notable exception is the work of Stephan, Bauchot and Andy,[2,3,4] whose quantitative studies of major brain structures in insectivores and prosimians are excellent examples of an important but under-utilised approach.

Most comparative studies involving prosimian brains have been purely descriptive and have dealt with gross external morphology alone. The best comparative study of prosimian brain morphology, both for the number of species treated and for the analysis of details, is that of Smith,[5] which, however, covered only 11 of the 17 commonly recognised genera. Recently,[6] I published figures of endocasts of 12 genera of recent prosimians, with some functional interpretation. The present study was undertaken firstly to survey the entire range of diversity of external brain morphology of extant prosimians so that data at least comparable to that published for ceboid primates would be available for prosimian studies,[7,8] and secondly to interpret functionally morphological differences and consider their relevance to phylogenetic interpretations. The use of endocranial casts made it possible to examine external brain morphology of all living genera of prosimians.

Methods

Although the value of endocasts for illustrating prosimian brain morphology was demonstrated a century ago,[9] little use has been made of them since then. As has been amply demonstrated, endocasts of prosimians, as of most mammals, reproduce in fine detail most of the external morphology of the brain, including all of the cerebral sulci, as well as the pattern of blood vessels, cranial nerve stumps, and often braincase sutures.[6,10] In addition, endocasts reproduce the shape and proportions of the major parts of the brain as they existed in life, undistorted by the shrinkage that usually accompanies brain fixation.

Endocasts were made from skulls representing all genera of living prosimians, using a latex casting technique which does not damage the skull.[6] Interpretation of prosimian external brain morphology

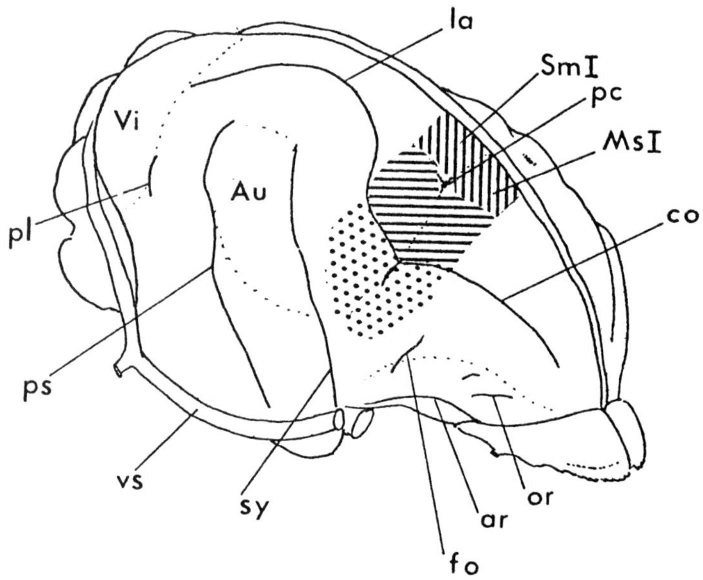

Figure 1. Endocast of *Lemur fulvus*, to show sulcal terminology used and to indicate cortical areas as inferred from references cited in text. Stippled area indicates head, horizontal hatching fore limb, and vertical hatching hind limb areas of somatic sensory and motor cortex. Abbreviations: ar, anterior rhinal fissure; Au, auditory cortex; co, coronal sulcus (? = sulcus rectus); fo, fronto-orbital sulcus; la, lateral sulcus (? = intraparietal); MsI, primary motor cortex; or, orbital sulcus; pc, postcruciate sulcus (transverse branch = central sulcus); pl, postlateral sulcus (= lunate sulcus); ps, postsylvian sulcus (= posterior suprasylvian, superior temporal or parallel sulcus); SmI, primary somatic sensory cortex; sy, sylvian sulcus (= anterior suprasylvian sulcus); Vi, visual cortex (Brodmann's area 17); vs, vascular sinus overlying posterior rhinal fissure.

was aided by the histological and physiological cortical mapping studies of Brodmann,[11] Campbell et al.,[12] Kanagasuntheram et al.,[13] Mott and Kelley,[14] Sanides and Krishnamurti,[15] and Zuckerman and Fulton.[16] Terminology for sulci used in this study, plus cortical areas as interpreted from the preceding references, are shown in Fig. 1. Examination of drawings of endocasts superimposed in proper position on skulls in many cases provided insights into the functional significance of differences in brain shape and proportions. To facilitate comparisons, all endocast-skull drawings were oriented with the floor of the medulla horizontal in lateral view and in the plane of the paper in dorsal view. Finally, the quantitative data provided by Stephan et al.[4] were used to check and supplement some of the ideas that resulted from study of the endocasts alone.

Observations

1. *Family Lemuridae*

(*a*) Subfamily Lemurinae: see Fig. 2. The three commonly recognised living genera of lemurines are *Lemur, Hapalemur* and *Lepilemur*, in order of decreasing size. Brains of the genus *Lemur* have the following sulci: coronal, postcruciate, lateral, postlateral, sylvian, postsylvian, orbital (usually represented by two dimples), and sometimes a fronto-orbital sulcus. A vascular sinus overlies the posterior rhinal fissure. The relationship between the caudal end of the sylvian and the postsylvian is variable. There is also considerable variation in the degree of overlap of the cerebellum and in the degree of convexity of the frontal lobes as seen in lateral view; the specimens of *Lemur variegatus* and *L. rubriventer* shown in Fig. 2 represent the extremes of variation of those features present in the sample available to me. My sample, even combined with figures from the literature, is too small to determine the extent to which species of *Lemur* may be distinguished by those or other neuro-anatomical features.

The brain of *Hapalemur* displays a sulcal pattern similar to that of *Lemur*, except that it lacks the postlateral sulcus. (The fronto-orbital sulcus is absent in the endocast I studied, but is present in a brain figured by Beddard.)[17] The one endocast available to me displays general proportions similar to those of *Lemur* species, except for being relatively narrower and in having the olfactory bulbs relatively more ventrally located than in most specimens of *Lemur*. (The *Lemur rubriventer* endocast figured here has the olfactory bulbs about as low as in *Hapalemur*, but differs in having a more convex frontal lobe).

Lepilemur, which is the smallest of the lemurines, has a sulcal pattern similar to that of *Lemur*, except that it lacks the postsylvian, postlateral and fronto-orbital sulci, and in having a less pronounced coronal sulcus. Brain shape and proportions fall within the range observed in *Lemur*.

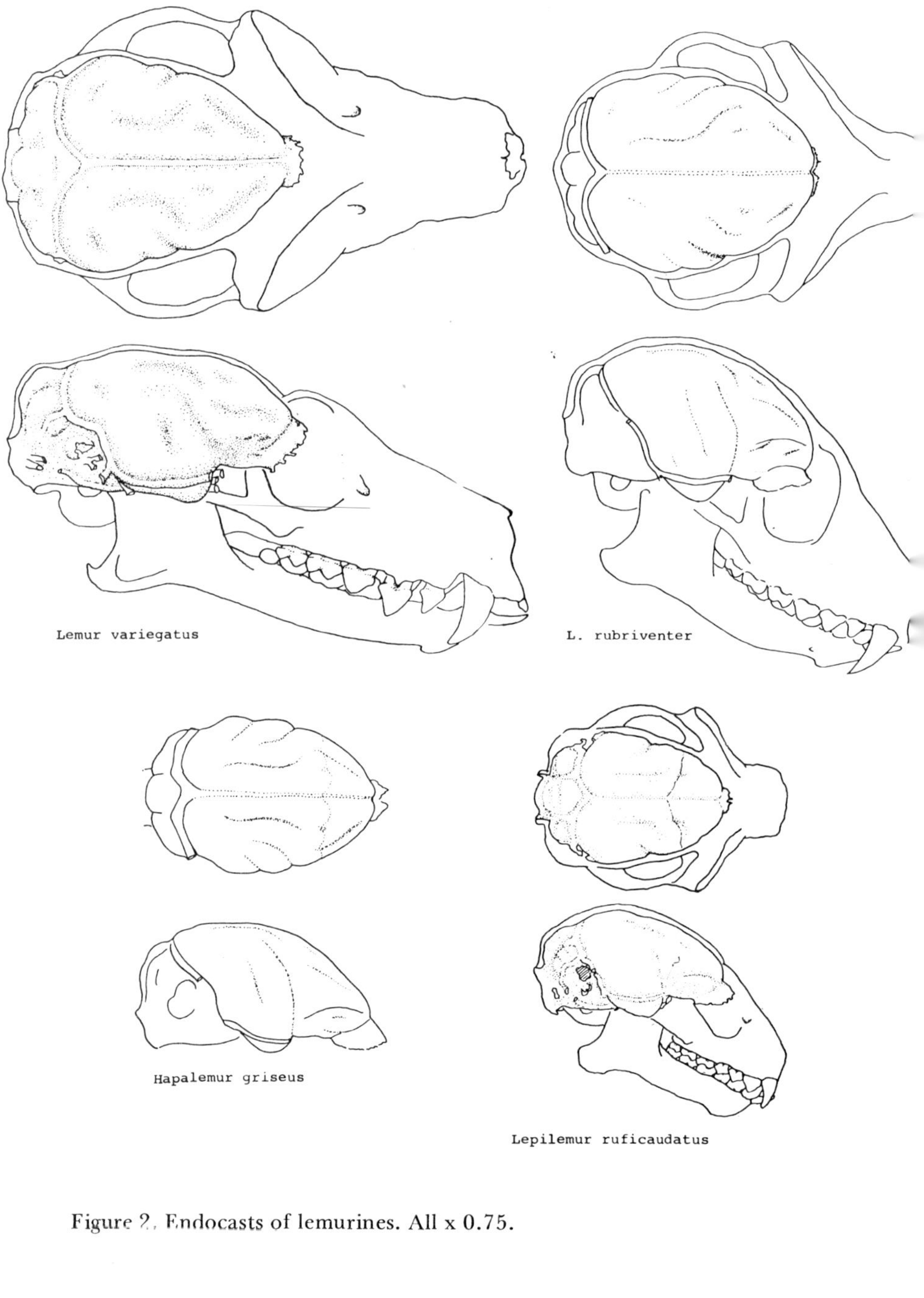

Figure 2. Endocasts of lemurines. All x 0.75.

(*b*) Subfamily Cheirogaleinae: see Fig. 3. The cheirogaleines include three genera, *Phaner, Cheirogaleus* and *Microcebus.* The brain of *Phaner*, the largest of the cheirogaleines, is similar in size to that of *Lepilemur*, the smallest lemurine, and has a similar sulcal pattern except that it lacks the postcruciate sulcus and possesses a short postsylvian. The brain of *Cheirogaleus* is similar to that of *Phaner* except that it is slightly smaller and lacks an orbital sulcus. *Microcebus*, the smallest of the lemurids, is one of the few prosimians for which the brain has been studied in any detail.[18] It differs from the preceding cheirogaleines in lacking all sulci except for the sylvian and a faint trace of a postsylvian. The olfactory bulbs are more compressed laterally than in the other lemurids, presumably a correlate of *Microcebus* being small and having relatively large eyes.

2. *Family Indriidae*

The family Indriidae (see Fig. 4) includes three living genera, *Indri, Propithecus* and *Lichanotus* (=*Avahi*), in order of decreasing body size. The brain of *Indri* is similar to that of comparable-sized *Lemur* specimens, except for having a more pronounced postlateral sulcus and broader frontal lobe, with the postcruciate sulcus better developed and, most striking, having a transverse branch that crosses the coronal sulcus. Related to the latter features, the motor area for the hand is larger and bulges out more in *Indri* than in *Lemur* (see

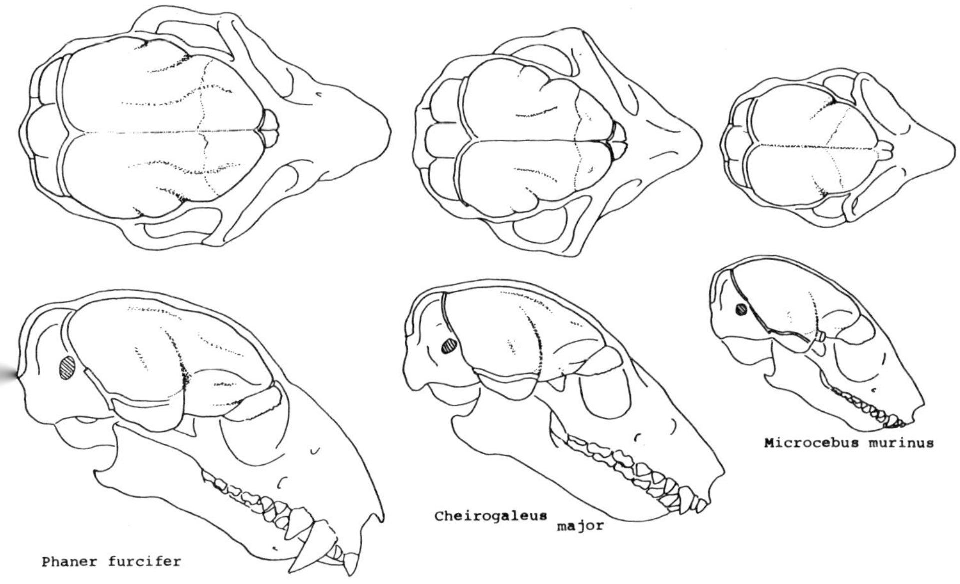

Figure 3. Endocasts of cheirogaleines. All natural size.

area marked 'x' in Fig. 4). The caudal end of the postsylvian is particularly variable in *Indri.*

The brain of *Propithecus* has a sulcal pattern similar to that of *Indri*, except that the sylvian sulcus extends further caudally than the lateral sulcus, usually terminating in a short caudomedially directed segment. Also, the transverse branch of the postcruciate sulcus, which is the homologue of the central sulcus, is more pronounced in *Propithecus* than in *Indri.* In shape, the brain of

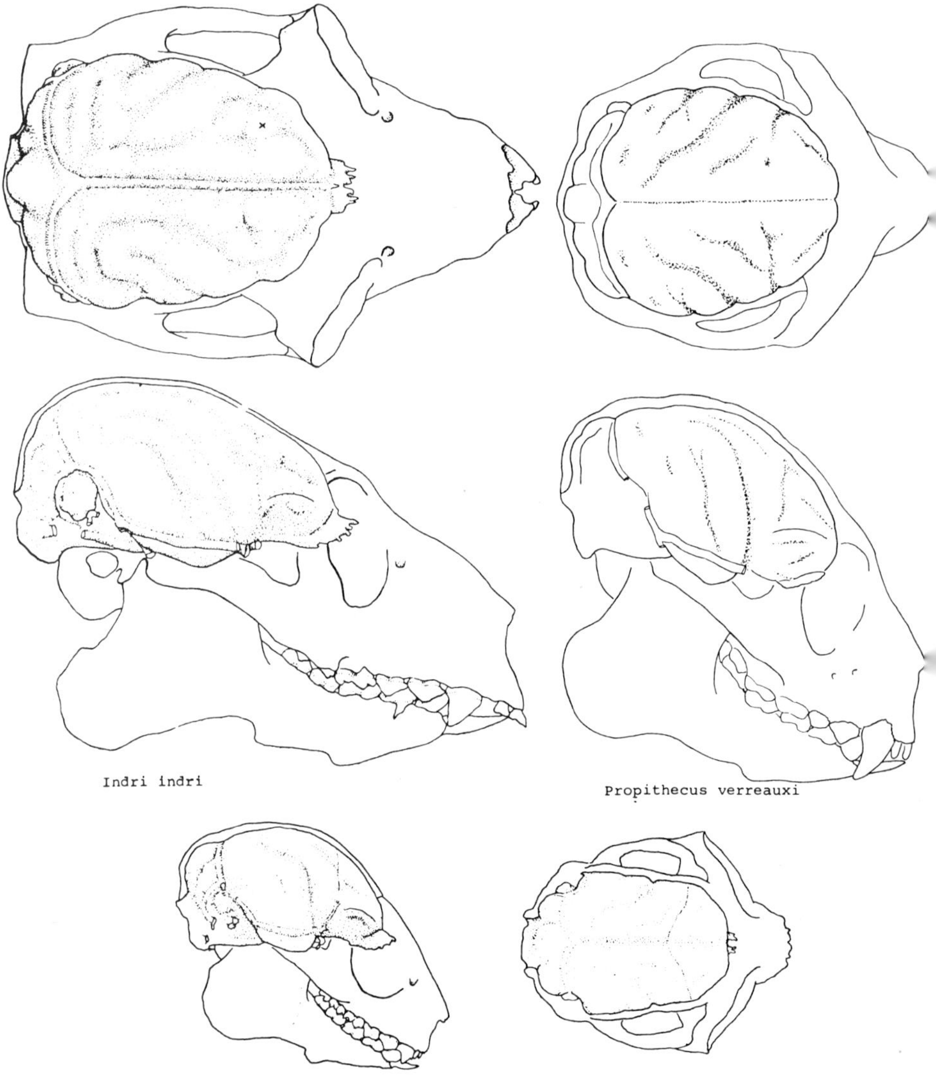

Figure 4. Endocasts of indriids. x (top left illustration) marks inferred hand area in primary motor cortex of *Indri.* All x 0.75.

Propithecus differs from that of *Indri* in having a more convex frontal lobe dorsal profile, with more downwardly rotated olfactory bulbs.

The brain of *Lichanotus* lacks the postcruciate and fronto-orbital sulci seen in its larger relatives, and the postsylvian and postlateral sulci are less developed. As in *Propithecus*, the sylvian extends far caudally, past the end of a shorter lateral sulcus. The frontal lobes are relatively broad, as in *Indri* and *Propithecus*, and intermediate in dorsal profile convexity between the latter two genera.

3. *Family Daubentoniidae*

The brain of *Daubentonia madagascariensis*, the sole surviving species in the family (see Fig. 5), is bizarre among prosimians and has been the subject of several investigations.[6,19–21] It is relatively high and narrow, and has a more convex and downwardly rotated frontal lobe than any comparable sized prosimian brain (compare with *Indri* and *Lemur*), and is further unique in lacking orbital impressions. The coronal sulcus is broken and variable in development, as is the postcruciate complex. The anterior suprasylvian is separated from the pseudosylvian, exposing an area of cortex that in all other prosimians lies buried in a sylvian sulcus. The postsylvian is variably developed and apparently there is no postlateral sulcus.

4. *Family Tarsiidae*

The brain of *Tarsius*, the sole living genus of this family (see Fig. 5), is relatively short and broad, and has relatively small frontal lobes, small olfactory bulbs and a wide occipital-temporal area (comprised of visual cortex), compared to other prosimians. The only sulcus present is the sylvian. The small frontal lobes are primitive and the large visual area and small olfactory bulbs are advanced characters.

5. *Family Lorisidae*

(*a*) Subfamily Galaginae: see Fig. 6. The genus *Galago* includes species differing more in size than do species of other prosimian genera, and with correspondingly greater differences in brain morphology. The largest species of *Galago, G. crassicaudatus*, has a brain close in size to that of the lemurine *Hapalemur*, and with a similar sulcal pattern, including coronal, lateral, orbital, sylvian, postsylvian and (occasionally present) a faint postcruciate sulcus. Six *G. crassicaudatus* brains examined by Haines et al. had a broken coronolateral sulcal complex, while in the two endocasts available to me that sulcal complex appeared to be continuous.[22] The olfactory bulbs are relatively larger in *Galago crassicaudatus* than in *Hapalemur*, and the frontal lobes are relatively narrower, although there is apparently considerable variation in that feature (see Fig. 6). The brain of *Galago senegalensis* is considerably smaller than that of *G. crassicaudatus*, with the sylvian and postsylvian the only clear sulci, and a faint trace of orbital and coronal sulci. The frontal lobes of *G. senegalensis* appear shorter and are downwardly rotated compared to those of *G. crassicaudatus*. An endocast of *Galago senegalensis* figured by Clark and Thomas,[23] is broader across the temporal lobes and has a more tapering frontal lobe profile in dorsal view than the endocast figured by me. An endocast of *Galago demidovii* reveals a brain somewhat smaller but otherwise similar to that of *G. senegalensis*.

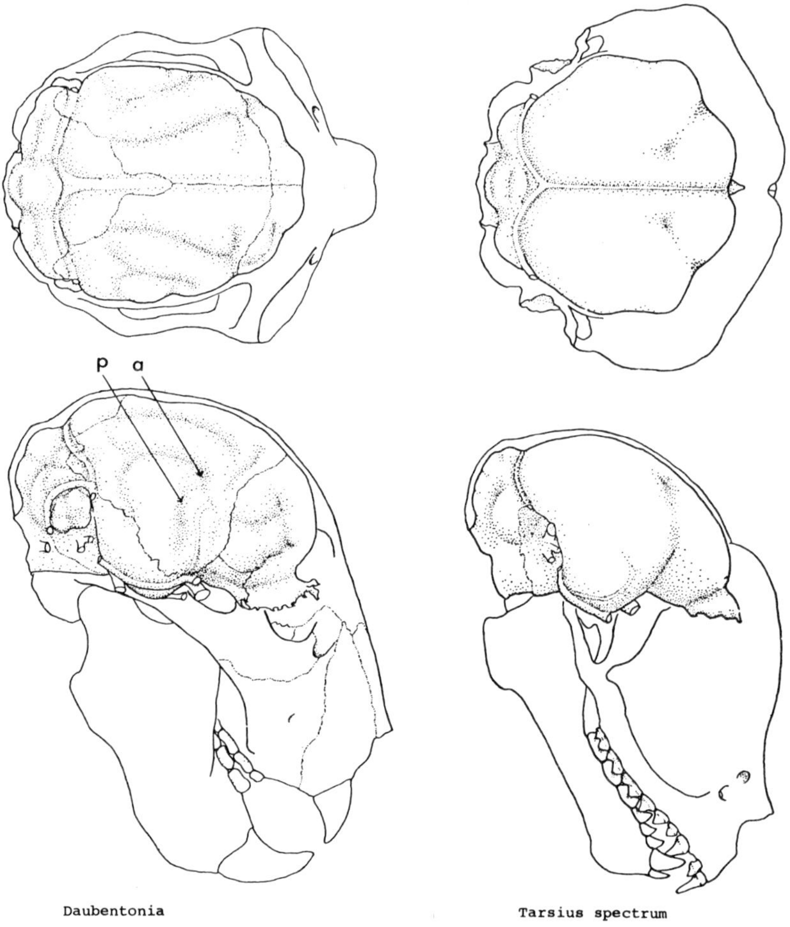

Figure 5. Endocasts of *Daubentonia* and *Tarsius*, the former x 0.75, the latter x 1.5. Abbreviations: a, anterior suprasylvian sulcus; p, pseudosylvian sulcus.

The brain of *Euoticus elegantulus* is slightly larger than that of *Galago senegalensis*, and has lateral, coronal and orbital sulci, in addition to the sylvian and postsylvian. The olfactory bulbs are less downwardly rotated than in *G. senegalensis.* The frontal lobes have the same narrow, tapering outline in dorsal view as in *Galago.*

(*b*) Subfamily Lorisinae: see Fig. 7. The lorisines include two African genera, *Perodicticus* and *Arctocebus*, and two genera confined to Asia, *Nycticebus* and *Loris.* The brain of *Perodicticus* is about the same size as that of *Galago crassicaudatus*, but with relatively broader frontal lobes and olfactory bulbs, smaller orbital impressions, and two major additions to the sulcal pattern: a short but prominent

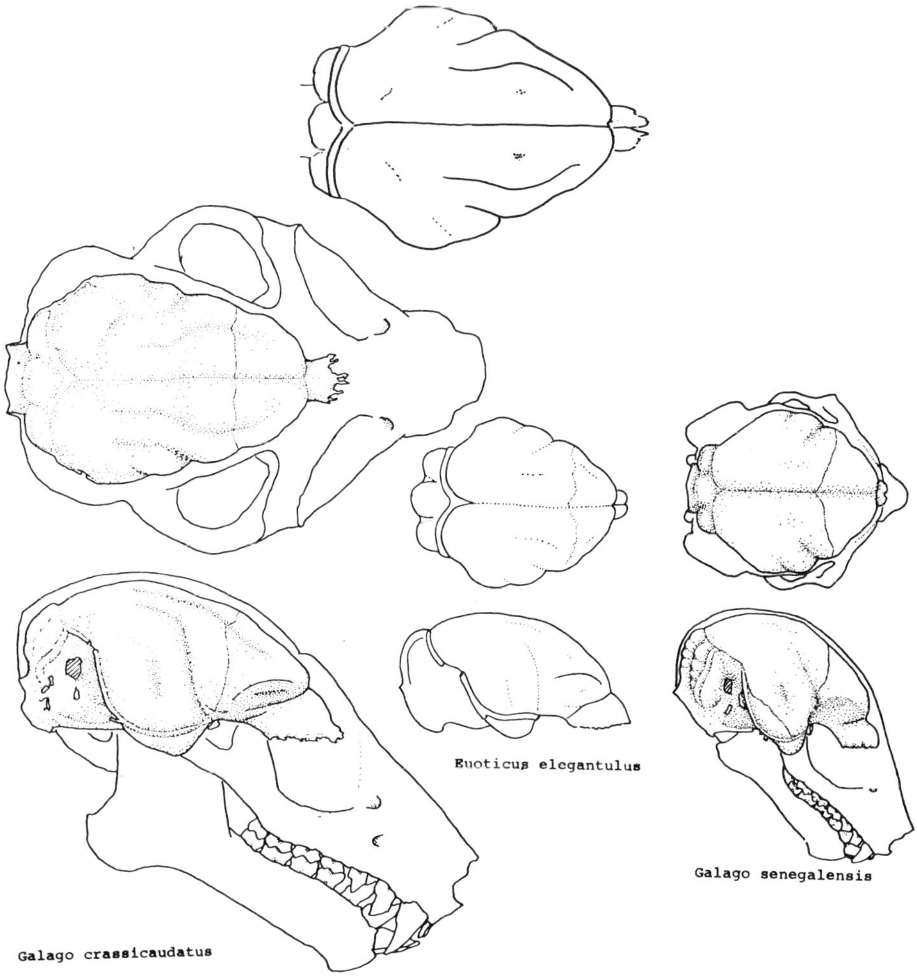

Figure 6. Endocasts of galagines. All natural size.

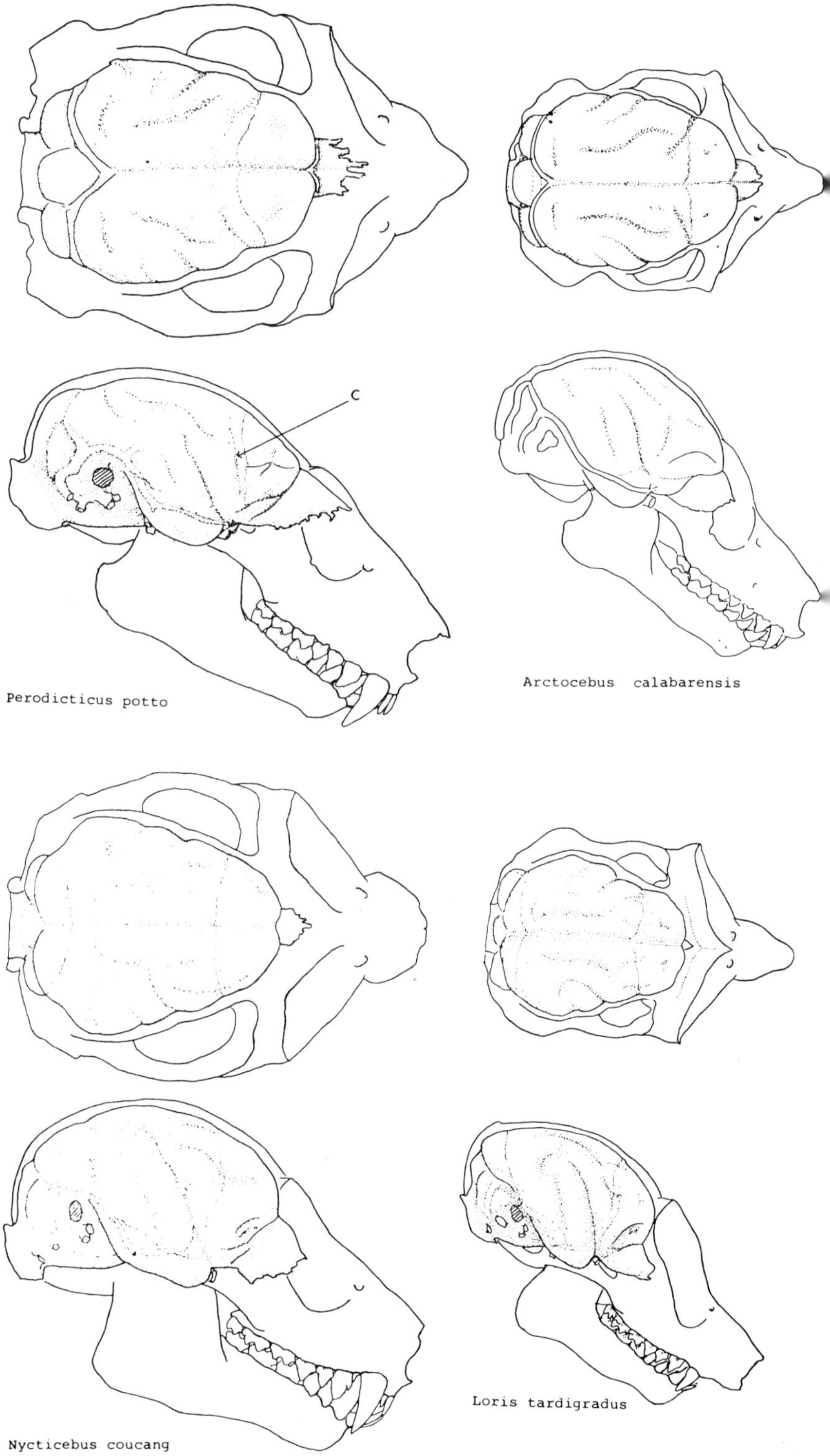

Figure 7. Endocasts of lorisines. All natural size. c = central sulcus.

postlateral sulcus and a long central sulcus (developed from the postcruciate complex) that cuts across the coronal sulcus and descends well down across the coronal gyrus. Those sulci are better developed in *Perodicticus* than in comparable-sized indriids or lemurids, and the central sulcus is better developed than in any other prosimian, regardless of size.

Arctocebus calabarensis has a brain smaller than that of *Perodicticus*, with a more convex dorsal profile and more downwardly rotated frontal pole, but with the same basic sulcal pattern, including the postlateral and central sulci. The posterior rhinal fissure is higher in *Arctocebus* than in the other prosimians examined.

Nycticebus coucang has a brain about the size of that of *Perodicticus potto*, but with a more convex dorsal profile. The sulcal pattern is unique in that there is a short portion of the central sulcus extending ventrally from the coronal sulcus, a longitudinally oriented postcruciate sulcus, and the sylvian sulcus rather than the coronal sulcus connects up with the lateral sulcus. The last is reminiscent of the condition in the indriids *Propithecus* and *Lichanotus*, but in these two genera the lateral sulcus is more extensive. Like *Perodicticus* the brain of *Nycticebus* has a short but marked postlateral sulcus. The olfactory bulbs are narrower in *Nycticebus* than in *Perodicticus*, being similar in relative width to those of the galagines. An endocast of *Nycticebus pygmaeus* is slightly smaller than that of *N. coucang* and displays the same sulcal pattern except for the apparent absence of a postlateral sulcus.

The brain of *Loris tardigradus* is smaller than that of *Nycticebus coucang*, with a more convex frontal lobe dorsal profile and more ventral and caudally displaced olfactory bulbs. The same sulci that are present in *Nycticebus* occur in *Loris*, including the orbital and postlateral sulci; the main difference is that in *Loris* the lateral joins the coronal in the normal prosimian manner, rather than the sylvian, as in *Nycticebus*.

Discussion

1. *Cortical folding*

A common kind of difference in external brain morphology observed between prosimian genera is in the number of neocortical sulci. Having a highly folded cortex is considered an advanced character, and in large part because of that the brain of *Microcebus*, for example, has been described as primitive.[8,24] However, it is clear from the above survey of prosimian brains that as a general phenomenon smaller brained forms have fewer sulci than their larger brained relatives. This is evident in examining brains of the lemurid genera arranged in order of size, in comparing the small indriid *Lichanotus* with *Propithecus* or *Indri*, or the small galagines *Galago senegalensis* and *G. demidovii* with the larger species *G. crassicaudatus*. The same allometric phenomenon may be observed in ceboid primates (see figures in Hershkovitz),[8] artiodactyl ungulates, and other groups of mammals, and was considered by Clark and others to result simply from the need to maintain an approximately constant proportion of gray cortex to internal white matter:[25] the thin sheet of cortex expands as a surface area, proportional to the square of a radius, while the white matter expands as a volume, proportional to the cube. Given that consideration, the lack of sulci in small prosimian brains cannot in itself be considered a primitive character; it merely reflects their small size. Only when brains of similar size are compared would fewer sulci suggest relatively less neocortex.

The quantitative data presented by Stephan and Andy, which must be interpreted with some caution owing to the small sample sizes and the fact that their indices are based on body weight, nevertheless provide the possibility of a check on inferences derived from endocast observations.[3] For example, the cerebral cortex of *Loris* is more highly folded than that of some similar sized or slightly larger-brained prosimians, such as *Cheirogaleus, Lepilemur* and *Lichanotus*, and the direct volumetric measurements of Stephan and Andy show *Loris* to have relatively more neocortex.[3] The same work shows the brain of *Microcebus* to have relatively more neocortex than that of *Cheirogaleus, Lepilemur, Hapalemur* and *Lichanotus*, all of which have more cortical sulci.

Details of sulcal pattern also differ between brains of various prosimian genera, and some of those differences can be interpreted by reference to the cortical mapping studies listed above and summarised in Fig. 1. For example, brains of *Indri* and *Propithecus* differ from *Lemur* brains in having a more prominent development of the transverse branch of the postcruciate sulcus and a larger area of cortex immediately rostral to that sulcus and medial to the coronal sulcus. That cortical area, marked 'x' in the *Indri* brain in Fig. 4, is the primary motor area for the fore limb; its relatively greater size in the large indriids suggests finer muscular control of the hand in *Indri* and *Propithecus* than in *Lemur*.

A postlateral sulcus delimits the rostral boundary of the visual cortex in all lorisine brains (except *Nycticebus pygmaeus*), but is not present in brains of comparable sized lemurids, indriids or galagines. By analogy with work that has been done on sulci of the somatosensory cortex, the presence of the postlateral sulcus in lorisine brains suggests expansion of the visual cortex (see Welker and

Campos for discussion of functional significance of sulci).[26] The volumetric measurements of visual cortex analysed by Stephan indicate that *Nycticebus coucang* and *Loris* do indeed have relatively considerably more visual cortex than those other prosimians.[2] However, the brain of *Perodicticus*, which is similar in size to that of *Nycticebus coucang* and has a similar development of the postlateral sulcus and occipital lobe, has considerably less visual cortex according to Stephan's data (measurements of the visual cortex of *Arctocebus* are not available).[2] This contradiction of an inference based on observation of sulcal pattern emphasises the need to interpret with caution and to check when possible, with direct measurements or experiments, hypotheses based on such observations.

A difference in sulcal pattern, the significance of which is not immediately obvious, is the presence of a well developed central sulcus in *Perodicticus* and *Arctocebus.* The dominance of that transverse sulcus, which separates primary motor cortex from primary somatic sensory cortex, over the longitudinally oriented coronal sulcus, which separates head from forelimb representations in both areas, distinguishes the brains of anthropoid primates from those of prosimians. Studies of the developing thalamus and cortex in *Perodicticus* and *Arctocebus* might yield insights into the question of why the central sulcus is so prominent in anthropoid primates.

Another difference in sulcal pattern which is equally puzzling is the caudal extension of the sylvian sulcus in *Propithecus, Lichanotus* and *Nycticebus.* A similar pattern occurs in some cebid monkeys, where the sylvian sulcus is confluent with the intraparietal.[27] In most but not all of the cebids that have that pattern, the overlap appears to have resulted from expansion of the tail representation area in somatic sensory cortex and consequent lateral displacement of the intraparietal sulcus. That particular explanation is not applicable to *Propithecus, Lichanotus* and *Nycticebus*, and since the cortical areas immediately involved are association areas, it is difficult to suggest a functional interpretation.

The peculiar sulcal pattern of *Daubentonia* appears to have resulted largely from changes in brain shape, and will be discussed below.

2. *Brain shape and proportions*

Another kind of difference between various prosimian genera is in shape and proportions of the brain. Endocasts are particularly well suited for revealing this kind of difference because they reproduce the shape and proportions of the major parts of the brain as they were in life. Examination of endocasts superimposed on the outline drawings of the skull may provide insight into the significance of differences in brain shape, since skull shape may affect brain shape.

A commonly occurring phenomenon is that the frontal lobe is more convex in smaller forms than in their larger relatives. This is seen in the following pairs: *Lemur rubriventer* and *L. variegatus; Propithecus* and *Indri; Galago senegalensis* and *G. crassicaudatus; Arctocebus* and *Perodicticus*; and *Loris* and *Nycticebus.* The more convex and downwardly rotated frontal pole, which is often correlated with a shorter and/or more ventrally flexed splanchnocranium, appears to be a general allometric phenomenon, and has been discussed in the literature (see Biegert and references cited therein).[28] In some cases, splanchnocranial deflection and the accompanying rotation of the frontal pole appears to arise from the need to accommodate larger orbits in a smaller face; *Tarsius* is an outstanding example of this.

Splanchnocranial deflection and consequent modification of brain shape may also occur in large forms, in response to other than allometric demands. The skull of *Daubentonia* appears to have been modified to resist stresses from biting into hard materials (the aye-aye bites into fruit stones and the wood of trees to get at boring insect larvae). Splanchnocranial rotation and rotation of the incisors has brought the long axis of the rostrum into line with the upper incisors, which allows more effective absorption and transmission of stresses than does the condition in other prosimians. The braincase is relatively high and narrow, which would also appear to facilitate resistance of stresses in the saggital plane.[29] The splanchnocranial flexion and changes in shape of the braincase, both in response to selective pressures relating to mastication, have resulted in the globose shape, with downwardly rotated frontal pole and olfactory bulbs, which characterises the brain of *Daubentonia.* Those modifications, perhaps coupled with what appear to be relatively smaller orbits in *Daubentonia* than in *Lemur* and *Indri*, probably account for the lack of orbital impressions in *Daubentonia.* The disrupted sulcal pattern of the frontal lobe appears to have resulted from those changes in frontal lobe shape, as may the separation of anterior suprasylvian and pseudosylvian sulci. Another suggestion to account for the latter is that *Daubentonia*, which uses hearing to locate its prey, may have an expanded auditory cortex (which lies between suprasylvian and pseudosylvian sulci).

The frontal lobes are relatively broader and less tapering in all indriids than in comparable-sized lemurids. Some of that difference might be the result of an expanded motor cortex (see above); but examination of endocasts suggests that motor cortex expansion alone is insufficient to account for the observed difference in frontal lobe shape. Examination of dorsal view drawings of endocasts superimposed on skulls reveals that indriids also have more forwardly facing orbits and that the anterolateral border of the frontal lobe approximately parallels the plane of the orbit in both groups (compare Figs. 2 and 4). This suggests that forward rotation of the orbits altered the shape of the frontal lobes in indriids, producing a relatively broader and less tapering outline.

The brain of *Nycticebus* is relatively globose for a prosimian of its size (compare with *Perodicticus, Galago crassicaudatus*, small indriids and lemurines), which may be correlated with its having relatively large and forwardly facing orbits. That effect on brain shape is even more marked in *Loris*, which has relatively even larger orbits. Similarly, the more forwardly facing orbits in all lorisines may be responsible for the smaller orbital impressions in those prosimians than in comparable sized galagines and lemuroids. Enormously expanded orbits probably account in part for the relatively short and wide brain shape of *Tarsius*, although expansion of the visual cortex probably also played a role in bringing about the very broad occipito-temporal region.

A greater degree of overlap of cerebrum over cerebellum has been noted by previous workers to distinguish brains of galagine and lorisine lorisids from those of comparable sized lemurids and indriids. This appears to be supported in the sample of endocasts I have studied, although the relatively great amount of intraspecific variation present in some taxa for which I have several specimens (*Lemur, Indri, Perodicticus*) suggests that it is a character that must be used cautiously. Quantitative data presented by Stephan and Andy on neocortical volumes reveal a relatively greater amount of neocortex in lorisines and galagines than in comparable-sized lemurids and indriids, which may be the explanation for the difference in cerebellar exposure.[3]

Another feature that should be examined carefully is relative size of olfactory bulbs. Appearances may be deceptive and, for example, measurements by Stephan and Andy of olfactory bulb size reveal that the olfactory bulbs of *Tarsius* are relatively larger than those of the indriids and *Hapalemur*, a difference that was not obvious from examination of endocasts.[3] The observation that lorisids have relatively larger olfactory bulbs than do lemuroids holds true for lemurines and indriines, but not for *Daubentonia, Cheirogaleus* and *Microcebus*, according to Stephan and Andy's data.[3,6]

3. *Phylogenetic implications*

Features of external brain morphology characterise various groups of prosimians; the families Lemuridae and Indriidae, the subfamilies Lorisinae and Galaginae, and the genera *Daubentonia* and *Tarsius*, represent six taxa that are clearly distinguishable on the basis of external neuroanatomy. While lorisines and galagines are currently placed in the same family (Lorisidae) by most primate taxonomists, the differences in external brain morphology between them are as great or greater than those that distinguish other families of prosimians, and in fact appear greater than the differences in brain morphology between galagines and lemurids. However, this is not to say that the latter two groups are related. Before neuro-anatomical features are used to support hypotheses of phylogenetic affinity, they should be carefully evaluated in light of their functional significance relative to that of other characters used to distinguish higher taxa of prosimians. At present I would only emphasise that neuro-anatomical data support evidence from other parts of the body that galagines and lorisines are more different from each other than are lemurids and indriids.

Simpson stated that *Loris* most closely resembles *Arctocebus*, while *Nycticebus* is closest to *Perodicticus*.[30] The brain morphology suggests, on the contrary, that *Loris* and *Nycticebus* are most closely related (the former appears to be a dwarfed version of the latter), while *Perodicticus* and *Arctocebus* are linked by the common possession of a central sulcus.

The differences in sulcal pattern that distinguish *Propithecus* from *Indri* and *Nycticebus* from *Loris* (continuity of sylvian and lateral sulci) should not, I believe, be given more than generic value, since in both cases they result from a change in the course of one sulcus in brains that are otherwise quite similar within each respective pair. Except for the differences in development of the central sulcus in lorisines, it appears that the features that distinguish brains of the prosimian genera within the major polygeneric groupings (Lemuridae, Indriidae, Galaginae, Lorisinae) can be attributed mainly to size differences.

It has been suggested by Le Gros Clark and others that similarities in external brain morphology indicate phylogenetic affinity between *Tarsius* and the Eocene prosimians *Tetonius* and *Necrolemur*.[24] However, as I have argued previously, I find the main source of special resemblance to arise from the failure of *Tarsius*, alone among the living prosimians, to have evolved frontal lobes larger than the Eocene condition, and such a similarity, i.e. retention of a primitive condition, cannot be taken as evidence of special phylogenetic affinity.[1] I see no features in the brains of any modern prosimians that suggest special relationships to any particular group of early Tertiary prosimians.

Advanced development of a central sulcus in *Arctocebus* and *Perodicticus* should not be taken to indicate special affinities to the anthropoid primates, for that sulcus also occurs, although less well developed, in the Asiatic lorisines and in indriids. It would not be surprising to find independent development of that sulcus in several groups of primates; parallel evolution of an equally prominent sulcus, the cruciate sulcus, has been demonstrated in the Carnivora.[31] Nor should the presence of a well developed coronal sulcus be considered to preclude from the ancestry of anthropoids any particular line of fossil prosimians, for the variable development of central and coronal sulci in various living prosimians suggests the possibility of the central sulcus replacing the coronal in prominence during brain evolution within a given lineage.

Acknowledgments

This work was supported in part by National Science Foundation Grant GB-31242 and a grant from the Abbott Memorial Fund, Anatomy Department, University of Chicago.

NOTES

1 Radinsky, L. (1970), 'The fossil evidence of prosimian brain evolution', in Noback, C.R. and Montagna, W. (eds.), *The Primate Brain: Advances in Primatology* 1, New York.

2 Stephan, H. (1969), 'Quantitative investigations on visual structures in primate brains', *Proc. 2nd. Int. Congr. Primat.* 3, 34-42.

3 Stephan, H. and Andy, O.J. (1969), 'Quantitative comparative neuroanatomy of primates: an attempt at a phylogenetic interpretation', *Ann. N.Y. Acad. Sci.* 167, 370-87.

4 Stephan, H., Bauchot, R. and Andy, O.J. (1970), 'Data on size of the brain and of various brain parts in insectivores and primates', in Noback, C.R. and Montagna, W. (eds.), *The Primate Brain: Advances in Primatology* 1, New York.

5 Smith, G.E. (1903), 'On the morphology of the brain in the Mammalia, with special reference to that of the lemurs, recent and extinct', *Trans. Linn. Soc. Lond. (Zool.)* 8, 319-432.

6 Radinsky, L. (1968), 'A new approach to mammalian cranial analysis, illustrated by examples of prosimian primates', *J. Morph.* 124, 167-80.

7 Anthony, J. (1946), 'Morphologie externe du cerveau des singes platyrrhiniens', *Ann. Sci. Nat. (Zool.)* 8, 1-150.

8 Hershkovitz, P. (1970), 'Cerebral fissural patterns in platyrrhine monkeys', *Folia primat.* 13, 213-40.

9 Gervais, P. (1872), 'Mémoire sur les formes cérébrales propres à l'orde des Lémures', *J. Zool.* 1, 5-27.

10 Bauchot, R. and Stephan, H. (1967), 'Encéphales et moulages endocraniens de quelques insectivores et primates actuels', *Colloq. Internat. C. N. R. S.* 163, 575-86.

11 Brodmann, K. (1908), 'Beiträge zur histologischen Lokalisation der Grosshirnrinde', 7: Die cytoarchitektonische Cortexgliederung der Halbaffen (Lemuriden)', *J. Psychol. Neurol.* 10, 287-334.

12 Campbell, C.B.G., Yashon, D. and Jane, J.A. (1966), 'The origin, course and termination of corticospinal fibres in the slow loris, *Nycticebus coucang* (Boddaert)', *J. Comp. Neurol.* 127, 101-12.

13 Kanagasuntheram, R., Leong, C.H. and Mahran, Z.Y. (1966), 'Observations of some cortical areas of the lesser bushbaby (*Galago senegalensis senegalensis*)', *J. Anat.* 100, 317-33.

14 Mott, F.W. and Kelley, A.M. (1908), 'Complete survey of the cell lamination of the cerebral cortex of the lemur', *Proc. Roy. Soc. Lond. B* 80, 488-506.

15 Sanides, F. and Krishnamurti, A. (1967), 'Cytoarchitectonic subdivisions of sensorimotor and limbic fields in slow loris (*Nycticebus coucang coucang*)', *J. f. Hirnforsch.* 9, 225-52.

16 Zuckerman, S. and Fulton, J.F. (1941), 'The motor cortex in *Galago* and *Perodicticus*', *J. Anat.* 75, 447-56.

17 Beddard, F.E. (1901), 'Notes on the broad-nosed lemur, *Hapalemur simus*', *Proc. Zool. Soc. Lond.* 1901, 121-9.

18 Le Gros Clark, W.E. (1931), 'The brain of *Microcebus murinus*', *Proc. Zool. Soc. Lond.* 1931, 463-86.

19 Hofer, H. (1956), 'Das Furchenbild der Hirnrinde von *Daubentonia madagascariensis* (Gmelin, 1788) und seine morphologische Bedeutung', *Zool. Anz.* 156, 177-94.

20 Hofer, H. (1958), 'Vergleichende Beobachtungen über die kraniocerebrale Topographie von *Daubentonia madagascariensis'*, *Morph. Jb.* 99, 26-64.

21 Starck, D. (1954), 'Morphologische Untersuchungen am Kopf der Säugetiere, besonders der Prosimier, ein Beitrag zum Problem des Formwandels des Säugerschädels', *Z. wiss. zool.* 157, 169-219.

22 Haines, D.E. et al., this volume.

23 Le Gros Clark, W.E. and Thomas, D.P. (1952), 'The Miocene lemuroids of East Africa', *Fossil Mammals of Africa* No. 5, British Museum of Natural History, London.

24 Le Gros Clark, W.E. (1971), *The Antecedents of Man*, 3rd ed., Chicago.

25 Le Gros Clark, W.E. (1947), 'Deformation patterns in the cerebral cortex', in Le Gros Clark, W.E. and Medawar, P.B. (eds.), *Essays on Growth and Form Presented to d'Arcy Wentworth Thompson*, Oxford.

26 Welker, W.I. and Campos, G.B. (1963), 'Physiological significance of sulci in somatic sensory cerebral cortex in mammals of the family Procyonidae', *J. Comp. Neurol.* 120, 19-36.

27 Radinsky, L. (1972), 'Endocasts and studies of primate brain evolution', in Tuttle, R. (ed.), *Functional and Evolutionary Biology of Primates*, New York.

28 Biegert, J. (1963), 'The evaluation of characteristics of the skull, hands and feet for primate taxonomy', in Washburn, S.L. (ed.), *Classification and Human Evolution*, Chicago.

29 Cartmill, M., this volume.

30 Simpson, G.G. (1967), 'The Tertiary lorisform primates of Africa', *Bull. Mus. Comp. Zool.* 136, 39-62.

31 Radinsky, L. (1971), 'An example of parallelism in carnivore brain evolution', *Evol.* 25, 518-22.

PART II SECTION D
Prosimian Locomotion

C. J. JOLLY and A. T. GORTON

Proportions of the extrinsic foot muscles in some lorisid prosimians

Interpreting the relationship between structural and functional diversity in the living world was a respectable occupation for naturalists long before Darwin substituted a mechanistic for a theistic explanation of the existence of the phenomenon. Today, however, confirmation of the universal existence of adaptation (or Divine Providence) is perhaps the least of our reasons for investigating the concordance between structure and function in particular cases. Much more important is the prominent role played by such deductions in two central procedures of evolutionary biology; determination of the phyletic relationships between species, and elucidation of the adaptive trends that have led to the array of relationships and characteristics observed in extant and fossil organisms. Whether one is concerned with living or extinct forms, and whether the operation is performed intuitively or with the aid of mathematical procedures, the construction of phyletic trees involves a critical assessment of the significance, or 'phyletic weight', of observed resemblances and differences. In this judgment, the functional interpretation of characters plays a vital part, since only thus can the potentially misleading products of parallelism, convergence, and rapid adaptive shifts be distinguished from phyletically significant heritage characters.

This rationale is implicit in recent reconsiderations of lorisid classification which question the phyletic unity of a subfamilial or familial group which includes both the Asiatic and African 'slow climbers' while excluding all forms of *Galago* and *Euoticus*.[1] To the extent that the resemblances between Asiatic and African lorisines can be attributed to a single morphological-functional pattern associated with common locomotor habits, they must be regarded as poor indicators of special phyletic relationship, although it is not unlikely that they are indeed heritage characters. (This interpretation does not, of course, prohibit the establishment of a phenetically-defined lorisine grouping on the grounds of adaptive resemblance.)[2]

On the other hand, the monophyletic status of the Lorisidae (*sensu lato*) as opposed to other extant prosimians is supported by a variety of details of dental and cranial structure to which high phyletic weight may be accorded.[3]

In drawing out correlations between features of structure and function in extant forms we are able to make use of two sets of data which are logically independent of each other: the comparative anatomy of the system under examination, and information about the naturalistic behaviour of the living animal. Each of these two sets may be used to derive hypotheses testable against the other. Since correlation, not causality, is the object, this process may proceed in either direction.

By contrast, when attempting a similar procedure in forms known only as fossils, our *independent* knowledge of their natural history is virtually limited to ambiguous information derived from their geological and taphonomic setting. Hypotheses about the behaviour of fossil forms (their locomotor repertoires, dietary emphases, and so on) thus rest largely upon the extent to which analogies can be drawn with living forms in which behavioural-structural correlations have been satisfactorily established. (It is perhaps worth emphasising that the most telling analogies are often derived from forms relatively remote, in the phyletic sense, from the species under consideration.)[4,5,6]

Isolated foot bones, especially of the tarsus, are among the postcranial elements most frequently recovered as fossils. Furthermore, there are few of the more contentious issues in prosimian evolution (the occurrence of vertical clinging and leaping adaptations in Eocene prosimians,[4,7] and the nature of the distinctive adaptation of basal primates,[8,9] are cases that come to mind), which do not involve to some extent the functional interpretation of hindlimb anatomy. Interpretation of skeletal elements, whether fresh or fossil, is sterile if it does not view them as parts of a total mechanical system including, in the living, muscles, cartilage, ligaments, membranes and tendons as well as bones. The present paper is an examination of variability among four lorisid species (*Galago senegalensis, G. crassicaudatus, Nycticebus coucang*, and *Perodicticus potto*) with respect to one aspect of a single part of this total system – proportional relationships among the weights of the extrinsic muscles of the foot. Parallel work on joint structure and the relationships between the bones, tendons, muscles and ligaments of the leg and foot is currently in progress. Since these factors, and others, profoundly influence the nature of the force exerted by a particular muscle,[10,11] the mechanical interpretations of muscular proportions which are offered in the discussion must be regarded as tentative and provisional. Regardless of their interpretation, however, the results presented below do seem to suggest that certain muscular ratios vary among the four in proportion to the importance of particular locomotor activities in the repertoire of each species.

Materials and methods

Material examined consisted of the legs and feet of two *Galago crassicaudatus* (three limbs), five *G. senegalensis* (six limbs), and one *Perodicticus potto* (one limb). All but one subadult lesser bushbaby were apparently mature animals. The galagos came from south-eastern Kenya, the potto from the Mabira Forest, Uganda. The former were feral animals which had been experimentally infected with hookworm prior to sacrifice; the latter had been held captive for a brief period before its death from unknown causes. None of the subjects showed any obvious pathology affecting the skeleton or musculature. All had been preserved in neutralised formaldehyde solution for five years prior to dissection. The skeleton exhibited no signs of demineralisation, and the musculature was apparently neither macerated nor differentially dehydrated in any specimen.[1 2]

After dissection of the leg, each of the extrinsic foot muscles (Table 1) was carefully removed from its origin, and its tendon severed at a standard point, usually immediately proximal to its passage beneath its most proximal retinaculum, if any. Muscle bellies and attached portions of tendon were then immersed in 10% formaldehyde solution for at least a week prior to weighing.

Each muscle was removed from the solution immediately before weighing, blotted quickly and without pressure to remove surface moisture, and weighed on a chemical balance legible to 0.003 g. The whole weighing procedure was then repeated. If the discrepancy between the two determinations exceeded 0.01 g., reweighings were repeated until a 'run' of three consistent weights was obtained. Specimens were always returned to the fluid and subsequently blotted between determinations. Weights were rounded to the nearest 0.01 g. After weighing, the saturated muscles of a represent-

Table 1 List of muscles discussed and abbreviations used in Tables 2-6

Gn	*Gastrocnemius*	Fdt	*Flexor digitorum tibialis*
So	*Soleus*	Fdf	*Flexor digitorum fibularis*
Pt	*Plantaris*	Pl	*Peroneus longus*
Edl	*Extensor digitorum longus*	Pb	*Peroneus brevis*
Ehl	*Extensor hallucis longus*	$\overline{X}$	Sample Mean
Ta	*Tibialis anterior*	Range obs.	Observed range in the sample
Tp	*Tibialis posterior*	S.D.	Estimate of population Standard Deviation
		C.L.	Upper and lower 95% Confidence Limits of the Population Mean

ative of each species were teased and interspecific comparisons made in terms of the architectural arrangement of their fibres and tendons.

Weight data were punched onto cards and a computer was used to calculate a number of ratios in each specimen, as well as a series of statistics (Mean, Standard Deviation, 95% Confidence Limits of the Mean) partially describing the distribution of each of the ratios in each species as represented by our samples. Ratios calculated fall into three groups:

1. The weight of each individual muscle expressed as a percentage of the total extrinsic foot musculature (referred to as 'Relative Weight') (Table 2).
2. The weights of individual muscles expressed in terms of the total musculature acting upon a particular joint or set of joints (Tables 3, 4, 5).
3. Ratios comparing the weight of certain pairs of muscles with partially similar action (Table 6).

Grand includes in his account of the anatomy of the leg and foot of *Nycticebus coucang* expressions of the weights of the extrinsic foot muscles of eight animals (nine limbs).[13] Although Grand weighed only to the nearest 0.1 g., used fresh rather than preserved material, and did not state how much of the tendon of insertion is included in the weight, his data comprise a valuable series for comparison with the African lorisids. Grand's data are presented in the form of ratios of the total musculature operating upon each of three joints (as in Tables 3, 4 and 5); but Relative Weights and ratios between pairs of muscles could be extracted from them by simple arithmetic manipulation.

Results

A general, qualitative account of lorisid limb musculature is to be found in the work of Jouffroy, who summarised her own and previous findings, comparing them with the lemuroid condition.[14] In general, the origins, insertions and gross structure of individual muscles were quite similar in the three species dissected. The following differences were noted:

Gastrocnemius in *Galago* has a large fabella in each tendon of origin; these are absent in *Perodicticus* and *Nycticebus*, as previously noted by Jouffroy.[14] Grand reports an origin from the fascia of flexor digitorum tibialis in the slow loris.[13] The bellies of the muscle are more slender and cylindrical, and their insertion into the common triceps tendon more fibrous, and less aponeurotic, in *Perodicticus* than *Galago*.

Soleus in the potto is similar in size and shape to the two heads of

Table 2 Relative weights of the extrinsic musculature of the foot

+ In this and subsequent tables indicates cases where there is no overlap in 95% confidence limits between the two *Galago* species.

* In this and subsequent tables indicates cases where *Perodicticus* falls more than 2.5 S.D. from the Mean for *Nycticebus.*

	Galago senegalensis				*G. crassicaudatus*				*Nycticebus coucang*				*P.potto*
	$\overline{X}$	Range obs.	S.D.	C.L.	$\overline{X}$	Range obs.	S.D.	C.L.	$\overline{X}$	Range obs.	S.D.	C.L.	
Gn	24.8	23.7-26.6	1.2	23.6-26.1	+18.8	17.7-19.3	0.9	16.5-21.0	5.2	3.8-87.4	1.3	4.2- 6.3	11.9*
Pt	7.7	5.8- 8.9	1.1	6.5- 8.9	6.3	5.9- 6.9	0.6	4.9- 7.7	1.0	0.6- 1.6	0.4	0.4- 1.5	1.5
So	3.8	3.1- 4.4	0.6	3.2- 4.4	5.0	4.6- 5.6	0.5	3.7- 6.3	9.4	8.1-10.6	0.7	8.7-10.0	8.8
Edl	3.5	1.2- 4.6	1.2	2.2- 4.7	2.8	2.6- 3.0	0.2	2.2- 3.3	4.2	3.6- 5.2	0.5	3.8- 4.6	5.4
Ehl	3.7	3.3- 4.1	0.3	3.4- 4.0	3.3	3.1- 3.4	0.1	3.0- 3.5	2.9	2.1- 3.8	0.5	2.6- 3.3	3.7
Ta	13.0	12.2-15.0	1.1	11.9-14.2	15.1	14.2-15.9	0.9	13.0-17.2	12.4	10.5-15.3	1.2	11.5-13.3	10.9
Tp	3.5	3.1- 4.1	0.4	3.1- 3.9	3.2	3.0- 3.4	0.2	2.8- 3.6	3.8	3.0- 5.2	0.8	3.2- 4.5	4.5
Fdt	6.7	5.1- 7.9	1.1	5.6- 7.9	8.7	7.7-10.0	1.2	5.8-11.7	32.5	29.2-40.3	1.6	31.2-33.7	20.6*
Fdf	14.8	13.1-16.9	1.6	13.1-16.5	18.9	17.3-21.6	2.4	13.0-24.7	14.3	12.4-16.5	0.8	13.6-14.9	15.0
Pl	11.7	10.3-13.0	1.0	10.6-12.8	13.5	13.2-13.9	0.4	12.5-14.5	11.4	10.8-12.5	0.6	10.9-11.9	12.3
Pb	6.8	5.8- 8.4	1.1	5.6- 7.9	+4.6	4.5- 4.6	0.1	4.3- 4.8	3.0	2.5- 3.6	0.2	2.8- 3.2	5.5*

Table 3 Weights of the individual talocrural muscles expressed as a percentage of the total for the group

	Galago senegalensis				*G. crassicaudatus*				*Nycticebus coucang*				*P. potto*
	$\bar{X}$	Range obs.	S.D.	C.L.	$\bar{X}$	Range obs.	S.D.	C.L.	$\bar{X}$	Range obs.	S.D.	C.L.	
Gn	57.3	55.3-61.7	2.3	54.8-59.7	52.0	50.7-53.5	1.4	48.6-55.5	23.0	17.8-30.7	4.3	19.4-26.6	38.0*
Pt	17.6	14.8-19.6	1.7	15.9-19.3	17.4	16.1-19.8	2.1	12.2-22.6	4.5	2.7- 7.3	1.9	2.1- 6.9	4.9
So	8.7	7.1- 9.9	1.2	7.4- 9.9	+13.9	13.2-15.1	1.0	11.2-16.4	42.3	38.4-48.2	3.0	39.7-44.8	28.2*
Edl	7.9	3.2-10.8	2.7	5.1-10.7	7.7	7.3- 8.0	0.4	6.8- 8.6	18.7	15.4-22.8	2.3	16.9-20.5	17.1
Ehl	8.5	7.7-10.6	1.1	7.3- 9.7	9.0	8.9- 9.0	0.1	8.9- 9.1	12.9	9.6-15.3	1.7	11.6-14.2	11.8
Gn+Pt+So	83.6	81.5-86.2	1.8	81.7-85.5	83.3	82.3-84.3	0.4	83.0-83.7	68.9	64.3-71.8	4.0	65.8-72.0	71.0

Table 4 Weights of individual invertors and evertors of the foot expressed as a percentage of the total for the group

	Galago senegalensis				*G. crassicaudatus*				*Nycticebus coucang*				*P. potto*
	$\bar{X}$	Range obs.	S.D.	C.L.	$\bar{X}$	Range obs.	S.D.	C.L.	$\bar{X}$	Range obs.	S.D.	C.L.	
Ta	31.3	28.4-36.2	2.8	28.4-34.1	33.5	32.7-34.1	0.7	31.7-35.2	19.6	16.6-22.6	1.9	18.1-21.1	20.2
Tp	8.5	7.5- 9.4	0.8	7.6- 9.3	7.1	6.7- 7.5	0.4	6.1- 8.0	6.1	4.7- 8.2	1.3	5.1- 7.0	8.3
Fdt	16.1	12.8-18.9	2.6	13.4-18.8	19.3	17.6-21.5	2.0	14.5-24.2	51.3	46.3-53.5	2.3	49.6-53.8	38.3*
Pl	27.9	25.6-29.9	1.5	26.4-29.5	30.0	28.5-32.1	2.0	25.1-34.8	18.0	16.5-19.6	1.0	17.3-18.8	22.9*
Pb	16.2	13.8-19.8	2.7	13.4-19.0	+10.1	9.6-10.6	0.5	8.8-11.4	4.7	3.9- 5.1	0.4	4.4- 5.0	10.2*
Ta+Tp+Fdt	55.9	50.8-58.5	2.8	52.9-58.8	59.9	57.4-62.3	2.4	53.8-66.0	77.1	76.1-78.7	0.8	76.5-77.8	66.9*

Table 5 Weights of individual flexors and extensors of the digits expressed as a percentage of the total for the group

	Galago senegalensis				*G. crassicaudatus*				*Nycticebus coucang*				*P.potto*
	$\bar{X}$	Range obs.	S.D.	C.L.	$\bar{X}$	Range obs.	S.D.	C.L.	$\bar{X}$	Range obs.	S.D.	C.L.	
Fdt	23.2	20.0-26.0	2.2	21.0-25.6	56.1	21.9-30.7	4.0	15.9-36.1	60.2	56.9-63.0	1.8	58.8-61.6	46.1*
Fdf	51.6	47.7-56.2	2.8	48.7-54.6	56.1	51.8-61.8	5.2	43.2-68.9	26.4	23.3-28.0	1.3	25.4-27.4	33.5*
Edl	12.2	4.1-15.9	4.3	7.7-16.7	8.3	7.3- 9.2	0.9	6.0-10.6	7.3	6.6-10.1	1.0	7.0- 8.6	12.0*
Ehl	12.8	11.4-14.6	1.3	11.5-14.2	+9.7	9.0-10.3	0.7	8.0-11.4	5.4	3.8- 7.1	0.9	4.7- 6.0	8.3*
Fdf	75.0	70.9-82.2	4.0	70.8-79.1	82.0	80.5-83.7	1.6	78.1-86.0	86.7	84.5-88.3	1.4	85.7-87.8	79.7*

Table 6 Intra-group ratios

	Galago senegalensis				*G. crassicaudatus*				*Nycticebus coucang*				*P.potto*
	$\bar{X}$	Range obs.	S.D.	C.L.	$\bar{X}$	Range obs.	S.D.	C.L.	$\bar{X}$	Range obs.	S.D.	C.L.	
Gn/so	6.70	5.70-7.75	0.85	5.80-7.60	+3.77	3.44-4.00	0.29	3.04-4.49	0.56	0.42-0.75	0.12	0.45-0.66	1.35*
Fdt/Fdf	0.45	0.39.0.53	0.05	0.40-0.50	0.47	0.35-0.58	0.12	0.18-0.76	2.28	2.04-2.65	0.18	2.15-2.42	1.38*
Ta/Fdt	2.54	2.00-3.27	0.57	1.94-3.13	2.11	1.90-2.26	0.19	1.64-2.57	0.50	0.42-0.66	0.07	0.45-0.56	0.75*
Pl/Pb	1.75	1.29-2.06	0.28	1.45-2.04	+2.96	2.89-3.03	0.07	2.78-3.13	3.36	3.34-4.96	0.52	3.46-4.26	2.23*

gastrocnemius. Distally it develops numerous tendinous fibres on its deep surface which extend directly into the common tendon of *triceps surae.* In *Galago* most of the fibres of the muscle insert fleshily on the deep surface of the gastrocnemius tendon, direct contributions by tendons developed in soleus being few. In all specimens of our series, soleus originated by a slender flat tendon from the head of the fibula; none showed the origin from the fibular shaft reported by Nayak.[14] *Plantaris* in our potto was distinct from gastrocnemius, though weakly developed. Its origin was femoral, not fibular as reported by Nyak.[14]

Flexor digitorum tibialis of galagos resembles its homologue in the lemuroids, a single-headed muscle arising from the lateral condyle and posterior surface of the proximal third of the tibia. In *Nycticebus* and *Perodicticus* it has a more extensive origin and is almost completely divisible into two heads. In *Nycticebus* the deep head, which is presumably the homologue of the single head of other prosimians, is supplemented by a superficial head with an extensive origin on the medial surface of the tibia, the medial surface of the distal end of the femur, and the medial aspect of the knee.[13] In our *Perodicticus*, the accessory, superficial head originated only from the medial tibial surface.

In *Galago* the relation between *tibialis anterior* and *extensor hallucis longus* is especially close, the tendon of the hallucial extensor passing between the fibres of the larger muscle, and lying in a superficial groove in its tendon under the extensor rectinaculum. In the potto the two muscles and their tendons are merely adjacent.

Extensor digitorum longus supplies no tendon to the reduced second digit in *Perodicticus*;[14] digits II, III, IV and V are supplied in *Galago.* In *Nycticebus*, according to Grand, it supplies 'each toe'.

The tiny *peroneus digiti quinti* was absent in our potto, as in some *Nycticebus*,[13] but occurred in all *Galago* specimens. When present it was weighed with peroneus brevis, although its contribution to the weight of that muscle was negligible.

The results of analysis of the relative and proportional weights of the eleven extrinsic foot muscles are presented in full in Tables 2-6.

In the following exposition of the salient points of interspecific difference, comparisons are first made between *Galago senegalensis* and *Nycticebus coucang*, which are represented by larger samples of material. *G. senegalensis* and *P. potto* are then compared with both.

Fig. 1 contrasts the profiles of relative muscle weight in *Galago senegalensis* and *Nycticebus coucang*, and demonstrates the major difference between them, namely the dominance of gastrocnemius in the bushbaby, and flexor digitorum tibialis in the slow loris. Within the group whose primary action is plantarflexion of the foot at the talocrural joint (Gn, So, Pt), gastrocnemius is by far the largest component in the bushbaby, and is supplemented by a substantial plantaris, another two-joint muscle crossing both knee and ankle. In

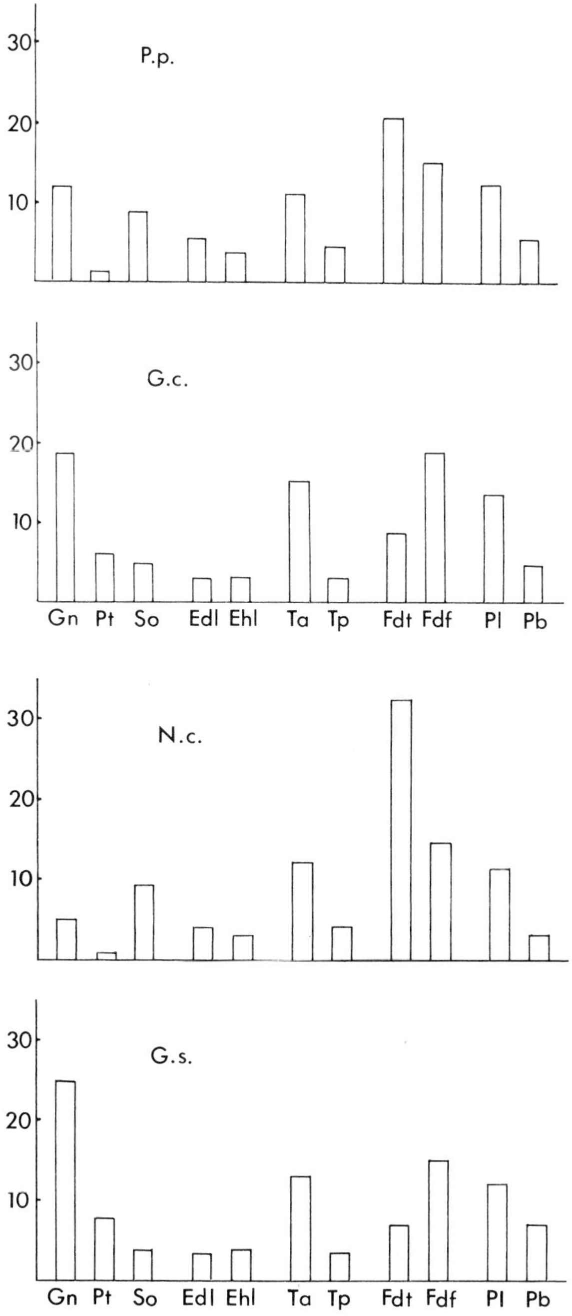

Figure 1. Relative weights of the extrinsic foot muscles.
G.s.: *Galago senegalensis* G.c.: *Galago crassicaudatus*
P.p.: *Perodicticus potto* N.c.: *Nycticebus coucang*
For other abbreviations see Table 1.

the slow loris both plantaris and gastrocnemius are very weak, and together weigh less than soleus. This contrast appears in the Gn/So ratio (Table 6), which averages 6.70 in the bushbaby, but only 0.56 in the loris. The long digital extensors, which are also dorsiflexors of the foot on the leg, are comparatively small and rather variably developed muscles which do not differ in relative size in the two forms. The plantarflexors therefore represent a higher proportion of the total talocrural musculature in the bushbaby (Table 3). Tibialis anterior and posterior, which are apparently primarily invertors of the foot, are rather similar in relative size in the two forms.

Flexor digitorum fibularis, which has little or no rotatory action on the foot, is very similar in relative size in the two forms, but the combined digital flexor and pedal invertor, flexor tibialis, is much larger by all standards in the loris, as well as exhibiting the specialised structure and long head of origin described above. With a relative weight averaging 32.5, it is much the largest of the extrinsic foot muscles, corresponding in importance to combined gastrocnemius and plantaris in the bushbaby. The importance of the combined flexor and invertor is very apparent when compared with that of the simple digital flexor (Fdt/Fdf, Table 6), or with that of the simple invertor (Ta/Fdt, Table 6), and produces the high ratio of invertors to evertors of the foot seen in the loris (Table 4).

The two major peronei are the principle evertors of the foot, and peroneus longus is, in addition, a powerful adductor of the mobile first metatarsal. In both species peroneus longus has a relative weight of about 11, but peroneus brevis is more than twice as large in the bushbaby (Table 2). The difference in relative size of the two muscles is further illustrated by the ratio Pl/Pb (Table 6).

The generalisations that can be made about the muscular proportions of *Perodicticus potto* on the basis of a single specimen are of course limited. Pending a study of more extensive material, attention may be drawn to those features in which the single specimen lies so far from the mean of the other series that a species-specific difference is clearly implied. In general, the profile of Relative Muscle Size in the potto differs from that of *Galago senegalensis* in the direction of *Nycticebus coucang* (Fig. 1). However in those characters in which the loris differs most strikingly from the galago, the potto is consistently less extreme in its deviation. Soleus, in the potto, is relatively as large a muscle as it is in the loris, but it is a much smaller part of the total talocrural musculature (Table 3) because of the relatively larger gastrocnemius. The difference between the two is brought out by the Gn/So ratio (Table 6). Since the relative weights of soleus and plantaris do not differ in the two lorisines (Table 2), the low Gn/So ratio in the loris is clearly due to a deficit in gastrocnemius. On the other hand, it is noteworthy that the talocrural plantarflexors are a similar proportion of the total talocrural musculature in the two lorisines (Table 3), both of which differ from the galago in this respect.

Flexor digitorum tibialis is large and bicipital in the potto. Although it exceeds *flexor fibularis* in weight, the disproportion between them is less pronounced than it is in the loris (Table 6). The more modest size of the muscle is reflected also in the lower invertor/evertor and digital flexor/extensor ratios in the potto (Tables 4 and 5).

Finally, peroneus brevis is relatively larger in our potto than in any of the loris series (Table 2), as is reflected in its lower Pl/Pb ratio (Table 6).

As with *Perodicticus*, our sample of three limbs of *Galago crassicaudatus* must be regarded with caution as representative of the species as a whole. They do, however, show some noteworthy differences from the other species, especially the lesser galago. The plantarflexors of the talocural joint comprise, as in the lesser bushbaby, about 83% of the total musculature acting upon the joint, but within the group the proportion contributed by gastrocnemius is much less (Table 3), giving a Gn/So ratio as close to that of the potto as to that of *G. senegalensis* (Table 6). No significant difference between the two galagos can be shown in the relative size of soleus or plantaris, so that (as in the case of the lorisines) the difference can be attributed to variation in the size of gastrocnemius. Again as in the lorisines, this reduction accounts fully for the smaller

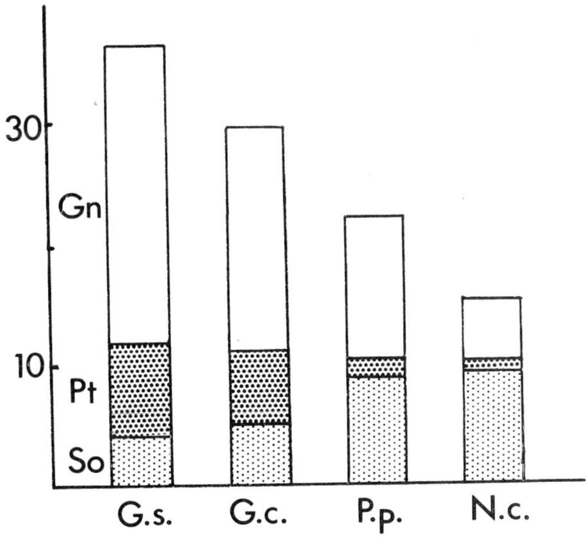

Figure 2. Relative weights of the talocrural plantarflexors.
Segments of each column represent the relative weights of soleus, plantaris and gastrocnemius.
Abbreviations as in Fig. 1.

relative size of the total plantarflexors (range 29.3-30.0) in *G. crassicaudatus* when compared with *G. senegalensis* (range 33.5-39.3) (Fig. 2).

Unlike the talocrural musculature, the proportions of the digital flexors of the greater bushbaby show no significant trend in a loris-like direction, the relative size of flexor digitorum tibialis and the ratio Fdt/Fdf showing values little different from those seen in *G. senegalensis*.

A significant difference does, however, appear in the peronei. Peroneus brevis is a smaller part of the total crural musculature, and of the total invertor-evertor group, than in the lesser galago, a difference that is also apparent in the ratio Pl/Pb (Table 6). In each of these expressions of the relative size of peroneus brevis *G. crassicaudatus* lies closer to the potto than it does to its smaller relative.

Discussion

The finding that gastrocnemius is a relatively large muscle in *Galago senegalensis* is not unexpected, in view of its importance, as a two-joint muscle, in the leap,[15] which is a common and characteristic gait in its repertoire.[16] Since leaping and hopping are reported to be less frequent in the behavioural repertoire of *G.s crassicaudatus*,[17-19] the reduced relative size of gastrocnemius in this form is explicable, as is its still smaller size in the lorisines. Clearly, the same explanation is insufficient to account for the difference between the two lorisines, neither of which has ever been observed to leap, hop or gallop.[20] It is possible that the femoral origin of flexor tibialis in the slow loris may partially replace the gastrocnemius as an accessory flexor of the knee, thus permitting its further reduction, but this action of the long head of flexor tibialis has not been demonstrated.

In contrast to gastrocnemius, the one-joint plantarflexor, soleus, is presumably able to exert its action at any position of the knee, which might account for its importance in the lorisines, whose mode of locomotion, especially on small branches, involves great variation in orientation of footholds.[13,21] The ability to plantarflex the foot with the knee fully extended may be an important factor since slow-climbing involves maintaining a grip with the trailing foot until the last possible moment. Plantaris, which varies almost exactly inversely to soleus in the four forms investigated here, is a two-joint flexor with, presumably, the additional function of tensing the strong plantar aponeurosis, and thus helping to support the elongated tarsus against the powerful internal forces of dorsiflexion which must be exerted upon it as the body weight is raised about a fulcrum at its distal end. This function is surely minimal in the short, hand-like foot of the lorisines, as is the development of the plantar aponeurosis itself.

The most striking muscular specialisation in the leg of the loris is the hypertrophy, bicipital structure and femoral origin of flexor digitorum tibialis. As a combined invertor and digital flexor this muscle performs two of the most important motions involved in deliberate climbing, especially when performed in a network of steeply-sloping, small-diameter footholds.[13] Its less specialised condition in the potto is less easy to interpret on the basis of present evidence about the natural history of the two forms. Perhaps future quantitative studies, modelled upon those of Charles-Dominique,[21] may indicate that the slow loris, while sharing the basic locomotor repertoire of the potto, may spend more of its time in the demanding small-branch milieu, and less on the larger boughs which, as footholds, do not require the same degree of pedal inversion. It is known that in primary forest 71% of the supports utilised by pottos exceeded 5 cm. in diameter, while in secondary forest 52% were in this size-range.[21]

Comparable data on *Nycticebus* are not available, although Grand[13] suggests that the form seems to prefer footholds about 3 cm. in diameter. Possibly, on the other hand, the femoral origin of flexor tibialis, and consequent reduction of gastrocnemius, represents a condition that would be adaptive for both species, but has arisen in the ancestors of *Nycticebus* by mutational events which by chance have not occurred in the *Perodicticus* stock. In any case, it would certainly be premature to interpret the apparently more primitive condition in the potto as evidence for the separate derivation of *Perodicticus* and *Nycticebus* from an ancestral stock whose locomotor repertoire did not include a predominant component of slow and deliberate climbing.

Variations in the peronei do not follow the galagine-lorisine dichotomy as do the other characters which vary in the four forms. However, the low Pl/Pb ratio seen in *G. senegalensis* may be associated with both its comparatively frequent descent to the ground,[16] and its small size, both of which are known to affect this ratio in cercopithecines.[10,22] The difference between the two lorisine species cannot be explained in terms of size differences, and perhaps may be regarded as an additional indication that the potto, though predominantly an arboreal animal, may not be as constantly dependent upon the pedal grip as the slow loris, either because it spends more time on large boughs, or because it descends more readily to the ground. Pottos in the wild are known to cross open ground between trees, especially where the forest is broken.[18,21] Their movement when placed on the ground is easy and as rapid as it is in the trees,[20] showing none of the discomfiture and difficulty in locomotion which Grand reports for *Nycticebus* forced to walk on a flat surface.[13]

It cannot be emphasised too strongly that the interpretations suggested in this section must be regarded as hypotheses to be tested against additional subsystems of the locomotor apparatus, further behavioural observations, and more species. Nevertheless they do support the view that despite the undoubted heuristic value of the concept of locomotor categories, the complex of adaptive characters distinctive of a particular species can be more fully elucidated by reference to its total locomotor repertoire, which is likely to comprise a unique combination of gaits and postures. They also illustrate the insight that may be gained by comparing groups of closely-related forms, and thus partly controlling the heritage component of inter-specific variation.[10,23]

NOTES

1 Groves, C.P., this volume.

2 Washburn, S.L. (1951), 'The new physical anthropology', *Trans. N.Y. Acad. Sci.*, ser. 2, 13, 198-304.

3 Le Gros Clark, W.E. (1959), *Antecedents of Man*, Edinburgh.

4 Cartmill, M. (1972), 'Arboreal adaptations and the origin of the order primates', in Tuttle, R. (ed.), *The Functional and Evolutionary Biology of Primates*, Chicago.

5 Cartmill, M., this volume.

6 Jolly, C.J. (1971), '*Hadropithecus*: A lemuroid small-object feeder?', *Man*, 5, 619-26.

7 Napier, J.R. and Walker, A.C. (1967), 'Vertical clinging and leaping – a newly recognized category of locomotor behaviour of primates', *Folia primat.* 6, 204-19.

8 Szalay, F.S. (1968), 'The beginnings of primates', *Evolution* 22, 19-36.

9 Martin, R.D. (1968), 'Towards a new definition of primates', *Man* 3, 376-401.

10 Tuttle, R. (1972), 'Relative mass of cheiridial muscles in catarrhine primates', in Tuttle, R. (ed.), *The Functional and Evolutionary Biology of Primates*, Chicago.

11 Hall-Craggs, E.C.B., this volume.

12 We are greatly indebted to Tom Miller and Alan Walker for the gift of cadavers used in this study, and to Fred Brett for his stalwart assistance with the computations.

13 Grand, T.I. (1967), 'The functional anatomy of the ankle and foot of the slow loris (*Nycticebus coucang*)', *Am. J. Phys. Anthrop.* 28, 168-82.

14 Jouffroy, F.K. (1962), 'La musculature des membres chez les lémuriens de Madagascar, étude descriptive et comparative', *Mammalia* 26, 1-386.

15 Hall-Craggs, E.C.B. (1965), 'An analysis of the jump of the lesser galago (*Galago senegalensis*)', *J. Zool.* 147, 20-9.

16 Doyle, G.A., this volume.

17 Walker, A.C. (1967), 'Locomotor adaptation in recent and fossil Madagascan lemurs', PhD thesis, University of London.

18 Kingdon, J. (1971), *East African Mammals*, Vol. 1, London.

19 Bearder, S.K. and Doyle, G.A., this volume.

20 Walker, A.C. (1969), 'The locomotion of the lorises, with special reference to the potto', *E. Afr. Wildl. J.* 7, 1-5.

21 Charles-Dominique, P. (1971), 'Eco-éthologie des prosimiens du Gabon', *Biol. Gabon.* 7, 121-228.

22 Jolly, C.J. (1967), 'The Evolution of the Baboons', in Vagtborg, H. (ed.), *The Baboon in Medical Research*, Austin, Texas.
23 Jolly, C.J. (1972), 'The large African monkeys as an adaptive array', in Napier, J.R. and Napier, P.H. (eds.), *Systematics of the Old World Monkeys*, New York.

F.K. JOUFFROY, J.P. GASC,
M. DÉCOMBAS and S. OBLIN

Biomechanics of vertical leaping from the ground in Galago alleni *: a cineradiographic analysis*

General observations

It is a well-known fact that galagos display a remarkable jumping ability; in height, as well as length, they can jump a distance fourteen times their body size (tail excluded). It is a record challenged among primates only by the tarsier and among the other mammals by the rat-kangaroos (Potoroinae) and the kangaroo-rats (Zapodinae).

Compared to the other galagos, *Galago alleni* displays the most specialised type of locomotion: 'The animal moves almost exclusively by vertical jumps between small trunks and vines, moving from one vertical starting place to another one.'[1] If the animal descends to the ground for feeding, selecting insects or gathering fallen fruits, at the least startling noise it jumps into the nearest tree. This type of jump is very common in captivity. In captivity, *Galago alleni* shows remarkable adaptability to its new 'environment', especially in its locomotor abilities. In the absence of tree trunks or vines (lianas), or whatever other vertical supports might be supplied, it will move easily on plane, usually horizontal, surfaces. On horizontal surfaces the animal is observed running quadrupedally, or moving by little bipedal hops like a jerboa, or even leaping from one horizontal platform to another. As in its native forest, it leaves the platform for a higher position at the slightest noise. Even after gathering food from the cage floor it will make the same movement. This jump is extraordinarily quick. The preparatory and launching phases together do not last more than a few tenths of a second.

Interpretation of photographs, hand and foot prints

Before our cinematographic and cineradiographic recording we conducted some preliminary investigations on the animal's locomotion by means of photographs and footprint registration with the animal running on smoked paper tracks.[2] The analysis of these prints (Fig. 1) shows that before leaping the animal is in a plantigrade position with its heels pressing strongly on the ground (Fig. 2*B*), but that when it runs quadrupedally it does so in a semidigitigrade manner with the tuber of the calcaneum slightly above the ground (Fig. 2*A*). The absence of hand prints just before the leap shows that the animal raises its trunk and jumps in a bipedal attitude. The distance between the hallux and the other digits,[3] as well as the general orientation of the feet, is quite variable. In the few cases where the landing places have been registered it was observed that the hands touched first (making smudged prints).

Cinematography and cineradiography

The vertical jump from the ground of *Galago senegalensis* has been studied by Hall-Craggs,[4,5] who filmed at 1000 frames/sec. vertical jumps of up to 7′ 4¾″. We shall quote some of his conclusions: the fore feet take little part; effective use is made of the length of the synchronously acting hind limbs (as a single structure); the force is applied to the ground in the region of the tarso-metatarsal joint. In an attempt to study more precisely the movements of the bones, we have analysed the same type of jump, but over a shorter distance, in *Galago alleni*, using for this purpose the cineradiographic technique.[6]

In cineradiography, the image is produced on a small screen that becomes fluorescent when subjected to X-rays, and the image is retained on movie film. Use of an image intensification system reduces the radiation necessary to produce an acceptable image. The quality and amount of information thus obtained depends not only upon radiological factors (e.g. number and speed of electrons, sensitivity of the fluorescent screen), and upon characteristics of the photographic system, (e.g. quality of the optics, frames per second), but, in addition, requires an awareness of both the ecology and behaviour of the animal studied. It is essential to obtain an animal's 'co-operation' in conforming as closely as possible to its natural behaviour, or at least (for a captive animal) to its habitual behaviour. In this study we used a species with which we were familiar and an animal which one of us had personally raised for seven years.

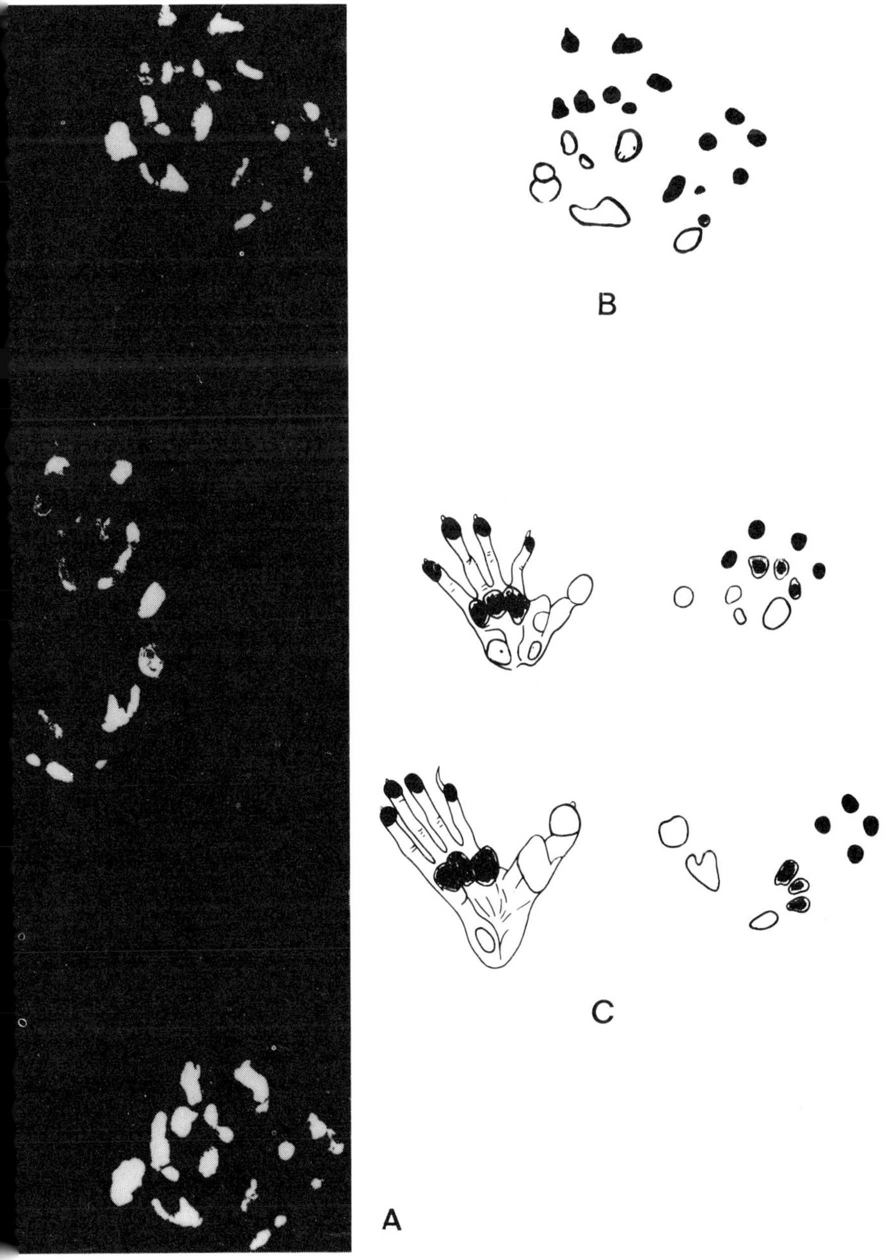

Figure 1.(*A*) Tracks of *Galago alleni* running on the ground at a moderate pace.(*B*) Schematic drawing of the top print.(*C*) Correspondence between the foot and hand prints (right) and the foot and hand pads (left).

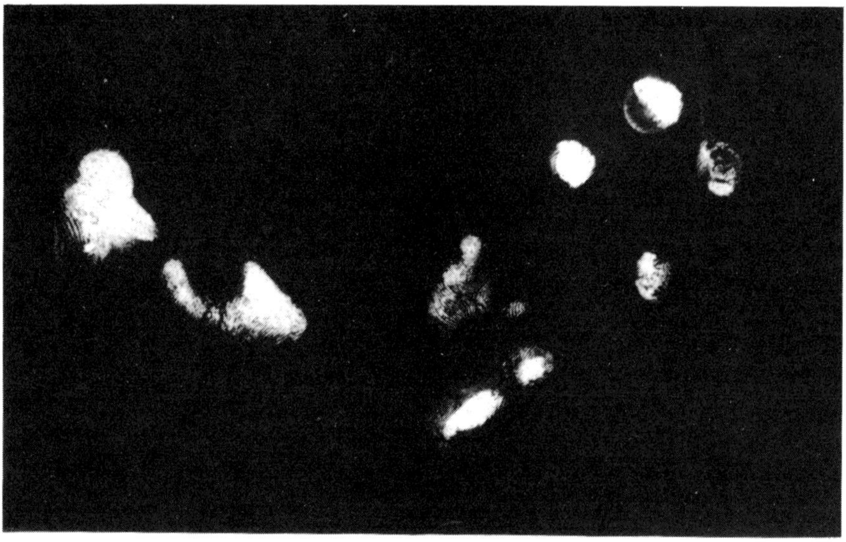

A

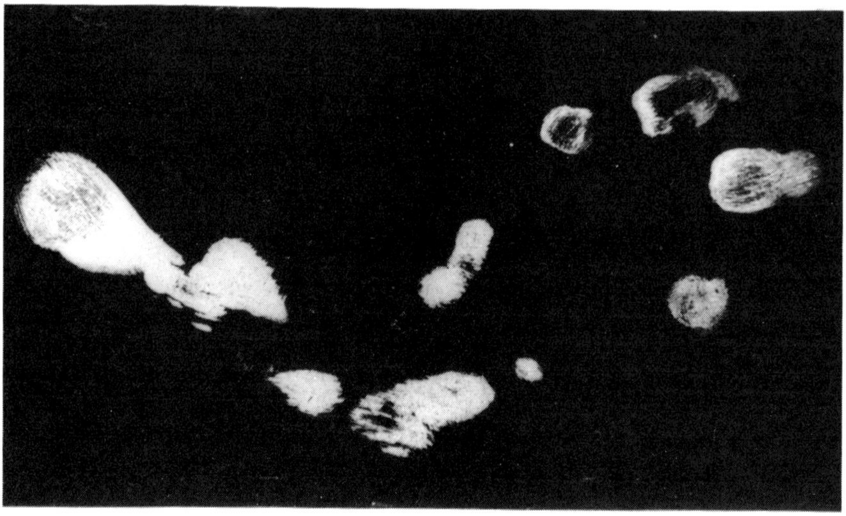

B

Figure 2. Right footprint: (*A*) while running in a quadrupedal posture on the ground; (*B*) a leap about to be launched.

Materials and methods

A Massiot Philips Medio 20 X-ray generator was employed at 60 KV. The milli-amperage was automatically regulated during the operation, but never exceeded 15 mA. The maximum length of uninterrupted recording under these conditions is 60 seconds. A pause of equal time is recommended before continuing with a new film sequence.

The operator, protected by a lead apron, follows the field on a fluoroscope reflected by a mirror. The cinefluoroscopic apparatus, mounted on a 'C' arm, is suspended by a motorised telescopic column which permits raising and lowering; this in turn is freely movable along the ceiling on a double system of perpendicular rails. Thus, although the diameter of the image intensifier is only 15 cm., the mobile suspension system makes it possible to centre the image rapidly on the movement of the animal.

For movements as rapid and sudden as the galago jump, the inertia of the apparatus, along with the time-lag between the observation of the event and the start and actual operation of the motors, makes centring of the apparatus a difficult task. Centring requires, besides rapid reflexes, sufficient familiarity with the tame animal to predict its direction of movement. In addition, the sharp noise made by the starting of the X-ray generator always startles the animal, interrupts its movement and, in most cases, provokes an escape reaction. For this reason we were unable to obtain any record of locomotion in another prosimian (*Cheirogaleus* sp.), which crouched on the ground in fright and attempted to hide itself in the smallest crevices. In the case of the galago, this escape reaction was used to advantage, since we were able to study the vertical or semi-vertical jump that the animal always employed when frightened, in an attempt to reach the security of its nestbox.

A 16 mm. Arriflex camera was run at 64 frames per second using Kodak Plus X, T.V. reversible film. When filming small animals the periphery of the field is brightest, obscuring the image definition on projection. For this reason we found it necessary to stop down the lens (Schneider-Kreznach Arriflex Cine Xenon 1:2/35) to a setting of f 5.6. The film was initially projected at 16 frames per second to permit observation of movements in slow motion. The most interesting sequences were then examined with the aid of a film editor. Two procedures were used to analyse these sequences frame by frame:

1. By printing all the frames on photographic paper with an enlarger. This extremely laborious process is only practicable for short sequences of the order of 40 frames.
2. By single frame projection onto a screen covered with tracing paper. In this way, the axes of various bones can be traced and successive images superimposed as required. Sequences of 160 frames (2.5 seconds duration) are readily analysed by this technique.

The animal

The subject of this study, a female *Galago alleni*, was a pet of one of the authors who captured it in primary forest in Gabon (May 1963: Mission biologique au Gabon, CNRS, France). It was brought back to France and raised first in the author's house and then, from 1968 until its death in 1971, in the laboratory.

The galago lived in a large-meshed, metallic cage measuring 30 x 45 cm. and 100 cm. high. A box, 18 x 16 x 15 cm., fastened 60 cm. above the floor, served as a site for sleeping, eating and retreat. In order to avoid upsetting the normal behaviour of the animal, it was placed during the recording sessions in a cage of similar dimensions to its own, but constructed with a plywood frame covered with netting and with the nestbox replaced by a wooden platform. The 'C' arm of the apparatus was positioned to deliver a horizontal X-ray beam.

Because of the restrictions of the technique, we analysed only the vertical or semi-vertical (within 30° of the vertical) jumps that the galago invariably made to reach its platform. For angular measurements we used sequences in which the animal faced the apparatus (frontal view) and also sequences in which the hindlimbs were seen in lateral view. Frontal views were most common because, by natural curiosity, the animal tended to watch the investigator and was therefore frontally positioned. Angular measurements thus recorded are not usually the actual angles between two bones, but instead represent a projection of the angles on to the frontal or parasagittal plane. Comparisons of angular variation are valid only during the time when the animal moves its limb segments in the same plane.

Slow motion analysis

Slow motion analysis at 16 frames per second yielded the following data: Of the seven jump sequences filmed, six were asymmetrical (i.e. all the body weight shifted to a single foot, the active one, before the jump), and one was symmetrical. In every case of an asymmetrical jump the active foot was always the left. The galago started from a quadrupedal posture, which in two cases was such that the animal was orientated laterally to the apparatus, and then, lifting its fore limbs, it was seen to be in a bipedal posture before pushing off with the active hind limb.

In the case of a symmetrical jump, three successive stages are distinguishable in slow motion: (1) straightening of the head and the beginning of realignment of the trunk; (2) lifting of the fore limbs with accompanying extension of the trunk, lowering of the pelvis and closure of the femoro-tibial angle; and (3) pushing off with both hindlimbs.

Single frame projection

1. *The symmetrical jump* (Figs. 3 and 5*A*). During this type of jump the different segments of the hindlimbs move alike and are strictly synchronised. The whole movement lasts 0.35 sec. (23 frames). It can be divided into a first, or preparatory phase (*phase d'appel*) of relatively long duration (0.26 sec.), and a very quick springing phase that lasts 0.09 sec.

   During the preparatory phase, the animal crouches with the feet in full plantigrade contact with the ground. The main movements occur at the femur and tibia levels. The femur undergoes a rotatory movement that diminishes the femoro-tibial angle in such a way that the pelvic/femoral articulations are lower than the knees. This movement covers the whole preparatory phase. About 0.05 sec. before the end of this phase, the tibia moves towards the ground, narrowing the tibio-tarsal angle. The outcome of this movement is that it narrows even more the femoro-tibial angle.

   The springing phase straightens, in a perfect alignment, all the hindlimb bone segments. 0.09 sec. before the launching, the heads of both the femurs begin the ascent; at 0.05 sec. before the launching, the tibiae follow the same ascending movement; and at 0.03 sec. before the launching the tarsus is raised, by dorsiflexion of the tarso-metatarsal articulation, before the digits finally leave the ground.

   This symmetrical type of jump, during which the propulsive strength is equally distributed between the two limbs, appears – at least in our film recordings – to be less frequent than the asymmetrical one.

2. *The asymmetrical jump* (Figs. 4 and 5*B*). In this type of jump, the right and left limbs play quite different roles, and this explains why no exact chronological correlations can be noted between the different bone segments of the two limbs.

   During the preparatory phase, the animal's weight is shifted on to one foot, while the other is lifted: the active foot (*pied d'appel*) and the inactive foot (*pied libre*), respectively. The left foot was always the active one during our film recording, probably as a result of the experimental conditions. The preparatory phase thus involved the shifting of the right foot, with all the articulations of the left limb flexing at the same time.

   Taking the right limb first, the foot is initially in contact with the ground and the tarso-metatarsal articulations are dorsiflexed. Just before lifting, successive oscillations of the tarsus are observed (measured by the tibio-tarsal angle). The position of the tibia, relative to the horizontal, remains almost constant.

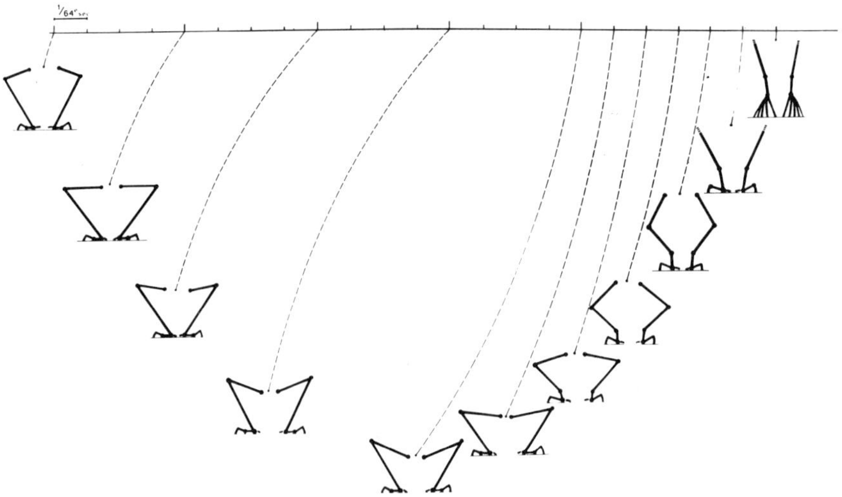

Figure 3. Schema of the movements of the hind limb segments during a symmetrical leap. Time elapsed: 1/64 sec.

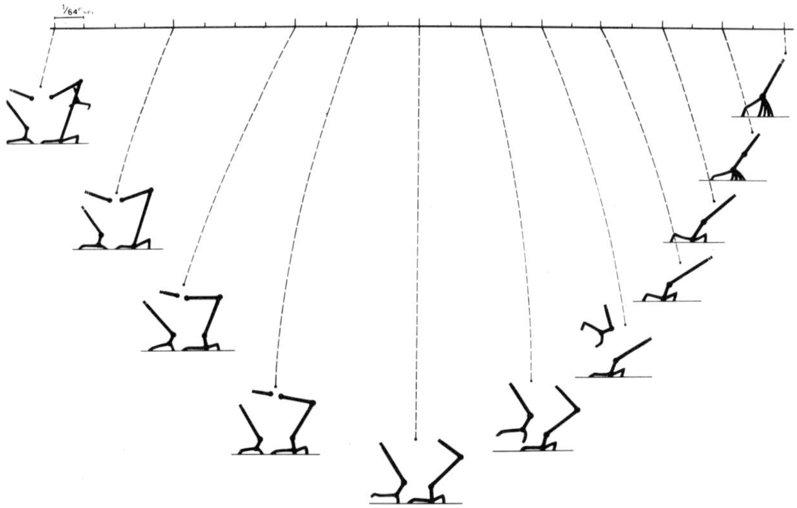

Figure 4. Schema of the movements of the hind limb segments during an asymmetrical leap. Time elapsed: 1/64 sec.

The oscillation amplitude is about 15°; its duration is from 0.04 to 0.08 sec. These oscillations last as long as the foot is not lifted. During this stage the pelvis is lowest, relative to the ground, and remains stable.

The left limb, meanwhile, remains immobile and flexed. This limb is entirely responsible for the launching phase. It supports the entire body weight. 0.34 sec. before launching, the springing phase begins simultaneously at two levels; the femur becomes

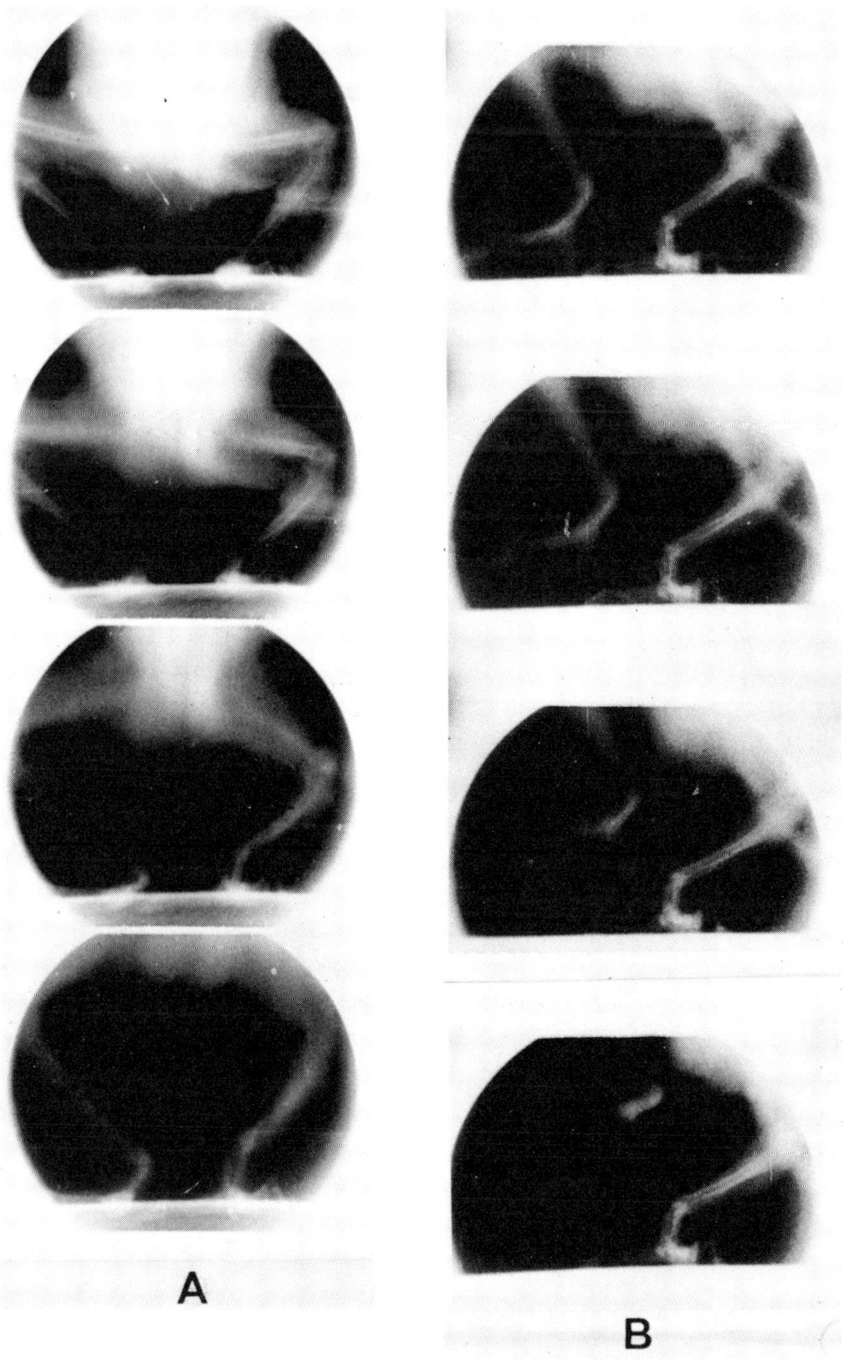

Figure 5. Cineradiographic frames showing two characteristic phases of a symmetrical (*A*) and an asymmetrical (*B*) launching. In each case, the time elapsed between two successive frames is 1/64 sec.

more vertical and the pelvis rises correspondingly; the tarsus rises and leaves the ground by tarso-metarsal dorsiflexion. The angle of the tibia relative to the horizontal remains constant during this stage. 0.28 sec. before launching, a preparatory movement of the tibia is observed: it reduces the tibio-tarsal angle by 40° (0.12 sec.) while the femur and the tarsus continue their upward motion. 0.09 sec. before the launch, the tarsus, which has previously become vertical in a parasagittal plane, inclines itself in the frontal plane (30°) to position itself in the direction of the jump. Finally, a slight push with the tip of the hallux takes place as the last contact with the ground.[7]

Discussion

The *Galago alleni* symmetrical jump appears to be exactly the same as the one recorded for *Galago senegalensis* by Hall-Craggs,[5] who concluded that 'each pair of limbs may be considered as a single structure'. The *G. alleni* specimen that we studied, however, used this type of jump far less than the asymmetrical one.

The high frequency of the asymmetrical jump is perhaps correlated with the ability of *G. alleni* to move in conditions that vary considerably, and to leap to any landing place. The symmetrical jump supposes a landing place situated vertically above, or directly in front of, the take-off point. The asymmetrical jump, on the other hand, enables the animal to choose a greater variety of landing places.

In both cases, a preliminary movement is observed that tends to flex the limbs before extension. This complex movement can be related to contractions of the muscle antagonists. The durations of the two phases, preparatory and launching, differ markedly. Under the conditions examined above, the launching phase (extension) began 0.34 sec. before the launch in the asymmetrical jump and only 0.09 sec. before the launch in the symmetrical jump.

In the case of the asymmetrical jump, the different bones follow each other harmoniously in their displacements, and the whole of this movement is easily analysed: firstly, the active limb femur becomes vertical, and secondly, the semi-plantigrade foot posture is attained before the tibia becomes more vertical. During the symmetrical jump, on the contrary, tibia and tarsus extend at the same time (0.03-0.04 sec. before the launch) and until this movement takes place the feet remain in a plantigrade position on the ground.

This preliminary analysis, which emphasises the precision of the data that can be recorded with the technique radiocinematography, should be completed by the study of other types of locomotion or movements such as grasping or mastication, for instance, which would give us a more thorough understanding and a more general knowledge of prosimian biology.

NOTES

1 Charles-Dominique, P. (1971), 'Eco-éthologie des prosimiens du Gabon', *Biol. Gabon.* 7, 121-228.

2 Décombas, M. (1970), 'Analyse comparée des pistes chez divers rongeurs et prosimiens', *Mém. D.E.A. Université de Paris*, 7, 26 (unpublished).

3 In more than 500 measurements of rodent and prosimian prints, the distance separating the hallux, or the thumb, from the other digits has always been found to be larger on the right side.

4 Hall-Craggs, E.C.B. (1964), 'The jump of the bushbaby, a photographic analysis', *Med. Biol. Ill.* 170-4.

5 Hall-Craggs, E.C.B. (1965), 'An analysis of the jump of the lesser galago (*Galago senegalensis*)', *J. Zool.* 147, 20-9.

6 Jouffroy, F.K. and Gasc, J.P. (1974), 'A cineradiographic analysis of jumping in an African prosimian (*Galago alleni*)', in Jenkins, F.A. (ed.), *Advances in Primatology*, 3, *Primate locomotion*, New York.

7 The film shown at the Research Seminar was produced by the Laboratory of Comparative Anatomy of the Muséum National d'Histoire Naturelle, Paris, and by the Audio-visual Department of the University of Paris.

E.C.B. HALL-CRAGGS

Physiological and histochemical parameters in comparative locomotor studies

Introduction

In previous communications the results of a study of the jump of *Galago senegalensis* have been published. In an initial paper an account was given of the method used to record the early stages of the jump using still and high-speed cinematography.[1] Later, osteometric data obtained from the hind limb skeleton of *Galago senegalensis* and a number of allied prosimian types were presented and their significance discussed.[2] Subsequently an attempt was made to combine the accumulated findings into an analysis of the jump as a whole.[3] Since this work, many additional and better studies have been made of locomotor patterns throughout the primate order including those of prosimians.[4] However, a number of these have been behavioural in nature or strictly concerned with musculoskeletal features that could be correlated with fossil evidence. No emphasis has been placed on the mechanical functions that individual muscles are required to perform, nor on the physiological characteristics that allow them to perform these functions. Over the last decade considerable advances have been made in the physiological and histochemical characterisation of limb muscles and their component fibres, and where possible these characteristics have been linked with the apparent functional calls made on the muscle. Because of this it is proposed to describe the results of some physiological experiments made on the locomotor muscles of *Galago senegalensis*, up to now only briefly reported,[5] and to discuss these in combination with a review of new knowledge of the differing contraction and metabolic properties of skeletal muscle.

Anatomical and physiological considerations

The Galaginae have a reputation for their striking saltatory mode of progression and this has been confirmed by observation of *Galago senegalensis* in its natural habitat. Despite the fact that it is a nocturnal creature, when disturbed in its arboreal habitat during the day it will make accurate jumps from branch to branch and if forced to the ground will proceed by a series of short hops in the upright position using only its hind limbs. When the jump is observed under controlled conditions it is found to be made only when a safe perch can be reached, and – using a chimney-like cage with smooth walls and an adjustable roof of wire netting – a maximum, almost vertical, jump of 7 feet 4¾ inches was recorded. In view of the fact that the animal's centre of gravity is probably only 1½ inches from the ground, this was considered by any standards a remarkable achievement. When the vertical and horizontal course of the jump were controlled by a shallow cage having a transparent front and a background ruled out in 5.0 cm. squares, it was possible to make a simultaneous photographic record of the height achieved in the jump, together with a slow motion study of the critical period when the animal had commenced the jump but still remained in contact with the ground. It is during this period that the height reached is determined, for it is related to only two variables – the angle and velocity of take off – while the weight of the animal and air resistance remain constant. Repeated observation of jumps made to differing heights under these conditions suggested that these variables were accurately gauged by the animal, which could only be persuaded to leap when the aiming point was kept in view. The take-off velocity, upon which so much depends, is attained by the animal while it is in contact with the ground and is related to the force applied to its centre of gravity either during this time or over the distance travelled by the animal during this time. A study of individual frames of the slow-motion film showed that the forelimbs contributed little or nothing to the jump and that the force used was equal and opposite to that applied by the hind-limbs against the ground. It was also possible to decide upon the moment at which the centre of gravity began to move, the moment at which the animal lost contact with the ground and the angle of take off. This information, combined with a knowledge of the frames exposed/sec. by the camera, allowed an estimate to be made both of the distance travelled by the centre of gravity and the time that had elapsed. Knowing the height reached in the jump and the angle of take-off, the take-off velocity could be calculated for each jump. Having established this, the acceleration required to reach this velocity over the observed time or distance could next be calculated, and this combined with the weight of the animal allowed a figure for the amount and direction of the reaction against the ground to be obtained from a force diagram. In one example a 69.0 cm. jump was found to require a force of 1328 g.wt. It is a fortunate fact that the two hind limbs are used in a symmetrical manner, for it could be further said that half this force was exerted by each limb.

A consideration of the anatomy of the hind limb shows that this force is largely developed by, and transmitted through, the two

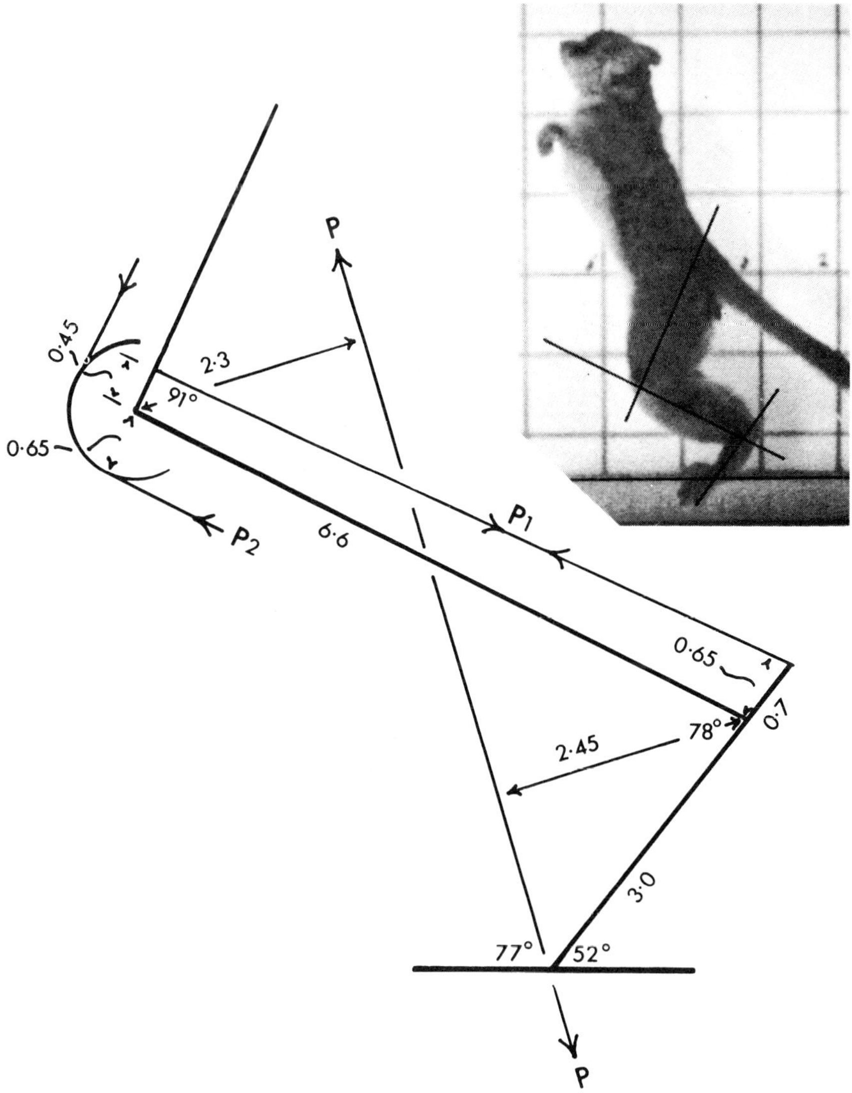

Figure 1. Diagram representing the lower limb skeleton as a series of rods and a pulley from which can be calculated the contribution of *triceps surae* (P1) and *quadriceps femoris* (P2) to the total force required to make the jump (P). Inset is a photograph from which the angles made by the rods were taken.

muscle groups *quadriceps femoris* and *triceps surae*. The *quadriceps femoris* extends the knee joint by acting around the axis of the joint over a pulley formed by the femoral condyles. In so doing, it is required to undergo considerable shortening. The *triceps surae* plantarflexes the foot at the ankle joint, acting as a lever whose fulcrum lies at the tarso-metatarsal joint. Its attachment in large part to the femur has, however, important consequences. Not only does it plantarflex the foot, but it also flexes the knee, and as a result the *quadriceps femoris* must counterbalance the force generated by the *triceps surae* before it is able to extend the knee. This 'two-joint' arrangement of *triceps surae* is not altogether as disadvantageous as it appears. From the scale drawing (Fig. 1) it can be seen that this muscle forms one side of a parallelogram, and that as the knee extends and the ankle plantarflexes the *triceps surae* needs to shorten little. The increased bulk necessary in the *quadriceps femoris* to compensate for this feature also has the effect of transferring the weight of the limb nearer to the centre of gravity of the animal, thus decreasing the inertia of the extremity, a feature observed in many 'fleet-footed' animals.

Turning again to the scale drawing (Fig. 1), which was prepared from a frame of the high-speed cine-film and measurements taken from X-rays of a comparable animal, it is possible, when the total force required to make the jump is known, to calculate estimates of the forces actually developed by these two groups of limb muscles. This was done at a number of points during the jump and the figures appear in Table 1. It can be seen that peak tensions of approximately 4.65 kg.wt. and 5.70 kg.wt. might be expected to be developed by the *triceps surae* and *quadriceps femoris* muscles respectively, and that, because of the changing arrangement of the limbs about the direction of the total force required to be applied to the centre of gravity, the estimated values fall off throughout the latter part of the jump. If these figures can be accepted, it becomes of interest to examine the anatomy of these two muscle groups and to know whether tensions of this order can be demonstrated to be developed in an experimental animal, especially in view of the considerable difference in size between the two. It would also be worthwhile to

Table 1 Estimates of tensions developed by *triceps surae* (P1) and *quadriceps femoris* (P2) at different intervals after the beginning of a jump of 69.0 cm.

Time elapsed	*P1*	*P2*
sec	*g wt*	*g wt*
.04	4454	4576
.05	4648	5313
.06	4058	5709
.07	3652	5514
.08	2503	4083
.09	1598	1660

see how the changes in tension needed are related to their physiological properties.

Thompson has said that in similarly constructed animals the height to which they are able to jump would be constant whatever the size of the animal.[6] However, differences in construction may well lead to differing performances. Of the possible differences, those in the proportion of the body weight devoted to jumping musculature might be expected to have a considerable effect.

In a small number of specimens of *Galago senegalensis* the body weights and weights of *triceps surae* and *quadriceps femoris* were obtained and compared with similar figures obtained from *Perodicticus potto* (Table 2). It was found that in *Galago* the combined weights of the two muscles formed 8-12 percent of the body weight, but in *Perodicticus* the same groups formed little more than 2 per cent of the body weight. These figures show that a considerably greater proportion of the body weight is devoted to jumping musculature in *Galago* than in *Perodicticus.*

A comparison was also made of the proportion that the weight of

Table 2 Body weights and muscle group weights in specimens of *Galago senegalensis* **and** *Perodicticus potto*

Specimen	*Body weight*	*Weight of quadriceps femoris + triceps surae*	*% of body weight*
	g	*g*	%
Galago senegalensis	300	24.5	8.2
,,	230	27.1	11.8
,,	168	20.9	12.4
,,	275	31.1	11.3
Perodicticus potto	1429	19.4	1.4
,,	1025	21.9	2.1

Table 3 Weight of *quadriceps femoris* **and** *triceps surae* **muscles in** *Galago senegalensis* **and** *Perodicticus potto*

Specimen	*Weight of triceps surae*	*Weight of quadriceps femoris*	*Weight of triceps surae as % of quadriceps femoris*
	g	*g*	%
Galago senegalensis	2.73	21.8	12.5
,,	2.96	24.1	12.3
,,	3.0	17.9	16.8
,,	2.6	23.2	11.2
,,	3.5	27.6	12.7
Perodicticus potto	6.2	13.2	47.0
,,	5.0	16.9	29.6

the *triceps surae* muscle forms of the weight of *quadriceps femoris* (Table 3). The evidence confirms the suggestion that there is a tendency in agile animals for the limb weight to be nearer to the centre of gravity, for in *Galago* the weight of *triceps surae* is only about 15 per cent of *quadriceps femoris* whereas in *Perodicticus* it is 30 per cent or more.

In order to explore the tensions that these muscle groups were in fact able to develop, isometric and isotonic recordings were made under controlled laboratory conditions. Under general anaesthesia, the lower limb was securely fixed by steel pins passed through the upper and lower ends of the femur and by a clamp holding the foot. The patellar tendon and *tendo calcaneus* were exposed, as were the nerves to each muscle group. The tendons were transfixed and ligated to steel hooks, which were in turn linked to either an isometric or isotonic lever recording on a revolving smoked drum. The nerves were divided and prepared for stimulation. All incisions were kept irrigated with liquid paraffin and muscle temperature was maintained by radiant heat controlled by a thermistor.

Only one series of satisfactory isotonic recordings was made, but these allowed both force-velocity (Figs. 2 and 3) and power-velocity curves to be constructed. The force velocity curves for each muscle showed the shape already established for muscles from many different types of animals.[7] The power-velocity curves showed that the peak output occurs at a higher rate of shortening in *quadriceps femoris* than in *triceps surae* – a fact which may be correlated with the distance over which each muscle is required to shorten during the jump; approximately 17 per cent of its initial length in the case of

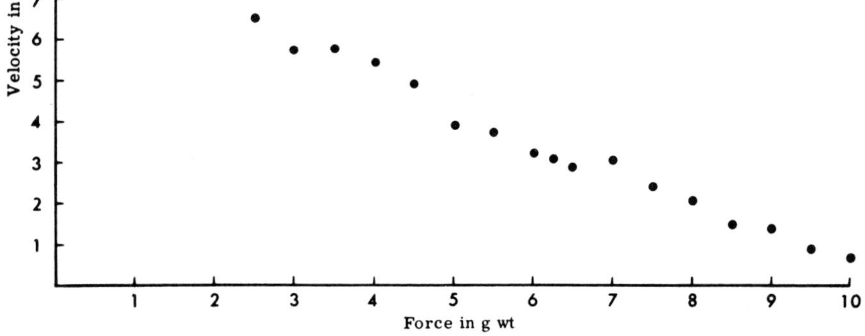

Figure 2. Force-velocity curve, *quadriceps femoris*.

quadriceps femoris and 5-6 per cent in the case of *triceps surae.* In terms of actual velocity of shortening and tensions developed, these results fell well below those that had been estimated as necessary to make the jump. This may have been for a variety of reasons, connected, for example, with the method of anaesthesia or inertia and friction in the mechanical method of making the recordings. The isotonic recordings were also made without any pre-loading, a situation which may in fact not be the ideal condition for the development of maximum tension (see later). The isometric recordings were, however, made over a range of muscle lengths more closely related to those found *in vivo.* This was achieved by taking an X-ray of each animal immediately before recordings were made. Before each tendon was divided, the position of a mark on the tendon was recorded when the limb was arranged in the fully extended position that it would take up at the end of the jump. The length of the muscle in this position was called the terminal length (see Fig. 4). The tendons were then divided and isometric recordings begun with the muscle at this length. These were repeated at increasingly greater muscle lengths up to a length just beyond that estimated from the X-rays to represent full flexion of the knee and full dorsiflexion of the ankle. This was called the initial length (Fig. 4). The difference

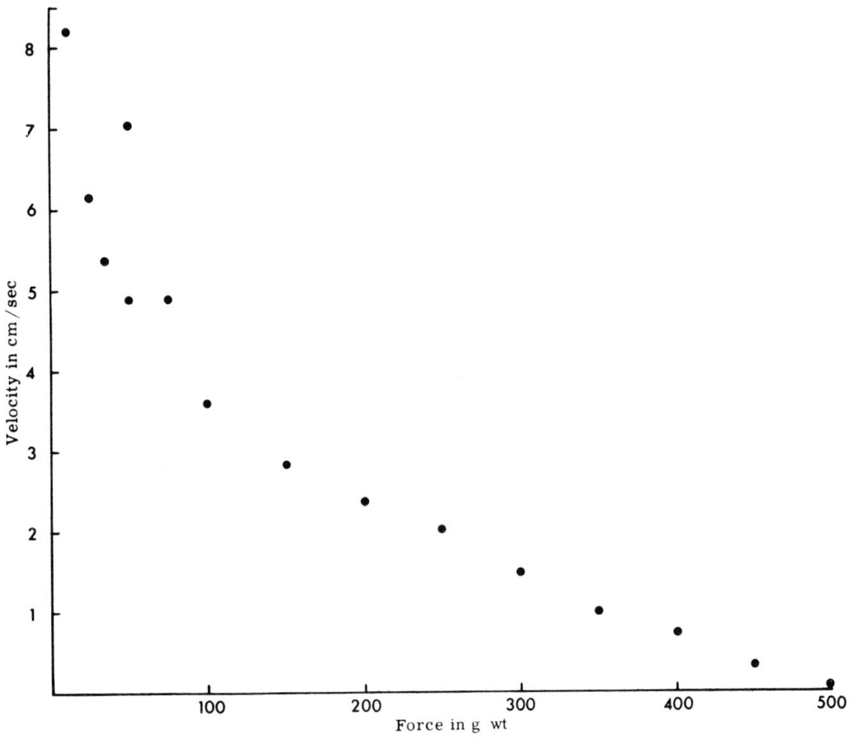

Figure 3. Force-velocity curve, *triceps surae.*

between these two lengths, as suspected from a consideration of the anatomy, proved to be small in the case of *triceps surae* – only 2.0 and 2.5 mm. in two animals – although the theoretical minimum and maximum lengths (Fig. 4) were 10.0 mm. above and below the range actually employed in jumping. It was found that stimulation at the terminal length produced a tension of just over 60 per cent of the maximum tension recorded (Fig. 5). This value was 3.3 and 3.375 kg.wt. in two experiments and was developed very close to a length estimated to exist at the beginning of the jump, i.e. the initial length.

The change in length of *quadriceps femoris* needed to execute the jump was calculated to be 1.25 cm. At the initial length the tension developed was approximately 4.75 kg.wt., but at this point the passive tension was 2.5 kg.wt. (Fig. 6). The distance over which *quadriceps femoris* must shorten is considerably greater than that required of *triceps surae* and as a result, while the maximum tension developed is greater than that developed by *triceps surae*, the tensions developed near the terminal length would appear to fall below those possible in *triceps surae.*

A further experiment was performed in which an attempt was made to reproduce as far as possible an *in vivo* situation. Using the isotonic lever, the mechanical leverage existing at the knee and ankle joints was more nearly simulated by attaching the muscle nearer to the fulcrum than the load. The load was also rigidly attached to the

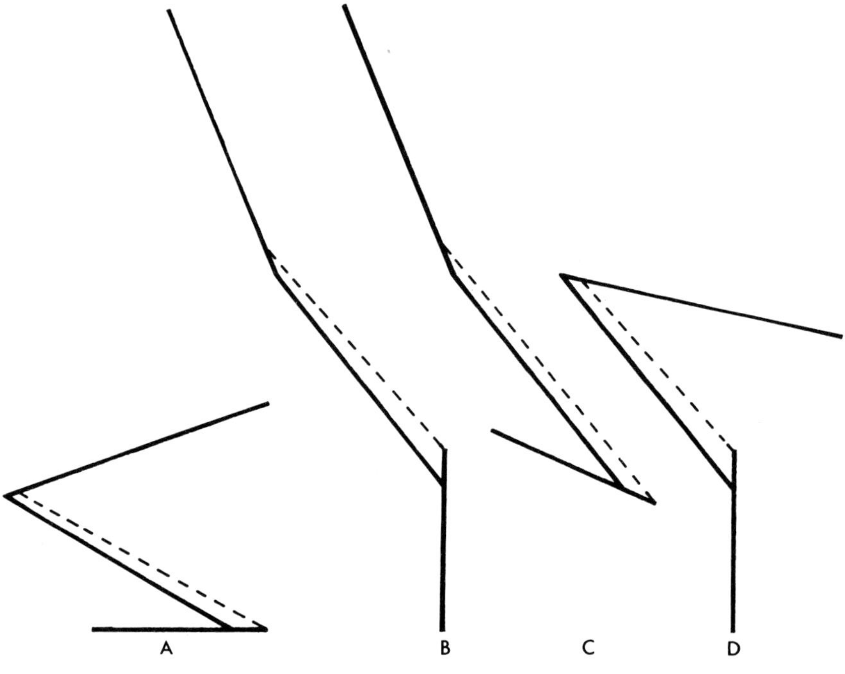

Figure 4. Diagram illustrating the 'initial length' of *triceps surae* (*A*) the 'terminal length' (*B*) and the theoretical 'maximum (*C*) and minimum (*D*) lengths'.

lever. The expected amount of shortening was again calculated from an X-ray and the muscles linked to the lever at their terminal lengths. From this position they were then extended by known amounts, and at each length stimulated while they were loaded with a range of weights. In this way, portions of their force-velocity curves were recorded at lengths believed to exist during the jump (Figs. 7 and 8). The results of this experiment showed, as might be expected, that at a given rate of shortening the force developed increased as the length of the muscle was increased. At its initial length *triceps surae* was now found to develop a force of 4.2 kg.wt. and the *quadriceps femoris* a force of 6.0 kg.wt. midway between its initial and terminal lengths. These tensions showed an increase over those previously recorded and would in fact have been sufficient to perform the jump originally described. However, at these peak tensions there remains a considerable discrepancy in the speed of shortening – a factor possibly related to the experimental conditions. Despite this, the geometry of the levers is such that the maximum force is required of each muscle when the speed of shortening would need to be least (Table 1) and falls off as the animal accelerates, thus allowing greater

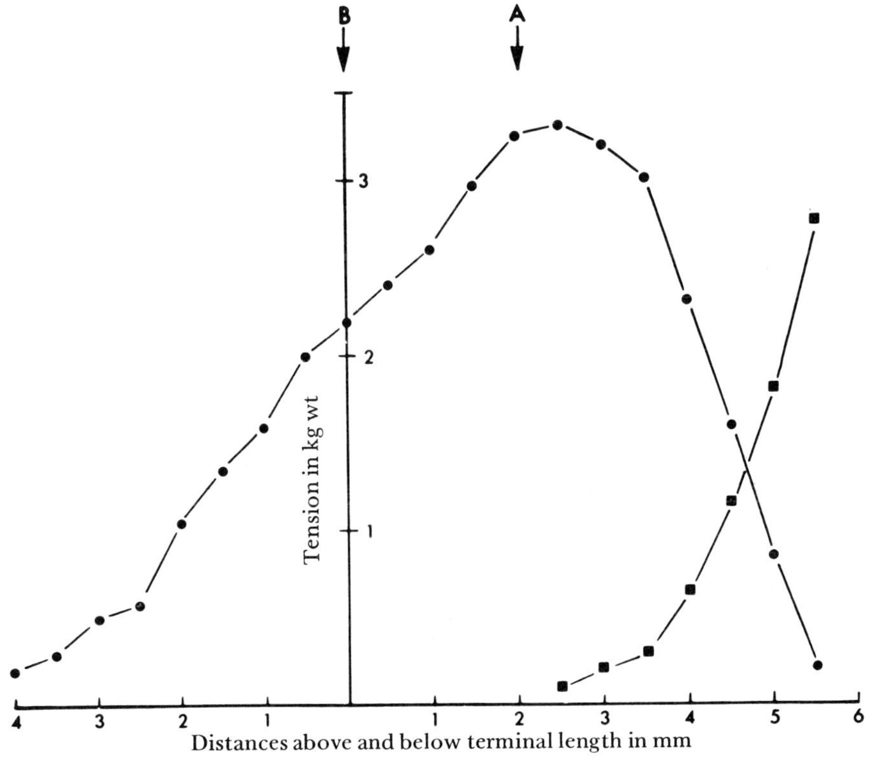

Figure 5. Length-tension curve, *triceps surae*. *A* indicates initial length, *B* indicates terminal length, ■ = passive tension, ● = developed tension.

speeds of shortening to be reached towards the end of the jump.

A further discrepancy also appears to exist between the maximum tensions developed by these muscles and their size. Following two of the experiments described, the two muscles were weighed and figures obtained for tension in kg.wt. per g. of muscle (Table 4). The figures for *quadriceps femoris* were in the same order as those obtained from rabbit *triceps surae* in two preliminary experiments and were well within the maximum figures of 0.88 kg.wt./g. quoted by Alder et al. for rabbit *tibialis anterior.*[8] The *triceps surae* of *Galago*, however, was found to develop between four and five times as much tension per g. as *quadriceps femoris.* A possible explanation for this lies in a differing fibre architecture, for the tension is related to the effective cross-sectional area of the fibres and (for muscles of the same weight) this will vary inversely with the length of their component fibres. Using a method described by Elliott and Crawford,[9] muscles were macerated and small fascicles containing perhaps half a dozen fibres were sorted into groups by length, dried and weighed. The weight of each group was divided by its mean length and the sum of the quotients gave an index of the relative fascicular cross-sectional area.

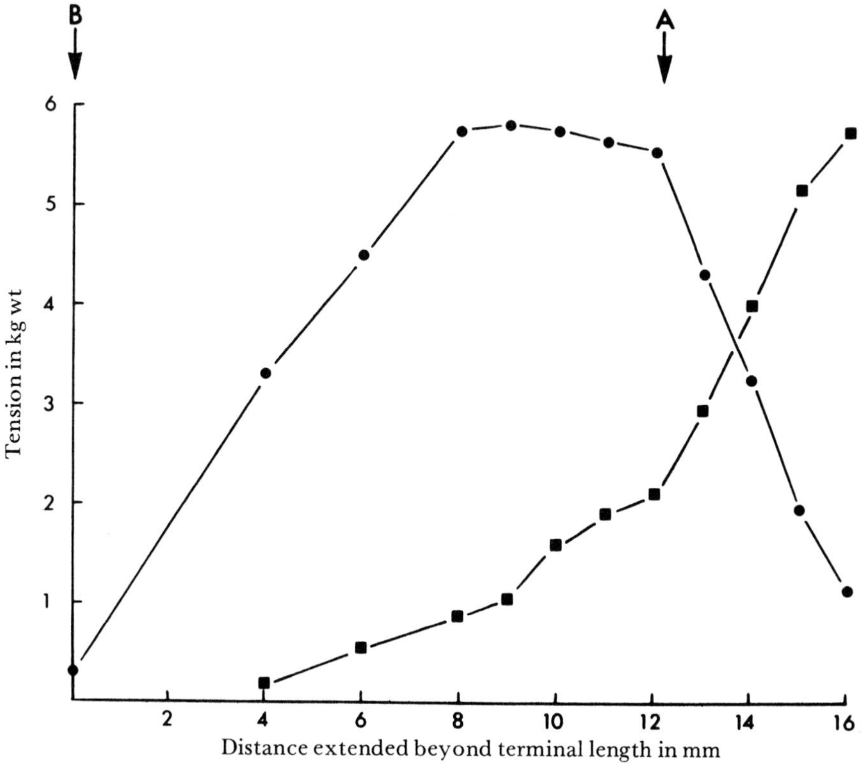

Figure 6. Length-tension curve, *quadriceps femoris. A* indicates initial length, *B* indicates terminal length, ■ = passive tension, ● = developed tension.

This figure was corrected for degree of extension by counting the number of sarcomeres per unit length in phase contrast microphotographs of specimen fibres. The final figure for relative fascicular area per gramme dry weight was found to be twice as great for *triceps surae* as for *quadriceps femoris* in both experiments. On these grounds, *triceps surae* could be expected to develop twice the tension per unit weight that can be developed by *quadriceps femoris*, thus accounting in part for the difference of four to five times that had been observed.

Table 4 Weights of muscle groups and maximum tensions recorded from them

Specimen	*Muscle*	*Weight* g	*Max. tension developed* *Kg wt*	*Tension developed per Kg of muscle* *Kg wt*
Galago	Quad. fem.	10.88	6.25	0.57
senegalensis	Tri. surae	1.37	3.49	2.56
,,	Quad. fem.	12.05	5.8	0.48
,,	Tri. surae	1.48	3.3	2.23
Rabbit	Quad. fem.	12.7	6.63	0.52
	Tri. surae	9.8	6.6	0.67

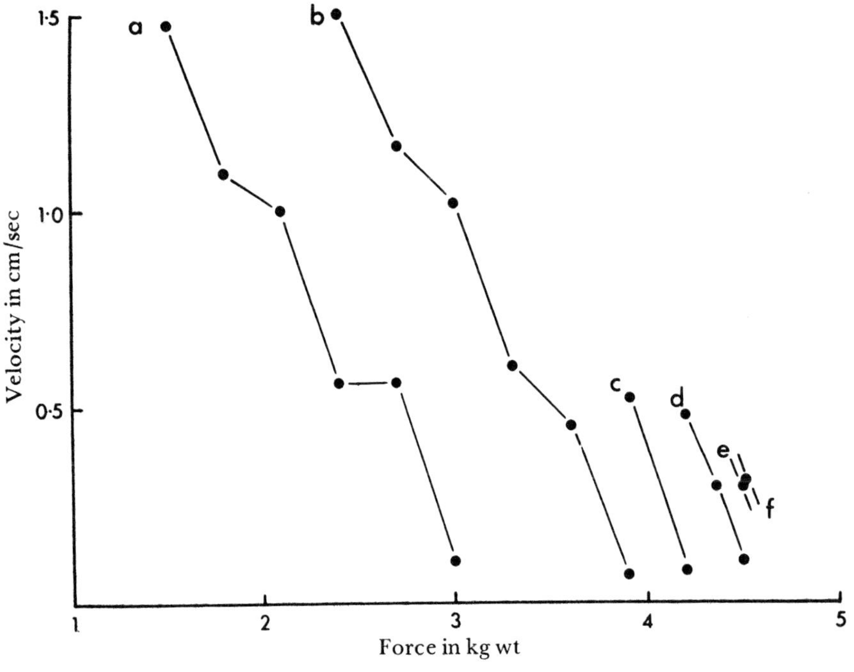

Figure 7. Portions of a series of force-velocity curves at different muscle lengths. The muscle was lengthened 1.0 mm. between *a* and *b* and 0.5 mm. between subsequent curves. The muscle length used at *c* corresponded to the initial length; *triceps surae.*

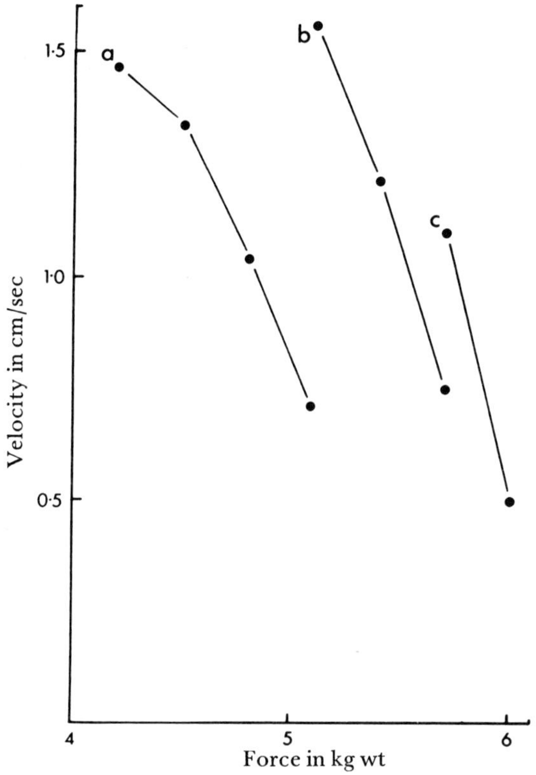

Figure 8. Portions of three force-velocity curves at different muscle lengths. The muscle was lengthened by 2 mm. between curves *a* and *b* and 1 mm. between curves *b* and *c*. The muscle length used at *c* lay midway between the initial and terminal lengths; *quadriceps femoris*.

Histochemical evidence

It has, of course, long been known that skeletal muscles differ from each other in both morphological and physiological detail. Ranvier was well aware of this when he demonstrated that a stimulus rate that produced tetanic fusion in a red muscle was insufficient to produce the same effect in a white muscle.[10] Over subsequent years the contraction characteristics of red and white limb muscles have been more fully analysed and a fairly wide spectrum of properties revealed.[11] Together with these studies, advances in histochemical techniques for demonstrating the nature of enzymes present in the individual fibres comprising these muscles have allowed some correlation to be made between metabolism and function. Stein and Padykula classified three distinct fibre types in the white *gastrocnemius* and red *soleus* of the rat.[12] This classification was based on the cytochemical distribution of the activity of succinic dehydrogenase (SDH). In the rat the white muscle was found to contain all three types (*a*, *b* and *c*, Fig. 9) and the *soleus* two (*b* and *c*, Fig. 10). A reciprocal relationship between the presence of oxidative enzymes and those of the glycolytic pathway has also been demonstrated (Figs. 11 and 12),[13] and differences revealed in the acid and alkaline stability of myofibrillar and intermyofibrillar adenosine triphosphatase (MATPase and IMATPase) have provided a further technique for characterising the fibres of fast and slow muscles (Fig. 13).[14]

Romanul applied a large number of histochemical tests to serial sections of the calf muscles of rats and was able to define eight fibre types.[15] From this more extensive study it emerged that there were: (1) a group of fibres with a high capacity to utilise glycogen, but with a low lipid and oxidative metabolism and a low myoglobin content; (2) an intermediate group with a moderate ability to metabolise glycogen and lipids and a high oxidative metabolism; and (3) a third type with a low capacity for glycogen breakdown, very active lipid and oxidative metabolism and a high myoglobin content.

The metabolic varieties of the fibres of limb muscles having been established, it remains to link these more closely with the physiology of the muscle and if possible with those of motor units. A number of workers have been able to record contraction properties of individual units either by using reflexly evoked stimuli or by the stimulation of single ventral root fibres.[16] Edström and Kugelberg found that repetitive stimulation of motor nerve fibres produced sufficient change in phosphosylase and glycogen levels to allow them to identify the fibres of the motor unit involved.[17] They were thus able to say that the fibre composition of units was largely uniform in fibre type and that the fatigability of units was closely correlated with their fibre types. More recently, by combining this method with a range of physiological determinations particularly aimed at the response of a fibre to fatigue, Burke et al. have described three varieties of motor unit in the cat *gastocnemius*.[18] These were a fast but rapidly fatigued unit, a fast unit resistant to fatigue and a slow and fatigue-resistant unit. Serial sections taken from the experimental muscles enabled them to show these units to be composed of fibres of low, moderate and high oxidative activity, respectively.

Little information appears to be available about fibre types in non-human primates. Beatty et al. have presented some histochemical and biochemical data from the Anthropoidea and in particular from the rhesus monkey.[19] As information about the Prosimii appears to be lacking, it seems worthwhile to report briefly on a number of preparations obtained post-mortem from members of this sub-order.[20] The quality of these is poor, owing to the time that elapsed between death and removal of the muscles, but in a number it is possible to identify the fibre type after incubation to demonstrate SDH.

Of the muscles examined in *Galago senegalensis*, the small *rectus femoris* showed the typical chequerboard pattern of a fast muscle

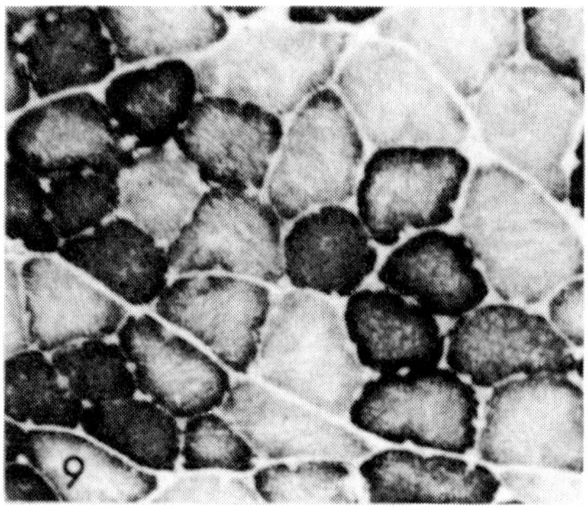

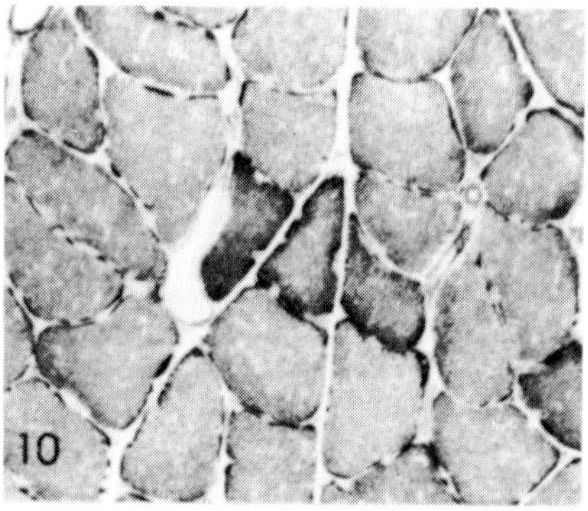

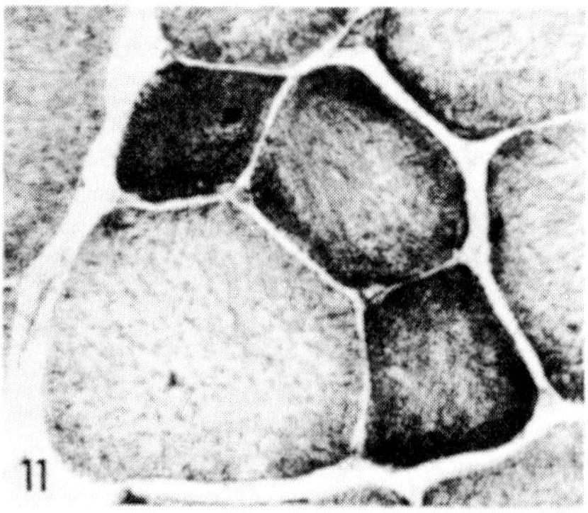

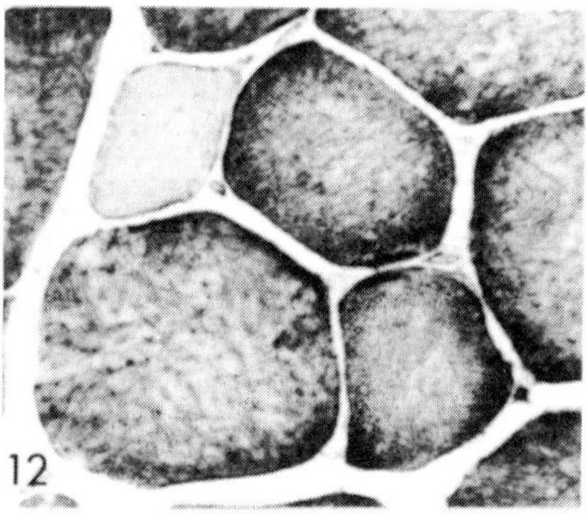

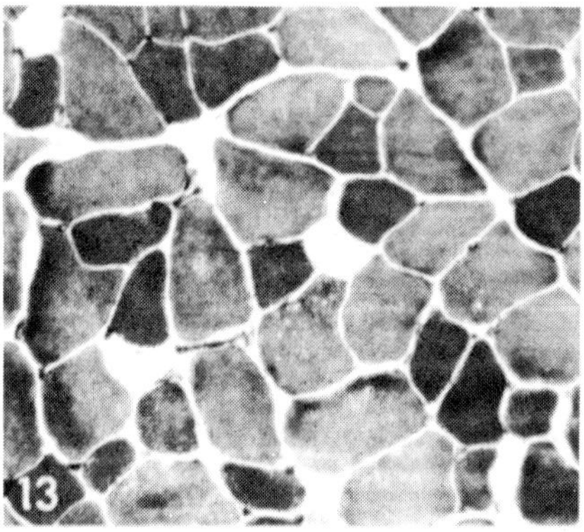

Figure 9. Rat *plantaris* muscle, SDH; *a, b* and *c* fibres are seen to be present.

Figure 10. Rat *soleus* muscle, SDH; only *b* and *c* fibres are present.

Figure 11 and 12. Serial sections of rat *plantaris* muscle. In Fig. 11 the oxidative enzyme reduced diphosphopyridinenucleotide-tetrazolium reductase is demonstrated, and in Fig. 12 the glycolytic enzyme α-glycerophosphate dehydrogenase. Comparison of the two figures shows the reciprocal relationship of these two enzymes.

Figure 13. Rat *plantaris* muscle, MATPase; using this reaction, three fibre types can also be distinguished: $\alpha\beta$ or *a* fibres (moderate staining), α or *c* fibres (dark) and β or *b* fibres (light).

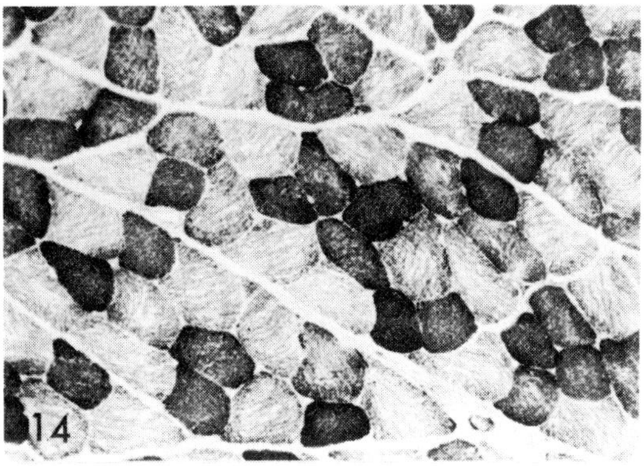

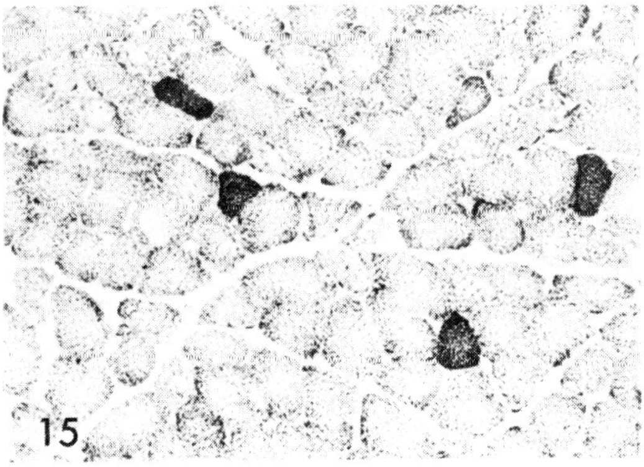

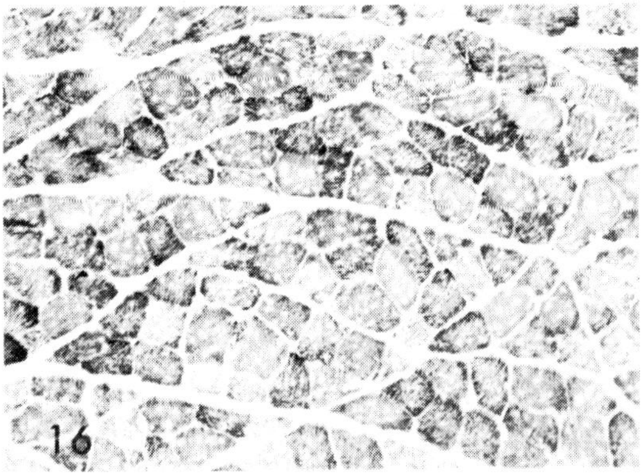

Figure 14. *Rectus femoris* muscle of *Galago senegalensis*, SDH; three fibre types are seen.

Figure 15. *Vastus medialis* muscle of *Galago senegalensis*, SDH; nearly all the fibres are Type *a*.

Figure 16. *Vastus medialis* muscle of *Loris tardigradus*, SDH; showing a high proportion of *b* and *c* fibres.

and all three types of fibre could be distinguished (Fig. 14). The two *vasti*, however, appeared to have a higher proportion of *a* fibres and in some regions of *vastus medialis* (Fig. 15) this was almost the only type seen. In contrast to this, the *vasti* of *Arctocebus* and *Loris tardigradus* showed a high proportion of *b* and *c* fibres (Fig. 16), the calibre of which often appeared to exceed that of the *a* fibres. The same picture was seen in the *gastrocnemii* and the *biceps* and *triceps brachii* of the slender loris.

It would be a mistake to reach any firm conclusions on this evidence, but one cannot avoid drawing attention to the presence of a high proportion of possibly fast *a* fibres in the *quadriceps femoris* of *Galago*, where speed of shortening is a paramount need, and the predominance of fatigue-resistant *b* and *c* fibres in the muscles of more slowly moving species that are known to maintain a posture for long periods without movement.

Summary

In summary it can be said that *Galago senegalensis* shows a number of features that appear to be positively advantageous to its saltatory mode of progression. Its long hind limb with a wide range of extension allows the maximum time or distance for the application of force against the ground. Against this must be placed the poor mechanical advantage of the leverage of the foot, which calls for great forces from the *triceps surae.* The geometry of the attachments of *triceps surae* have been seen to limit the shortening needed, thus increasing the possible cross-sectional area of the muscle fibres per unit weight when compared with the *quadriceps femoris.* This feature is reflected in the length-tension diagrams which show that *triceps surae* only uses the optimum portion of the curve when shortening during the jump. The attachment of this muscle to the femur also has the effect of transferring the weight of the limb nearer to the centre of gravity and reducing the inertia near the moment of take-off. It would appear that the *quadriceps femoris* may be largely composed of muscle fibres having a high speed of shortening, a fact that again is seen to be necessary from a study of its mechanical action.

Since the work described here was completed, major advances in the investigation of the physiological and histochemical characteristics of skeletal muscle have been made. Many of these new techniques could be applied to comparative locomotor studies in the Prosimii, which present such a wide range of locomotor patterns, and it is thought that if these were to be combined with the type of anatomical and mechanical data presented here a significant contribution to knowledge in both the field of primatology and muscle biology could be made.

NOTES

1 Hall-Craggs, E.C.B. (1964), 'The jump of the bushbaby – a photographic analysis', *Med. Biol. Ill.* 14, 170-4.

2 Hall-Craggs, E.C.B. (1965), 'An osteometric study of the hind limb of the Galagidae', *J. Anat.* 99, 119-26.

3 Hall-Craggs, E.C.B. (1965), 'An analysis of the jump of the lesser galago (*Galago senegalensis*)', *J. Zool.* 147, 20-9.

4 e.g. Grand, T.I. (1967), 'The functional anatomy of the ankle and foot of the slow loris (*Nycticebus coucang*)', *Am. J. Phys. Anthrop.* (n.s.) 26, 207-18. Napier, J.R. and Walker, A.C. (1967), 'Vertical clinging and leaping – a newly recognized category of locomotor behaviour of primates', *Folia primat.* 6, 204-19. Simons, E.L. (1967), 'Fossil primates and the evolution of some primate locomotor systems', *Am J. Phys. Anthrop.* (n.s.) 26, 241-54.

5 Hall-Craggs, E.C.B. (1966), 'Muscle tension relationships in *Galago senegalensis*', *J. Anat.* 100, 699-700.

6 Thompson, D'A.W. (1942), *On Growth and Form*, Cambridge.

7 Wilkie, D.R. (1956), 'The mechanical properties of muscle', *Brit. Med. Bull.* 12, 177-82.

8 Alder, A.B., Crawford, G.N.C. and Edwards, R.G. (1958), 'The growth of the muscle *tibialis anterior* in the normal rabbit in relation to the tension-length ratio', *Proc. Roy. Soc. B.* 148, 207-16.

9 Elliott, D.H. and Crawford, G.N.C. (1963), 'The relationship between the size of muscles and their tendons in the adult rabbit', *J. Anat.* 97, 139.

10 Ranvier, L. (1873), 'Propriétés et structures différentes des muscles rouges et des muscles blancs, chez les lapins et chez les raies', *C.R. Hebd. Séanc. Acad. Sc. Paris* 77, 1030-4.

11 Denny-Brown, D.E. (1929), 'The histological features of striped muscle in relation to its functional activity', *Proc. Roy. Soc. B* 104, 371-411. Cooper, S. and Eccles, J.C. (1930), 'The isometric responses of mammalian muscles', *J. Physiol.* 69, 377. Buller, A.J., Eccles, J.C. and Eccles, R.M. (1960), 'Differentiation of fast and slow muscles in the cat hind limb', *J. Physiol.* 150, 399-416. Close, R. (1967), 'Properties of motor units in fast and slow skeletal muscles of the rat', *J. Physiol.* 193, 45-55.

12 Stein, J.M. and Padykula, H.A. (1962), 'Histochemical classification of individual skeletal muscle fibres of the rat', *Am. J. Anat.* 110, 103-23.

13 Dubowitz, V. and Pearse, A.G.E. (1960), 'Reciprocal relationship of phosphorylase and oxidative enzymes in skeletal muscle', *Nature* 185, 701-2.

14 Guth, L. and Samaha, F.J. (1969), 'Qualitative differences between actomyosin ATPase of slow and fast mammalian muscle', *Exp. Neurol.* 25, 138-52.

15 Romanul, F.C.A. (1964), 'Enzymes in muscle. I: Histochemical studies of enzymes in individual muscle fibres', *Arch. Neurol.* 11, 355-68.

16 Henneman, E. and Olson, C.B. (1965), 'Relations between structure and function in the design of skeletal muscles', *J. Neurophysiol.* 28, 551-98. Close, R. (1967), 'Properties of motor units in fast and slow skeletal muscles of the rat', *J. Physiol.* 193, 45-55.

17 Edström, L. and Kugelberg, E. (1968), 'Histochemical composition, distribution and fatiguability of single motor units: Anterior tibial muscle of the rat', *J. Neurol. Neurosurg. Psychiat.* 31, 424-33.

18 Burke, R.E., Levine, D.N., Zajac, F.E., Tsaris, P. and Engel, W.K. (1971), 'Mammalian motor units: physiological-histochemical correlations in three types in cat gastrocnemius', *Science* 174, 709-12.

19 Beatty, C.H., Basinger, G.M., Dully, C.C. and Bockek, R.M. (1966), 'Comparison of red and white voluntary skeletal muscles of several species of primates', *J. Histochem. Cytochem.* 14, 590-600.

20 The author is indebted to the Zoological Society of London for making specimens available.

PART II SECTION E
Chromosomes, Proteins and Evolution

N. A. BARNICOT

The molecular and cytogenetic approaches to prosimian phylogeny

Although inherited variation is a cornerstone of the theory of natural selection very little was known about the mechanisms of heredity in 1859. Following the rise of Mendelian genetics a re-examination of evolutionary theory was initiated by mathematical geneticists in the 1920s. This led to clearer concepts in which gene mutation and gene frequency change were seen as the fundamental processes involved in evolutionary divergence. It follows that changes in the structure of the genome (i.e. the genetic material as a whole) should provide the most basic information about evolutionary divergence; but the chemical nature of the genome was not known and it was impossible to compare the genomes of distantly related species by classical breeding methods.

This impasse was resolved barely 20 years ago by developments in molecular biology. The crucial advances were the discovery of the structure of DNA and its role in protein synthesis, and the invention of methods for determining the complete amino acid sequences of protein chains. At about the same time work on the three-dimensional structure of certain proteins by X-ray crystallographers made it possible to understand how the amino acid sequence is related to the functional properties of the molecule.

These discoveries opened new prospects for evolutionary biology which were soon exploited, notably by Margoliash and his colleagues in their remarkable studies on the amino acid sequences of the enzyme cytochrome *c* in a very wide range of organisms.[1] It was no doubt reassuring to find that a scheme of phylogenetic relationships based on this single protein, which represents only a minute fraction of the whole genome, was reasonably consistent with prevailing taxonomic views. But it should be noted that there is little or no fossil evidence for many invertebrate groups and comparative anatomy is not much help; these parts of the cytochrome *c* phylogeny are therefore difficult to check against other types of evidence.

I do not wish (after this appalling example of potted history) to detail the biochemical facts on which the molecular approach to phylogeny is based. This is ably done by Dr Doolittle and other authors in the papers that follow. I believe that amino acid sequence data on several different proteins in a well-chosen spectrum of primates will go far to solve problems of phylogenetic relationship. But it is worth remembering that relationship is only one, and perhaps not even the most interesting, facet of the evolutionary puzzle. A diagram of the course of evolutionary divergence with a correct branching pattern and correct distances along the branches is certainly well worth having; but it is merely a description of what happened. The much more difficult problem of explaining why evolution took that particular course remains, and it will certainly need many other approaches to reach plausible solutions.

The study of proteins in an evolutionary context has generated some ideas that are disturbing to orthodox belief in the all-pervasive power of natural selection. The question is whether natural selection has in fact been responsible for most of the amino acid substitutions that have been incorporated in proteins over the course of time. Kimura pointed out that if the rates of change in the few proteins that had then been examined could be taken as typical of the thousands of proteins that must be coded by the genome, they would imply an intolerable load of genetic deaths if the changes were due to selection.[2] One way out of this difficulty is to suppose that many of these mutations are selectively neutral. A small fraction of those that occur then become fixed in the population by random processes (genetic drift). Selection still operates, but mainly in a negative sense, to weed out unfavourable mutants. The wide variation in the evolutionary rates of different proteins could be accounted for by differences in their tolerance of mutational change in structure. Various arguments for and against this theory of protein evolution by neutral mutations have been advanced, and the controversy continues.[3–5] Perhaps it should be remarked that neither Kimura nor his protagonists have denied the importance of natural selection at levels of organisation higher than protein molecules, so that there is no cause for general alarm.

Polymorphic variation of proteins is remarkably frequent in man and in a few other organisms that have been closely examined.[6] The neutral mutation theory sees this polymorphism as a stage in the passage of neutral alleles towards fixation by drift, and it enables one to predict the numbers and frequencies of such alleles to be expected in populations of given effective size. Studies on biochemical polymorphism in non-human primates, such as Hewett-Emmett and I have attempted (this volume), may therefore help to check the validity of this theory if they can be combined with field observations on population size and breeding structure. Although non-human primates are not ideal for work of this kind, they do have the advantage that considerable effort is now being devoted to studying their behaviour in the wild. Perhaps some of the smaller and more abundant prosimians that are not too difficult to locate and catch, such as galagos or mouse-lemurs, would provide favourable material.

One attraction of the neutral mutation theory is that it predicts constant rates of change in proteins provided that mutation rates remain constant. If this is the case, and if the theory is substantially right, then proteins can serve as evolutionary clocks. Approximately constant rates of change have in fact been claimed for some proteins, but Goodman et al. (1971) have postulated declining rates for anthropoid haemoglobin chains and there are various other facts in the literature which suggest that rates are not always constant.[7]

A striking example has been presented by Staehlin (1972), who studied the sequences of the hormone calcitonin in man, pig, ox, sheep and salmon.[8] He found 18 differences between man and pig in a chain of 32 residues and only 15 differences between man and salmon. He suggested that major changes in the function of this calcium-controlling hormone may have occurred early in the evolution of artiodactyls.

To estimate rates of change we need an acceptable phylogenetic tree, measures of distance (i.e. divergence) along the branches and reliable dates for the nodes. Some of the problems of constructing phylogenetic trees from protein sequence data are discussed by Cook and Hewett-Emmett in this volume. It is easy to forget the assumptions on which these procedures are based and to relax into the pleasant belief that one need only feed the data into a computer for infallible phylogenies to emerge. As a rule no measure of the reliability of a dendogram is given and it is tempting to attach undue weight to details of the branching pattern that may in fact be supported by very slight evidence.

Palaeontological dates for branching points in primate phylogeny are often approximate. This is not so much because dating of the geological deposits is uncertain, but because we are unlikely to find fossils that are true common ancestors (i.e. lie exactly on nodes in the phylogenetic tree). We therefore have to extrapolate to a greater or lesser extent; and, although I hesitate to mention it for fear of rousing either Professor Simons, Dr Szalay or both, it is not unknown for palaeontologists to disagree in their attributions of ancestors.

There is much evidence pointing to changes of evolutionary rates at the morphological level.[9] Adaptation to new ecological situations may at first be relatively rapid and later slow down. It would be surprising if major adjustments of habit are not reflected in structural and functional change of at least *some* proteins. But it may not always be possible to get evidence of rate changes in the evolution of a protein. If we have, say, only two living descendants (or two groups of closely related descendants) from some early common ancestor, we cannot tell whether the sequence differences between them that we see today accumulated at a steady rate, or whether divergence was more rapid at some restricted period. To investigate this we need data on the survivors of lines that branched from these lineages at some suitable stage in their divergence. Even if such branches ever existed it may be that extinction will have deprived us of the evidence.

If we are interested in the structural divergence of genes it would be best to analyse and compare the genes themselves, because most amino acid changes in proteins do not give unequivocal information about the precise nature of the gene change and because some gene changes are not manifested in proteins at all. In general it is still impossible to isolate and analyse particular genes, but it is worth mentioning that Min Jou et al. have now reported the complete base sequence of the gene coding for the coat-protein of phage MS 2.[10] Admittedly the genome of this virus is much smaller than that of higher organisms, and the genetic material is RNA, not DNA, but the work serves to show how rapidly molecular genetics is advancing. In any case it is already possible to make a *general* comparison of the genomes of higher organisms by means of DNA hybridisation methods. These techniques test the overall structural similarity of large segments of DNA containing perhaps many thousands of genes. Some work has been done on primates,[11] and it is a pity that we were unable to include a paper on this approach, since it is a valuable complement to studies on single proteins.

Despite technological advances amino acid sequence analysis is still a slow (and expensive) business and quicker, if less informative, methods still command attention. Immunological comparisons of primate proteins were initiated by Nuttall over fifty years ago and the methods have been refined and extended since then.[12] Insofar as amino acid substitutions alter the molecular suface to which antibodies stick, we should expect serological and chemical comparisons to give similar results. This seems to be the case, though very few proteins have been intensively studied by both methods. However, discrepancies sometimes occur if the same antigenically potent substitution happens to have taken place in two distantly related species (see Nisonoff et al. on cytochrome *c*).[13] Bauer gives us a clear account of the rationale of the serological approach to phylogeny and both he and Goodman present massive new data on primates. It is interesting to see that Bauer's estimate of the time since divergence of man and chimpanzee, based on a technique different from that of Sarich and Wilson,[14] and on an examination of many serum proteins rather than one, is substantially longer.

Lastly, with the contributions of Chiarelli, Egozcue and Rumpler we step from the molecular to the microscopic world of chromosomes. The distinction is of course arbitrary and essentially one of method, because re-arrangements of the genetic material also occur at submicroscopic levels. Comparisons of protein sequences have shown the importance of gene duplication for the genesis of novel proteins, and hybrid protein chains, such as those of the Lepore haemoglobins, testify to breaks and errors of reunion within the confines of structural genes.

The karyotypes of a substantial proportion of the Prosimii have now been described and some striking differences both within and between species have been recorded. But, as Chiarelli and Egozcue point out, chromosome numbers, lengths and centromere positions provide only limited clues to the cytological events that produced karyotype divergence. Small deletions and translocations, and even quite large inversions, may be virtually undetectable in somatic cells by the older methods of examination. Even if meiotic material is available, the amount of evolutionary change is sure to be underestimated.

In the last few years new staining techniques have revealed patterns of banding in the arms of mammalian chromosomes. These patterns are certainly very gross in comparison with the exquisite banding of Dipteran giant chromosomes, but they suffice to make the identification of each member of the chromosome set much more reliable. They also make it easier to detect structural deviations and to give a precise account of the changes that occurred in karyotype divergence. In a recent study using these procedures De Grouchy et al. concluded that at least five chromosome mutations must have been incorporated during the divergence of man and chimpanzee.[15] Curiously enough, four of them are pericentric inversions, a type of change which is liable to generate abnormal meiotic products. Pearson et al. have shown that the intense fluorescent staining which marks the distal part of the long arm of the human Y-chromosome is also present in the chimpanzee and gorilla, but not in the orang and gibbon.[16] Human Y-chromosomes show considerable individual variations in length, apparently due to loss of material in this fluorescent region, and in the chimpanzee it is conspicuously small. It seems that material from certain regions of chromosomes can be lost without causing deleterious effects. It may well be that these are regions in which particular sequences are repeated hundreds or thousands of times. There is also evidence that such repetitious DNA is concentrated near the centromeres of many chromosomes and this may be one reason why 'centric-fusions' between acrocentric chromosomes are such a conspicuous feature of karyotype evolution. Naturally the anthropoid apes have received preferential attention from workers using these new methods and prosimians will have to wait their turn in the queue. It need hardly be added that many prosimian karyotypes differ from one another far more than do these of man and chimpanzee and it will probably be correspondingly more difficult to unravel the changes that have taken place over such relatively long periods of divergence.

Obviously genes are ideally precise markers of regions of the chromosomes. If we had sufficiently good linkage-maps they might give us a much more detailed picture of the rearrangements of chromosomal material that occurred in evolution than we can get by microscopy. This may seem utopian, because fine linkage-mapping by classical methods is hopelessly tedious in slow-breeding species. But there are other possibilities. A good deal of work is now being done with cell cultures, though the long-cherished hope that somatic crossing-over can be used for gene-mapping has not yet been fulfilled. Nevertheless cell cultures (see Ruddle for a recent review)[17] are beginning to be useful in assigning particular genes to particular chromosomes. This approach depends on being able to detect the specific product of a gene in the cells of a tissue that is easy to culture, and biochemical genetics has given us numerous enzyme variants that are good markers for such experiments. The technique of fusing cells from different species (e.g. mouse and man), which results in the progressive elimination of some of the chromosomes from the hybrid cell lines, has been very important in this work. Attempts have also been made to use radio-labelled messenger RNA as a specific 'stain' to reveal the sites of particular genes, but the results are still controversial. I mention these developments simply to show that the prospects of comparative cytogenetics in the next few decades are distinctly attractive.

When we are dealing with protein sequences it is easy to derive quantitative measures of evolutionary divergence by simply counting amino acid substitutions. In the case of karyotypes there is no great difficulty, apart from human fatigue, in measuring chromosome arms. But to express karyotype differences effectively we also need to take into account re-associations of arms, or parts of them, due to translocations, as well as changes (i.e. paracentric inversions) that do not alter arm lengths. The best measure of divergence would seem to be the number of mutational events (i.e. translocations, deletions, inversions, etc.) accounting for the observed differences between two species. To make such estimates we must be able to homologise chromosome arms. The resolution of our technique will put an upper limit to the number of observable changes and we should have to assume that invisible changes are in proportion to those we can actually see. Since at present we can identify points of breakage only very crudely, there may be an element of doubt as to whether a given structural change present in two species is really due to the same mutational event.

As in the case of protein evolution we should obviously like to go beyond descriptions of what happened in evolution and to understand why karyotype divergence took a particular course in certain lineages. Presumably karyotypes diverge because, like gene mutations, breaks and errors in reunion of chromosome strands are going on all the time and sometimes the products spread through a population. Many of these products must surely be eliminated quite rapidly because they entail loss of essential material or lead to uncompensated meiotic troubles that lower fitness. We now have many examples of more or less deleterious chromosome changes in man. On the other hand we know from classical studies on *Drosophila* that some inversions persist in wild populations and it is supposed that they are favourable because they ensure tight linkage of co-adapted blocks of genes. At present our knowledge of the organisation of genes in chromosomes and its functional significance is somewhat rudimentary, at least for higher organisms. Advances in this area of cell biology will surely have much to offer to students of karyotype evolution.

NOTES

1 Fitch, W.M. and Margoliash, E. (1967), 'Construction of phylogenetic trees', *Science* 155, 279-84.

2 Kimura, M. (1968), 'Evolutionary rate at the molecular level', *Nature* 217, 624-6.

3 King, J.L. and Jukes, T.H. (1969), 'Non-Darwinian evolution', *Science* 164, 788-98.

4 Maynard-Smith, J. (1968), ' "Haldane's dilemma" and the rate of evolution', *Nature, Lond.* 219, 114-16.

5 Clarke, B. (1970), 'Darwinian evolution of proteins', *Science* 168, 1009-11.

6 Harris, H. (1969), 'Enzyme and protein polymorphism in human populations', *Brit. Med. Bull.* 25, 5-13.

7 Goodman, M., Barnabas, J., Matsuda, G. and Moore, G.W. (1971), 'Molecular evolution in the descent of man', *Nature* 233, 604-13.

8 Staehlin, M. (1972), 'The calcitonins: an example of unusual evolution', *J. Molec. Evol.* 1, 258-62.

9 Simpson, G.G. (1944), *Tempo and Mode in Evolution*, New York.

10 Min Jou, W., Haegeman, G., Ysebaert, M. and Fiers, W. (1972), 'Nucleotide sequence of the gene coding for the bacteriophage MS 2 coat protein', *Nature* 237, 81-8.

11 Kohne, D.E., Chiscon, J.A. and Hoyer, B.H. (1972), 'Evolution of Primate DNA sequences', *J. Hum. Evol.* 1, 627-44.

12 Nuttall, G.H.F. (1904), *Blood Immunity and Blood Relationship*, London.

13 Nisonoff, A., Reichlin, M. and Margoliash, E. (1970), 'Immunological activity of cytochrome *c*. II: Localisation of a major antigenic determinant of human cytochrome *c*, *J. Biol. Chem.* 245, 940-6.

14 Sarich, V.M. and Wilson, A.C. (1967), 'Immunological time scale of hominid evolution', *Science* 158, 1200-3.

15 De Grouchy, J., Turleau, C., Robin, M. and Klein, M. (1972), 'Evolutions caryotypiques de l'homme et du chimpanzé: Etude comparative du topographie des bandes après dénaturation ménagée', *Ann. Genet.* 15, 79-84.

16 Pearson, P.L., Bobrow, M., Vosa, C.G. and Barlow, P.W. (1971), 'Quinacrine fluorescence in mammalian chromosomes', *Nature* 231, 326-9.

17 Ruddle, F.H. (1973), 'Linkage analysis in man by somatic cell genetics', *Nature* 242, 165-9.

J. EGOZCUE

Chromosomal evolution in prosimians

Introduction

The study of primate chromosomes has experienced great expansion in the past few years. Unfortunately, data on prosimian chromosomes remain scarce. With the exception of the Lemurinae, which have been fairly well studied, the gaps in our knowledge of prosimian chromosomes are wider than in any other primate group. Since the publication of our review of primate chromosomes,[1] only a few papers have dealt with the problem of prosimian karyology; most of them (Rumpler and Albignac, Rumpler, Egozcue)[2-4,5,6] are devoted to the genus *Lemur*, and only two deal with the genera *Galago* and *Nycticebus* (Egozcue, Hayata et al.).[7,8] In spite of these shortcomings, the steps of chromosomal evolution can be traced in some prosimian genera. Following previous practice (Egozcue),[1] the charts of chromosomal evolution presented in this paper apply only to structural chromosome changes, and it is not pretended that they have phylogenetic implications. For comparison of chromosomal formulae, the chromosomes are classified into submetacentrics and acrocentrics; the sex chromosomes are always excluded. Karyotypes are compared only when numerical and morphological similarities warrant it.

Lemuroidea

While extensive data are available on the genus *Lemur*, chromosome studies have been almost entirely lacking for other Lemuroidea. Although the preliminary program of the Third Conference on Experimental Medicine and Surgery in Primates (Lyon, June 21-23, 1972) includes a report on the chromosomes of the Indridae (Rumpler and Albignac),[9] and a paper on the chromosomes of the

Table 1 Chromosomes of Lemuroidea

Species	*2N*	*S*	*A*	*X*	*Y*
Lemur mongoz	60	4	54	A	A
L. fulvus fulvus	60	4	54	A	A
L. f. rufus	60	4	54	A	A
L. f. albifrons	60	4	54	A	A
L. f. sanfordi	60	4	54	A	A
L. f. collaris	60	4	54	A	A
L. catta	56	8	46	M	A
L. catta	56	10	44	M	—
L. f. collaris	52	12	38	A	A
L. f. collaris	48	16	30	A	—
L. variegatus	46	18	26	M	A
L. macaco	44	20	22	A	A
Hapalemur griseus	54	10	42	A	A
H. griseus	58	6	52	—	—
Cheirogaleus major	66	—	64	A	A
Microcebus murinus	66	2	64	—	—
Propithecus verreauxi verreauxi	48	—	—	—	—
P. v. coquereli	48	—	—	—	—

Lemuridae, Indridae and Daubentoniidae has been presented by Rumpler in this volume,[10] no details of the karyotypes are known to me at the time of this writing, except for the incomplete data published by Chu and Bender on *Propithecus*.[11]

Recently, we have studied the chromosomes of two female *Microcebus* in blood cultures.[12] The diploid number is 2N = 66, and the karyotype is composed of one pair of submetacentric chromosomes and 32 pairs of acrocentrics.

Table 1 summarises the diploid numbers and chromosomal formulae of the Lemuroidea. While data on the different species of the genus *Lemur* have been confirmed by several authors, the chromosomal complements described so far in the genera *Hapalemur* (Chu and Swomley),[13] *Cheirogaleus* (Chu and Swomley, cited by Chu and Bender),[11] and *Propithecus* (Chu, cited by Chu and Bender)[11] have to be considered as tentative until Rumpler's data are published in full.[10]

The possible pathways of chromosomal evolution in the Lemurinae are shown in Fig. 1. Only confirmed data have been included. However, the diploid numbers and chromosomal formulae of *Hapalemur* also fit the pathways illustrated in Fig. 1, and correspond to intermediate steps in the chromosomal evolution of the Lemurinae (Fig. 2).

In the genus *Lemur*, hybrids have been obtained between species with very pronounced chromosomal differences. Gray lists some of these hybrids and qualifies the crosses as unlikely.[14] However, such hybrids do occur, and some of them have been studied by Rumpler and Albignac. The chromosomal complement of the hybrids

Table 2 Chromosomes of lemur hybrids

Species	*2N*	*S*	*A*	*X*	*Y*
L. f. fulvus x *L. f. albifrons* (1)	60	4	54	A	A
L. f. fulvus x *L. f. rufus* (2)	60	4	54	A	A
Hybrid 1 x Hybrid 2	60	4	54	A	—
L. macaco x *L. f. albifrons*	52	12	38	A	A
L. macaco x *L. f. rufus*	52	12	38	A	—
L. f. collaris[a] x *L. f. collaris*[b]	54	10	42	A	—

[a] *L. f. collaris* with 2N=60
[b] *L. f. collaris* with 2N=48

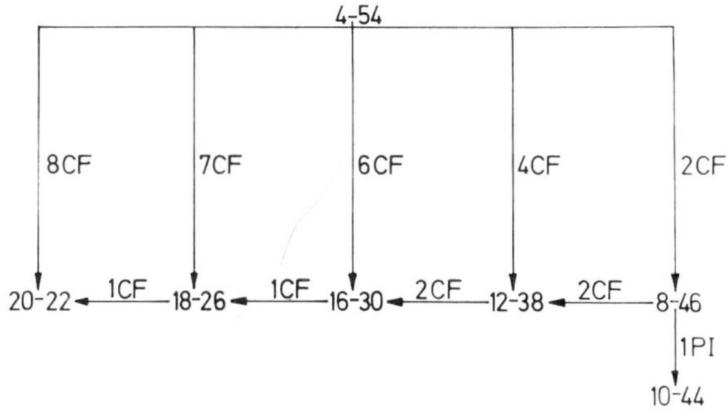

Figure 1. Chromosomal evolution in the genus *Lemur*. CF stands for Centric Fusion, PI for Pericentric Inversion. (Numbers indicate submetacentrics and acrocentrics, respectively — cf. Table 1.)

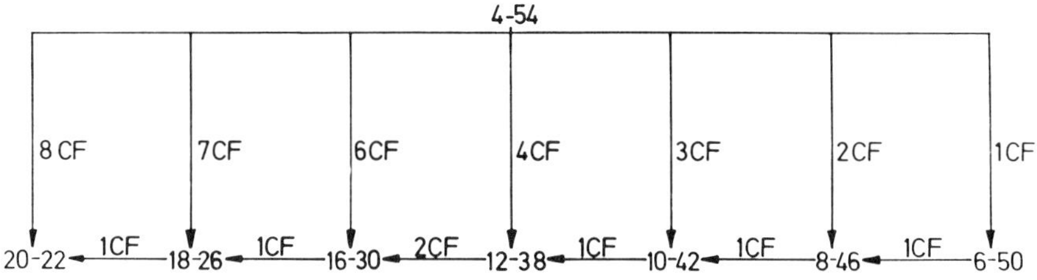

Figure 2. Chromosomal evolution in the genera *Lemur* and *Hapalemur* (see caption to Fig. 1).

(Table 2) is made up of the haploid set from each parent.[6] That such hybrids can be produced and are viable suggests that, in spite of the considerable structural changes that have taken place in the chromosomes of Lemuridae, a high degree of genetic homology must exist between the different species concerned.

Lorisoidea

Table 3 shows the chromosome numbers of the Lorisoidea. Some of the results have recently been confirmed by De Boer.[15] In the Lorisidae, a case of intraspecific polymorphism has been described in *Nycticebus coucang*, where diploid numbers of 2N = 50 and 2N = 52 are found. Of special interest are the striking similarities between the karyotypes of *Nycticebus coucang* and *Arctocebus calabarensis*. There is some discrepancy regarding the chromosomal formula of *Loris tardigradus* (Manna and Talukdar).[16] However, as we have stated before,[6] some differences may be due to the subjective criterion used by an author in classifying subterminal chromosomes as acrocentric or submetacentric, rather than to actual polymorphism in the species. Since all animals studied in this group belong to different genera, no attempt has been made to establish any pathways of chromosomal evolution, in spite of the fact that from a strictly morphological point of view it would not be difficult to establish a relationship among the chromosomal complements of the Lorisidae.

In the Galagidae, different results have been published for *Galago senegalensis* and *Galago crassicaudatus*. As far as *Galago senegalensis* is concerned, the only study conducted on suitable material revealed a karyotype composed of 30 submetacentrics and 6 acrocentrics.[7] In *Galago crassicaudatus*, these results have been confirmed by Hayata et al.[8] The chromosomal complement described by Chu and Bender for *Galago crassicaudatus* has not been confirmed, but it provides a

Table 3 Chromosomes of Lorisoidea

Species	*2N*	*S*	*A*	*X*	*Y*
Lorisidae					
Perodicticus potto	62	24	36	S	A
Loris tardigradus	62	34	26	S	S
L. tardigradus	62	38	22	S	S
Arctocebus calabarensis	52	50	—	S	S
Nycticebus coucang	52	2	48	S	S
N. coucang	50	48	—	S	S
Galagidae					
Galago crassicaudatus	62	6	54	S	A
G. crassicaudatus	62	30	30	S	A
G. senegalensis	38	30	6	S	—

chromosomal link between the greater and the lesser bushbabies.[17] The possible pathways of chromosomal evolution in the Galagidae are shown in Fig. 3.

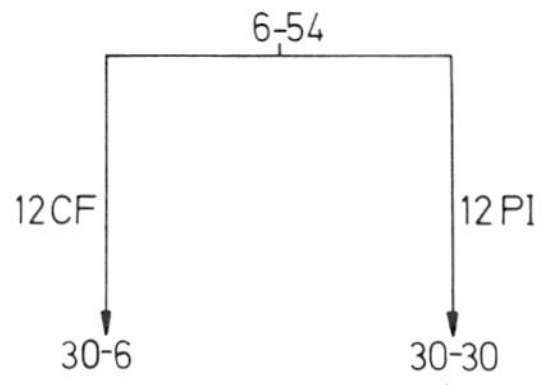

Figure 3. Chromosomal evolution in the Galagidae (see caption to Fig. 1).

Tarsioidea

Data on the chromosomes of Tarsioidea are extremely scarce. The diploid number of 2N = 80, described by Klinger in *Tarsius bancanus*,[18] was confirmed by Chiarelli and Egozcue in testicular material of *Tarsius syrichta*,[19] and is the highest diploid number of any primate. The karyotype of *Tarsius bancanus* is composed of 14 submetacentrics and 66 acrocentrics. Since the only animal studied was a female, no information is available on the sex chromosomes (Table 4).

Table 4 Chromosomes of Tarsioidea

Species	*2N*	*S*	*A*	*X*	*Y*
Tarsius bancanus	80	14	66	—	—
T. syrichta	80	—	—	—	—

Discussion

In spite of the scarcity of data on the chromosomes of prosimians, it is still possible to trace the structural changes that may have given rise to new species. This is permissible because of the fact that the chromosomal mechanisms that may lead to speciation are monotonously similar in mammals, and almost always correspond to centric fusion or pericentric inversion events. In other words, at the present time mammals seem to be in a process of chromosome stabilisation characterised by centric fusions and pericentric inversions.[21] Both mechanisms increase the number of submetacentric chromosomes in a given karyotype; furthermore, centric fusions increase chromosomal size while reducing the diploid number of a species. Since there is some evidence that smaller chromosomes and acrocentric chromosomes tend to have a lower number of chiasmata during meiosis I, and thus show a tendency to separate precociously into univalents and provoke meiotic problems,[20] it seems logical that any mechanisms that increase the two main morphological factors of chromosome stability in meiosis, i.e. chromosomal size and metacentricity, will have an important role in chromosome evolution. This is not to deny the importance of centric fission (Todd),[21] a mechanism that may have been predominant at some stages of mammalian evolution.

From a taxonomic point of view, chromosomal data may be of importance, as in the case of the proposed classification of *Lemur fulvus* in a single species with *L. macaco* (a suggestion not supported by chromosomal studies). In other cases, karyological studies are of no assistance, for example in deciding whether to classify *L. catta* in the genus *Hapalemur*, or whether to give *L. variegatus* separate generic status (*Varecia*), as would apparently be warranted by data from other sources.

Recently, the new techniques of quinacrine fluorescence and Giemsa banding of chromosomes have developed to provide an extraordinarily useful tool for comparative cytogeneticists. The fact that both the Q (fluorescent) and G (Giemsa) bands permit detection of chromosome homology will be useful for establishing interspecific relationships on a more secure basis. I have not had the opportunity to apply these techniques to prosimian chromosomes; but the studies carried out so far in our laboratory suggest that banding patterns are extremely stable, and that they thus provide an excellent means of establishing chromosome homologies between closely related species.

Acknowledgments

Most of our studies on prosimian chromosomes were conducted at the Oregon Regional Primate Research Center, with the skillful technical assistance of Fay Hagemenas. I thank R.D. Martin and N.A. Barnicot, University College, London, for supplying the samples of *Microcebus*.

NOTES

1 Egozcue, J. (1969), 'Primates', in Benirschke, K. (ed.), *Comparative Mammalian Cytogenetics*, New York.

2 Rumpler, Y. and Albignac, R. (1969), 'Etude cytogénétique de quelques hybrides intraspécifiques et interspécifiques de lémuriens', *Ann. Sci. Univ. Besançon* 6, 1.

3 Rumpler, Y. and Albignac, R. (1969), 'Etude cytogénétique de deux lémuriens, *Lemur macaco macaco*, Linné 1766, et *Lumur fulvus rufus* (Audebert 1800) et d'un hybride *macaco macaco/fulvus rufus*', *C. R. Soc. Biol.* (Paris) 163, 1247-50.

4 Rumpler, Y. and Albignac, R. (1969), 'Existence d'une variabilité chromosomique intraspécifique chez certains lémuriens', *C. R. Soc. Biol.* (Paris) 163, 1989-92.

5 Rumpler, Y. (1970), 'Etude cytogénétique du *Lemur catta*', *Cytogenetics* 9, 239.

6 Egozcue, J. (1972), 'The chromosomes of lemurs', *Folia primat.* 17, 171-6.

7 Egozcue, J. (1970), 'The chromosomes of the lesser bushbaby (*Galago senegalensis*) and the greater bushbaby (*Galago crassicaudatus*)', *Folia primat.* 12, 236-240.

8 Hayata, I., Sonta, S., Itoh, M. and Kondo, N. (1971), 'Notes on the karyotypes of some prosimians, *Lemur mongoz, Lemur catta, Nycticebus coucang* and *Galago crassicaudatus*', *Japan. J. Genet.* 1, 61.

9 Rumpler, Y. and Albignac, R. (1972), 'Cytogenetic study of the Indridae', *Third Conf. on Exptl. Med. and Surg. in Primates*, Lyon, June 21-23.

10 Rumpler, Y., this volume.

11 Chu, E.H.Y. and Bender, M.A. (1962), 'Cytogenetics and evolution of primates', *Ann. N.Y. Acad. Sci.* 102, 253-66.

12 Egozcue, J. (1972), unpublished data.

13 Chu, E.H.Y. and Bender, M.A. (1961), 'Chromosomes of lemurine lemurs', *Science* 133, 1925-6.

14 Gray, A.P. (1954), *Mammalian Hybrids*, Technical Communication No. 10 of the Commonwealth Bureau of Animal Breeding and Genetics, Alva.

15 De Boer, L.E.M. (1972), 'Chromosome studies in some primates', *Mammal. Chromos. Newsl.* 13, 4.

16 Manna, G.K. and Talukdar, M. (1968), 'An analysis of the somatic chromosome complements of both sexes of two primates, the slender loris, *Loris tardigradus*, and rhesus monkey, *Macaca mulatta*', *Mammalia* 32, 118.

17 Chu, E.H.Y. and Bender, M.A. (1961), 'Chromosome cytology and evolution in primates', *Science* 133, 1399-405.

18 Klinger, H.P. (1963), 'The somatic chromosomes of some primates (*Tupaia glis, Nycticebus coucang, Tarsius bancanus, Cercocebus aterrimus, Symphalangus syndactylus*)', *Cytogenetics* 2, 140-51.

19 Chiarelli, B. and Egozcue, J. (1968), 'The meiotic chromosomes of some primates', *Mammal. Chromos. Newsl.* 9, 85-6.

20 Egozcue, J. (1969), 'Aneuploidia cromosómica: nuevas observaciones sobre la separación precoz de bivalentes en meiosis I', *Sangre* (Barcelona) 14, 442.

21 Todd, N.B. (1967), 'A theory of karyotypic fissioning, genetic potentiation and eutherian evolution', *Mammal. Chromos. Newsl.* 8, 268.

Y. RUMPLER

Cytogenetic contributions to a new classification of lemurs

Introduction

The systematic relationships of the Malagasy lemurs are still a matter of controversy, as is testified by the successive different classifications that have been proposed. In recent years, however, new studies have led to a certain degree of convergence of the conclusions of various authors, permitting us to attempt a revision of the classification, and in this paper we propose a classification of lemurs based on a systematic cytogenetic study of almost all of the Malagasy lemur species. For this revision, we have utilised our cytogenetic data to modify one of the most recent classifications (Hill)[1] in cases where our results generally accord with the conclusions of recent publications.

Results

We propose to raise the existing subfamily Cheirogaleinae Gregory, 1915 to the rank of the family Cheirogaleidae, with type genus *Cheirogaleus* E. Geoffroy, 1812. This family includes the Cheirogaleinae Gregory, 1915 and a new subfamily Phanerinae, with type genus *Phaner* Gray, 1870. The family Lemuridae is modified to include two subfamilies: the Lemurinae Mivart, 1864 and the Lepilemurinae, with type genus *Lepilemur* I. Geoffroy, 1851, corresponding to the tribe Lepilemurini Stephan and Bauchot, 1965.

Discussion

Study of the chromosomes of species within the previously recognised subfamily Cheirogaleinae indicates that there are two separate groups: (*a*) *Cheirogaleus* and *Microcebus* species; (*b*) *Phaner* species. *Cheirogaleus* and *Microcebus* have the same diploid numbers (2N = 66) and fundamental numbers (FN = 66). The diploid number (2N = 48) and particularly the fundamental number (FN = 62) of *Phaner* are markedly different from those of *Cheirogaleus* and *Microcebus* (Rumpler and Albignac).[2] Other characters which have been reported recently, such as the dermatoglyphic patterns (Rakotosamimanana and Rumpler),[3] the presence of a specific marking gland on the anterior aspect of the neck in *Phaner* (Rumpler and Andriamiandra),[4] and the peculiar behaviour of *Phaner* (Petter et al.),[5] all support the chromosomal evidence in favour of separation of the two groups. On the basis of cranial and behavioural characters, Petter has already proposed that the subfamily Cheirogaleinae should be raised to family level,[6] and recent parasitological investigations similarly underline the specificity of the nematode parasites of members of this group,[7] which are quite distinct from those of species within the family Lemuridae. This point of view has been accepted in the present investigation, and the new family Cheirogaleidae has already been described as containing the 2 subfamilies Cheirogaleinae and Phanerinae (Rumpler and Albignac).[2] We have left *Allocebus* in the subfamily Cheirogaleinae along with *Cheirogaleus* and *Microcebus* (see Petter-Rousseaux and Petter),[8] although at the present time we have not been able to study the karyotype of this genus and therefore cannot provide evidence to support this provisional allocation.

The subfamily Lemurinae includes 3 genera: *Lemur, Hapalemur* and *Varecia.* The 3 genera have the same fundamental number (FN = 64), which indicates robertsonian evolution within this group. There are no cytogenetic characters indicating that the variegated lemur should be placed as a separate genus, rather than as a species within the genus *Lemur* (i.e. as *Lemur variegatus*). However, other recently described characteristics, such as the quite distinctive dermatoglyphic patterns (Rakotosamimanana and Rumpler).[9] the presence of a marking gland on the neck (Rumpler and Andriamiandra),[4] and various cranial features,[10] support J.-J. Petter's suggestion that this animal should be placed in the independent genus *Varecia.*

The subfamily Lepilemurinae contains quite distinctive animals which were previously placed in the genus *Lepilemur* and classified together with *Lemur* and *Hapalemur.* Stephan and Bauchot have already proposed that the sportive lemurs should be more clearly distinguished, by placing them in a separate tribe Lepilemurini (as distinct from the Lemurini).[11] The extremely distinctive karyotypes of the sportive lemurs, with diploid numbers varying between 2N = 20 and 2N = 38, and fundamental numbers ranging from FN = 44 to FN = 38 (Rumpler and Albignac),[19] have induced us to raise them to the level of a subfamily, Lepilemurinae. Other recently described characters of the sportive lemurs, in particular the dermatoglyphs (Rumpler and Rakotosamimanana),[13] haematological characters,[14,15] and parasitological features (Chabaud, Brygoo and Petter),[7] along with behavioural peculiarities (Petter, Charles-Dominique and Hladik),[16,17] also argue in favour of this allocation. These recently recognised characteristics, when viewed along with the very distinctive karyotypes of the *Lepilemur* group, might even indicate that they should in fact be classified in a separate family, Lepilemuridae.

The 3 genera of the family Indriidae are clearly distinguished both by their karyotypes and by their fundamental numbers.[18] In particular, the genus *Avahi* is clearly distinct from the other two genera, *Indri* and *Propithecus.* Other characters, such as the marking glands (Rumpler and Andriamiandra),[4] emphasise the distinctiveness of each of these 3 genera. Despite the differences which exist between the 3 genera in the palmar and plantar pads, however, there is marked homogeneity within the family Indriidae as a whole (Rakotosamimanana and Rumpler).[3]

The family Daubentoniidae is represented by the sole genus *Daubentonia.* The diploid number (2N = 30), the karyotype and the fundamental number (FN = 54) are all extremely distinct from those of other lemurs and would provide a further argument – if such were necessary – for classifying the aye-aye quite separately.

We therefore propose the following new classification:

Superfamily LEMUROIDEA (Mivart, 1864)
Family CHEIROGALEIDAE Rumpler & Albignac, 1972
1° – Subfamily CHEIROGALEINAE Gregory, 1915
genera: *Microcebus* E. Geoffroy, 1828
Cheirogaleus E. Geoffroy, 1812
8 (see news article, Mar. 1990) *Allocebus* Petter-Rousseaux & Petter, 1967
2° – Subfamily PHANERINAE Rumpler & Albignac, 1972
genus: *Phaner* Gray, 1870

Family LEMURIDAE Gray, 1821
1° – Subfamily LEMURINAE Mivart, 1864
genera: *Lemur* Linnaeus, 1758
Hapalemur I. Geoffroy, 1851
Varecia Gray, 1863
2° – Subfamily LEPILEMURINAE Rumpler & Rakotosamimanana, 1972
genus: *Lepilemur* I. Geoffroy, 1851
Family INDRIIDAE Burnett, 1828
genera: *Avahi* Jourdan, 1834
Propithecus Bennett, 1832
Indri E. Geoffroy & Cuvier, 1795
Family DAUBENTONIIDAE Gray, 1870
genus: *Daubentonia* E. Geoffroy, 1795

NOTES

1 Hill, W.C.O. (1953), *Comparative Anatomy and Taxonomy of the Primates*, vol. 1: *Strepsirhini*, Edinburgh.

2 Rumpler, Y. and Albignac, R. (1972), 'Cytogenetic study of the endemic Malagasy lemurs subfamily Cheirogaleinae Gregory, 1915', *Am. J. Phys. Anthrop.* 38, 261-4.

3 Rakotosamimanana, B.R. and Rumpler, Y. (1972), 'Evolution of palm, sole pads and dermatoglyphic patterns in the Malagasy lemurs', in press.

4 Rumpler, Y. and Andriamiandra, A. (1971), 'Etude histologique des glandes de marquage de la face antérieure du cou des lémuriens malgaches', *C.R. Soc. Biol.* 165, 436-44.

5 Petter, J.-J., Schilling, A. and Pariente, G. (1971), 'Observations éco-éthologiques sur deux lémuriens malgaches nocturnes: *Phaner furcifer* et *Microcebus coquereli*', *Terre et Vie* 25, 287-327.

6 Petter, J.-J. (1962), 'Recherches sur l'écologie et l'éthologie des lémuriens malgaches', *Mém. Mus. Nat. Hist. Nat.* (n.s.) Sér. A, *Zoologie* 27, 1-146.

7 Chabaud, A.G., Brygoo, E. and Petter, A. (1965), 'Les nématodes parasites des lémuriens malgaches. VI: Description de six espèces nouvelles et conclusions générales', *Ann. Parasit. Hum. Comp.* 40, 181-214.

8 Petter-Rousseaux, A. and Petter, J.-J. (1967), 'Contribution à la systématique des Cheirogaleinae', *Mammalia* 31, 574-82.

9 Rakotosamimanana, B.R. and Rumpler, Y. (1970), 'Etude des dermatoglyphes et des coussinets palmaires et plantaires de quelques lémuriens malgaches', *Bull. Ass. Anat.* 148, 493-510.

10 Mahé, J., personal communication; thesis in preparation: 'Crâniométrie des lémuriens'.

11 Stephan, H. and Bauchot, R. (1965), 'Hirn-Körpergewichts-Beziehungen bei den Halbaffen (Prosimii)', *Acta. Zool.* 46, 1-23.

12 Rumpler, Y. and Albignac, R. (1972), 'Etude cytogénétique de *Varecia variegata* et du *Lemur rubriventer*', *C.R. Soc. Biol.* 165. 741-7.

13 Rumpler, Y. and Rakotosamimanana, B.R. (1972), 'Coussinets palmo-plantaires et dermatoglyphes des représentants des Lemuriformes malgaches', *Bull. Ass. Anat.* 154.

14 Richaud, J., Rumpler, Y. and Albignac, R. (1971), 'Quelques données hémobiologiques de la sous-famille des lémurinés malgaches', *Arch. Inst. Pasteur Madagasc.* 40, 137-44.

15 Buettner-Janusch, J., Washington, J.L. and Buettner-Janusch, V. (1971),

'Hemoglobins of Lemuriformes', *Arch. Inst. Pasteur. Madagasc.* 40, 127-36.
16 Petter, J.-J. (1960), 'Remarque sur la systématique du genre *Lepilemur*', *Mammalia* 24, 76-86.
17 Charles-Dominique, P. and Hladik, C.M., 'Le *Lépilemur* du sud de Madagascar: Ecologie, alimentation et vie sociale', *Terre et Vie* 25, 3-66.
18 Rumpler, Y. and Albignac, R. (1973), 'Etude cytogénétique des Indriidae', International Congress of Anthropological and Ethnological Sciences, Chicago, in press.
19 Rumpler, Y. and Albignac, R. 'The cytogenetic study of the subfamily Lepilemurinae', in preparation.

B. CHIARELLI

The chromosomes of the prosimians

In many different animal groups, recently acquired knowledge on number and morphology of the chromosomes has furnished important data for reconstruction of their phylogenies and has provided us with one more criterion for taxonomic organisation. The group of the prosimians, which are so heterogeneous in terms of their remote evolutionary relationships and their geographical isolation, seems particularly suitable for a karyological approach to their phylogeny (Fig. 1). The phylogenetic organisation of the species currently assigned to this group has in fact been the object of continuous discussion, notably with respect to classification above the family level.

Knowledge of chromosomes is important in the study of phylogeny and taxonomy because these structures are the direct carriers of the genetic information, which is itself organised in a stable arrangement on the chromosome. For each species of Eukaryota, the structural affinity between homologous chromosomes is consistently controlled by their pairing at meiosis in each individual. Incomplete chromosome pairing at meiosis suggests the presence of structural changes in the organisation of the genetic information which is linearly distributed along the chromosomes. If such structural changes are not restricted, they cease to be compatible with the functional organisation of the genetic information. The result of such incompatibility is the inability of homologous chromosomes to pair at meiosis, with consequent disruption of the meiotic process.

In all higher organisms, pairing of homologous chromosomes at meiosis plays the rôle of a filter through which only a functionally patterned genetic system can pass. Furthermore, in natural populations meiosis represents a barrier which prevents exchange between diverging genetic systems, and it should therefore be considered as a basic mechanism involved in the division of populations into new species.

For any given species, constancy of karyotype appearance (which we can appreciate at the level of each chromosome) is therefore a consequence of the meiotic filter. On the other hand, the existing differences between karyotypes of different species represent the product of variation in the somatic chromosomes of the germinal cell line which were able to filter through the meiotic sieve and become

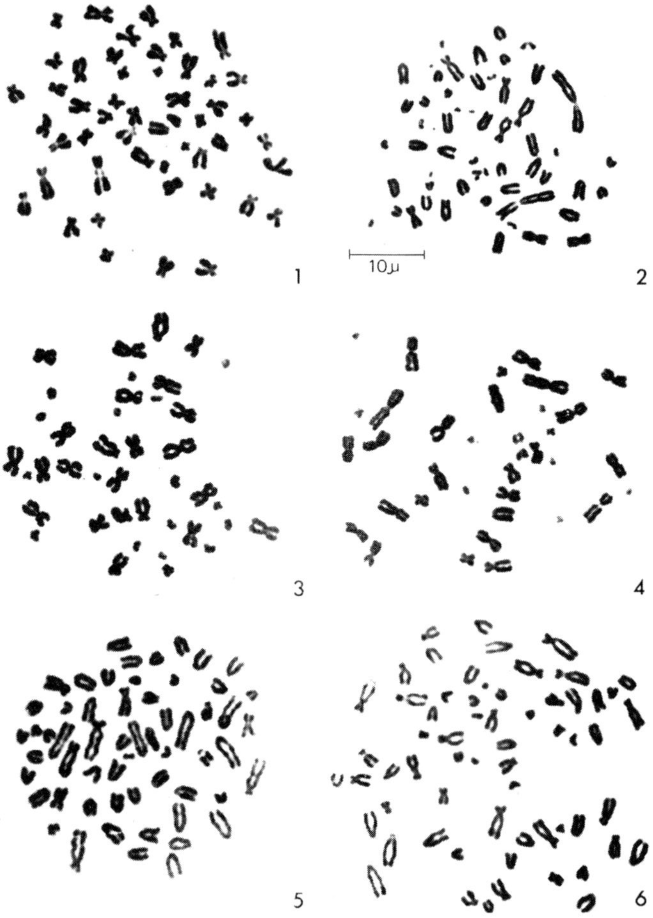

Figure 1. Metaphase plates of six prosimian species.
1 – *Nycticebus coucang* (female; 2N = 50)
2 – *Lemur catta* (male; 2N = 56)
3 – *Galago alleni* (female; 2N = 40)
4 – *Galago senegalensis* (female; 2N = 36)
5 – *Galago demidovii* (male; 2N = 58)
6 – *Galago crassicaudatus* (male; 2N = 62)
(These metaphase plate photographs were kindly placed at our disposal by L.E.M. de Boer, Institute of Genetics, University of Utrecht, The Netherlands.) The magnification is the same for all figures (1000 x), and the actual scale is indicated by the 10 μ line on photograph 2.

established because of some selective advantage.

On the basis of hypotheses about the possible steps by which the chromosomes of two related species became differentiated, one can construct a phyletic line connecting them and establish their taxonomic affinities.

However, before one can conclude that there is chromosome identity between individuals of different species it is important to be sure of their homologous genetic content. The only control one can have at the moment is the observation of the behaviour of the chromosomes during meiosis (expecially diakinesis) in individuals derived from cross-breeding of two related species (F_1). In cyto-taxonomic research, therefore, data on hybrids deserves particular consideration.

The purpose of the present study is to analyse how far the karyological information so far available for the prosimians is of value in arriving at a systematic and phylogenetic organisation of this group.

The taxonomic nomenclature and order followed here is the traditional one.[13] A synthesis of the information available for the chromosomes of those species which have been studied to date is presented in Table 1.

From this table it can be seen:

1. That there is a total lack of karyological information for many important genera: *Dendrogale, Ptilocercus, Avahi, Indri.*
2. That for many other genera only preliminary information is available (*Propithecus, Microcebus, Cheirogaleus, Phaner, Lepilemur, Daubentonia*).
3. That great variability exits in the number of chromosomes for all of those genera which have been more extensively studied (e.g. *Galago* and *Lemur*).

Disregarding, for the moment, the intrageneric chromosomal variation mentioned in point 3, (which will be discussed later) a more general question may be asked: Can the phylogenetic relationship between families be established on the basis of variations in the chromosome number between the genera belonging to each family within this group of Primates?

Mere statement of the diploid number of somatic chromosomes excludes one of the most easily occurring types of chromosome mutation; namely, centric fusion. This difficulty can be partially overcome by resorting to Mattheys's fundamental number (FN), since it takes into consideration solely the number of arms. The use of the fundamental number is, moreover, based on the assumption that centric fusion (or misdivision of the centromere) is one of the more successful types of mutation in passing the filter of meiosis. Such a mutation would not in fact interfere with the organisation of

Table 1 Numerical and morphological data on the chromosomes of Tupaioidea, Lorisoidea, Lemuroidea and Tarsioidea
(2n = diploid chromosome number; M = metacentric, S = submetacentric, A = acrocentric; x and y = sex chromosomes)

Taxa	*2n*	*M-S*	*A*	*X*	*Y*	*References* (see Notes)
1. Tupaiidae						
Tupaia glis	60-62	14-12	44-48	M-S	A	1,2,7,17,27,34,35,32
Tupaia montana	52-68			M	A	1,32
Tupaia minor	66					1,31,32
Dendrogale sp.						
Urogale everetti	44					2,8,17,19
Ptilocercus lowii						
2. Lorisidae						
Loris tardigradus	62	34-38	26-22	S	S-A	28,39
Nycticebus coucang	50-(52)	48		S	S	3,17,30,33,35
Nycticebus pygmaeus	50					17
Arctocebus calabarensis	52	50		S		29
Perodicticus potto	62	24	36	S	A	2,8,16,23
3. Galagidae						
Galago senegalensis	36-38	22-24-30	14-12-6	S	S-A	4,16,22,23,24,38,40,49,50
Galago crassicaudatus	62	6-30	54-30	S	A-S	16,17,23,24,33,38
Galago alleni	40					4
Galago demidovii	58					5

4. Lemuridae						
Microcebus murinus	66		64	A	A	16,18,42,44,45,46
Cheirogaleus major	66		64	A	A	17,18,42,44,45,46
Phaner furcifer	48					42
Hapalemur griseus	54-58	10-6	42-50	A	A	16,18,42,44,45,46
Lemur catta	56	10-14	44-50	S-A	A	2,10,11,16,17,18,21,33,41,13,20,42,44,45,46
Lemur variegatus	46	18	26	S	A	10,16,17,18,21,20,42,44,45,46
Lemur macaco	44-48-52-58-60	20-16-4	22-30-52-54	A	A	2,10,11,16,18,21,23,43,12,9,14,20,42,44,45,46
Lemur mongoz	60	4	54	A	A	10,16,18,21,33,12,9,20
Lepilemur mustelinus	22-38					42
5. Indridae						
Propithecus verreauxi	48					17,42,44,45,46
Avahi laniger	—					
Indri indri	42					48
6. Daubentoniidae						
Daubentonia madagascariensis	30					42
7. Tarsiidae						
Tarsius syrichta	80					15
Tarsius bancanus	80	14	66			35

the genetic information on the chromosomes. The only change which would result would be a reduction or increase in randomnesss of the distribution of genetic information in the offspring. Reduction or increase in the number of chromosome units certainly has adaptive advantages for an organism, in that it reduces or increases the potential variability in a population. In addition, there must be a relation between chromosome morphology and chiasma frequencies, although at the moment we do not have enough information to definite this relationship exactly.

(N.B. Calculation of the fundamental number (FN) is made by counting a metacentric or submetacentric as 2 and an acrocentric or sub-acrocentrix as 1 in any general karyotype).

Data of this type are shown in the fourth column of Table 2. Matthey's fundamental number is 70 to 84 in Tupaioidea; 87 to 102 in Lorisidae; 61 to 94 in Galagidae; 62 to 70 in Lemuridae; 94 in Tarsiidae and 54 in Daubentoniidae.

Karyological data elaborated in this way lend themselves to more reliable analysis and yield more useful taxonomic information.

No problem exists with regard to the taxonomic, and hence

Table 2 Basic karyological information for the prosimians

Genera	*Known/Studied*	*2n*	*Fundamental number*	
Tupaia	11/3	52-60-62-66-68	70-72-74-76	70
Dendrogale	2/0			
Urogale	1/1	44	84-80	↓
Ptilocercus	1/0			84
Loris	1/1	62	98-101	87
Nycticebus	2/2	50	100	
Arctocebus	1/1	52	102	↓
Perodicticus	1/1	62	87	102
Galago	6/4	38-62	61-64-69-75-94	61→94
Microcebus	2/1	66	68	62
Cheirogaleus	2/1	66	68	
Phaner	1/1	48	62	
Hapalemur	2/2	54-58	64	
Lemur	6/6	44-46-48-52-56-58-60	62-64-66-70	↓
Lepilemur	1/1	22-38	42-36	70
Propithecus	2/1	48	?	
Avahi	1/0			
Indri	1/1	42	?	
Daubentonia	1/1	30	54	54
Tarsius	3/2	80	94	94

phylogenetic, separation of the Tupaioidea with respect to the Lorisoidea and Lemuroidea. However, difficulties do arise with these two latter superfamilies, whose relationships are more controversial. A ready interpretation is provided by the data on fundamental numbers available to us. The Lorisidae can, in fact, be sharply distinguished from Lemuridae because their fundamental numbers are different and do not in any way overlap (87-102 vs. 64-70). The taxonomic position of the Galagidae, whose fundamental numbers (FN : 61-94) provide a kind of bridge between the Lemuridae and Lorisidae, is particularly interesting.

The strikingly large difference in the diploid chromosome number between the two *Galago* species previously studied (*Galago senegalensis* with 2N = 38, and *G. crassicaudatus* eith 2N = 62) provides a tempting basis for speculation about a possible polyploid mechanism as the origin of this variation.

However this hypothetical indication of polyploidy has been definitely eliminated by the recent findings that:

(*a*) There is polymorphism in the chromosome number of *Galago senegalensis*, due to a mechanism of centric fusion (Ying and Butler).[50]

(*b*) There is a diploid chromosomal number of 58 in *Galago demidovii* and one of 40 in *Galago alleni* (De Boer, personal communication).

(*c*) There is an identical DNA content in the two *Galago* species (7.54 ± 0.09 a. u. in *G. senegalensis* and 7.26 ± 0.09 a. u. in *G. crassicaudaus*); Manfredi-Romanini et al.[38]

These findings provide specific support for a relatively close taxonomic relationship of the Galagidae to the Lorisidae, and at the same time open the field to extensive speculation about the adaptive advantages of such chromosomal polymorphism in the Galagidae. The fact that extensive polymorphism due to centric fusion also exists in the genus *Lemur* (see Table 3) with production of karyological subspecies, as has recently been underlined by Rumpler and Albignac,[46] enormously increases our interest in the adaptive advantage of these variations and in such a peculiar mechanism of speciation.

Another aspect of particular interest concerns the relationship of the living prosimians to other groups of primates, especially the South American monkeys. According to one classic theory, the South American monkeys originated from a group of now extinct North American prosimians (Omomyidae), representatives of which migrated from North to South America. However, recent re-evaluation of the theory of continental drift makes room for speculation about direct migration of some early stock of primates directly from Africa to South America before the end of the Eocene. This hypothesis was recently set out by Lavocat,[36] who conducted a

Table 3 Chromosome polymorphism in the genus *Lemur* (after Rumpler et al. 1971)

Species	*Chromosomes*						
	2n	*M*	*S*	*A*	*X*	*Y*	*FN*
L.m. macaco	44	12	8	22	A	A	64
L.f. fulvus	48	10	6	30	A	A	64
L.f. fulvus	58	—	4	52	A	A	62
L.f. fulvus	60	—	4	54	A	A	64
L.f. albifrons	60	—	4	54	A	A	64
L.f. sanfordi	60	—	4	54	A	A	64
L.f. collaris (barbe blanche)	52	8	4	38	M	A	66
L.f. collaris a (barbe rouge)	60	—	4	54	A	A	64
L.f. collaris b (barbe rouge)	48	8	8	30	A	—	64
L.f. collaris c (barbe rouge)	52	6	6	38	A	A	64
L.f. collaris a x *L.f. collaris* b	54	4	6	42	A	—	64
L.m. macaco x *L.f. rufus*	52	7	5	38	A	—	64
L.m. macaco x *L.f. albifrons*	52	7	5	38	A	A	64
L.m. macaco x *L.f. collaris* b	46	10	8	26	A	—	64
L.f. fulvus x *L.f. albifrons*	58	—	4	52	A	—	62
L.f. fulvus x *L.f. albifrons* (1)	60	—	4	54	A	A	64
L.f. fulvus x *L.f. rufus* (2)	60	—	4	54	A	—	64
Hybrid *F2* : (1) x (2)	60	—	4	54	A	—	64
Lemur variegatus	46	14	4	26	M	A	66
Lemur catta	56	2	6	46	M	A	66

comparative study of the African Miocene rodents and the South American caviomorphs and came to the conclusion that the close morphological relationship between these two groups can only be explained on the basis of such migration. A research programme with the aim of finding chromosomal evidence for a possible relationship between some group of living prosimians and South American monkeys is now being carried out in our Institute.

Apart from the general morphological detail of the chromosomes, another important parameter to take into consideration in developing hypotheses about phylogeny is, as has been mentioned above, the DNA content. Unfortunately only limited data are available so far (Manfredi-Romanini),[37] and the subject is mentioned here in the hope that this might lead to a cooperative effort in obtaining materials for such analysis.

NOTES

1 Arrighi, F.E., Sorenson, M.W. and Shirley, L.R. (1969), 'Chromosomes of the tree shrew (Tupaiidae)', *Cytogenetics* 8, 199-208.

2 Bender, M.A. and Chu, E.H.Y. (1963), 'The chromosomes of Primates', in

J. Buettner-Janusch (ed.), *Evolutionary and Genetic Biology of Primates* 1, 261-310, New York.

3 Bender, M.A. and Mettler, L.E. (1958), 'Chromosome studies of primates', *Science* 128, 186-90.

4 de Boer, L.E.M. (1972), 'The karyotype of *Galago alleni* Waterhouse 1837 (2N = 40) compared with that of *Galago senegalensis* Geoffroy 1796 (2N = 38) (Primates, Prosimii: Galagidae)', *Genetica* 43, 183-9.

5 de Boer, L.E.M. (1972), 'The karyotype of *Galago demidovii* Fischer 1808 (2N = 58) (Primates, Prosimii: Galagidae)', *Genen en Phaenen* 15, 19-22.

6 Borgaonkar, D.S. (1968), 'Additions to the lists of chromosomes in the Insectivores and Primates', *J. Hered.* 58, 211-13.

7 Borgaonkar, D.S. (1969), 'Insectivora cytogenetics', in Benirschke, K. (ed.), *Comparative Mammalian Cytogenetics*, New York.

8 Buettner-Janusch, J. (1963), 'An Introduction to the Primates', in Buettner-Janusch, J. (ed.), *Evolutionary and Genetic Biology of Primates*, vol. 1, New York.

9 Chiarelli, B. (1961), 'Ibridologia e sistematica in Primati. I: raccolta di dati', *Atti Assoc. Genetica Ital.* 6, 213-20.

10 Chiarelli, B. (1961), 'Some chromosome numbers in Primates', *Mammal. Chrom. Newsl.* 6, 3-5.

11 Chiarelli, B. (1963), 'Primi risultati di ricerche di genetica e cariologia comparata in Primati e loro interesse evolutivo', *Riv. Antropol.* 50, 87-124.

12 Chiarelli, B. (1966), 'Ibridologia e sistematica in Primati: deduzioni sul cariotipo degli ibridi', *Riv. Antropol.* 53, 113-17.

13 Chiarelli, B. (1972), 'New data on prosimian chromosomes', *J. Human Evol.* 1, 61-3.

14 Chiarelli, B. and L. Barberis (1966), 'Some data on the chromosomes of Prosimiae and of New World monkeys', *Mammal. Chrom. Newsl.* 22, 216.

15 Chiarelli, B. and Egozcue, J. (1968), 'The meiotic chromosomes of some Primates', *Mammal. Chrom. Newsl.* 9, 85-6.

16 Chu, E.H.Y. and Bender, M.A. (1961), 'Chromosome cytology and evolution in Primates', *Science* 133, 1399-405.

17 Chu, E.H.Y. and Bender, M.A. (1962), 'Cytogenetics and evolution of Primates', *Ann. N.Y. Acad. Sci.* 102, 253-66.

18 Chu, E.H.Y. and Swomley, B.A. (1961), 'Chromosomes of lemurine lemurs', *Science* 133, 1925-6.

19 Dodson, E.O. (1960), cited by Bender, M.A. and Chu, E.H.Y. (1963), 'The chromosomes of Primates', in Buettner-Janusch, J. (ed.), *Evolutionary and Genetic Biology of Primates*, vol. 1, New York.

20 Eckhardt, R.B. (1969), 'A chromosome arm number index and its application to the phylogeny and classification of lemurs', *Am. J. Phys. Anthrop.* 31, 85-8.

21 Egozcue, J. (1967), 'Chromosome variability in the Lemuridae', *Am. J. Phys. Anthrop.* 26, 341-8.

22 Egozcue, J. (1968), 'The meiotic chromosomes of the lesser bushbaby (*Galago senegalensis*)', *Mammal. Chrom. Newsl.* 9, 92-3.

23 Egozcue, J. (1969), 'Primates', in Benirschke, K. (ed.), *Comparative Mammalian Cytogenetics*, New York.

24 Egozcue, J. (1970), 'The chromosomes of the lesser bushbaby (*Galago senegalensis*) and the greater bushbaby (*Galago crassicaudatus*)', *Folia primat.* 12, 236-40.

25 Egozcue, J. (1972), 'Chromosomal evolution in prosimians', this volume.

26 Egozcue, J. (1972), 'The chromosomes of lemurs', *Folia primat.* 17, 171-6.

27 Egozcue, J., Chiarelli, B., Sarti-Chiarelli, M. and Hagemenas, F. (1968), 'Chromosome polymorphism in the tree shrew (*Tupaia glis*)'. *Folia primat.* 8, 150-8.

28 Egozcue, J., Ushijima, R.N. and Vilarasau de Egozcue, M. (1966), 'The chromosomes of the slender loris (*Loris tardigradus*)', *Mammal. Chrom. Newsl.* 22, 204.

29 Egozcue, J. and Vilarasau de Egozcue, M. (1966), 'The chromosomes of the angwantibo (*Arctocebus calabarensis*)', *Mammal. Chrom. Newsl.* 20, 53.

30 Egozcue, J. and Vilarasau de Egozcue, M. (1967), 'The chromosome complement of the slow loris (*Nycticebus coucang*)', *Primates* 7, 423-32.

31 Elliot, O.S. and Lisco, H. (1969), cited by Borgaonkar, D.S. (1968), in Benirschke, K. (ed.), *Comparative Mammalian Cytogenetics*, New York.

32 Elliot, O.S., Wong, M. and Borgaonkar, D.S. (1969), 'Karyological study of *Tupaia* from Thailand', *J. Hered.* 60, 153-7.

33 Hayata, I., Sonta, S., Itoh, M. and Kondo, N. (1971), 'Notes on the karyotype of some prosimians, *Lemur mongoz, Lemur catta, Nycticebus coucang* and *Galago crassicaudatus*', *Japan. J. Genet.* 1, 61-8.

34 Hsu, T.C. and Johnson, M.L. (1963), 'Karyotypes of two mammals from Malaya', *Amer. Nat.* 97, 127-9.

35 Klinger, H.P. (1963), 'The somatic chromosomes of some Primates: *Tupaia glis, Nycticebus coucang, Tarsius bancanus, Cercocebus aterimus, Symphalangus syndactylus*', *Cytogenetics* 2, 140-51.

36 Lavocat, R. (1969), 'La systématique des rongeurs hystricomorphes et la dérive des continents', *C.R. Acad. Sci.* (Paris) (Série D) 269, 1496-7.

37 Manfredi-Romanini, M.G. (1972), 'Nuclear DNA content and area of primate lymphocytes as a cytotaxonomical tool', *J. Human Evol.* 1, 23-40.

38 Manfredi-Romanini, M.G., de Boer, L.E.M., Chiarelli, B. and Tinozzi Massari, S. (1972), 'DNA content and cytotaxonomy of *Galago senegalensis* and *Galago crassicaudatus*', *J. Human Evol.* 1, 473-6.

39 Manna, G.K. and Talukdar, M. (1968), 'An analysis of the somatic chromosome complements of both sexes of two Primates, the slender loris, *Loris tardigradus*, and Rhesus monkey, *Macaca mulatta*', *Mammalia* 32, 118-24.

40 Matthey, R. (1955), 'Les chromosomes de *Galago senegalensis* Geoffroy (Prosimii – Lorisidae – Galaginae)', *Rév. Suisse Zool.* 62 (suppl.), 163-206.

41 Rumpler, Y. (1970), 'Etude cytogénétique de *Lemur catta*', *Cytogenetics* 9, 239.

42 Rumpler, Y., this volume.

43 Rumpler, Y. and Albignac, R. (1969), 'Etude cytogénétique de deux lémuriens, *Lemur macaco macaco*, Linné 1766, et *Lemur fulvus rufus* (Audebert 1800) et d'un hybride *macaco macaco/fulvus rufus*', *C.R. Soc. Biol.* 163, 1247-50.

44 Rumpler, Y. and Albignac, R. (1969), 'Etude cytogénétique de quelques hybrides intraspécifiques et interspécifiques de lémuriens', *Ann. Scien. Univ. Besancon* 6, 1-3.

45 Rumpler, Y. and Albignac, R. (1969), 'Existence d'une variabilité chromosomique intraspécifique chez certains lémuriens', *C.R. Soc. Biol.* 163, 1989-92.

46 Rumpler, Y. and Albignac, R. (1971), 'Etude cytogénétique des Cheirogaleidae', *IVe congrés international de Primatologie*, Beaverton, in press.

47 Rumpler, Y. and Albignac, R. (1972), 'Etude cytogénétique de *Varecia variegata* et de *Lemur rubriventer*', *C.R. Soc. Biol.* 165, 741-5.

48 Rumpler, Y. and Albignac, R. (1972), 'Cytogenetic study of the Indriidae', *Third Conf. on Exptl. Med. and Surgery in Primates*, Lyon, 21-23 June, in press.

49 Ying, K.L. and Butler, H. (1969), 'The somatic chromosomes of a lesser bushbaby (*Galago senegalensis*)', *Mammal. Chrom. News.* 10, 21.

50 Ying, K.L. and Butler, H. (1971), 'Chromosomal polymorphism in the lesser bushbabies (*Galago senegalensis*)', *Can. J. Genet. Cytol.* 13, 793-800.

M. GOODMAN, W. FARRIS Jr.,
W. MOORE, W. PRYCHODKO,
E. POULIK and M. SORENSON

Immunodiffusion systematics of the primates II: findings on Tarsius, *Lorisidae and Tupaiidae*

Introduction

The rabbit and chicken antisera used in the present study are listed in Table 1. These antisera were produced against plasma proteins from *Tarsius syrichta, Galago crassicaudatus, Nycticebus coucang, Loris tardigradus, Perodicticus potto, Tupaia chinensis*, and *Urogale everetti.*

Using these antisera, and taking the sera and plasmas from the species listed in Table 2 as test antigens, 800 trefoil Ouchterlony plate comparisons were carried out. A trefoil Ouchterlony plate consists of three wells surrounding a centre field of agar. Antiserum

Table 1 The antisera used in this study and the number of plate comparisons performed with each antiserum type

ANTISERUM					
Produced in	*Donor species*	*Antigenic preparation*	*No. of lines**	*No. of antisera*	*No. of plate comparisons*
Rabbit	*Tarsius syrichta*	plasma	8.3	1	56
Rabbit	*Galago crassicaudatus*	plasma	6.6	2	43
Rabbit	*Nycticebus coucang*	plasma	6.3	1	15
Rabbit	*Loris tardigradus*	plasma	6.8	3	160
Rabbit	*Urogale everetti*	plasma	6.9	3	171
Rabbit	*Tupaia chinensis*	plasma	7.4	2	202
Chicken	*Perodicticus potto*	plasma Fr 12	3.3	2	51
Chicken	*Perodicticus potto*	plasma Fr 14	3.1	2	42
Chicken	*Perodicticus potto*	plasma Fr 13	4.4	2	32
Chicken	*Perodicticus potto*	plasma	4.5	2	28

*Average number of precipitin lines which the antiserum developed in plate comparisons with the homologous antigenic preparation.

Table 2 Type and number of primate and other mammalian samples (individual or pooled sera and plasma) used as test antigens in this study

Scientific name	*No. of Samples*
HOMINOIDEA	
Homo sapiens	10
Pan troglodytes	3
Gorilla gorilla	2
Pongo pygmaeus	1
Hylobates lar	2
CERCOPITHECOIDEA	
Macaca mulatta	1
Macaca fuscata	6
Cercopithecus aethiops	1
Papio anubis	2
Papio hamadryas	1
Theropithecus gelada	1
Erythrocebus patas	1
Cercocebus torquatus	1
Presbytis eristatus	2
CEBOIDEA	
Ateles geoffroyi	1
Ateles panisus	1
Ateles sp.	1
Saimiri sp.	2
Saimiri sciureus	2
Cebus albifrons	1
Cebus sp.	2
Cacajao rubicundua	1
Aotes trivirgatus	3
TARSIOIDEA	
Tarsius syrichta	2
LORISOIDEA	
Galago crassicaudatus	6
Galago senegalensis	1
Galagoides demidovii	3
Nycticebus coucang	6
Loris tardigradus	3
Perodicticus potto	3
Arctocebus calabarensis	1
LEMUROIDEA	
Lemur fulvus	3
Lemur fulvus albifrons	1
Lemur catta	1
Lemur mongoz	2
TUPAIODEA	
Urogale everetti	1
Tupaia glis	4
Tupaia chinensis	3
Tupaia longipes	1
Tupaia montana	1
Tupaia minor	1
Tupaia palawanensis	1

NON-PRIMATES	
Rhynchocyon sp.	1
Petrodromus sultan	3
Nasilio brachyrhynchus	1
Atelerix sp.	1
Hemiechinus auritus	1
Tenrec ecaudatus	1
Scapanas aquaticus	1
Sorex cinereus	1
Eptesicus sp.	1
Pipistrellus hesperus	1
Dasypus novemcinctus	3
Bradypus tridactylus	1
Myrmecophaga tridactyla	1
Manis pentadactyla	1
Bos taurus	3
Loxodonta africana	1
Canis familiaris	2
Potos flavus	2
Ursus arctos	1
Eumetopias jabatus	1
Rattus sp.	1
Cavia sp.	1
Citellus mexicanus	1
Suncus murinus	1
Elephantulus myurus	1

is put into the bottom well and antigen preparations from two different species are placed in each of the two top wells. The reactants diffuse towards each other in the limited area of agar, producing precipitin lines. If antigen from the species against which the antiserum was produced (the homologous species) is in one of the top wells and antigen from a heterologous species in the other top well, the precipitin line of the homologous reaction will extend beyond the precipitin line of the heterologous reaction, provided that the two species do not share all of the antigenic sites against which there are anti-bodies. This extension is called a spur. The fewer the antigenic sites shared by the two species, the longer the spur. When one heterologous species is compared to another, the one sharing more antigenic sites with the homologous species yields the larger net spur. Spurs are numbered according to length and intensity, on a scale from *one*, which denotes a reaction of identity when antigen from both the right and left top wells show precipitin lines, to *five*, the largest spur possible. Photographs of several species comparisons (tarsier vs. man and man vs. galago) developed by rabbit antiserum to tarsier serum in 'Y' trefoil Ouchterlony plates are shown in Fig. 1, with the net spur sizes recorded for the man vs. galago comparison given in the legend to Fig. 1. The spur size results from the 800 plate comparisons gathered in the present study were recorded on IBM coding forms and then converted by a computer

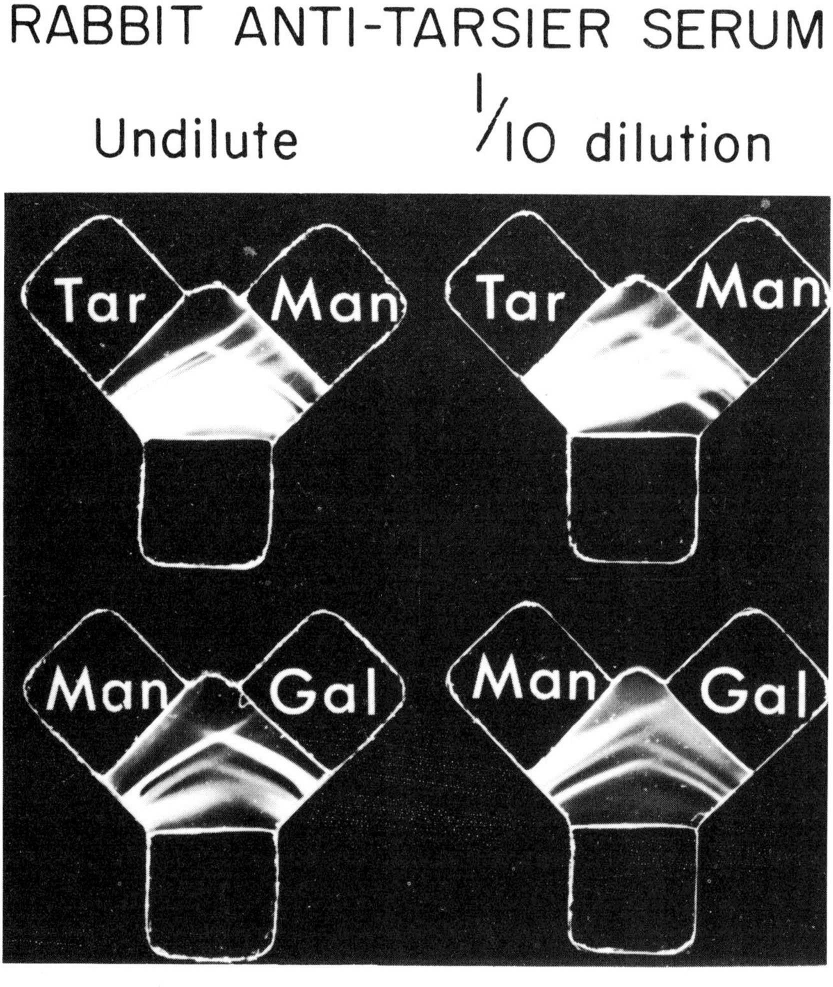

Figure 1. In the top two plates, the homologous antigen, tarsier serum, developed 8 distinct precipitin lines, each of which yielded strong intense spurs against the heterologous antigen, human serum. In the bottom two plates, in which there is comparison of two heterologous antigens, human and galago sera, the crossreacting serum proteins (diffusing into the agar from each antigen well) developed about three precipitin lines, with the human proteins developing positive net spurs against the galago proteins. In this heterologous to heterologous comparison, recording of the first two precipitin lines from the plate with 1/10 dilutions of human and galago serum, and of the third precipitin line from the plate with undilute human and galago serum, yielded the following net spur sizes:

	Human - Galago	*Net spur size*
Line 1	2 – 1	1
Line 2	3 – 1	2
Line 3	4 – 3	1

program IMMDF into the Taxonomic Distance Tables shown in Tables 3-12. The methods for producing the antisera, carrying out the trefoil Ouchterlony plate comparisons, and converting the plate results into Taxonomic Distance Tables are thoroughly described in a previous paper.[1]

These taxonomic distance data (Tables 3-12) provide evidence for the following conclusions about the cladistic relationships of different prosimian lineages.

Table 3 Taxonomic distance table: rabbit anti-*Tarsius syrichta* plasma

Species	*Antigenic distance*
Tarsius syrichta	0.000
Presbytis cristatus	4.128
Theropithecus gelada	4.188
Erythrocebus patas	4.249
Pan troglodytes	4.249
Gorilla gorilla	4.249
Pongo pygmaeus	4.249
Cercocebus torquatus	4.249
Hylobates lar	4.249
Cercopithecus aethiops	4.249
Papio hamadryas	4.249
Macaca fuscata; M. mulatta	4.533
Homo sapiens	4.650
Saimiri sp.	4.896
Aotes douroucouli	4.992
Cebus albifrons; C. sp.	5.037
Ateles sp.	5.158
Galago crassicaudatus	5.320
Tupaia chinensis	5.364
Cacajao rubicundua	5.400
Lemur fulvus	5.485
Perodicticus potto	5.509
Loris tardigradus	5.616
Nycticebus coucang	5.673
Citellus mexicanus	6.169
Dasypus novemcinctus	6.169
Bos taurus	6.411
Bradypus tridactylus	6.775
Petrodromus sultan	6.775
Atelerix sp.	6.956

Table 4 Taxonomic distance table: rabbit anti-*Galago crassicaudatus* plasma

Species	*Antigenic distance*
Galago crassicaudatus	0.000
Galago senegalensis	0.114
Galagoides demidovii	0.275
Nycticebus coucang	2.242
Perodicticus potto	2.400

Arctocebus calabarensis	2.626
Loris tardigradus	2.926
Lemur mongoz	6.236
Homo sapiens	6.800
Saimiri sciureus	6.951
Tupaia glis	7.515

Table 5 Taxonomic distance table: rabbit anti-*Nycticebus coucang* plasma

Species	*Antigenic distance*
Nycticebus coucang	0.000
Loris tardigradus	1.395
Galago crassicaudatus	2.727
Galago senegalensis	2.864
Arctocebus calabarensis	3.250
Perodicticus potto	3.422
Lemur mongoz	7.685
Homo sapiens	8.021
Saimiri sciureus	8.100
Tupaia glis	8.830

Table 6 Taxonomic distance table: rabbit anti-*Loris tardigradus* plasma

Species	*Antigenic distance*
Loris tardigradus	0.000
Nycticebus coucang	1.430
Galagoides demidovii	1.930
Galago crassicaudatus	1.996
Galago senegalensis	2.186
Perodicticus potto	2.441
Lemur fulvus	4.295
Lemur catta	4.576
Lemur sp.	4.743
Lemur mongoz	4.764
Ateles panisus	4.839
Homo sapiens	4.978
Macaca fuscata	5.027
Ateles geoffroyi	5.186
Pipistrellus hesperus	5.270
Gorilla gorilla	5.591
Papio anubis	5.594
Erythrocebus patas	5.627
Macaca fuscata	5.700
Pan troglodytes	5.810
Hylobates lar	5.957
Tupaia chinensis	6.156
Presbytis cristatus	6.206
Potos flavus	6.303
Aotes trivirgatus	6.350
Citellus mexicanus	6.505
Bos taurus	6.509
Atelerix sp.	6.895

Bradypus tridactylus	6.919
Dasypus novemcinctus	7.059

Table 7 Taxonomic distance table: rabbit anti-*Urogale everetti* plasma

Species	*Antigenic distance*
Urogale everetti	0.000
Tupaia minor	1.302
Tupaia longipes	1.392
Tupaia glis	1.399
Tupaia montana	1.434
Tupaia chinensis	1.441
Tupaia palawanensis	1.611
Lemur fulvus	3.453
Galago crassicaudatus	3.690
Loris tardigradus	3.690
Nycticebus coucang	3.987
Macaca fuscata	3.987
Potos flavus	4.038
Homo sapiens	4.058
Dasypus novemcinctus	4.080
Canis familiaris	4.228
Aotes trivirgatus	4.279
Petrodromus sultan	4.287
Tenrec ecaudatus	4.342
Perodicticus potto	4.352
Cavia sp.	4.494
Atelerix sp.	3.579
Scapanas aquaticus	4.652
Elephantulus myurus	4.895
Eptesicus sp.	4.980

Table 8 Taxonomic distance table: rabbit anti-*Tupaia chinensis* plasma

Species	*Antigenic distance*
Tupaia chinensis	0.000
Tupaia longipes	0.406
Tupaia glis	0.439
Tupaia minor	0.666
Tupaia montana	0.717
Tupaia palawanensis	1.141
Urogale everetti	1.331
Tarsius syrichta	5.882
Macaca fuscata fuscata	6.108
Lemur fulvus	6.114
Galago crassicaudatus	6.128
Aotes trivirgatus	6.222
Lemur fulvus albifrons	6.336
Bradypus tridactylus	6.368
Homo sapiens	6.369
Macaca fuscata	6.408
Loxodonta africana	6.423

Perodicticus potto	6.538
Ursus arctos	6.629
Potos flavus	6.726
Nycticebus coucang	6.819
Manis pentadactyla	6.901
Canis familiaris	6.961
Eumetopias jabatus	7.105
Dasypus novemcinctus	7.161
Petrodromus sultan	7.244
Rattus sp.	7.309
Tenrec ecaudatus	7.377
Cavia sp.	7.377
Rhynchocyon sp.	7.513
Myrmecophaga tridactyla	7.581
Nasilio brachyrhynchus	7.784
Scapanas aquaticus	7.852
Atelerix sp.	7.852
Eptesicus sp.	7.920
Suncus murinus	7.920
Sorex cinereus	8.192

Table 9 Taxonomic distance table: chicken anti-*Perodicticus potto* plasma Fr 12

Species	*Antigenic distance*
Perodicticus potto	0.000
Galago crassicaudatus	2.108
Arctocebus calabarensis	2.367
Nycticebus coucang	3.158
Loris tardigradus	3.275
Galago senegalensis	3.492
Lemur mongoz	5.158
Tupaia glis	5.758
Macaca irus	5.833
Cebus apella	5.908
Tarsius syrichta	6.058
Homo sapiens	6.583

Table 10 Taxonomic distance table: chicken anti-*Perodicticus potto* plasma Fr 14

Species	*Antigenic distance*
Perodicticus potto	0.000
Arctocebus calabarensis	0.995
Loris tardigradus	1.676
Nycticebus coucang	1.774
Galago crassicaudatus	1.891
Galago senegalensis	2.052
Lemur mongoz	4.242
Tarsius syrichta	4.306
Macaca irus	4.375
Homo sapiens	4.415
Cebus apella	5.210

Table 11 Taxonomic distance table: chicken anti-*Perodicticus potto* plasma Fr 13

Species	*Antigenic distance*
Perodicticus potto	0.000
Arctocebus calabarensis	1.250
Galago crassicaudatus	2.235
Galago senegalensis	2.500
Nycticebus coucang	2.500
Loris tardigradus	2.538
Lemur mongoz	5.346
Tarsius syrichta	5.790
Cebus apella	5.255
Macaca irus	6.385
Homo sapiens	6.645
Macaca mulatta	6.710

Table 12 Taxonomic distance table: chicken anti-*Perodicticus potto* plasma

Species	*Antigenic distance*
Perodicticus potto	0.000
Arctocebus calabarensis	1.704
Nycticebus coucang	2.574
Galago crassicaudatus	2.981
Galago senegalensis	2.981
Loris tardigradus	3.093
Lemur mongoz	5.241
Tupaia glis	6.796

Major ancestral branchings

At the base of the Primates, the tree shrew separates, or has already been separated, from the line to lemuroids, lorisoids, *Tarsius*, and Anthropoidea. (The question of whether or not the tree shrews should be treated as Primates is left unanswered by our present data.) In the early Primates, four major lines branch apart, separating the ancestors of lemuroids, lorisoids, *Tarsius* and Anthropoidea, with the lorisoids and lemuroids sharing a slightly more recent common ancestor with each other, and, similarly, *Tarsius* and Anthropoidea sharing a slightly more recent common ancestor before branching apart. (Taxonomic distance tables for anti-ceboid sera reveal the Ceboidea to be closer to catarrhines than to any living prosimian group.)

Subsequent branchings

1. Lorisidae

Three lines separate at the base of the family, one leading to *Galago* and *Galagoides*, another to *Nycticebus* and *Loris*, and the third to *Perodicticus* and *Arctocebus*. The *Nycticebus, Loris* and *Galago* lines may have slightly closer cladistic relationships with each other than with the *Perodicticus/Arctocebus* line. However, the latter line may simply have diverged more from the ancestral state than the former, thus accounting for the slightly smaller antigenic distances between the former lines. The splitting between *Galago* and *Galagoides* appears to be much more recent than either the *Nycticebus-Loris* split or the *Perodicticus/Arctocebus* split.

2. Tupaiidae

After the ancestral branching apart of *Urogale* and *Tupaia, T. palawanensis* appears to separate from the remaining *Tupaia* complex, then the line or lines to *T. montana* and *T. minor*, and finally quite small divergences separate *T. glis* and *T. longipes* from *T. chinensis*. Additional antisera to either *T. minor* or *T. montana* are needed to determine whether these two species represent a distinct monophyletic group within the genus *Tupaia*.

Acknowledgments

This research was supported by NSF grant GB7420. We thank Francis Vincent and Alan Walker for contributing blood samples from *Galagoides demidovii* and also John Buettner-Janusch and William Montagna for contributing blood samples from several lorisoids and lemuroids. The help of Mary Etta Hight in obtaining samples is also appreciated.

NOTES

1 Goodman, M. and Moore, G.W. (1971), 'Immunodiffusion systematics of the Primates. I: the Catarrhini', *Syst. Zool.* 20, 19-62.

N. A. BARNICOT and
D. HEWETT-EMMETT

Electrophoretic studies on prosimian blood proteins

Introduction

A simple way of comparing proteins from different species or individuals is to study their migration under the influence of an electric potential. Filter paper, cellulose acetate paper or agar gel can be used as supporting media, but better resolution is usually obtained with fine-pored media such as starch or acrilamide gels. If the protein is sufficiently concentrated its position in the gel at the end of the run can be revealed by applying some non-specific dye; but many proteins, notably enzymes, have to be located by special staining methods that depend on their characteristic activities. Many enzymes from blood or other tissues show more than one band when examined by standard methods. In some cases these *isoenzyme* bands are due to the presence of more than one form of the enzyme, each controlled by a different genetic locus, and sometimes to the presence of their interaction products as well. In other cases the reasons are less clear.

It is obvious that these electrophoretic procedures give much less information about the structure of the protein than do amino acid sequence analyses. As a rule, only amino acid changes that result in alteration of the electric charge of the molecule will be detected. Identical mobilities in the medium by no means prove identity of structure and differences of mobility may be due to a change in one amino acid or may be the resultant of many differences. Despite these limitations, the method does allow many specimens of many different proteins to be examined quite rapidly, and in interspecies comparisons regularities of taxonomic interest often emerge (see Tashian;[1] Barnicot and Cohen;[2] Barnicot and Hewett-Emmett).[3]

Electrophoresis is also very useful in detecting inherited variations of a given protein within a species; that is, biochemical polymorphism. It has been found that in man and a few other species that have been closely studied, about one third of the proteins examined are polymorphic. It is known that most of the numerous haemoglobin variants that have been described in man are due to mutations producing a change in a single amino acid in one or other of the two chains composing the molecule. The problem of why there is such widespread protein polymorphism has been touched on in the introduction to this section of the symposium.

We have been able to examine certain red cell and serum proteins in a fair range of prosimian species (Table 1) and we are very grateful to the workers, listed at the end of this paper, who have been good enough to send us material. In this paper we wish to draw attention to some of the main findings; a detailed report will be published elsewhere.

Table 1 Material examined

	Individual samples
Tarsius bancanus	1
Microcebus murinus	7
Cheirogaleus major	2
Lemur catta	5
Lemur mongoz	6 (parents and 4 offspring)
Lemur macaco	3 (parents and 1 offspring)
Hapalemur griseus	9 (includes 2 parents and 1 offspri
Galago senegalensis	8
Galago crassicaudatus	19
Galago demidovii	7
Galago alleni	2
Euoticus elegantulus	2
Arctocebus calabarensis	2 (one examined for Hb only)
Perodicticus potto	1
Nycticebus coucang	3
Tupaia tana	3
Tupaia belangeri	9 (4 of them sibs)
Tupaia minor	9

Since we are concerned with patterns of bands, varying in intensity and position, it would no doubt be possible to make more precise numerical descriptions and to combine the results from different proteins to give measures of overall resemblance by the multivariate methods that are now so much in vogue. So far we have not felt that the added labour would be worthwhile.

A key to the lettering of the diagrams is given in the caption to Fig. 1.

1. Haemoglobins (Hb)

In Fig. 1 the haemoglobins of prosimians, roughly arranged in taxonomic groups, are compared with the common human adult phenotype Hb-A. We did not detect any clear examples of haemoglobin polymorphism. It will be noted that the major Hb of *Tupaia* (three species) is conspicuously slow (i.e. cathodal or nearer the negative pole). This confirms a previous report on *T. glis.*[4] The major Hbs of both *Microcebus* and *Cheirogaleus* are somewhat anodal to those of the other Madagascan species examined, in agreement with Buettner-Janusch et al.[5] These authors also found that all specimens of *Lemur catta* and *Hapalemur griseus* had two haemoglobins in roughly equal concentrations, whereas *L. mongoz* and *L. macaco* had only one;[4,5] this was confirmed by our investigations.

The presence of two major haemoglobins in all specimens of *Galago senegalensis* and *G. crassicaudatus* has been reported by several workers.[4-6] We found this to be so in these two species, but only one major Hb was present in *G. demidovii* (N = 7) and *Euoticus elegantulus* (N = 2). The results for *G. alleni* are uncertain, since the specimens were very dilute. Nute et al.[7] reported two major Hbs in *G. demidovii* and it seems possible that there may be regional variation in this species. The presence of two major haemoglobins in

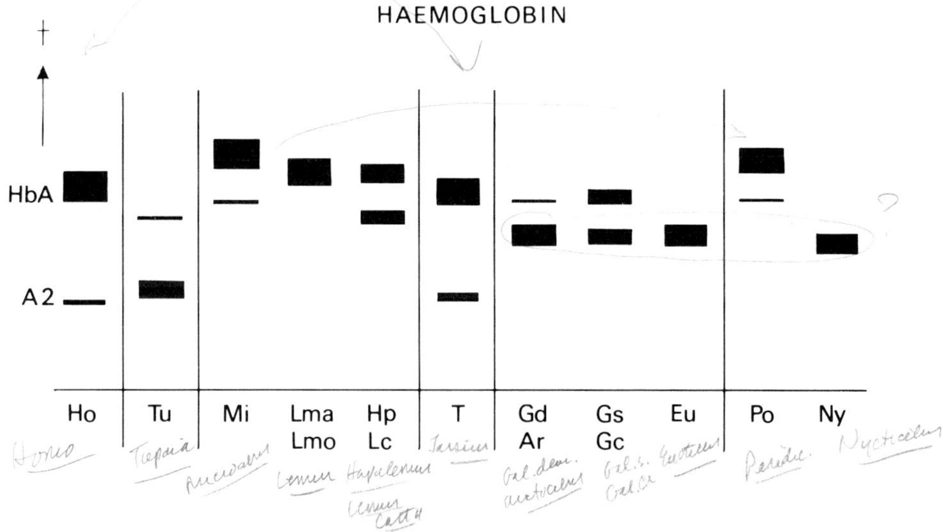

Figure 1. *Haemoglobins* (pH 8.6 tris-EDTA-borate buffer system).
Arrow indicates direction of the anode.
In this and subsequent figures: HO = human
Tu = *Tupaia*, 3 species Mi = *Microcebus*
Lma = *Lemur macaco* Lmo = *Lemur mongoz*
Lc = *Lemur catta* Hp = Hapalemur T = *Tarsius*
Gd = *Galago demidovii* Gs = *G. senegalensis*
Gc = *G. crassicaudatus* Eu = *Euoticus* Ar = *Arctocebus*
Po = *Perodicticus* Ny = *Nycticebus*
The horizontal line at the bottom in this and subsequent figures indicates starting point of run.

virtually every individual of certain prosimian species is perhaps best explained by postulating gene duplications at very high frequencies; similar situations are known in certain anthropoids (see Sullivan).[8]

A minor Hb component Hb-A_2 is shown in the diagram (Fig. 1) of a human specimen. This contains a chain (δ) similar to the β-chain and probably derived from it by duplication. A similar and probably homologous minor component occurs in pongids and gibbons. The only prosimian in which we found a minor Hb of comparable mobility to Hb-A_2 was *Tarsius*, but its concentration was considerably higher than that of Hb-A_2. The chain composition is not yet known and there is therefore no evidence as to whether it contains a homologue of the δ-chain.

A conspicuous minor component slightly cathodal to the major one is also seen in *Microcebus*, but its strength seems to vary in different individuals. Weak bands anodal to the major Hb, or in roughly the position of Hb-A_2, are sometimes seen in certain other prosimian species after o-tolidine/peroxide staining. It may be that some of these components are polymers or other secondary products of the major Hb, but this needs further study.

Buettner-Janusch and Twichell showed that in some prosimians (and also in certain other mammals) the haemoglobin is more resistant to denaturation by alkali than is human Hb-A.[9] We also found that *Galago crassicaudatus* Hb shows marked alkali resistance and that the two bands are not differentially affected. Resistance was also high for *Microcebus* and *Cheirogaleus* Hbs, but did not differ appreciably from human Hb-A in *Tarsius* and was only slightly raised in tree-shrews.

2. *Phosphoglucomutase (PGM)*

This red cell enzyme shows several isoenzyme bands. A common human type (PGM2) is shown on the left in Fig. 2. It is know that in man the more cathodal pair of bands and the anodal set are controlled by different gene loci.

The tree-shrews, represented in the diagram by *T. belangeri*, have very strong anodal bands. There is a clear polymorphism of these bands in this species; one of nine animals showed a four-banded pattern interpretable as a heterozygote for the two types shown in Fig. 2.

There was also polymorphism of the more cathodal bands in *Microcebus* (see Fig. 2) and probably in *Cheirogaleus*. The patterns in these two species are rather distinct from those found in the lemurine material. In some specimens of *Lemur mongoz* and *L. macaco* faint bands in a very anodal position were sometimes, but not consistently, detected.

The pattern in *Tarsius* is unusual and difficult to interpret. Since

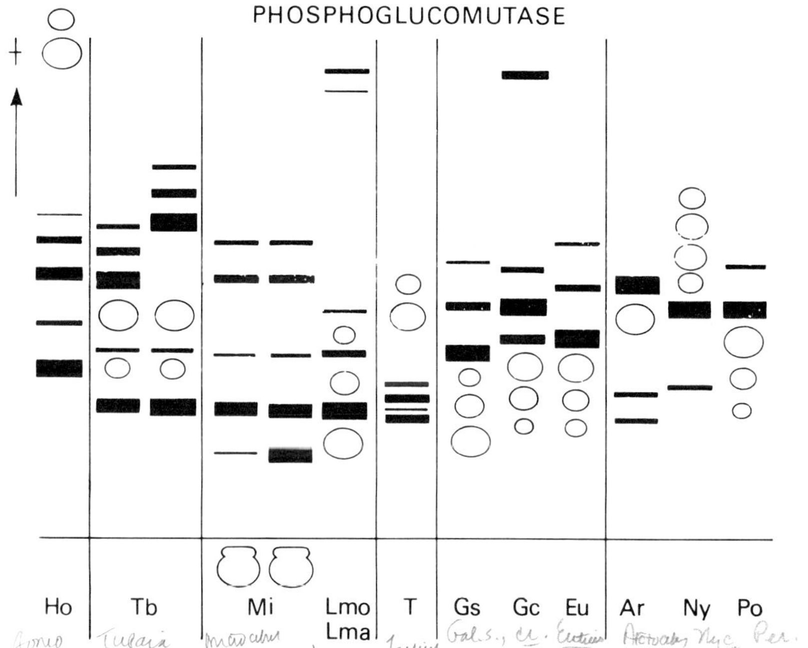

Figure 2. *Phosphoglucomutase* (black) and achromatic bands (white). Tb = *Tupaia belangeri.* Two phenotypes of *Microcebus* and *T. belangeri* are shown.

the specimen was taken some time after death the possibility of deterioration cannot be excluded.

It is not clear whether the set of bands in the galagines corresponds to the anodal set or the cathodal ones found in other species. Even very strong staining failed to show more cathodal bands, but the presence of achromatic bands (see below) in this region may interfere; this may also be true for *Perodicticus*, though further study is needed. The pattern shown for *G. crassicaudatus* suggests a heterozygous type, but it was seen in all the specimens. A very anodal band, which may sometimes be double, is consistently found in this species and faint bands rather less anodally situated were detected in some *G. senegalensis* specimens.

3. Achromatic bands

Fig. 2 also shows (white circles) the locations of an oxidase which prevents darkening of the gel background by the tetrazolium stain and therefore shows up as an unstained area. Wide variations in the mobility of this enzyme are seen in higher primates.[2,3] It will be noted that: (1) the cheirogaleines have more cathodal bands than the lemurines;[15] (2) the gradient of band strengths in galagines (except *G. senegalensis*) is the inverse of that in the lemurines; (3) the bands are rather more anodal in the lorisines than in the galagines, notably in *Nycticebus*. The pattern in this latter species suggests heterozygosity, but it was observed in all six animals.

4. Adenylate kinase (AK)

In the buffer system described by Fildes and Harris,[10] human homozygous types show a series of bands decreasing in strength anodally. The prosimian patterns are similar to this (Fig. 3), though in some species the set of bands is much more anodal and in some there are faint cathodal bands in addition. We may note again a clear difference between the two cheirogaleines and the lemurines. Clear examples of polymorphism were seen in *Galago senegalensis* and *G. crassicaudatus* (Gs, Gc in Fig. 3).

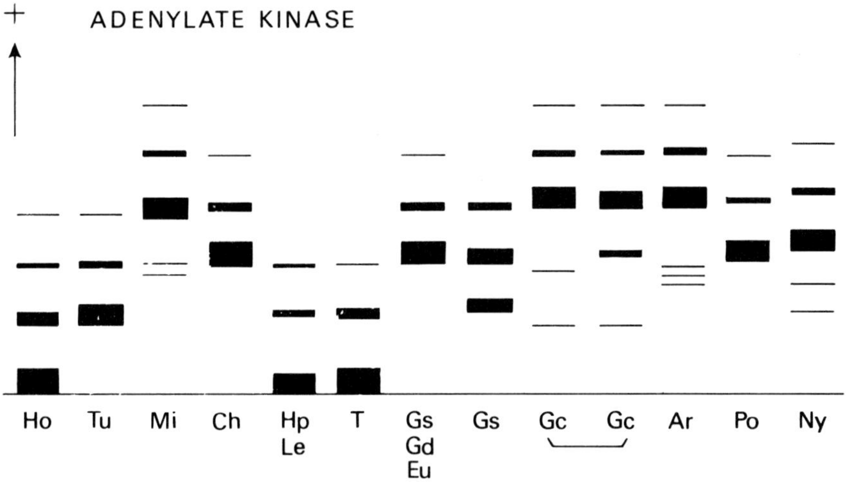

Figure 3. *Adenylate kinase.* Ch = *Cheirogaleus*, Le = *Lemur* (3 species). Note that two phenotypes are shown both for *Galago senegalensis* (Gs) and for *G. crassicaudatus* (Gc).

5. Acid phosphatase (AP)

The patterns show wide interspecific, and in some cases intraspecific, variations. Since many of the prosimian patterns lie well outside the range of the human control, cross-comparison of different gels is inexact and we show only (Figs. 4 and 5) comparisons based on individual gels.

All the tree-shrews showed a rather characteristic pattern of two widely separated bands. The more anodal band was clearly seen even in very fresh material and is therefore unlikely to be due to deterioration.

The bands in *Microcebus* and *Cheirogaleus* were substantially cathodal to those in the other Madagascan forms. Due to the small

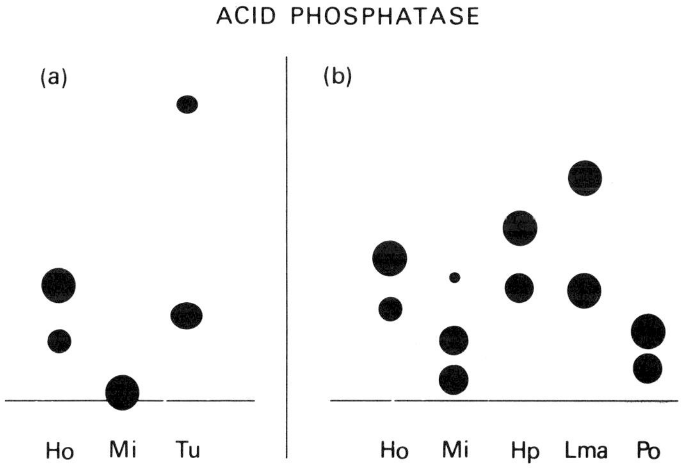

Figure 4. *Acid phosphatase.* Diagram of comparisons on two different gels (a) and (b). Gel (b) was run at increased voltage with cooling plates. The human control is an AB type.

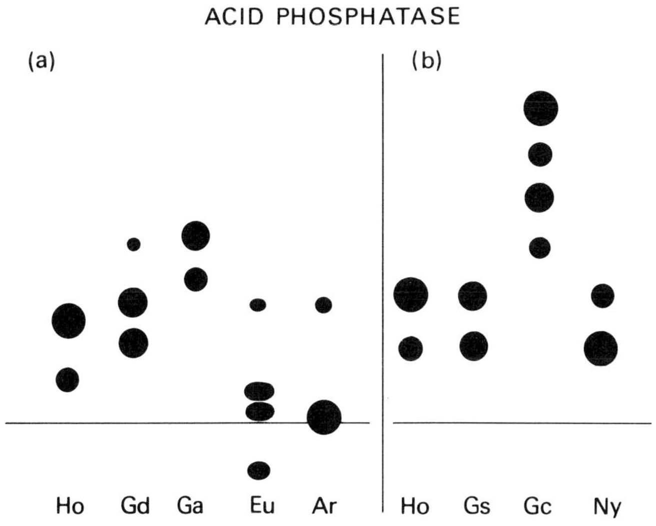

Figure 5. *Acid phosphatase.* Comparisons on two different gels (a and b) as in Fig. 4. The phenotype shown in 5b for *Galago crassicaudatus* is a heterozygous type.

migration distance, only one band is usually seen in the former; but it can be resolved into two by higher voltage runs using cooling-plates (Fig. 4*b*).

There is considerable variation between galagine species (Fig. 5*a*). Clear polymorphism was found in *G. crassicaudatus* (Fig. 5*b*), the types probably corresponding to those described in families of this species by Buettner-Janusch and Wiggins.[11]

6. *Malate dehydrogenase (MDH: cytoplasmic form)*

Tariverdian, Ritter and Schmitt have shown that this enzyme is more anodal in prosimians than in anthropoids.[12] We have examined some half dozen prosimian species (Fig. 6) and our results confirm this. It may be significant that the bands were relatively anodal in *Nycticebus* and *Perodicticus* and also in the single galagine examined so far.

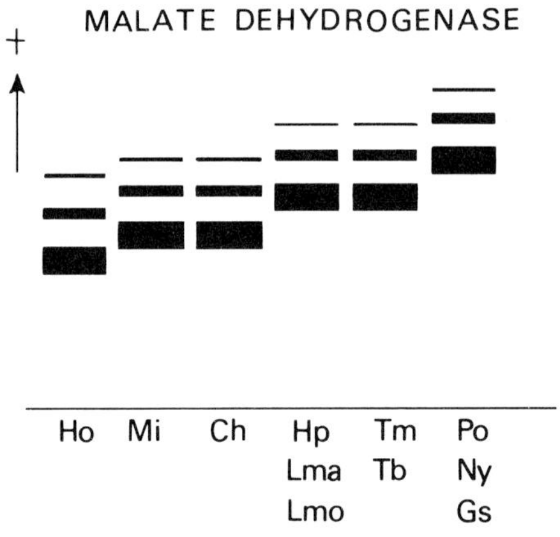

Figure 6. *Malate dehydrogenase.*

7. *Phosphohexose isomerase (PHI)*

This is an interesting enzyme since, as Tariverdian et al. have shown, there is a mobility gradient with hominoids at one end, prosimians at the other and monkeys in between.[13] However, we do not advocate a return to the concept of the Great Chain of Being on this account. We have not completed our studies, but it is already evident that there are wide mobility differences among prosimians. *Microcebus* and *Tupaia* show relatively cathodal bands and *Nycticebus* is at the opposite extreme. Mobility differences are also seen between

Hapalemur and *Lemur catta*, and between *Galago senegalensis* and *G. crassicaudatus.*

8. Serum proteins

(*a*) *Albumin:* The albumins of most of the prosimians are anodal to human albumin at pH 8.6, *Tarsius* being the most notable exception. There were some interspecies differences in the lemurines and tree shrews. An interesting feature is the presence of two albumin components of unequal strengths in all the galagines (*G. alleni* was excluded since the serum was in poor condition) and also in *Arctocebus* (Fig. 7). Simply inherited variations of albumin are known in man and various other mammals; but heterozygotes for such variants generally have two bands of roughly equal strength. Some experiments on the serum of one individual of *Galago crassicaudatus* showed that: (*a*) the albumin was single in runs on agar at pH 8.6, suggesting that the components do not differ in charge, (*b*) that a single band is also obtained if the pH of starch gels is lowered to 7.5, (*c*) that, at this pH, addition to the serum of mercaptoethanol (a disulphide bond-splitting reagent) results in two bands even at this lower pH, while at pH 8.6 the slower band is enhanced. Possibly the albumin in these species tends to undergo a conformational change, or to bind some relatively large molecule in the higher pH range; but the explanation is still unclear.

(*b*) *Transferrin* (Tf): The position of the iron-transport protein, transferrin, is also shown in Fig. 7. Some species show polymorphism; for simplicity we have illustrated only one phenotype in such cases, choosing a presumed heterozygote to show the mobility range of the protein.

The tree shrews and all the prosimians have transferrins anodal to the common human type TfC, and the lorisids tend to be most extreme in this respect.

Marked polymorphic variation was found in the lemurines, in agreement with Nute and Buettner-Janusch,[14] each of the four species showing three different transferrins. Three transferrins were also seen in our small sample of *Microcebus murinus.* There was clear polymorphism in *Tupaia belangeri, T. minor, Nycticebus* and *Galago demidovii*; but in these cases the phenotypes were less easy to interpret due to the presence of minor bands, which are clearly seen after autoradiography with radioactive iron (^{59}Fe). This complication was noted by other authors in *Lemur* species and was attributed to variations in the numbers of sialic acid residues attached to the protein.[14] Our

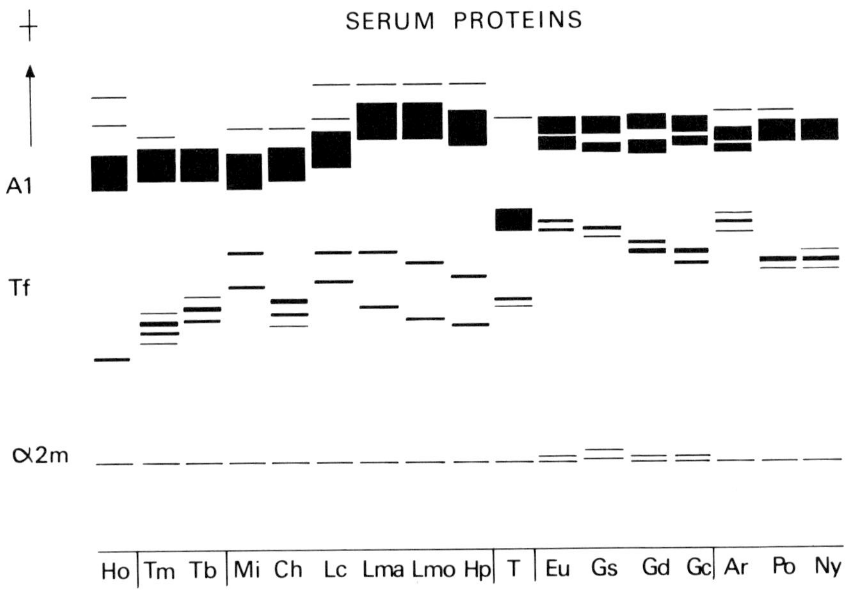

Figure 7. *Serum proteins*
Al = albumin (heavy black areas)
Tf = transferrin
α^2 m = α^2 macroglobulin
Thin lines anodal to albumin are prealbumin bands.
Tm = *Tupaia minor.*

own lemurine material took up the isotope only weakly, so that minor bands were not seen. It may be noted, however, that minor bands were not seen in *Microcebus* specimens, some of which took up ^{59}Fe strongly.

Discussion

Bearing in mind the methodological limitations mentioned in the introduction and the dangers of characterising a species from small samples, the following points are worth making.

The tree shrews, represented by three *Tupaia* species, differ from the prosimians most conspicuously in their Hb and AP patterns.

The two cheirogaleine species resemble one another quite closely but differ from the lemurines with regard to many proteins, notably in their Hb, AP, AK, achromatic band and PGM patterns. The comparison should be extended by the examination of the Indriidae.

The resemblances between *Tarsius* and man in Hb and AK patterns is interesting in view of morphological, and especially embryological, evidence of relationship between tarsioids and anthropoids. However these similarities are offset by marked differences in albumin and PGM. Since we had only one specimen of *Tarsius*, which had been dead for some hours before the blood was taken, the results clearly need confirmation.[16] At this level of analysis it would not be particularly surprising to find phenotypic resemblances for a few proteins even in forms that are in fact only distantly related.

Perhaps the most striking and consistent features of the galagines are the double albumin and transferrin bands. However, the former occurs in *Arctocebus*, and multiple transferrin bands, though somewhat different in pattern, are found in lorisines, Madagascan lemurs and tree shrews.

No useful generalisations can be made about polymorphism from such small samples, except that it certainly exists in several of the species. Variations of AK, AP, PGM and Tf were seen in some of the forms examined and our data confirm the impression from earlier work that transferrin may be very polymorphic in some species, especially lemurines. It is hoped that such variations may be put to good use in studies on the population genetics of prosimians.

Acknowledgments

We are very grateful to the following for providing us with material:

R. Albignac	*Hapalemur, L. catta*
P. Charles-Dominique	*G. alleni, G. demidovii, Euoticus, Arctocebus*
F.C. Jones and F.G. Garcia	*G. crassicaudatus, G. senegalensis, Nycticebus*
Ilaar Muul and Lim Boo Liat	*Nycticebus*
R.D. Martin	Tree-shrews, *Microcebus, Cheirogaleus, G. demidovii*
J.-J. Petter	*L. macaco, L. mongoz*
A. Petter-Rousseaux	*Microcebus, Perodicticus*
J. Rollinson	*Tarsius*
P.V. Tobias	*G. senegalensis*
D.A. Valerio	*G. crassicaudatus*
A. Voller	*G. crassicaudatus*

We wish also to acknowledge a grant from the Medical Research Council in support of this work, and to express our thanks for the technical assistance of Carol O'Reilly and Jackie Hillier.

NOTES

1 Tashian, R.E. (1965), 'Genetic variation and evolution of the carboxylic esterases and carbonic anhydrases of primate erythrocytes', *Am. J. Hum. Genet.* 17, 257-72.

2 Barnicot, N.A. and Cohen, P. (1970), 'Red cell enzymes of primates (Anthropoidea)', *Biochem. Genet.* 4, 41-57.

3 Barnicot, N.A. and Hewett-Emmett, D., (1971), 'Red cell and serum proteins of Talapoin, Patas and Vervet monkeys', *Folia primat.* 15, 65-76.

4 Buettner-Janusch, J. and Buettner-Janusch, V. (1964), 'Haemoglobins in Primates', in Buettner-Janusch, J. (ed.), *Evolutionary and Genetic Biology of Primates*, New York.

5 Buettner-Janusch, J., Washington, J.C. and Buettner-Janusch, V. (1971), 'Haemoglobins of Lemuriformes', *Arch. Inst. Pasteur Madagascar* 40, 127-36.

6 Hoffman, H.A. and Gottlieb, A.J. (1969), 'Investigations of non-human primate electrophoretic variation', *Ann. N.Y. Acad. Sci.* 162, 205.

7 Nute, P.E., Buettner-Janusch, V. and Buettner-Janusch, J. (1969), 'Genetic and biochemical studies of transferrins and haemoglobins of *Galago*', *Folia primat.* 10, 276.

8 Sullivan, B. (1971), 'Comparison of the haemoglobins in non-human Primates and their importance in the study of human haemoglobins', in Chiarelli, A.B. (ed.), *Comparative Genetics in Monkeys, Apes and Man*, London and New York.

9 Buettner-Janusch, J. and Twichell, J.B. (1961), 'Alkali resistant haemoglobins in prosimian primates', *Nature* 192, 669.

10 Fildes, R.A. and Harris, H. (1966), 'Genetically determined variation of adenylate kinase in man', *Nature* 209, 261.

11 Buettner-Janusch, J. and Wiggins, R.O. (1970), 'Haptoglobins and acid phosphatases of *Galago*', *Folia primat.* 13, 166.

12 Tariverdian, G., Ritter, H. and Schmitt, J. (1971*a*), 'Zur transspezifischen Variabilität der NAD-Malatdehydrogenase (E.C.:1.1.1.37) der Primaten', *Humangenet.* 11, 339.

13 Tariverdian, G., Ritter, H. and Schmitt, J. (1971*b*), 'Transpecific variability of phosphohexose isomerase (E.C.:5.3.1.9) in Primates', *Humangenet.* 12, 105.

14 Nute, P.E. and Buettner-Janusch, J. (1969), 'Genetics of polymorphic transferrins in the genus *Lemur*', *Folia primat.* 10, 181.

15 The specimens of *Microcebus* represented in Fig. 2 are both *M.m. murinus*. Further work subsequent to the conference has shown oxidase bands somewhat anodal to the origin in *Cheirogaleus* and in four individuals of *M.m. rufus*.

16 Subsequent to the Research Seminar meeting, we received an additional sample of *Tarsius bancanus* blood, by courtesy of Dr Carsten Niemitz. Once again we identified two haemoglobins, with the same mobilities as before; the minor band represented 18% of the total quantity. The unusually slow albumin was also confirmed.

F. M. BUSH, D. E. HAINES and K. R. HOLMES

Haemoglobin of the lesser bushbaby, Galago senegalensis: *starch gel electrophoresis and alkali-resistance*

Introduction

Heterogeneity of haemoglobin has been found in the suborder Prosimii. In a survey of some Lorisiformes, Buettner-Janusch and Twichell mentioned the presence of at least two haemoglobin components in a number of species of Galaginae, including *Galago crassicaudatus crassicaudatus, G. c. agisymbanus*, and *G. senegalensis senegalensis.*[1] Later, Buettner-Janusch described the variability in electrophoretic patterns of haemoglobins for five specimens of *G. c. crassicaudatis.*[2] Among 19 specimens of *G. c. crassicaudatus*, ten had two haemoglobin components and nine had three components (Buettner-Janusch and Buettner-Janusch).[3,4] In another subspecies, *G. c. panganiensis*, electropherograms of 23 adult and 13 young (over two months old) exhibited the presence of three haemoglobin components, two major bands and one satellite band of variable intensity (Stormont and Cooper).[5] The significance of this variability among the species so far studied has been minimised by the discovery that aged haeomolysates pooled from six specimens of *G. senegalensis* and *G. crassicaudatus* had two major and one minor electrophoretic bands, and that the minor band was absent in fresh samples (Hoffman and Gottlieb).[6] This phenomenon exhibited by aged haemolysates has been confirmed by Nute, Buettner-Janusch and Buettner-Janusch,[7] who observed the occurrence of two major components in fresh samples and also satellite bands in aged samples when they compared the electrophoretic patterns from 57 specimens of *G. c. crassicaudatus* and *G. c. argentatus*, 12 of *G. demidovii*, and nine of *G. senegalensis*. The satellite bands formed from polymerisation which could be reversed by treatment of the aged samples with 2-mercaptoethanol.

The two major haemoglobin components have been isolated by

DEAE-Sephadex column chromatography from individual samples in 13 specimens of *G. crassicaudatus* (Nute et al.).[7] Subsequent separation of both components in urea gels revealed the occurrence of identical alpha-chains and different beta-chains. Similar patterns in urea gels for unfractionated haemolysate samples in single specimens of *G. senegalensis* and *G. demidovii* suggested a similar polypeptide substructure for *Galago* haemoglobins. Accordingly, Nute et al. proposed an $alpha_2 beta_2{}^1$ polypeptide substructure for the component migrating slowest toward the anode on starch gels and an $alpha_2 beta_2{}^2$ polypeptide substructure for the rapidly-migrating electrophoretic band.[7]

Measurement of alkali-resistant haemoglobin reveals significant differences between the *Galago* species studied. Mean levels were found to be considerably higher in *G. crassicaudatus* than in *G. senegalensis.* Five specimens of adult *G. c. agisymbanus* had a mean level of 36.0% (range, 32.0-46.8) and 18 *G. c. crassicaudatus*, 36.5% (25.3-59.1);[3] 19 *G. crassicaudatus* spp., 27.7% (19.2-36.3);[3] and four *G. s. senegalensis*, 8.4% (7.2-10.1).[1]

The paucity of information about the haemoglobin in other colonies of the lesser bushbaby, particularly during ontogeny, prompted the present investigation. This paper reports on the results of additional studies on neonate, juvenile, and adult *G. senegalensis* haemoglobin, including studies on alkali-resistance, polypeptide chain composition, and electrophoresis in alkaline and acid buffer systems.

Materials and methods

The adult bushbabies used in the present study have all been maintained in captivity either at the Medical College of Virginia or at Michigan State University for periods of two to four years. The neonates were offspring of adults in the respective colonies. All animals were on a 12-hour day/night light cycle and received water and a standard diet of bananas, monkey chow, and a vitamin supplement. Blood samples were obtained from 34 different animals and were drawn from a superficial vein on the caudal aspect of the lower leg, designated as the short saphenous vein in *Tarsius.*[8] All animals were sampled at least twice. Each sample was drawn in a heparinised syringe between 8.30 and 12.00 a.m.

The procedure for preparation of red cell haemolysates was the same as described by Bush and Townsend.[9] Plasmas were separated after centrifugation of the sample at 2,000 g for 10 minutes. Red cells were washed at least three times with physiological saline kept at 4°C. The cells were lysed with distilled water and toluene and then the solution was exposed to carbon monoxide to form the CO-haemoglobin derivative. Each sample was centrifuged at 10,000 g for one hour and the clear red haemoglobin layer was filtered to remove any traces of cell debris. Small quantities of each haemolysate were further diluted either by adding approximately three volumes of distilled water or by making up to 0.1 M in 2-mercaptoethanol, a procedure followed by Nute et al.[7]

Haemolysate samples were subjected to vertical starch gel electrophoresis according to the method of Smithies.[10] Gels were prepared with the Tris-EDTA-borate buffer, pH 8.5 (Manwell, Baker and Betz).[11] Electrophoresis was for seven hours at 4°C, with power of 280 V supplied by a Beckman Duostat. Gels were slit longitudinally; one half was stained for haemoglobin by the standard peroxidative procedures involving H_2O_2 and either benzidine or *o*-dianisidine (Chernoff and Pettit),[12] and the other half was stained with amido black 10B in methanol: distilled water: acetic acid (50:50:10) solution. Some haemolysate samples were subjected to electrophoresis in a 0.1 M formate buffer, pH 1.8 (Muller) for chain separation;[13] individual components of other haemolysates were cut from the Tris-EDTA-borate gels and either incubated with 6 M urea for 30 minutes and transferred into the sample slots of acid gels, or transferred directly to the acid gels. These samples were subjected to electrophoresis at 150 V for seven hours and then stained as described above.

Alkali-resistant haemoglobin was determined by the method of Szakacs.[14] The haemoglobin concentration for each sample was adjusted to 10 g haemoglobin per 100 ml. The modified Wong method (MacFate) was used to determine the iron content in haemoglobin.[15] For each sample, the haemoglobin concentration was:

$$\text{g haemoglobin} = \frac{\text{mg iron/100 ml}}{3.4}$$

All colorimetry was accomplished with a Beckman D. U. Spectrophotometer.

Results

Electrophoresis of freshly prepared haemolysates or aged haemolysates treated with 2-mercaptoethanol produced identical patterns, indicating that the erythrocytes of all of these bushbabies contained two major components (Figs. 1 and 2). We refer to these as the rapidly (R)- and slowly (S)-migrating haemoglobins. In most animals, the S haemoglobin had a slightly greater stain intensity than did the R haemoglobin. These same two-band patterns were found in haemoglobin samples of neonate and juvenile animals between the ages of five days (Fig. 3) and six months (Fig. 2), and for all adult samples. No sexual dimorphism in electrophoretic patterns was detected.

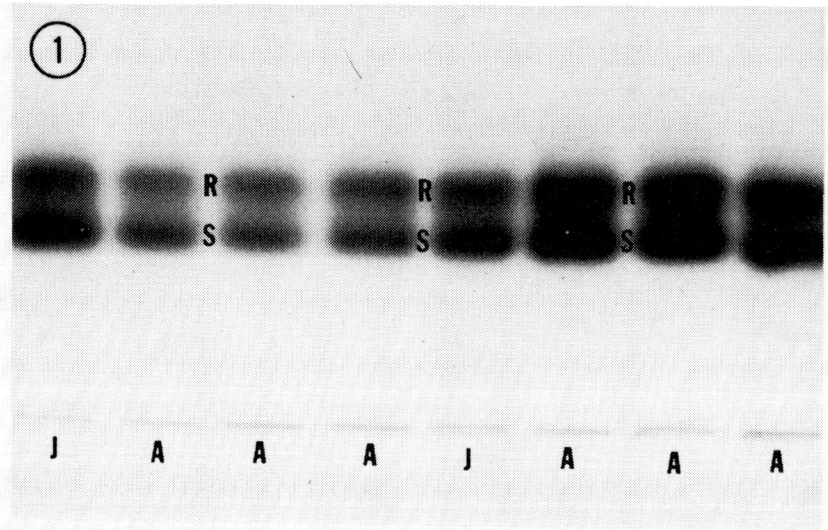

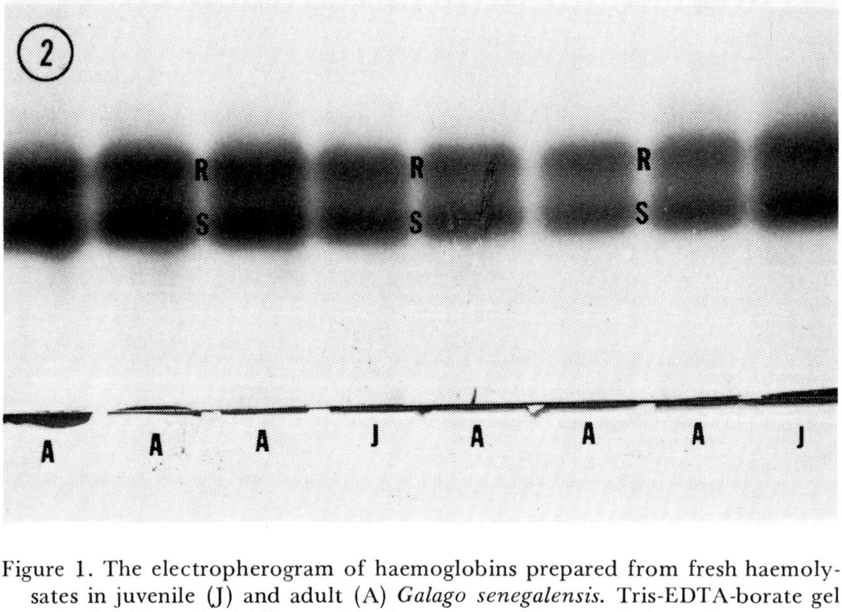

Figure 1. The electropherogram of haemoglobins prepared from fresh haemolysates in juvenile (J) and adult (A) *Galago senegalensis.* Tris-EDTA-borate gel stained with *o*-dianisidine and hydrogen peroxide. Anode is at the top. R is the rapidly-migrating band and S is the slowly-migrating band.

Figure 2. The other half of the electropherogram of *Galago senegalensis* haemoglobins, stained with amido black.

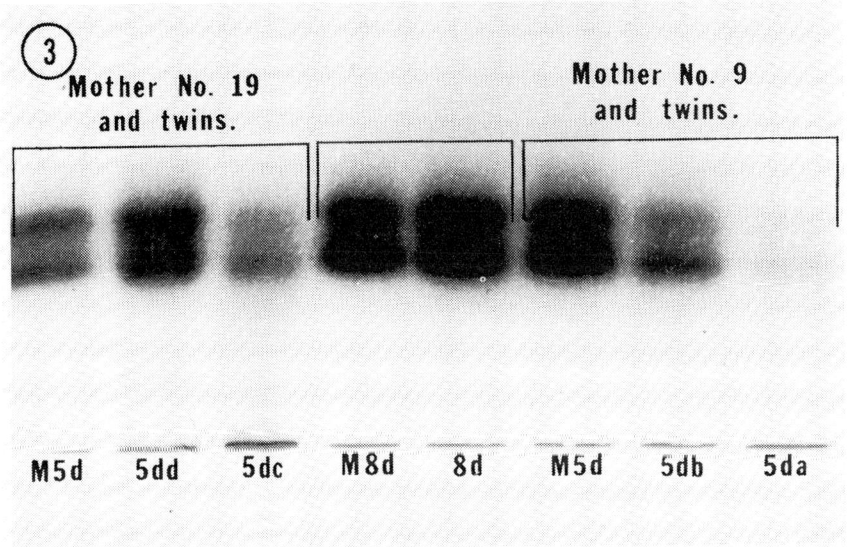

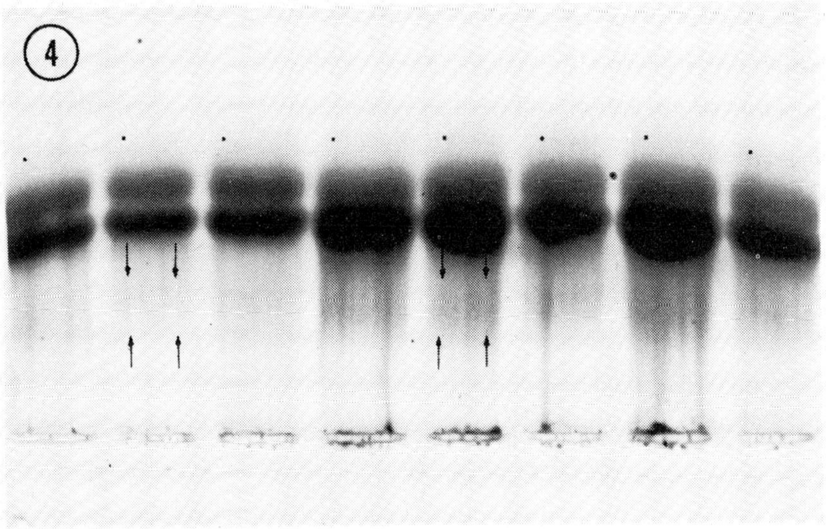

Figure 3. The electropherogram of *Galago senegalensis* haemoglobins prepared from haemolysates of neonates and their respective mothers. Patterns stained with *o*-dianisidine. Abbreviations: 5da and 5db = young of mother 9; 5dc and 5dd = young of mother 19; 8d = young of mother 8.

Figure 4. The electropherogram of *Galago senegalensis* haemoglobins resolved in formate buffer, pH 1.8, 1.2 M. Anode is at the bottom. Arrows denote diffuse zone in amido black patterns. Dots indicate non-haeme protein.

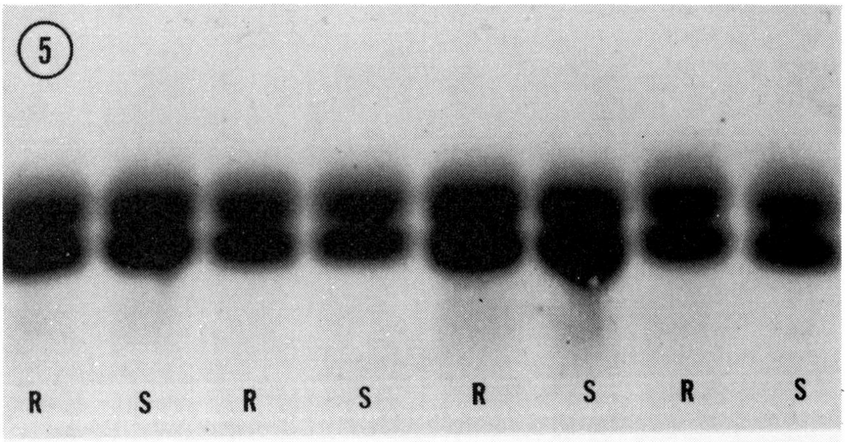

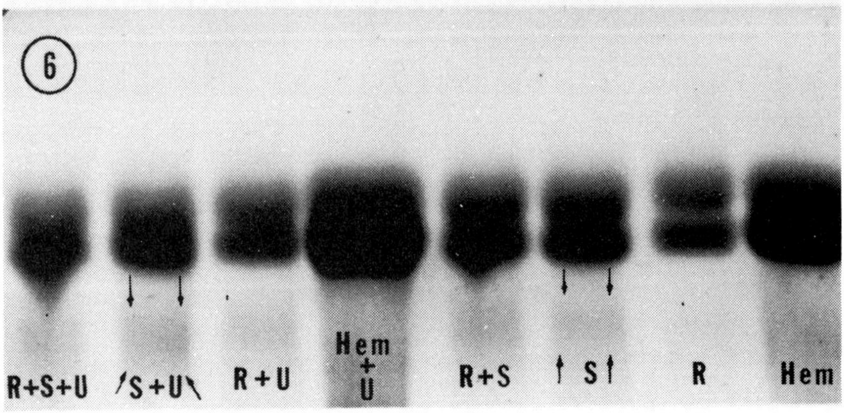

Figure 5. The electropherogram of R and S components in *Galago senegalensis* cut from the starch gels and subjected to electrophoresis in formate buffer. Anode is at the bottom of amido black pattern.

Figure 6. The electropherogram of the R and S components and haemolysates (Hem) subjected to electrophoresis in formate buffer, and the same components incubated with 6 M urea (U) and then subjected to electrophoresis. Anode is at the bottom of amido black pattern.

The formate gel patterns of unfractionated haemolysates treated with 2-mercaptoethanol were characterised by two major components: one anodic satellite band, and a more slowly migrating diffuse zone (Fig. 4). The satellite band proved to be a non-haeme protein, because it failed to stain with benzidine. When the two major components were cut from the starch gels and then subjected to electrophoresis in formate gels, the segments containing the proteins with the mobilities of the R and S haemoglobins resolved during electrophoresis into a four-band pattern of benzidine-positive components (Fig. 5). Each of the R and S haemoglobins separated into two bands. The more anodic band of the R haemoglobin had the same mobility as the more anodic band of the S haemoglobin. The second bands of each haemoglobin were also of similar mobility. Treatment of the haemolysate and the two bands separately or in combination with 6 M urea did not appreciably alter the patterns (Fig. 6).

Tests for the presence of haemoglobin resistant to denaturation by alkali are shown in Table 1. The mean level was 9.3% (range 6.6-13.0) for the 31 animals studied. Females had slightly, but not significantly (for *t*-test, $p>0.05$), higher levels than were found in

Table 1 Alkali-resistant haemoglobin in the lesser bushbaby, ***Galago senegalensis***

(a) MALES

Animal number	*Age*	*Per cent alkali-resistant*
4 Michigan		9.0
30	Up to 4	9.6
32	years in	12.9
41 B	captivity	13.0
2 Virginia		7.1
3 A		9.5
36		10.4
38		7.6
42		7.7
46		8.9
Juvenile	6 weeks	6.6
	6 months	7.4
Neonate	8 to 10 days	7.3
	Mean ± S.E. = 9.2 ± 0.7	

(b) FEMALES

Animal number	*Age*	*Per cent alkali-resistant*
5		9.4
25		9.8
37		12.0
41		7.3
41 C		10.9
1 Virginia		7.1
3		8.9
5		9.6
39		7.7
45		11.2
47		8.8
51		9.4
53		10.2
55		10.9
57		11.0
59		7.2
63		10.8
Juvenile	6 weeks	10.8
	6 months	11.4
Neonate	8 to 10 days	8.3
	Mean ± S.E. = 9.5 ± 1.2	

males: females = 9.5 (7.1-12.0), males = 9.2 (6.6-13.0). Percentages for neonate and juvenile bushbabies were within the range of adult levels. Measurements of alkali-resistant haemoglobin taken at six weeks and at six months from the same juveniles revealed no significant change in the level of pigment. Although the number of young tested was small, family data failed to reveal any appreciable difference between the levels of parents and those of their offspring. For example, animals 41 B and 41 C, born at different intervals to female 41, had higher levels than their mother, but their levels fell close to the higher values found in some other adults.

Discussion

The starch gel patterns from haemolysates in our colony of bushbabies confirm the results previously reported by Nute et al. for the few specimens so far sampled in this species.[7] The reproducibility of the two-band pattern in our animals, the findings by Nute et al. in *G. senegalensis* and in over 100 specimens of *G. crassicaudatus*,[7] and the report by Barnicot and Hewett-Emmett in the present seminar of two major components in all of eight specimens of *G. senegalensis* and 19 specimens of *G. crassicaudatus*,[16] indicates a remarkable homogeneity of haemoglobin phenotype for these species of *Galago* from different colonies. Such homogeneity suggests homozygosity, although it is difficult to generalise from such small samples. The uniformity in patterns for neonate, juvenile, and adult haemoglobins in our animals agrees with the statement made by Hoffman and Gottlieb,[6] who mentioned the similarity in the electrophoretic mobilities at pH 9.1 of haemoglobin from adult *Galago* blood and from foetal *Galago* cord blood.

Our results of unfractionated haemolysates and segments cut from the starch gels and resolved in formate buffer do not refute the evidence by Nute et al. of the three-band pattern for the unfractionated sample of haemolysates in one specimen of *G. senegalensis* and the DEAE-Sephadex fractionated samples of several *G. crassicaudatus* separated in urea gels.[7] However, the separation of the R and S haemoglobins cut from the starch gels and resolved with the formate buffer into four benzidine-positive components gives rise to some problem in interpretation. The presence of a fast-migrating band in the R component with the same electrophoretic mobility as a fast-migrating band in the S component agrees with the concept of alpha-chains with the same mobilities. The presence of a slowly-migrating band in the S component with similar electrophoretic mobility as the slowly-migrating band of the R component, even in samples treated with urea, would not agree with the results of beta-chains with different mobilities. Such a discrepancy could arise by several means; for example, the charges on the beta-chains may be similar enough in the acid system even though the amino acid compositions of the two components are different, or there may be incomplete separation of both components that migrate so close together in alkaline gels, or buffer interaction with the haemoglobin. The presence of the non-haeme protein as well as the diffuse zone in amido black patterns that stains intensely with benzidine, the existence of satellite bands in aged haemolysates, and the findings of a two-band pattern, including a diffuse zone, in urea gels by Nute et al. for the single unfractionated haemolysate in *G. demidovii*, all show the technical problems in handling haemoglobins from these animals.[7] In man, Muller reported a slow band of undetermined nature for both fractionated and unfractionated globins in acid gels that possibly represented agglomeration of single chains.[13] Because most of these investigators stained the haemoglobins only with amido black, it is possible that this diffuse staining in acid gels represents the separated haeme that would be benzidine-positive.

The mean level of 9.3% for alkali-resistant haemoglobin in our animals is slightly higher than the mean level found previously by Buettner-Janusch and Twichell for four specimens of this species.[1] Our discovery of similarities in levels of alkali-resistant haemoglobin for neonates, juveniles, and adults is consistent with the similarities in electrophoretic patterns for these animals of different ages. This similarity in levels corresponds to the results found previously by Buettner-Janusch in the young Ringtail Lemur, *Lemur catta.*[2] In that species, the percentage of alkali-resistant haemoglobin was nearly the same in the four-month old animal as that found in the adult.

Knowledge of the amino acid sequence in *Galago* haemoglobins does not seem to account for the differences found between alkali-resistant haemoglobins from *G. senegalensis* and *G. crassicaudatus.* A partial sequence of both alpha- and beta-chains has been determined for the prosimians *G. crassicaudatus, Lemur fulvus*, and *Tupaia glis* (Hill; see Sober).[17] Both *G. crassicaudatus* and *L. fulvus* have fairly high levels of alkali-resistant haemoglobin, while *T. glis* has a lower level comparable to that found in *G. senegalensis.*[1] Haemoglobins of *G. crassicaudatus* and *T. glis* have a number of residues in common, particularly in the beta-chain; whereas, there are fewer corresponding residues in either chain of *G. crassicaudatus* and *L. fulvus.*

It appears unlikely that this difference in alkali-resistance in the two *Galago* species relates directly to the oxygen-binding affinities of the haemoglobins. Compared with anthropoid haemoglobins, prosimian haemoglobins have lower oxygen affinities in the physiological range, a large acid Bohr effect, and a larger alkaline Bohr effect (Sullivan).[18] Sullivan showed that the haemoglobin oxygen affinity in *G. senegalensis* was similar to that of most lorisiform haemoglobins, that a slight increase and shift of the alkaline Bohr effect occurred at higher pH values, and that a slightly greater increase in the base-level oxygen affinity was found than in *G. crassicaudatus.* The oxygen affinities, but not the alkaline Bohr effects, were disparate between *G. senegalensis* and *T. glis*, while the alkaline Bohr effects were different in the two species of bushbabies. Furthermore, the increased base-level oxygen affinity and decreased alkaline Bohr effect were found in *L. fulvus*, and neither was characteristic in *G. crassicaudatus.*

We found no association between the percentage of alkali-resistant haemoglobin and the stain intensity of electrophoretic bands. Buettner-Janusch and Buettner-Janusch reported that no obvious correlation existed between the level of alkali-resistant pigment in *Galago* and the number of bands on starch gels; they did not identify the alkali-resistant haemoglobin with either band.[3,4] No specific relationship was found between the proportion of alkali-resistant haemoglobin and haemoglobin types in six species of macaques (Ishimoto et al).[19] Dissimilarities in alkali-resistant haemoglobin and the type of placentation in prosimian and anthropoid species have been discussed by Buettner-Janusch and Buettner-Janusch, who hypothesised that the shift from the epitheliochorial placenta in lower mammals to the haemochorial placenta in higher Primates affected the haemoglobin advantageously in the latter species.[4] Interestingly, the mean levels of alkali-resistant haemoglobins in the lesser and greater bushbabies are far apart; but both species possess the diffuse, non-deciduate epitheliochorial type placenta (Gérard; see Hill).[20] A more valid comparison between this blood property and placental structure might be permissible if the level of alkali-resistant haemoglobin were known in *G. demidovii*, which has an endotheliochorial type placenta (Gérard, see Hill),[20] closer to the type characteristic of higher Primates. Thus, the present body of evidence does not appear to be sufficient to disclose whether or not either the presence of alkali-resistant haemoglobin or two major haemoglobin components in *Galago* species have any adaptive significance. Recently, Sullivan[18] has implied a possible difference in function between the two major bands in *G. crassicaudatus*, after study of the haemoglobin oxygen affinity in this species.

Acknowledgments

We thank H.C. Dessauer (Department of Biochemistry, Louisiana State University) E.S. Kline (Department of Biochemistry) and J.I. Townsend (Program in Genetics, Virginia Commonwealth University) for assistance with techniques or criticisms of the manuscript. M.F. Dolwick, G.E. Goode, and D. Wright, Virginia Commonwealth University, helped with the typing or provided useful comments during the course of this research.

NOTES

1 Buettner-Janusch, J. and Twichell, J.B. (1961), 'Alkali-resistant haemoglobins in prosimian primates', *Nature* 192, 669.

2 Buettner-Janusch, J. (1962), 'Biochemical genetics of the primate haemoglobins and transferrins', *Ann. N.Y. Acad. Sci.* 102, 235-48.

3 Buettner-Janusch, J. and Buettner-Janusch, V. (1963), 'Haemoglobins of *Galago crassicaudatus*', *Nature* 197, 1018-19.

4 Buettner-Janusch, J. and Buettner-Janusch, V. (1964), 'Haemoglobins in primates', in Buettner-Janusch, J. (ed.), *Evolutionary and Genetic Biology of Primates*, New York.

5 Stormont, C. and Cooper, R. (1965), 'Haemoglobin and serum proteins in *Galago crassicaudatus panganiensis*', *Fed. Proc.* 24, 532.

6 Hoffman, H.A. and Gottlieb, A.J. (1969), 'Discussion paper: Investigations of nonhuman primate haemoglobin–electrophoretic variation', *Ann. N.Y. Acad. Sci.* 162, 215-238.

7 Nute, P.E., Buettner-Janusch, V. and Buettner-Janusch, J. (1969), 'Genetic and biochemical studies of transferrins and haemoglobins of *Galago*', *Folia primat.* 10, 276-87.

8 Hill, W.C.O. (1955), *Primates, Comparative Anatomy and Taxonomy*, vol. 2: *Haplorhini: Tarsioidea*, Edinburgh.

9 Bush, F.M. and Townsend, J.I. (1971), 'Ontogeny of haemoglobin in the house sparrow', *J. Embryol. Exp. Morph.* 25, 33-45.

10 Smithies, O. (1959), 'An improved procedure for starch-gel electrophoresis: further variations in the serum proteins of normal individuals', *Biochem. J.* 71, 585-7.

11 Manwell, C., Baker, C.M.A. and Betz, T.W. (1966), 'Ontogeny of haemoglobin in the chicken', *J. Embryol. Exp. Morph.* 16, 65-81.

12 Chernoff, A.I., and Pettit, N.M., Jr. (1964), 'Some notes on the starch-gel electrophoresis of haemoglobins', *J. Lab. Clin. Med.* 63, 290-6.

13 Muller, C.J. (1960), 'Separation of the α- and β-chains of globins by starch gel electrophoresis', *Nature* 186, 643-4.

14 Szakacs, J.E. (1964), 'Measurement of alkali-resistant haemoglobin in blood (method of Singer et al.), in Sunderman, F.W. and Sunderman, F.W. Jr., *Haemoglobin: Its Precursors and Metabolites*, Philadelphia.

15 MacFate, R.P. (1964), 'Measurement of haemoglobin in blood. II: Determination of iron concentration in haemoglobin solution', in Sunderman, R.W. and Sunderman, F.W. Jr., *Haemoglobin: Its Precursors and Metabolites*, Philadelphia.

16 Barnicot, N.A. and Hewett-Emmett, D., this volume.

17 Sober, H.A. (1968), *Handbook of Biochemistry*, Cleveland.

18 Sullivan, B. (1971), 'Structure, function, and evolution of primate haemoglobins. II: A survey of the oxygen-binding properties', *Comp. Biochem. Physiol.* 40 B, 359-80.

19 Ishimoto, G., Tanaka, T., Nigi, H., and Prychodko, W. (1970), 'Haemoglobin variation in macaques', *Primates* 11, 229-41.

20 Hill, W.C.O. (1953), *Primates: Comparative Anatomy and Taxonomy*, vol. I: *Strepsirhini*, Edinburgh.

K. BAUER

Comparative analysis of protein determinants in primatological research

Introduction

There are two main ways of establishing relationships between different species. The first is to examine a number of morphological parameters comparatively and to draw conclusions from the resulting data. This approach can be termed the classical anatomical and palaeontological method. The second technique is to investigate the same relationships at the molecular level.

Since the genetic information which is expressed in the phenotype of an organism is coded for in the DNA sequence, this might at first sight appear to be the level most suited for this type of investigation. Unfortunately, technical difficulties have inhibited progress in this field, and it is not possible at present to determine the numbers of mutational changes separating homologous DNA samples isolated from different species.

The translations of structural genes in the form of amino acid sequences have been amenable to experimental analysis since Sanger's pioneering work:[1] most of our knowledge about basic mechanisms of information-transfer and modification (both from generation to generation and inside the cell) originate from work at this level.

Single proteins contain only very small fractions of the information coded in the whole genome, so it is not possible to employ them as adequate representatives. In other words: to draw phylogenetic conclusions regarding a whole species, a panel of various such proteins must be examined. However, actual elucidation of polypeptide sequences is only possible with rather sophisticated and time-consuming methods. Therefore, this approach poses technical problems if it is to be applied to phylogeny. A more rapid, simpler technique for comparing proteins is urgently needed for this field of research.

This paper describes an immunological method based upon a comparative analysis of the determinant patterns reacting in pairs of cross-reacting protein homologues. No isolation procedure is needed for the analysis proper, which uses unfractioned blood plasma or serum. The method permits detection of mutations separating protein homologues from various species.

We have applied this approach to primate phylogeny, deriving a time-scale of primate evolution from data on a minimum of 15 different plasma proteins per species. The resulting data on the ages of the last common ancestors for man and various subhuman primates are consistent with estimations based on classical palaeontological work.

The principle of comparative determinant analysis (CDA)

Genetic information, coded in a structural gene, is translated into an amino acid sequence, which thus contains only information from the structural gene. The information transfer from DNA to protein levels is characterised by partial loss due to the degeneracy of the genetic code. However, all information present in a polypeptide must come from the corresponding cistron. Moreover, any substitution through which two polypeptide chains differ must have its equivalent in an accepted mutation at the DNA level distinguishing the two genes. But it is important to remember that not all mutations established as the nucleotide base level are apparent at the level of amino acid sequence.

With respect to information transfer, there exists a very similar relationship between the polypeptide chain level and the level of antigenic structure of a protein. Of all the amino acids present, only a certain fraction takes part in the formation of the antigenic structure of the molecule. These amino acids must:

1. be situated at the surface of the native molecule, in order to be accessible to the antibody,
2. form part of an antibody-binding site.

The second condition depends on the antigen-producing species (in our experiments: man) in relation to the antibody-producing species (in our experiments: the rabbit). For example, if human transferrin is injected into a rabbit, antibodies will be synthesised against those parts of the human molecule which are different from the rabbit transferrin molecule. The immunisation procedure thus becomes, in a certain sense at least and with certain limitations, a process of comparison between the two (sets of) structural genes (in our example, human and rabbit genes) coding for transferrin. Those parts of the human molecule against which antibodies are produced and with which they are able to react are the *antigenic determinants.*

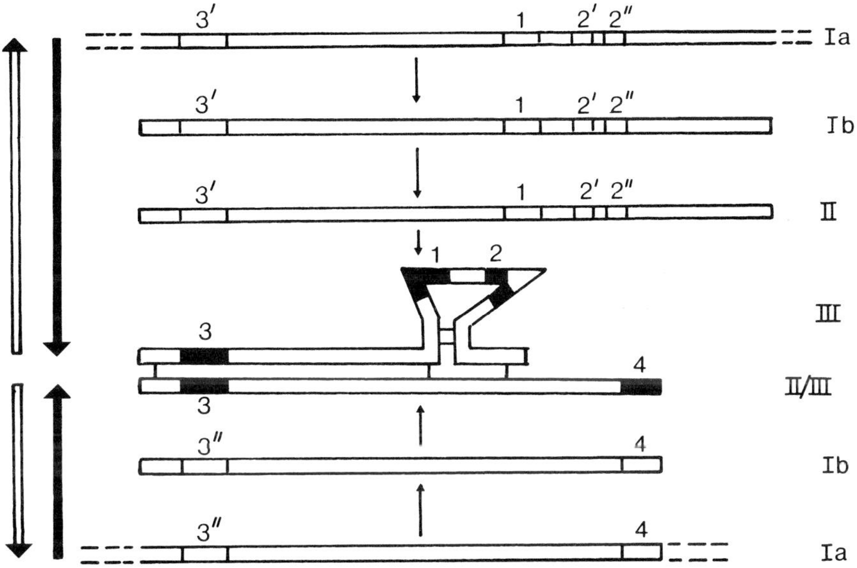

Figure 1. Diagram illustrating the three levels of potential investigation of genetic information; *nucleic acid* (DNA: I*a*; m-RNA: I*b*), *polypeptide chain* (II) and the *antigenically active polypeptide*, either folded (upper chain) or unfolded (lower chain). Immunogenetic comparative determinant analysis (CDA) utilises the third level (III). The flow of information is indicated by the black arrows, whilst the white arrows indicate the direction of the analytical approach. Black areas shown at level III show the determinants: 1 and 2 represent CDs, 3 is a CD coded for on two different structural genes, 4 is an SD. In most cases, no differentiation between the structural genes is possible through CDA, the main exception to the rule being immunoglobulin chains, for which specific antisera are available.

The antigen-antibody-reaction in such a system is therefore one between antigenic determinants, located on the surface of an otherwise inert residual molecule, and a mixture of determinant-specific antibodies; in short between a determinant carrier and a number of antibody populations.

A single determinant consists of a small number of amino acids which either represent a section from a polypeptide chain (*sequence determinant* – SD) or are only brought together by the molecule's conformation (*conformation determinant* – CD). It is apparent that both types of determinant are directly coded for in the corresponding structural gene, and that DNA sequence, polypeptide, and determinant pattern are colinear structures connected by information transfer from the genome. A model of this relationship is given in Fig. 1. Evidently, all determinants are coded in the structural gene itself, while in the actual definition of the single determinant – which takes part in the immunisation process – the genes from the immunised species also take part in the manner described.

If determinant patterns of a pair of homologues are analysed comparatively with the same antiserum, any mutations accepted in those parts of the structural genes corresponding to a determinant will be observable through techniques which allow us to state whether one of them carries more determinants than the other. Appropriate techniques will be described below. By their use for any pair of cross-reacting homologous proteins, one of the following statements can be made (Fig. 2):

1. homologue 1 is identical to homologue 2,
2. homologue 1 carries more determinants than does homologue 2,
3. homologue 1 carries less determinants than does homologue 2.

There are two limitations of such comparative determinant analysis (CDA):

1. Since any one determinant is formed by a *group* of amino acids, one is dealing with larger units in CDA than in sequence analyses, so that the structure examined only partly reflects the information involved.
2. Since only part of the whole genetic information coded in a gene takes part in the formation of the molecule's antigenic structure, only this fraction of the information is actually investigated. The residue, especially that corresponding to internal segments of the protein, is beyond the reach of CDA.

In spite of this loss of information, results obtained by CDA can be employed as an estimation of mutational differences between pairs of homologous proteins, because, among all changes occurring,

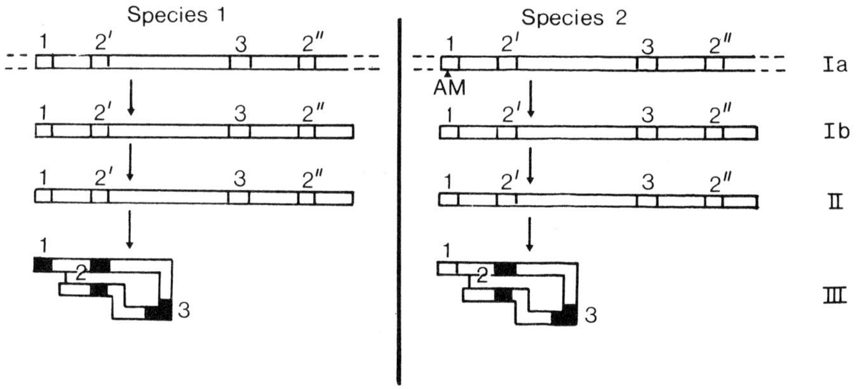

Figure 2. Diagrammatic representation of a CDA comparison between two structural genes (homologues) in species 1 and species 2. The antibody-binding capacity of the determinants in the final, native proteins, shown in the figure at the third level (III) is the property which is analysed. Because of the presence of an accepted mutation (AM) in the gene of species 2, determinant 1 loses its capacity to bind the antibody specific for the original determinant. If an Ouchterlony test of the two homologues is conducted, species 1 will 'spur over' species 2 (see Fig. 3*b*). For key, see Fig. 1.

there is no apparent departure from randomness in the selection of mutational events detected in the determinant structure. Since all species commonly used as antiserum producers (e.g. rabbit, sheep, horse, goat) are related to man via the eutherian mammal common ancestor, no difference would be expected if antiserum producers were exchanged. This expectation is borne out by our finding that antisera (to human proteins) from the species mentioned do not differ systematically in the range of animals with which they react.[2] There is one more possible limitation, which might greatly interfere with CDA and its application to primatological problems. Many proteins – among them the majority of plasma proteins on which our investigation is based – contain prosthethic groups, which are mainly carbohydrate. The whole principle of CDA, however, only applies to semantides, i.e. informational macro-molecules, connected directly with genetic information.[3] It is only for semantides, which include nucleic acids and polypeptides, that the relationship of *direct homology* can exist at all.[4] Episemantides, on the other hand, may be related by *indirect homology*, in cases where the enzymes taking part in their synthesis are homologous. In most cases, it is difficult to decide from the experimental findings alone whether two carbohydrate determinants occurring in two different species are – indirectly – homologous or analogous. This renders the use of episemantides in immunogenetic work very problematic whenever more than one species is involved. A very good example is provided by the classic carbohydrate determinants, ABO-blood group antigens. If experimental data on *these* determinants alone were used, certain intestinal bacteria would have to be considered as close relatives to primates (Table 1). This observation clearly demonstrates that phylogenetic immunogenetic research should be restricted to semantide determinants.

In order to exclude the use of episemantide determinants in our own analysis of proteins we have conducted the following experiments:

1. Denaturation by various agents, the effect of which should be confined to the polypeptide portion of the protein, destroyed the antibody-binding capacity of the plasma proteins in our test panel either totally or to a large degree. There were two exceptions to this rule, one containing only 0.8% and the other 38% carbohydrate (pre-albumin and acid α_1-glycoprotein, respectively).
2. The use of various glycolytic enzymes in no case interfered with antigenicity or destroyed any single determinant used for our study.[5]

While additional experiments, especially on the structure of carbohydrate moieties of plasma proteins, are desirable, these results indicate that the determinants of plasma proteins in our test panel

Table 1 Comparison of bacterial surface antigens and blood group determinants in various species of intestinal bacteria and of primates

The occurrence of these 'blood group' agents, which are carbohydrate in structure, is apparently independent of taxonomic relationships of species. The lack of such relationships is due to the episemantide nature of carbohydrates. Only semantides, such as polypeptides, are suited for this type of phylogenetic work. (Data compiled from ref. 28 – see Notes.)

Species (O-antigen)	*Location of determinant*	*Specificity*
E. coli (86)	O-antigen	B
E. coli (127)	O-antigen	H (O)
E. coli (128)	O-antigen	H (O)
S. poona (13, 22)	O-antigen	H (O)
S. grumpensis (13, 22)	O-antigen	H (O)
S. atlanta (13, 22)	O-antigen	H (O)
S. berkeley (43)	O-antigen	B
Arizona (9)	O-antigen	H (O)
Arizona (21)	O-antigen	B
H. sapiens	Erythrocytes	A, B, H (O)
P. troglodytes	Erythrocytes	A, H (O)
G. gorilla	Erythrocytes	B
P. pygmaeus	Erythrocytes	A, B
Hylobates lar	Erythrocytes	A, B
Papio	Erythrocytes	A, B
Erythrocebus	Erythrocytes	A

(see later: Table 5) are generally semantide in nature (i.e. are amino acid groupings).

It has been demonstrated that it is possible to compare two homologous proteins (viz. their structural genes) to determine the occurrence of established mutations by investigating their determinant structures. Homology can be demonstrated immunologically by the existence of cross-reaction itself, i.e. the reaction of a protein other than the one used for immunisation (the immunogen), if appropriate controls for the specificity of the reaction are conducted. Of course, single proteins comprise too small a fraction of the information present in the genome of higher organisms to permit any meaningful analysis. Buettner-Janusch states in this connection: 'Haemoglobin is . . . nice . . . to work with . . . but I would like to examine 30 proteins before I made a taxonomic decision.'[6] The minimal condition that we have employed was 15 proteins per primate species, and in most cases more proteins were included in the investigation. Such a panel of proteins is taken as a statistical sample of the total genetic information present.

The proteins investigated in our laboratory included only blood proteins (a few other proteins examined by other authors were also considered), and it may be argued that this selection is not random in the statistical sense. On the other hand, there are good reasons for our choice. If intracellular proteins are analysed, complicated isolation procedures must precede the actual investigation. These procedures may well interfere with the conformation of the proteins, so that certain determinants are altered or destroyed and escape detection, in spite of their presence in the native molecule. It is difficult to standardise the isolation procedure for all species examined to the extent that such possible deterioration effects are at least kept equal (assuming that the different homologues are equally sensible to denaturation). The same changes must be expected to a still higher degree if structural proteins, such as keratin or myosin or cell wall proteins, must be solubilised before the investigation. By contrast, plasma proteins comprise a wide variety of proteins as regards physicochemical properties (molecular weight, electrophoretic mobility) and functional activities (enzymatic, antigenic, immunoglobulin, transport, coagulation etc.). In all of these respects, blood proteins appear to be a good choice if a whole spectrum of properties is to be included in such an investigation. We therefore believe that deviation from a random sample of all proteins coded is not very large in the case of blood proteins. It should be mentioned, however, that an ideal genome information sample should include structural as well as other types of genes, and also those DNA sequences coding for classes of RNA other than m-RNA.

If the concept described here is accepted for the estimation of evolutionary rates from the CDA of a panel of plasma proteins, the next step (calculation of other parameters, especially times of divergence, leading to the different lines of descent) becomes feasible with familiar methods established in biochemical phylogenetical studies.[3] Various aspects of this calculation will be discussed below (see section on 'Evaluation').

The technique

From the range of immunological methods available, only those which *directly compare numbers of determinants* can be employed for CDA. This criterion is *not* fulfilled by quantitative techniques which measure the amount of precipitate formed in an antigen-antibody-reaction, or which are based on indirect quantitative determinations, such as the various modifications of complement fixation tests. The reason for this limitation is that determinants are to some degree interchangeable in the reaction, as far as the amount of reaction product is concerned.

The minimum requisite for a precipitation reaction is the presence of two determinants per molecule, regardless of their specificity.[7] If the assumption is made that a protein carries two determinants and the antibodies specific for them are abundant in the reaction medium, a maximum precipitation should occur even without marked participation of additional determinants. Therefore, in all immunological techniques measuring the amount of precipitate, great influence is exerted by one group of parameters which is normally unknown and not easy to estimate: the relative quantities of the various determinant-specific antibodies. Without a knowledge of the composition of the antiserum, which is practically very difficult to obtain, it is impossible to distinguish the various determinants and their participation in the total reaction. Moreover, the binding constants for each determinant-antibody-reaction should be known in order to estimate the composition of the reaction product. Therefore, this type of method is unsuited for elucidation of determinant structures, which are the only features of a protein molecule indicating substitutions at the base level with appropriate immunological methods. This fact is one reason for the failure of attempts to derive a time-scale of primate evolution with techniques not connected with determinant structures.[8] For further critical discussions of these authors' approach and results, we refer to papers by Simons,[9] Read and Lestril,[10] and Bauer.[11]

CDA was especially adapted from standard immunological methods for use in relationship to palaeontological studies, and it was first applied to the primate order.[11] It consists of the following steps:

1. Comparative Ouchterlony analysis

The gel diffusion technique developed by Ouchterlony[12] detects differences in determinant patterns of pairs of cross-reacting antigens. The three reactants, the antiserum and the two antigens, are poured into small reservoirs punched into an agarose gel covering a microscopic slide. The reactants are allowed to diffuse into the gel and to react with each other as and where they meet. If the antiserum contains appropriate antibodies, specific for at least two determinants per molecule of the two antigens, a precipitation line develops between each of the sample wells and the central trough containing the antiserum. Where these lines meet, various patterns can develop which are indicative of the relationship of the antigenic structures present in the two samples. Fig. 3 shows the main patterns. In general, we followed the technique described by Clausen;[13] our own modification is summarised below.

(*a*) *Gel*: 2.2 ml of a 1% solution made of agarose (Behringwerke, Marburg, Germany) in a pH 8.6 barbiturate buffer were poured on microscopic slides and allowed to cool. They were kept overnight at 4°C. The trough pattern shown in Fig. 3 was punched into the layer and the superfluous gel removed.

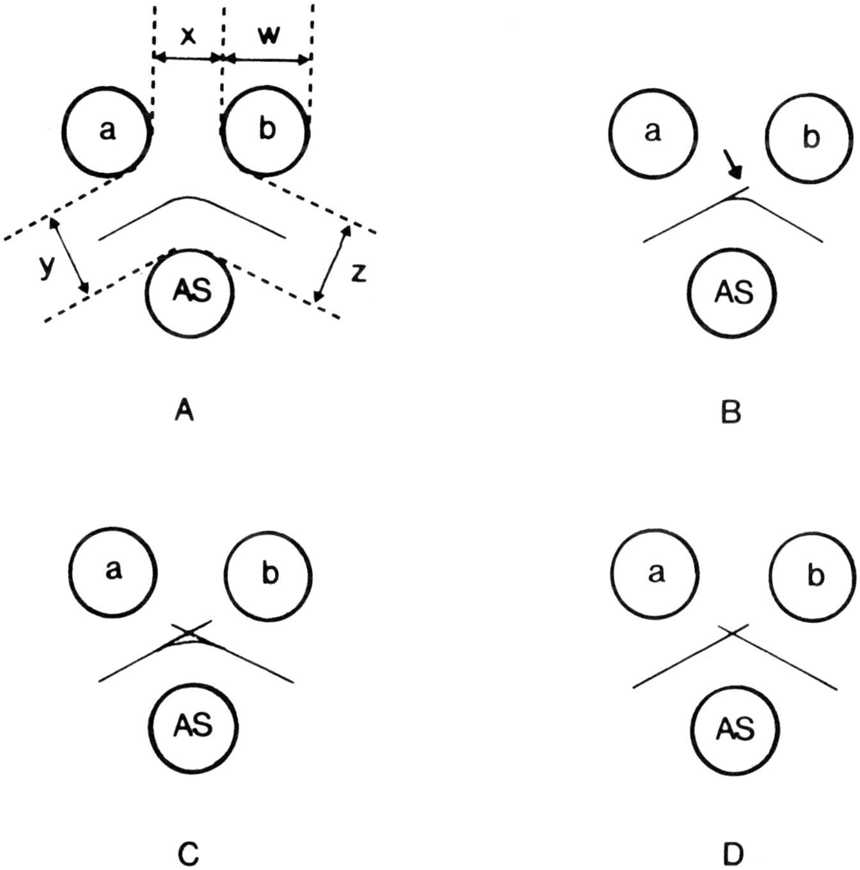

Figure 3. Diagram illustrating various patterns of precipitation lines between paired samples of homologous protein antigens, reacting with an antiserum (AS).

A – identity of determinant patterns.
B – sample *a* contains more determinants than sample *b*.
C – common determinants for *a* and *b* and one determinant, lacking on the other homologue, for both *a* and *b*.
D – *a* and *b* carry no common determinants.
In B, the arrow indicates a 'spur'.

(*b*) *Test*: The troughs were filled with the samples and the antiserum as shown in Fig. 3. Then the slides were kept at 37°C in a moist chamber. Evaluation followed after 1, 2 and 3 days of diffusion, by inspection of the precipitation lines for the occurrence of spurring phenomena (Fig. 3).

In order to apply comparative Ouchterlony analysis to the relationship between different primate species, we have systematically investigated – by use of antisera to the human homologues of some 25 different plasma proteins – pairs of plasmas from the following species:

Homo sapiens	*Erythrocebus patas*
Pan troglodytes	*Cercocebus galeritus*
Gorilla gorilla	*Cebus albifrons*
Pongo pygmaeus	*Galago crassicaudatus*
Macaca mulatta	*Nycticebus coucang*
Cercopithecus aethiops	*Perodicticus potto*

For comparative purposes, the plasmas of *Bos taurus, Ovis aries, Equus caballus, Cavia, Cricetus cricetus, Felis domestica, Lupus canis* were also examined. It must be emphasised that not all plasma proteins included did cross-react with all homologues mentioned. These differences between the various proteins as to their ranges of cross-reaction are observable without exact analyses of determinant structures. They allow a distinction of immunological evolution groups (IEG) of human proteins, which show parallels between ontogenetic and phylogenetic development.[14,15] (Fig. 4) A survey of the IEG system is given in Table 2.[16]

As a typical example of the results obtained by paired CDA, we show our data on the distribution of human fibrinogen determinants among subhuman primates (Table 3).[17] The determinants themselves are elucidated by combining all relations between pairs of homologues a and b of the types $a \equiv b$, $a > b$, $a < b$, or a (X) b (symbol adopted for reaction pattern C of Fig. 3).

The question of controls has assumed great importance since

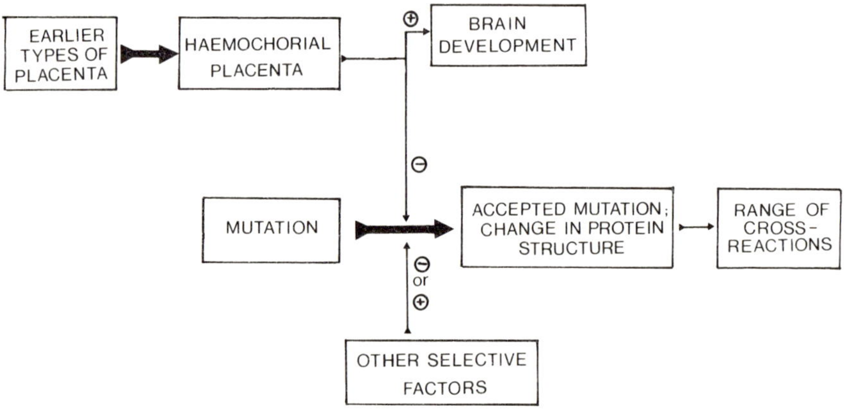

Figure 4. Diagram illustrating the theoretical concept developed by Goodman,[15] on the assumption that the emergence of the haemochorial placenta was essential for brain evolution. As a side effect, the acceptance of immunologically relevant mutations is affected in all proteins which are synthesised during early foetal life. This parallelism between ontogenetic and phylogenetic development has been reaffirmed in our laboratory.[14]

discrepancies were observed between determinant patterns of human α_2-macroglobulin, reported by James,[18] and by Hässig's group in Switzerland.[19] Because of the great interest in the immunogenetic treatment of primatological and other palaeontological problems we have re-examined this protein in our laboratory,[20] and obtained results reaffirming James' data. Both sets of findings are at variance with those of Roulet et al.[19] Apart from the fact that the same determinant formulas were estimated in two different laboratories, we have conducted extensive control experiments. The discrepancy demonstrates in general the importance of running appropriate controls with each and every experiment involving comparative

Table 2 The system of immunological evolution groups (IEG) of human plasma proteins

A stepwise increase in the range of cross-reactions observed is typical for the IEG system.[16]

Groups of species cross-reacting with human plasma protein homologue	*IEG* *I*	*IIa*	*IIb*	*IIc*	*III*	*IV*
Pongids	+	+	+	+	+	+
Catarrhines	—	+	+	+	+	+
Simians	—	—	+	+	+	+
Primates	—	—	—	+	+	+
Eutheria	—	—	—	—	+	+
Mammalia	—	—	—	—	—	+

Table 3 The results of comparative Ouchterlony tests[17] with antiserum to human fibrinogen

The determinants derived from the reaction patterns are indicated for each pecies. Determinants which are present in non-primates as well as in primates are termed A, B . . .; those restricted to primates are designated P_1, P_2. . . The guinea pig was selected because it possesses the largest number of human determinants among all the non-primates examined.

Species	*Determinants*	*Reaction of Plasma from* *Ho*	*Go*	*Po*	*Ma*	*Ce*	*Ga*	*GP*
H. sapiens (Ho)	AB $P_1 P_2 P_3$	≡	≡	≡	≡	>	>	>
G. gorilla (Go)	AB $P_1 P_2 P_3$		≡	≡	≡	>	>	>
P. pygmaeus (Po)	AB $P_1 P_2 P_3$			≡	≡	>	>	>
M. nemestrina (Ma)	AB $P_1 P_2 P_3$				≡	>	>	>
C. albifrons (Ce)	AB $P_1 P_2$					≡	>	>
G. crassicaudatus (Ga)	AB P_1						≡	>
Guinea pig (GP)	AB							≡

KEY: ≡ identical determinant structure (Fig. 3*a*); > Spurring indicating a higher number of determinants on the homologue from the species to the left, compared with the one at the top (Fig. 3*b*).

Ouchterlony analysis, in order to exclude erroneous results. These can be caused by technical errors, evaluation faults and the phenomenon of 'false spurs' (spur formation which occurs in spite of the presence of equal determinant structures if the examination is conducted under conditions of extreme concentration).

To avoid such errors we have used three controls: firstly, only data obtained in three independent experiments were accepted; secondly, absorption studies were run after the Ouchterlony tests; and finally, immunoelectrophoresis was used as a further criterion that the cross-reactions of the subhuman primates were actually due to homologue activity and not to defective antisera. The latter test introduces a non-immunological parameter, electrophoretic mobility, to characterise the homologues.

Only the combination of Ouchterlony analysis and absorption control (and where possible immunoelectrophoresis) is accepted as a CDA.

2. *Absorption control*

We have mentioned that protein antisera are mixtures of determinant-specific antibodies. If one adds to the antiserum a homologue of the immunogen (human protein) carrying less determinants than this, all antibodies which find their respective determinants on the homologue will be bound and precipitated; the supernatant, 'absorbed' antiserum contains the rest of the antibodies and will react only with the corresponding determinants. Absorption with the immunogen (in this study: human plasma) totally destroys the

Table 4 Absorption test of anti-human fibrinogen serum from rabbit used for the Ouchterlony analysis shown in Table 3

The same stepwise decrease of cross-reactions with increasing taxonomic distance from man is apparent. The absorption is conducted exhaustively, i.e. for as long as any residual reaction with plasma from the species used for the absorption is observed. The first and the last line taken together indicate the specificity of the cross-reaction: Antibody populations the same as those effecting the cross-reaction are responsible for the reaction of human plasma.[28]

Antiserum to human homologue of	*Absorbed exhaustively with plasma of*	*Reaction with plasma from*						
		Ho	*Go*	*Po*	*Ma*	*Ce*	*Ga*	*GP*
fibrinogen	*H. sapiens* (Ho)	–	–	–	–	–	–	–
fibrinogen	*G. gorilla* (Go)	–	–	–	–	–	–	–
fibrinogen	*P. pygmaeus* (Po)	–	–	–	–	–	–	–
fibrinogen	*N. nemestrina* (Ma)	–	–	–	–	–	–	–
fibrinogen	*C. albifrons* (Ce)	+	+	+	+	–	–	–
fibrinogen	*G. crassicaudatus* (Ga)	+	+	+	+	+	–	–
fibrinogen	Guinea pig (GP)	+	+	+	+	+	+	–
fibrinogen	saline control	+	+	+	+	+	+	+

ability of the antiserum to react with any homologue, including the human, all residual reactions (if any) being unspecific in nature. Table 4 provides an example for such an absorption test.

3. *Immunoelectrophoresis*

Because of a limited supply of several subhuman primate plasmas and as a result of the somewhat lower sensitivity of this method, this test was not always conducted. Using the technical procedure described elsewhere,[13] in all cases examined electrophoretic mobilities were identified in the same range for the human and subhuman primate samples.

Evaluation

Determinant analysis of single proteins as described in the preceding paragraphs has several implications. Firstly, one can estimate the minimal number of determinants with different specificity on the human protein. At the same time, a certain characterisation and definition of these determinants is achieved (e.g. a certain determinant is always restricted to certain homologues and is missing in all other species). Secondly, the distribution of human determinants among proteins of other primates (and non-primates) is examined.[21] In the present connection, the third implication is the most important; namely, that it is possible to estimate evolutionary rates and subsequently to calculate the age of Latest Common Ancestors (LCAs).

Every spur in CDA (Fig. 3*B*) corresponds to at least one accepted mutation separating the two species compared. This is a direct consequence of the colinearity of antigenic structure and the respective structural gene. Those substitutions observable by immunological means are evidently a random sample of all substitutions occurring. It is thus possible to conclude from the resulting data, which have been compiled in Table 5, that differences in determinant structures tend to increase with taxonomic distance between the species involved. However, it is a matter for discussion whether, and to what extent, the number of mutations fixed is a function of time elapsed. Yet it is a precondition for the use of CDA for the estimation of numerical relationships among various species that such a function exists. The main argument against such a dependence upon time elapsed is that the acceptance of mutations is influenced by natural selection. The selective forces must have varied almost unceasingly during evolution, so that no continuous acceptance or acceptance rate would have resulted. This problem was faced by biochemical geneticists long before immunogeniticists, and the latter can therefore happily make use of biochemical arguments for

Table 5 Compilation of the determinants present on primate molecules of plasma proteins, detected by antisera to human homologues. (In addition to our own results, data from the literature have been employed. An earlier paper[11] provides references, while our experimental findings on the chimpanzee will be published elsewhere.)[29] Only P-determinants are considered.

	Gm, γ_3-chain	Plasminogen	Gm, γ_1-chain	β-lipoproteins	μ-chain (IgM)	β_2-glycoprotein I	Fibrinogen	Ceruloplasmin	α_1-lipoproteins	C4	C5	Inv-determ. (κ-chain)	Xh-factor	Myoglobin	Prealbumin	α-chain (IgA)	Transferrin	α_1-trypsin inhibitor	γ-chain (IgG)	C_1-esterase inhibitor	Haptoglobin	β_2-glycoprotein II	β_2-glycoprotein III	α_2-macroglobulin	Albumin	Lp (a) (β-lipoproteins)	Inter-α-Trypsin inhibitor	Cholinesterase	δ-chain (IgD)	Acid α_1-glycoprotein	α_{2HS}-glycoprotein	Haemopexin	Catalase (E.C.1.11.1.6)	GOT (E.C. 2.6.1.1)	Gc-globulin	Isf (1) (γ-chain)	Lt (a) (β-lipoproteins)
Homo sapiens	9	6	5	4	4	3	3	3	3	3	3	3	3	3	2	2	2	2	2	2	2	2	2	2	2	2	1	1	1	1	1	1	1	1	1	1	1
Gorilla gorilla	4	5	2	4	4	2	3	2	3	3		0	3	3	2	2						2	2				1	1	1						1	0	1
Pongo pygmaeus	4	4	1	4	3	3	3	3	3	3	2	0		3	2											2	1		1						1	0	
Macaca mulatta	3	3	0	3	2	2	3	3		2	1	0	1	2	1	1	1	1	1	1	1	1	1	1	1	1	1	1	1	1	1	1	1	1		0	
Cercopithecus aethiops	0	3	0	3	2	2	3	3		2	1	0		2	1		1	1	1		1	1	1	1						1	1	1	1	1		0	
Erythrocebus patas	3	3	0	3	2	2	3	3	3	2	1	0			1		1	1	1		1	1	1	1						1	1	1				0	
Cercocebus galeritus		3		3	2	2	3	3		2	1	0			1		1	1	1		1	1	1	1						1	1	1					
Cebus albifrons	0	2	0	2	2	2	2		2	1		0				0						0	0				0	0	0								
Galago crassicaudatus	0	1	0	1	1	1	1		1	1		0				0		0			0	0	0				0	0	0	0	0	0					
Nycticebus coucang	0		0			1						0						0			0										0	0					
Perodicticus potto	0		0		1					1		0				0		0													0						
Non-primate eutherians	0	0	0	0	0	0	0	0	0	0	0	0	0	0	0	0	0	0	0	0	0	0	0	0	0	0	0	0	0	0	0	0	0	0	0	0	0

an evolutionary rate which is constant over long periods of time.

1. It is possible to estimate evolutionary rates for two of the proteins (haemoglobins and cytochromes c) most thoroughly examined by sequence analyses, as well as for other proteins. It has been observed that the number of substitutions between homologues from pairs of taxonomic groups (e.g. mammals and birds) are more or less of the same magnitude. From findings of this kind, it is possible to calculate either 'unit evolutionary periods' – times for one substitution to occur – or changes per 100 residues and per 100 million years by simple comparison with palaeontologically estimated times of divergence among major species groups.[3,22,23]
2. New observations appear to indicate that neutral mutations may have played a more important role during evolution than would be expected as a result of purely selective forces,[24] and that the amino acid composition of an average vertebrate protein more or less reflects the genetic code.[25]

The calculation of the ages of Latest Common Ancestors for man and several other primates from CDA data. Our evaluation is based on the same procedure as that described for biochemical methods. We have used those determinants restricted to primate homologues and their distribution in the primate order, assuming a direct linear relation between time elapsed and numbers of immunologically detected mutations. The formula used was

$$T_i(\mathrm{RTU}) = \frac{100}{n} \cdot \sum_{j=1}^{j=n} \frac{f_{ji}}{f_{j\,HS}} \qquad \text{– I}$$

where T_i (RTU) is the time since the eutherian common ancestor (T_i = O) to the LCA of man and species i in relative time units (RTU), defined as T_i = 100 (RTU) for man. n is the number of proteins included, and f_{ji} and $f_{j\,HS}$ are the numbers of (human) determinant present on the i and the human homologue of protein j. The relative immunological time scale (*RITS*) of primate evolution, measuring time in RTU, can be transformed into an absolute one (*AITS*) by taking

$$100\ \mathrm{RTU} = x \text{ million years} \qquad \text{– II}$$

Since x (which is equivalent to the age of the eutherian common ancestor) cannot be determined exactly by classical means, uncertainties not inherent in the immunogenetical method of CDA must enter into an absolute scale as compared with the relative scale. Therefore we prefer the latter in cases where only the molecular relationships are concerned. If absolute values are needed, equation I can be transformed accordingly. With x = 70 million years:

$$T_i(MY) = \frac{70}{100} T_i\,(RTU) = \frac{70}{n} \sum_{j=1}^{j=n} \frac{f_{ji}}{f_j\,HS} \qquad \text{– Ia}$$

It should be mentioned that Zuckerkandl and Pauling propose an exponential function to be used for the same purpose (using biochemical data) in order to take account of multiple substitutions at the same site.[3] We believe however, that a linear relationship is appropriate for periods of time as 'short' (by phylogenetic standards) as that involved in primate evolution.

Results

The results obtained with equations I and Ia from CDA results are compiled in Table 6. All scale values prove to follow the classically established traditional pattern as to the successive divergence of prosimians, New World monkeys, Old World monkeys and the pongids from the line leading to man. In the same Table, classically determined values for times of divergence are given, and these are in the same range as are our CDA data. This concordance of results derived by two methods as different as the morphological and the molecular immunogenetic approaches further reaffirms the theoretical concepts discussed above.

Table 6 The T_i values for the primate species included in our time scale of primate evolution[11,29] compared with data derived from morphological criteria

Species	*Immunogenetically determined*			*Classically derived data after*	
	RITS in RTU	*AITS*	*age (in 10⁶ years)*	*Pilbeam*[26]	*Simons*[9]
H. sapiens	100.0	70.0	0.0	—	—
P. troglodytes	86.2	60.3	9.7		
G. gorilla	82.6	57.9	12.1	15	14-20
P. pygmaeus	77.5	54.3	15.7	(10-20)	
Old World monkeys (mean of 4 species)	59.6	40.4	28.3	40 (30-50)	30 (27-33)
New World monkeys (*Cebus albifrons*)	22.9	16.1	54.9	55 (45-60)	50-55
Prosimians (*Galago crassicaudatus*)	9.5	6.7	63.3	60-80	70
LCA of eutherians	0.0	0.0	70.0	—	—

1. *Prosimians*

Three species were included in our study: *G. crassicaudatus, N. coucang* and *P. potto*, all three belonging to the Lorisiformes. Only the first-mentioned species was investigated according to our criterion of 15 different proteins per species, because of the limited supply of plasma available in the other two cases. However, in all cases where *N. coucang* or *P. potto* homologues were actually investigated, they proved to be identical to *Galago* proteins in their antigenic structure. Therefore, it may be justifiable to estimate the same value of 6.7 million years after the origin of the scale as the time of divergence of all three prosimians (*Nycticebus, Potto* and *Galago*) from the line of descent leading to higher primates.

The *Galago* time of divergence is the oldest observed by us in the primate order. On the other hand, all proteins examined in this species shared more determinants with man than did any non-primate. In this connection, investigation of insectivores and of other groups of prosimians regarded as diverging very close to the origin might provide interesting conclusions concerning the positions of prosimians relative both to other primates and to other eutherians.

2. *New World monkeys*

Only one species, *Cebus albifrons*, was investigated. In all cases, determinants present on *Cebus* homologues were also observed in all higher primate protein molecules, while the *Cebus* molecules carried at least the same number of determinants as the *Galago* proteins, and usually exhibited more.

3. *Old World monkeys*

Four species were included in the scale. They were in all cases found to share identical antigenic structures in their molecules.* Several more Old World monkeys (OWM) investigated showed the same behaviour, including *M. nemestrina* and *P. hamadryas*. OWM represent the second group of primates which is remarkably homogeneous as to the 'human' determinants their members possess. This finding is quite contrary to the pongid determinant patterns (see below) and this at first recalls Landsteiner's idea of *serological perspective.*[27] This 'perspective' analogy was employed to explain the observation that antisera from closely related species detected small antigenic differences which were otherwise not observable between homologues of other closely related species. However, on the basis of modern molecular immunogenetics,[28] we have been able to offer an alternative concept to this 'perspective effect', introduced by

* One of the values indicated in the literature (Gm factors of γ_3-chains) provides an exception to this rule, however, as has been confirmed in our own experiments.

classical serologists into early theory. With respect to the appearance of both prosimians and OWM as homogenous groups in terms of their patterns of 'human' determinants, we propose the following explanation: all antisera employed were raised in the rabbit, which recognises those surface areas of introduced protein molecules which are sufficiently substituted as determinants† in relation to its own homologue. Evidently, these determinants are not recognised as 'human' but as 'non-self'. On the other hand, the existence of nodal sequences for proteins of a common evolutionary group should have a parallel in antigenic structures. Therefore we suggest that determinants which are present in all primates, including prosimians, are actually *common primate determinants*, those found in all primates except prosimians are *common simian determinants*, those observed in all higher primates are *common catarrhine determinants.*

4. Pongids

With few exceptions, all pongids included in this study carried either more or an equal number of 'human' determinants on their plasma proteins than do OWM homologues. A general regularity observed, in addition, was that the orang-utan possesses less or an equal number of determinants of this group than does the gorilla, which in turn bears the same relationship to the chimpanzee. This pattern allows us to distinguish three more subgroups of 'human' determinants and – if any – species-specific human determinants proper. The number of the latter group was astonishingly small in comparison with structures common with pongids, especially the chimpanzee. The consequent implications for the evolution of pongids and man will be discussed in greater detail elsewhere.[29]

Conclusions and future aspects

The foregoing description of concepts and results of CDA in the study of primate evolution can be summarised as follows: it is possible by immunological techniques to measure mutations established during phylogeny and to derive for the LCAs of man and major subhuman primate groups time relations which are consistent with classical palaeontological evidence.[11] Of the three groups in which several species were investigated, prosimians and OWM are remarkably homogenous, while pongids show stepwise approximation to human antigenic structures from orang to gorilla to chimpanzee. This observation is explained by a distinction of common determinant groups (CDG), common to primates (present as the only determinants on prosimian molecules), common to

† The antiserum to C_1-esterase inhibitors was raised in the goat.

simians (on New World monkey proteins) and common to catarrhines (present in addition to other, primate and simian, determinants on OWM homologues). Similar determinant groups can be distinguished in pongids. The relationships thus established for the primate order by the immunogenetical approach are indicated in Fig. 5. It is apparent that much additional research is needed before results of comparable volume can be used to establish phylogenetic relations among primates on the basis of sequence analyses. It would therefore seem promising to investigate further primate species by CDA. Key species which were omitted from the present study because of lack of plasma samples are the gibbon (a more primitive pongid) and prosimians such as *Tarsius* or a representative of the Tupaiiformes. Study of these forms would yield additional information on the relationships of these primates, with their special branching positions at the root of the whole order.

A major advantage of CDA is that the same technique and the same antiserum can be applied to problems like this. Difficulties arise, however, whenever antisera to proteins from other species than man are employed. If our above-mentioned criteria are applied, 15

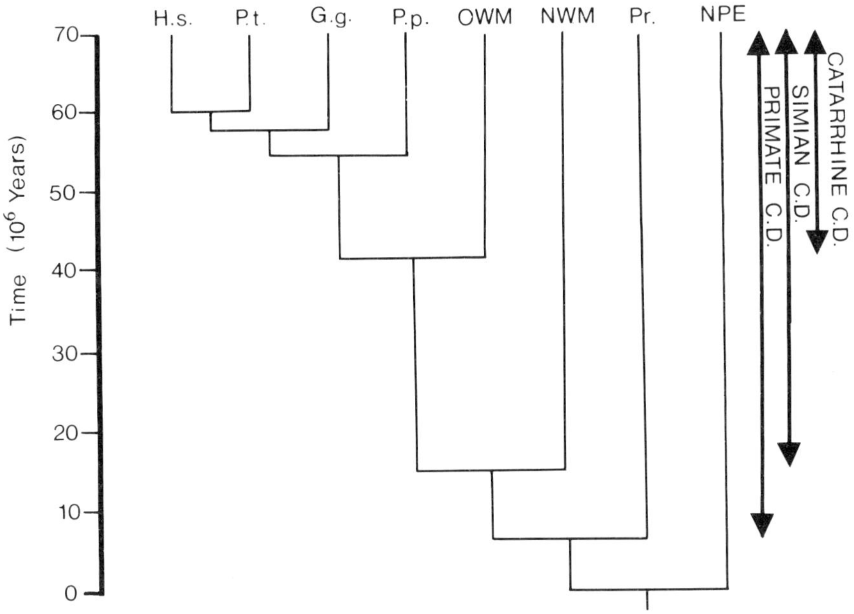

Figure 5. An evolutionary pedigree (AITS) derived from the CDA data given in Table 6. The arrows at the right indicate the ranges over which common primate, simian and catarrhine determinants are found. Key: H.s. = *Homo sapiens*; G.g. = *Gorilla gorilla*; P.p. = *Pongo pygmaeus*; OWM = Old World Monkeys; NWM = New World Monkeys (*Cebus albifrons*); Pr. = Prosimians; NPE = non-primate eutherian mammals. Also included are our results on the chimpanzee (P.t. = *Pan troglodytes*).[29]

different proteins of, say, *Galago* must be isolated and purified, antisera must be prepared and absorbed properly (since 'pure' proteins are rarely a reality), *before* the actual analysis can start. However, such antisera might eventually be employed to provide control over our present results. Moreover, they may be expected to distinguish more readily between the closer relatives of the species producing the immunogen. In the case of anti-human sera, differences between proteins from man and the various pongids become particularly apparent; accordingly a panel of anti-*Galago* sera might be used to investigate relations between various prosimian species.

In this connection, one further aspect should be briefly mentioned, though it is outside the subject of the present symposium, and that is the use of CDA outside the primate order. Human plasma proteins do not cross-react to a sufficient degree to permit phylogenetic conclusions to be drawn outside the primate order. But here again other antigen-producers might be used. Another possibility would be to employ as antigens proteins which show a wider range of cross-reactions than do plasma proteins. Possible candidates for such proteins are the eye-lens proteins investigated by Manski.[30] Interesting results for the new field of immunogenetics which has been created by CDA of homologous proteins may be expected as soon as the various components of this protein group are better characterized; at present it is already established that the range of cross-reactions for eye-lens proteins comprises all vertebrates.

The medical use of antisera to a wide variety of human plasma proteins has, however, led to a preponderance of investigations exploiting the practical availability of anti-human sera. One more application which deserves mention, as it may prove valuable in primate research, is the use of antisera to human proteins for the quantitative determination of a subhuman primate homologue. If, for example, human albumin cross-reacts strongly with its *Galago* homologue, it is also possible to determine the latter quantitatively in the blood of the animal, by making use of the anti-human serum or even of commercially available sets of Mancini plates which have been developed and are used in clinical diagnostics.[31]

In conclusion, it would seem that the construction of two time-scales of primate evolution consistent with fossil evidence, and the selection of likely further applications of cross-reactions (particularly through CDA) among proteins, adequately demonstrate that the approach described here should prove to be a powerful tool in primatological research.

Acknowledgments

The experiments described in this paper have been sponsored as part of a study of the determinant structures of human plasma proteins supported by the Deutsche Forschungsgemeinschaft, Bonn-Bad Godesberg, Germany. The skilful technical assistance of U. Reinle and R. Rauschnabel is gratefully acknowledged.

NOTES

1 Sanger, F. (1965), in Green, D.E. (ed.), *Currents in Biochemical Research*, New York.

2 Bauer, K. (1974), 'Cross-reactions between human and animal plasma proteins. V: The role of the antiserum and immunological specificity in relation to the phenomenon of cross reaction', *Humangenetik*, 21, 179.

3 Zuckerkandl, E. and Pauling, L. (1965), 'Evolutionary divergence and convergence in proteins', in Bryson, V. and Vogel, H.J. (eds.), *Evolving Genes and Proteins*, New York.

4 Florkin, M. (1966), *A molecular approach to Phylogeny*, Amsterdam.

5 Bauer, K. and Rehberger, D. (1970), unpublished results.

6 Buettner-Janusch, J. (1965), comment in the discussion of part III of Bryson, V. and Vogel, H.J. (eds.), *Evolving Genes and Proteins*, New York.

7 Pauling, L., Pressman, D. and Campbell, D.H. (1944), 'The serological precipitation of simple substances. VI: The precipitation of a mixture of two specific antisera by a dihaptenic substance containing the two corresponding haptenic groups. Evidence for the framework theory of serological precipitation', *J. Am. Chem. Soc.* 66, 330.

8 Sarich, V.M. and Wilson, A.C. (1967), 'Immunological time scale of hominid evolution', *Science* 158, 1200.

9 Simons, E.L. (1969), 'The origins and radiation of the primates', *Ann. N.Y. Acad. Sci.* 167, 319.

10 Read, D.W. and Lestril, P.E. (1970), 'Hominid phylogeny and immunology: a critical appraisal', *Science* 168, 578.

11 Bauer, K. (1970), 'An immunological time scale for primate evolution consistent with fossil evidence', *Humangenetik* 10, 344.

12 Ouchterlony, O. (1962), 'Diffusion-in-gel methods for immunological analysis – II', *Progr. Allergy* 6, 30.

13 Clausen, J. (1969), *Immunochemical Techniques for the Identification of Estimation of Macromolecules*, Amsterdam and London.

14 Bauer, K. (1970), 'Cross-reactions between human and animal plasma proteins. II: The influence of structural and functional factors', *Humangenetik* 10, 1.

15 Goodman, M. (1964), 'The specificity of proteins and the process of primate evolution', *Protides of the Biol. Fluids* 1964, 70.

16 Bauer, K. (1971), 'Cross-reactions between human and animal plasma proteins, IV: Non-eutherian Mammalia', *Humangenetik* 13, 49.

17 Bauer, K. (1970), 'Investigations on the antigenic heterogeneity of human fibrinogen and plasminogen', *Klin. Wschr.* 48, 443.

18 James, K. (1965), 'A study of the alpha$_2$-macroglobulin homologues of various species', *Immunology* 8, 55.

19 Roulet, D.L.A., Gugler, E., Rosin, S., Renaud, N.M. and Hässig, A. (1960), 'Untersuchungen über die Determinantenstruktur menschlicher Serumproteine', *Vox Sang.* 5, 479.

20 Bauer, K. (1970), 'The antigenic determinants of several human plasma proteins: their determination from the investigation of the cross-reactions between human and other mammalian plasmas', *Intern. J. Prot. Res.* 2, 137.

21 Bauer, K. (1969), 'The occurrence of antigenic determinants of human blood proteins in mammalian plasmas', *Humangenetik* 8, 27.

22 Margoliash, E., Fitch, W.M. and Dickerson, R.E. (1971), 'Molecular expression of evolutionary phenomena in the primary and tertiary structures of cytochrome c', in Schoffeniels, E. (ed.), *Biochemical Evolution and the Origin of Life*, Amsterdam and London.

23 Dayhoff, M.O. (1969), *Atlas of Protein Sequence and Structure*, Silver Spring: Nat. Biomed. Res. Found.

24 Kimura, M. and Ohta, T. (1971), 'Protein polymorphism as a phase of molecular evolution', *Nature* 229, 467.

25 King, J.L. and Jukes, T.H. (1969), 'Non-Darwinian evolution', *Science* 164, 788.

26 Pilbeam, D. (1970), personal communication.

27 Landsteiner, K. (1947), 'The specificity of serological reactions', Cambridge, Mass.

28 Bauer, K. (1973), *Einführung in die Immungenetik*, Stuttgart.

29 Bauer, K. (1973), 'Age determination by immunological techniques of the last common ancestor of man and chimpanzee', *Humangenetik* 17, 253.

30 Manski, W. (1959), 'On the biological basis of organ-specific cross-reactions of tissue antigens', *Intern. Convoc. Immunol.*, Buffalo, N.Y., Basel.

31 Bauer, K. (1974), 'Cross-reactions between human animal plasma proteins VI: An assay method for ape and monkey plasma proteins using antihuman antisera', *Humangenetik* 21, 273.

C. N. COOK and D. HEWETT-EMMETT

The uses of protein sequence data in systematics

Introduction

Amino acid sequences, besides defining the structure and hence function of proteins, could provide an extremely large pool of taxonomic characters of equal weight. As such, they could be used to refine our current inferred phylogenetic trees to a greater extent than any other single type of character, except the base sequence of the DNA itself. The uniformity of these characters, derived from the fact that each amino acid is coded by three nucleotide units of chromosomal DNA, should eliminate the need for subjective weighting (or equally subjective lack of weighting) associated with classical character sources. We need: (1) to discuss the validity of the above postulates; (2) to ensure that the maximum amount of phylogenetic information is extracted from the data; (3) to examine the relevance of the methods to systematics as a whole. The rationale behind the use of characters at the molecular rather than the organismal level has already been adequately argued by Zuckerkandl and Pauling.[1]

Before considering the methods currently in use, it is desirable to discuss the relationship between phylogenetic trees and topologies. We follow Eck and Dayhoff and define a topology as a branched structure which can represent the relationships between the N taxa under consideration.[2,3] Proposed evolutionary events can be marked along the branches; but a topology has no time direction. A topology can acquire a time direction and be converted into a phylogenetic tree by folding it about one of its (2N-3) branches, such that all the branches progress forward in time from the point of folding, which will represent the common ancestor of all the taxa involved. It is arguable whether a phylogenetic tree, so produced, can have a time scale, although certain points on the tree might be fixed by use of dated fossils. To apply a time scale to a phylogenetic tree based on protein sequence data would require the anchoring in time of some of the ancestor nodes by use of fossils of a reliable nature, and knowledge either of the constancy of the rate of mutation of the protein or of those sites in the protein sequence which are free to change randomly because they are not subject to selection.

The problem of deciding where the topology should be folded is a common one in classical taxonomy and usually manifests itself as argument about the primary sub-division of an order, for example whether the Carnivora should be divided into Pinnipedia and Fissipedia or whether a division into Feloidea and Canoidea-Pinnipedia is more appropriate. Closer to home, there is still discussion about whether the Primates are best split into Prosimii and Simii or Strepsirhini and Haplorhini, i.e. whether the tarsiers should be included with the lemur-loris group or with the monkey group (Fig. 1). These are folding problems and not topological questions.

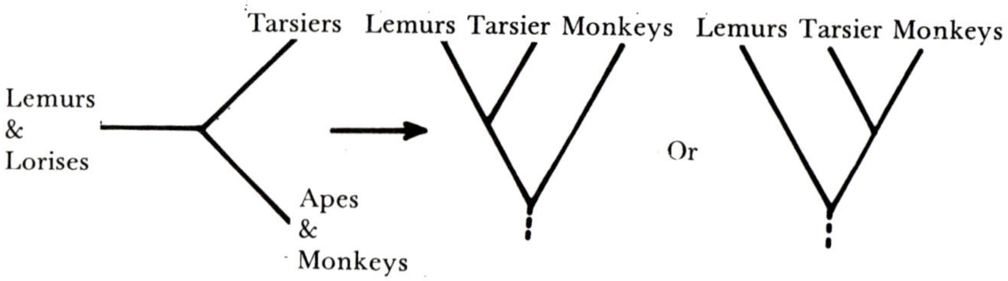

Figure 1. Derivation of alternative phylogenetic trees from a single topology.

The ancestral node of all the taxa being compared (i.e. the folding point of the topology) is not strictly determinable, but two approaches can give an approximation. The first invokes the somewhat dubious principle of non-reversibility of evolution, which might hold if a large number of complex characters were involved. This method cannot be used for protein sequences without bringing in functional considerations and, to date, we have very little data to associate function with the variable parts of sequences (apart from the physiological studies of Sullivan, and the crystallographic comparisons of human variant haemoglobins undertaken by Greer and Perutz).[5,6] In fact, some would say that these variable sites have no functional significance and are therefore not subject to natural selection (e.g. Kimura; King and Jukes).[7,8] The second method attempts to find the mid-point of the topology, such that the average branch lengths from this point are equal. This method is well suited to uniform characters such as sequence data; but it involves the assumption that the rate of evolution is more or less constant. If there is a radiation of branches near this point, its localisation could be suspect; but this problem may be side-stepped by the addition of an extra distantly related taxon, for instance to compare a group of primates with a clear non-primate such as a bird. In this way, a distinct ancestral primate node could be computed and the topology could be safely folded along the bird-ancestral primate branch using both of the above criteria.

The number of dichotomous topologies (T) which can be fitted to N taxa ($N \geq 3$) conforms to a product series, viz:

$$T = \prod_{x=o}^{N-3} 2(N-x)-5$$

T therefore rises very rapidly as the number of taxa is increased. As examples, 10 taxa can be arranged in 2,027,025 topologies, while the 51 recognised 'prosimian' species[9] can be arranged in nearly 10^{76} topologies: to get the number of trees, these figures are multiplied by (2N-3). It is essential, therefore, that some selection of promising structures be imposed and that the folding of the topology be delayed to a late stage in the work. The methods discussed all build a best fit topology and then refine the structure until no further improvement in the fit can be attained.

Matrix methods

A matrix method for amino acid sequences was first proposed by Fitch and Margoliash,[10] and used with slight modification by Barnabas, Goodman and Moore.[11] It derives from the application of cluster analysis to numerical taxonomy (Rao; Sokal and Sneath; Edwards and Cavalli-Sforza).[4,12,13]

The set of protein sequences are aligned so that supposedly homologous sites are compared, and a table of differences between the taxa is constructed. As a refinement, each difference is scored according to whether one, two or three nucleotide base mutations (it is necessary to distinguish base mutations at the DNA level from phenetic mutations, e.g. amino acid changes) would be required to interconvert the two amino acids using the genetic code (see Fig. 2).[14] This is called a minimum mutation distance. Since deletions have occurred over time, these distances are corrected according to the number of amino acids in each sequence to give a triangular or half-matrix of 'relative distances'.

The shortest distance, d_{AB}, in the matrix between any two taxa (A and B) is selected, and these taxa are assumed to form a terminal pair of branches. All other taxa are assigned to set C and the average of the distances from A and B, respectively, to the members of C are calculated to give d_{AC} and d_{BC}. It is now possible to solve the three simultaneous equations to obtain the lengths of branches a, b and c, as shown in Fig. 3. Whereas lengths a and b will correspond to actual branch lengths, c will be the average length of all the branches from the node to the taxa of set C.

The next shortest distance is selected and the branch lengths

First Nucleotide	Second nucleotide: U	C	A	G	Third Nucleotide
U	Phe	Ser	Tyr	Cys	U
	Phe	Ser	Tyr	Cys	C
	Leu	Ser	terminate	terminate	A
	Leu	Ser	terminate	Trp	G
C	Leu	Pro	His	Arg	U
	Leu	Pro	His	Arg	C
	Leu	Pro	Gln	Arg	A
	Leu	Pro	Gln	Arg	G
A	Ile	Thr	Asn	Ser	U
	Ile	Thr	Asn	Ser	C
	Ile	Thr	Lys	Arg	A
	Met	Thr	Lys	Arg	G
G	Val	Ala	Asp	Gly	U
	Val	Ala	Asp	Gly	C
	Val	Ala	Glu	Gly	A
	Val	Ala	Glu	Gly	G

Figure 2. The Genetic Code.

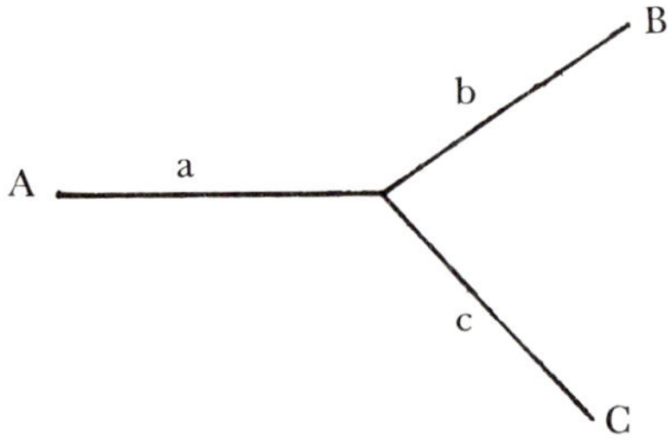

$$a = \tfrac{1}{2}(d_{AB} + d_{AC} - d_{BC})$$
$$b = \tfrac{1}{2}(d_{AB} + d_{BC} - d_{AC})$$
$$c = \tfrac{1}{2}(d_{AC} + d_{AC} - d_{AB})$$

Figure 3. Calculation of average branch lengths.

calculated. The branch lengths about an internal node of the topology may all be averages (see Appendix 1 for details) and the average distance AB, for example, will be the average of all the combinations of comparisons between all the members of sets A and B. This is an unweighted pair group method of clustering. Finally only two clusters of taxa will remain, and the overall common ancestor will be determined by finding the average mid-point such that $a = b = \tfrac{1}{2}d_{AB}$ (see Appendix 1). The lengths of the internal branches are found as follows: the length of any branch between two adjacent internal nodes x and y equals the average length from x to all its terminal elements, via y, minus the average length from y to these same elements. Examples of these calculations are given in Appendix 1.

Having assigned all the branch lengths, the original matrix may be reconstructed and the 'average per cent standard deviation' calculated:

$$\text{'Average \%S.D.'} = 100 \times \sqrt{\frac{1}{\frac{1}{2}N(N-1)-1} \sum_{i>j} \left(\frac{d_{ij}\ \text{original} - d_{ij}\ \text{reconstructed}}{d_{ij}\ \text{original}} \right)^2}$$

where N = number of taxa
dij = mutation distance (original and reconstructed) between taxa i and j

The fraction before the sum is the reciprocal of the number of elements in the half matrix minus one. The 'average per cent standard deviation' is not in fact a standard deviation as used in statistics. It can be used as an arbitrary figure indicating the fit of the reconstructed matrix to the original matrix. By slightly relaxing the criteria for choosing the next pair of branches to be joined, the data may be fitted to alternative topologies, until no improvement of fit can be obtained by further alterations.

If there is to be an exact fit of data and reconstructed matrices, one must formulate the axiom that there exists a topology such that, for all pairs of taxa, the sum of the branch lengths between any two taxa i and j is always equal to the value of the matrix element of i and j.

In the event that this axiom does not hold (in the real case this could be due to the presence of convergence or multiple mutations at a single site), there will not be complete agreement and a positive 'standard deviation' will be noted.

The trees constructed using matrix methods appear to be in fairly good general agreement with the course of evolution as at present conceived on independent grounds. The major anomaly that may be observed is the occasional presence of branches with negative length. If the axiom formulated above holds good, then the presence of a negative branch would indicate that the clustering was at fault, and that an alternative tree should be tried.[15] If it is impossible to get an exact fit, then negative branches may have to be accepted; but it is difficult to see what they could represent in evolutionary terms. It is probably best to consider them as manifestations of the fact that the data do not conform to the initial axiom, i.e. that a matrix distance should be equal to the sum of the corresponding branch lengths. It might be more appropriate to load against negative branches by making them either zero or positive for the purposes of reconstructing the matrix. In Appendix 2, we show the results of constructing a few of the many possible trees from some α-chain haemoglobin sequences, and applying the criteria discussed above. It may be noted that the 'better-fitting' trees lacked or had insignificant negative branches.

The method described above uses an unweighted cluster analysis, since the length from a node along a branch to a group of taxa is the mean of all the lengths from that node to each of the taxa, irrespective of their arrangement on more terminal branches. Many biologists feel intuitively that one should give equal weight to the two branches which are being joined, irrespective of the number of taxa giving rise to these branches.

Hence in Fig. 4 the average length from node X is:

(i) *weighted* = ½ d + n + ½ [c + m + ½(a + b)]
(ii) *unweighted* = ¼ d + (n + c) + (n + m + b) + (n + m + a)

A weighted clustering is performed initially as in an unweighted method; but when the two branches A and B are joined, the two rows of the matrix elements of A and B are replaced by a single row of (AB) elements, which are the average of A and B. This reduced matrix is taken as the starting-point for the next clustering cycle. The disadvantage of this weighted method is that for the same tree different values may be obtained for the branch lengths depending upon the order in which available pairs of branches are joined. The unweighted method generates a unique set of branch lengths for any given tree, so that, on balance, we prefer this method. Dayhoff (personal communication) has developed a weighted matrix method which does not suffer from this disadvantage. Both Fitch and Margoliash and Barnabas, Goodman and Moore use an unweighted method,[10,11] but in a later paper Fitch, Margoliash and Gould use a weighted method.[16]

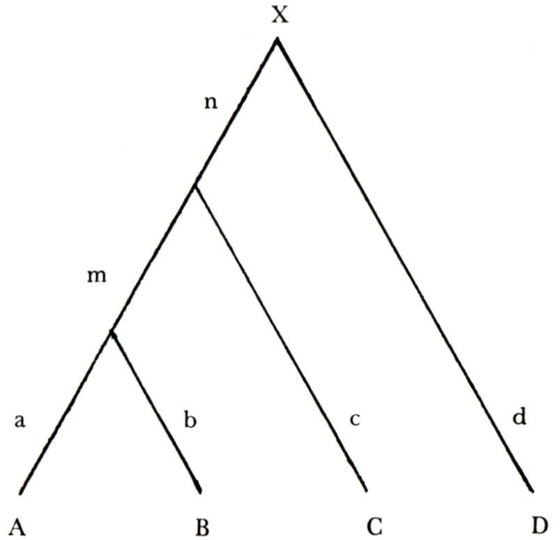

Figure 4. Tree illustrating effect of weighting branch lengths.

When, as often happens, two homologous proteins differ not by an amino acid change, but by a site having been deleted, it is difficult to compare them directly. To make allowance for this both Fitch and Margoliash and Barnabas, Goodman and Moore corrected their minimum mutation distances.[10,11] Fitch and Margoliash multiply their distances by the ratio of amino acids in each protein (with the larger number as numerator of the fraction) to give a relative distance. Barnabas, Goodman and Moore prefer to divide their distances by the number of shared amino acids and multiply by 100 to give a 'percentage mutational divergence'. Both these corrected values may give valid measures of affinity (that of Barnabas, Goodman and Moore being probably the better of the two), but it is clear that neither leaves the data with comparable units. We cannot therefore expect to obtain branch lengths in distinct units (i.e. mutations) either. It may also be noted that the amino acid sites in a set of sequences may be divided between the variable and so-called invariant groups. The latter provide us with no taxonomic information and in this context are superfluous and therefore should not be incorporated in the above correction factors. The only solution to the problem of deletions is to compare only those sites which are common to all the taxa under study.

The question of testing the fit of the tree to the data must remain a central one. We have already stated that the 'average per cent standard deviation' is not in this case a statistically valid one. It is clear that the data, if they can be used to produce a tree of affinities, cannot be considered as at all independent; for example, 20 'independent' sequences would generate 190 matrix comparisons, from which would be calculated the lengths of only 35 independent branches. Hence this type of statistical analysis cannot be invoked. Other workers have fallen into the trap of making non-independent comparisons when trying to compute rates of protein evolution, e.g. Air et al. and Kimura.[17,7] Having established the non-statistical nature of the test of fit, we may ask whether it may nevertheless be a useful arbitrary figure of merit. If the intention of the method is to obtain a tree which produces a reconstructed matrix as similar as possible to the original data matrix, then the 'average per cent standard deviation' is probably as good an indication as any. If, however, the intention is to build a minimum evolution tree from the data, then it may not provide the best answer. Consider the case (see Fig. 9 in Appendix 1) of an average deviation from the data by one unit in the terminal branch between S and node 1. This branch length would be used in computing six distances of the reconstructed matrix. A similar unit deviation on the internal branch between node 2 and node 3 would appear in twelve distances of the reconstructed matrix. In a 20-taxon tree, a single branch value could be used in reconstructing between 19 and 100 of the 190 elements of the matrix. It is clear that there will be a bias towards selecting a tree in which deviations from the data are found in the more terminal branches, even though that tree may not be the minimum evolution tree. A comparison between the trees produced by Eck and Dayhoff, and Fitch and Margoliash, for cytochrome-c showed that roughly equivalent middle branches were lower, and terminal branches greater in length in the tree of Fitch and Margoliash.[2,10]

A further point which should be noted is that the distances in the data matrix are *minimum* mutation distances. It is therefore axiomatic that the elements of the reconstructed matrix should all exceed or be equal to the corresponding data elements; yet an examination of the two matrices of Fitch and Margoliash shows the presence of a large number of negative deviations from the data, confirming that the best fit contains distortions.[7] The best test of fit for the minimal evolution model would be to find that tree which gave the lowest sum of the absolute branch lengths. In this test, each branch is used only once and, by taking absolute values, negative branches would be discouraged and occur only rarely in better fit trees.

Minimum path methods

This method attempts to find a minimum evolution fit at each homologous site, among all the taxa under consideration simultaneously,[2,18,19] rather than to try to fit all the data simultaneously to pairs of taxa as the matrix methods do. To achieve this, the method makes use of three computer routines:

Routine 1. For a given topology, the routine generates for each homologous site the nodal amino acids (as far as this is possible) and counts the mutations (i.e. phenetic changes) required to interconvert all members. This is repeated for the whole set of data and the total number of mutations required is calculated. This total is the 'minimum mutation value' for the topology.

Routine 2. This is a topology-building routine.

Routine 3. This is a topology-refining routine.

This method is best illustrated by working a simple example. Consider the following sequences of 5 amino acids from 5 species (in fact, the data are positions 8, 35, 68, 71 and 111 from haemoglobin α-chains selected purely to illustrate the method). Four of these species are chosen at random and Routine 2 sets out the three possible topologies (Fig. 5*a*):

Routine 1 fits the data to each topology in turn and, by reconstructing the most likely ancestral sequence at each node, counts the substitutions required to interconvert each node. It does this by listing all the possible amino acids that could occur on each

	Haemoglobin α – chain sites				
	8	35	68	71	111
Chicken	Asn	Ser	Asn	Glu	Ala
Human	Thr	Ser	Asn	Ala	Ala
Rabbit	Thr	Gly	Lys	Gly	Ser
Pig	Ala	Gly	Lys	Gly	Val
Horse	Thr	Gly	Leu	Gly	Val

Note: Ala – Asn, Leu – Lys, Leu – Asn, Val – Ser are double base mutations, others are single steps

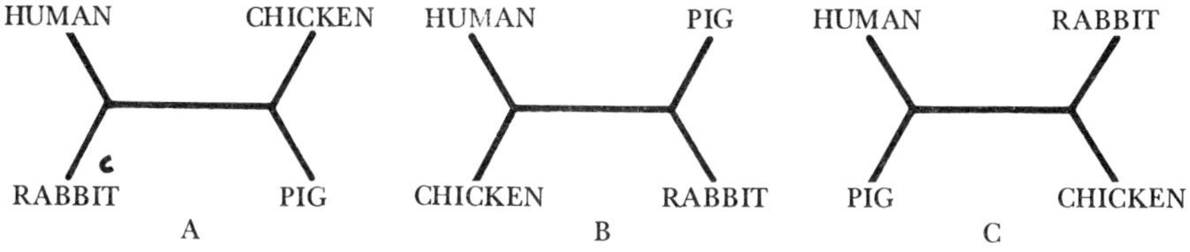

Figure 5*a*. Example used to illustrate Eck and Dayhoff method.

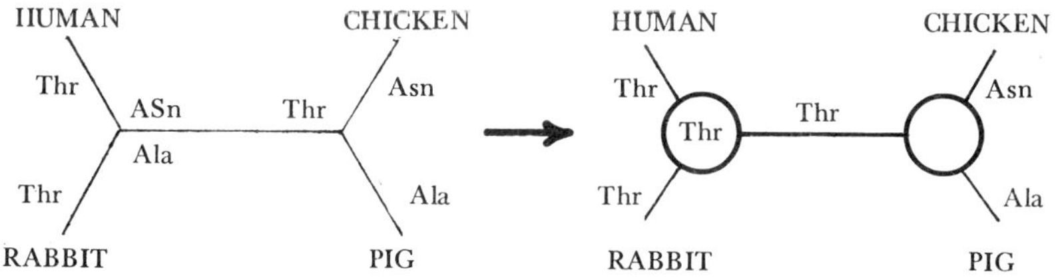

Figure 5*b*. Site 1 (α8) amino acids in topology A.

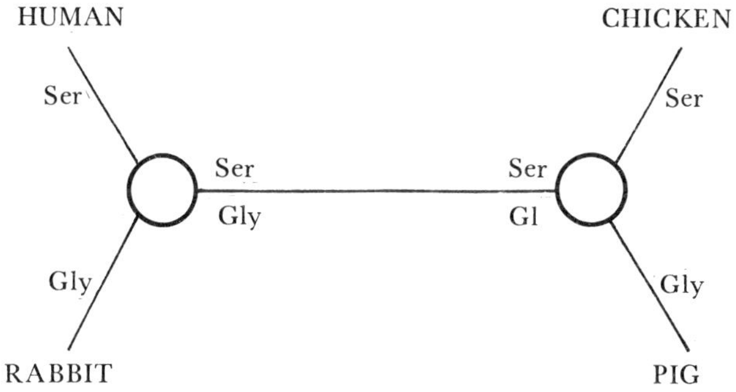

Figure 5*c*. Fit of site 2 (α35) amino acids to topology A.

of the three branches about a node and choosing the most abundant one as the nodal amino acid. For example, fitting the data from the first site (α8) to the branches about the 'left hand' node of topology A shows that two of the branches are Thr, which is therefore the nodal amino acid (see Fig. 5*b*). The 'right hand' node has three different amino acids, and therefore this site (α8) requires, for topology A, two of the three changes Thr-Ala, Thr-Asn, and Ala-Asn.

Site 2 (α35) presents the same situation at each node; each nodal amino acid being indeterminate, since there is an equal choice between glycine and serine (see Fig. 5*c*). The mutation value will be a minimum of two, there being two Gly-Ser changes (the exact branches on which they occur being indeterminate). The third site (α68) is an identical situation to site 2 (α35), requiring a minimum of two mutations, Ans-Lys, but with indeterminate nodal amino acids. The fourth (α71) and fifth (α111) sites both require the theoretical minimum of two mutations to interconvert the three amino acids. Thus, for topology A, the minimum number of mutations required to fit the sequences is 10. Topology C also requires 10 mutations. With topology B we find a different situation when the sequences are fitted. The species are now arranged such that only one Gly-Ser change is required for site 2 (α35), the nodes now being firmly allocated, and only one Asn-Lys change for site 3 (α68). The score for this topology is thus 8, which is also the theoretical minimum.

Routine 2 then takes this best fit topology and adds a fifth taxon, the new branch being added to each of the 5 existing branches, generating 5 topologies. The sequences are then fitted to each topology in turn by routine 1 as before. With larger numbers of taxa, the method described may leave several blank nodes, so the routine has been constructed so that the 3 neighbouring nodes to a blank node are examined. If two of these are the same, then the blank node can be filled in, and this cycling is repeated until no further nodes can be assigned. Any remaining blank nodes are left indeterminate. Routines 2 and 1 thus allocate branches and assess the resultant topologies until all the taxa have been added.

It is possible, however, that a chance convergence of the data has caused us to choose a topology which is not the best fit for the data when all the taxa have been added in. Routine 3 systematically cuts each branch and applies the 'cut end' to all the branches in the other part of the topology (see Fig. 6) and Routine 1 tests the fit until no topology with lower mutation value can be found. The nodal sequences are then determined (as described above) and the mutations interconverting them assigned to the branches.

Finally, the topology may be folded to convert it into a phylogenetic

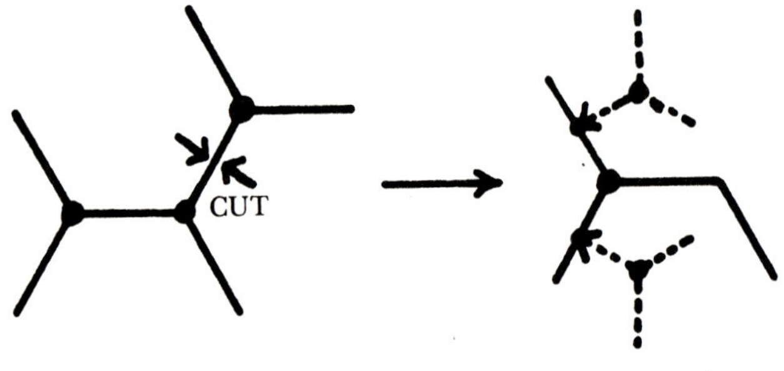

Figure 6. Showing how a topology may be cut (by routine 3) and alternative topologies produced.

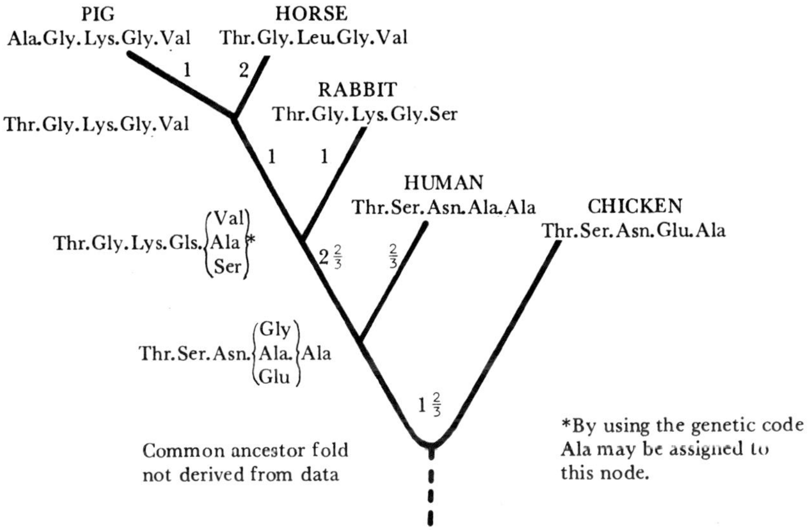

Figure 7. Best fit tree for five haemoglobin α-chain sites used to illustrate the method of Eck and Dayhoff.[2]

tree using either the criterion of constant rate of mutation or other criteria not derived from the data. Fig. 7 represents the tree produced from the 5 amino acid sequences of the 5 species used to explain the method.

Although the computer programme has not yet been published, the 3 routines would seem to provide a fully efficient method for arriving at a minimum evolution topology, using amino acids as unit characters.

In particular, Routine 3 represents an important advance in refining taxonomic structures, but a slight doubt must always remain that the final 'minimum' topology may not be the true minimum, but a local 'hollow' (to use a geographical analogy). This is unlikely, but cannot be proved not to occur unless all possible topologies are tested.

Examination of the genetic code (Fig. 2) reveals that a single amino acid mutation may require up to three nucleotide *base* mutations. This suggests that a greater precision might be obtained by inferring nucleotide base sequences from amino acid sequences, and processing these character states by the above method. In Fig. 7, the fifth site (α111) ancestral node for rabbit-pig-horse is ambiguous when amino acid unit characters are employed. However, the genetic code shows that a valine-serine mutation requires a double base mutation and that alanine is a possible intermediate. Hence alanine may be assigned to this site.

Fitch has produced an algorithm to fit inferred nucleotide sequences to a selected tree.[20] Although he has solved the problem posed by the alternative codons (triplets) for serine, leucine and arginine, his technique suffers in that it operates on a tree and not on a topology, and hence the potential work-load is increased by a factor of (2N-3).

Discussion

In comparing the matrix methods with the minimum path method of Eck and Dayhoff,[2] one is attracted to the latter in that it portrays events which could have taken place in the course of evolution of the proteins. The Eck and Dayhoff method handles variable amino acid sites as single characters which could equally well be included as additional characters in a numerical analysis of morphological character states. It is because of this approach that the method can be fully efficient in a minimum evolution model. This cannot be said of the matrix methods. In particular, the presence of convergence and/or multiple substitution sites will, with matrix methods, be realised as errors from the best fit. Such characters are acceptable to the minimum path method as long as their effects on the topology are not too localised or too extensive, as is also the case with classical studies.

Matrix methods are, however, the only ones that may be used for serological data. If the experimental data for a protein can be transformed into distance data, a matrix can be tabulated which could be considered as an approximation to a distance matrix of sequence data. We would recommend that in these circumstances an unweighted analysis (or a weighted one giving a unique solution) be used with a minimum total path length test of fit.

It might be worthwhile to point out that all the methods described have a tree-building programme which produces only an approximation to the best fit. This is because cluster analysis relies on an assumption of an even rate of evolution, and the computer cannot be programmed with hindsight. Accepting this as the case, any reasonable approximation such as one based on classical methods could be used to initiate the refining programme in the Eck and Dayhoff method and, in addition, this could be adopted in the matrix methods, provided that alternative trees were examined systematically in much the same manner.

Having devised adequate methods of finding a minimum evolution fit of the sequences, it would be nice to say that all we needed now were a lot of variable proteins, accurately sequenced, which could be lined up to generate a phylogeny, with high accuracy for all those species in which we were interested.

Although this is *the* major requirement, there remains one more problem which is still somewhat intractable – that of sequence homology. When proteins from two species are examined, although we may be sure that they are homologous (in that they were derived from a common ancestral *protein*), we may not be able to say whether that common ancestral protein was present in the most recent common ancestral species. For example, if we compared a human haemoglobin α-chain with a baboon β-chain, we would find ourselves faced with the ludicrous conclusion that the two species separated in the late pre-Cambrian! This is an extreme case; but others are more subtle. Boyer et al. have found haemoglobin chain variants in both the chimpanzee and gorilla.[21] Both of these variant chains differ from the chimpanzee (= human) α^A chain by at least 9 mutations, but from each other by only 2 mutations, suggesting the presence of an α-chain gene duplication which predated the divergence of the two species but which has remained 'silent' except in these two individuals. Boyer et al. have also examined the haemoglobins of ceboid monkeys.[22] In this group, all species examined contain a minor haemoglobin which shows resemblances to hominoid haemoglobin A_2 and consists of two α^A-chains and two δ-like chains. This type of minor haemoglobin component is not found in the Cercopithecoid monkeys.[23] Comparison of the δ-chain sequences from the two groups (Ceboidea and Hominoidea) makes it equally likely that they are either the result of a β-chain gene duplication before the separation of those two lineages, or the product of independent duplications after the superfamilies had separated. The significance of this situation is that if the β-duplication occurred separately (Fig. 8*a*), β-β, δ-δ and intergroup β-δ comparisons (including the cercopithecoid β-chains) may be made for the purpose of studying the superfamilial relationships. However, if the β-duplication occurred before the simian radiation (Fig. 8*b*), a β-δ

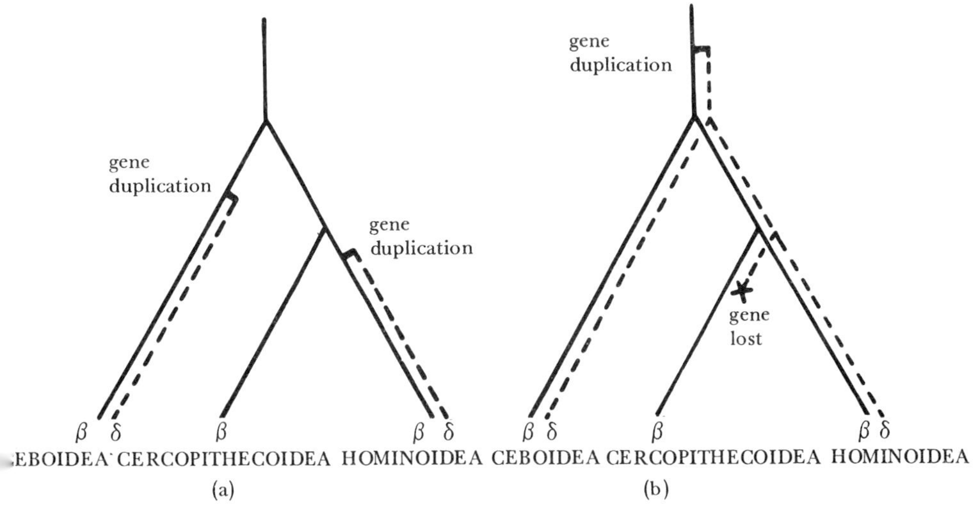

Figure 8. Alternative explanations of the origin of δ-like chains in the Anthropoidea.

comparison would lead to the erroneous assessment that there was an early separation of the superfamilies. The only slight hope that this problem might be resolved would be to examine the δ (or δ-like) chains of more species, particularly those with the more ancient ancestral relationships within each superfamily.

Buettner-Janusch and Twichell, Buettner-Janusch and Hill, and Goodman et al. have all suggested that the non-α chains of lemuroids are likely to be homologues of the human γ-chain (foetal chain), since lemuroid haemoglobins show several properties analogous to simian foetal haemoglobins: no separate foetal haemoglobin has been found in the lemuroids, and the sequences of the non-α globin chains of *Lemur fulvus* and *Propithecus verreauxi*, though not yet adequately published, are claimed by Goodman et al. to cluster with human γ and kangaroo non-α in their best globin phylogenetic tree.[24–28]

Duplications are also known in the chains of the buffalo, goat, mouse, *Macaca fascicularis* and horse.[29–34] The studies of the two latter species well illustrate the complexities that arise when both gene duplications and allelic variations are present. Lack of serial homology in the sequence set of a particular protein may normally be detected by considering sequence sets from other types of proteins, i.e. it is essentially a sampling problem, as in classical evolutionary studies.

Goodman et al. have also suggested that, in view of its marked difference from other mammalian haemoglobin α-chains, *Tupaia glis* α-chain should be considered as the survivor of an ancient duplication, but it could equally well be used as further evidence that the tupaioids are not primates.[26,27,35]

In Appendix 2 are some examples of trees which we have computed using the methods described in this paper. The data are sequences of haemoglobin α-chains from four prosimians, a macaque, man and representatives of some other orders. These trees illustrate several features of interest, some of which are relevant to tree building with more complex characters.

It will be seen that there is not a complete correspondence between the two criteria of fitness used. Conversely, there are several different trees giving very similar low values, so that a final choice of tree is not easy to make. The tree shrew, although initially placed as the most isolated mammal by the cluster analysis, gave a much better fit when placed with the mouse. (However, bearing in mind the range of species involved, this may not be taken as a suggestion that the tupaioids are members of the Rodentia!) This latter move also eliminated or gave insignificant negative branches.

Most of the internal branches are short, having very few mutations assigned to them. Although part of the reason for this is undoubtedly the short time duration of the branches, the longer branches computed by the Eck and Dayhoff method suggest that procedural factors may also be involved. Whatever the reason, they suggest that only a small proportion of the data are relevant to the determination of the branching structure of the trees. We feel that these short branches justify the use of the genetic code to gain some additional precision and also indicate that the weights applied to classical characters may be more critical than generally thought. Many of the changes are species-specific so that, although they show the distinctness of the taxa, they contribute nothing to elucidation of their relationships; within the context of the species chosen they are superfluous.

The trees in Appendix 2 are drawn to scale and not, as most authors have done, brought to a common 'now' plane. This format shows very clearly that there has been a very wide variation in the apparent evolutionary rates of these proteins. We therefore reject the use of protein sequences as 'evolutionary clocks'. A similar conclusion may be drawn from the fibrinopeptide data of Doolittle.[36,37] While the inconstancy of evolutionary rates in general is accepted by most biologists, it is often forgotten that constancy of rates is enshrined in the principle that phenetic similarity is an accurate measure of degree of kinship, which is the basis of most of our numerical methods.

The method of Eck and Dayhoff allows us to distribute the changes of character states upon the branches of a given tree. This implies a partition of characters into ancestral and group-specific, an approach proposed for morphological characters by Remane,[38] which allows a better assessment of the status of the higher taxa that we have erected. It may be suggested, for instance, that the Lemuridae are at present defined by mainly ancestral prosimian characters, and that the failure to discriminate between ancestral and group-specific characters has been responsible in the past for at least part of the controversies over the relationships of the higher apes with man. This approach makes it evident that further characters will be required to refine our present trees. It is likely that protein sequencing will be of great assistance in this task.

We have described how sequences may be fitted to a minimum evolution model. Although we feel intuitively that the minimum evolution model is the best approximation to the actual evolutionary tree, we must not regard the final phylogenetic tree that we compute as necessarily reflecting the precise course of the evolution of our proteins. For example, from an analysis of apparent single, double and triple mutations, Jukes has concluded that there is more multiple substitution than the minimum model shows.[39] The principle of minimum evolution has not yet been formally demonstrated to be the best estimator of the actual course of evolution.

We hope that this paper has shown that in practice the use of protein sequences in taxonomy is not so straightforward as the simple statement in our introductory remarks had suggested.[40] It appears, in fact, that the use of sequences is prone to many of the problems that beset those who use more complex polygenic characters, with the exception that the character states are of equal weight. The advantage of this uniformity should not be underrated. It allows us to quantify our analytical methods, and in so doing we are forced to re-consider the mechanisms involved in the act of reconstructing phylogeny.

We dance round in a ring and suppose
But the secret sits in the middle and knows

Robert Frost

Acknowledgments

We thank N.A. Barnicot, A.W.F. Edwards, R.D. Martin, and D.A. Coleman for stimulating discussion, and we are very grateful to M.O. Dayhoff and N.A. Barnicot for reading the manuscript and making valuable suggestions.

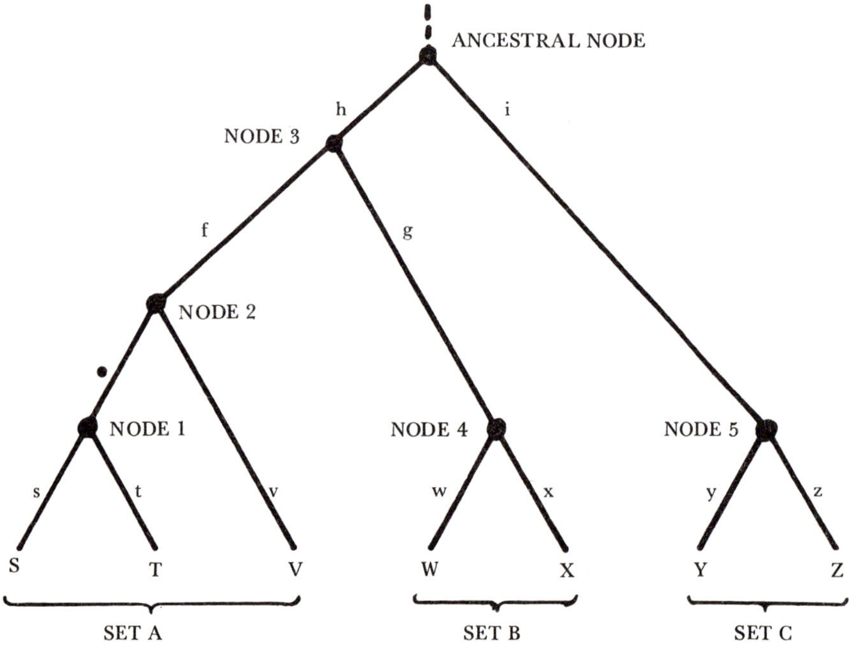

Figure 9. Hypothetical tree, labelled for use with the equations in Appendix 1.

Appendix 1

Derivation of branch lengths by the unweighted matrix methods used by Fitch and Margoliash (1967)[10], and Barnabas, Goodman and Moore (1971).[26] See Fig. 9.

1. Average distance d_{AB} = average distance between SET A and SET B

$$= \tfrac{1}{6}\,[d_{SW} + d_{SX} + d_{TW} + d_{TX} + d_{VW} + d_{VX}]$$

where d_{IJ} = a matrix element between taxa I and J.

2. Average branch length l_{A3} = average of NODE 3 to SET A.

$$= \tfrac{1}{2}\,[d_{AB} + d_{AC} + d_{BC}]$$

3. Internal branch length $f = l_{A3} - \tfrac{1}{3}\;[v + (s+e) + (t+e)]$

f will also equal average distance from NODE 2 to SETS B and C minus the average from NODE 3 to SETS B and C.

Thus $f = \tfrac{1}{2}\,(l_{B2} + l_{C2}) - \tfrac{1}{2}\,(l_{B3} + l_{C3})$

4. ANCESTRAL NODE is at $\dfrac{d_{(AB)C}}{2} = \tfrac{1}{2}\,[\tfrac{1}{10}(d_{SY} + d_{SZ} + d_{TY} + d_{TZ} + d_{VY} + d_{VZ} + d_{WY} + d_{WZ} + d_{XY} + d_{XZ})]$

Hence $h = \dfrac{d_{(AB)C}}{2} - \tfrac{1}{5}\,[(e + f + s) + (e + f + t) + (f + v) + (g + w) + (g + x)]$

and $i = \dfrac{d_{(AB)C}}{2} - \tfrac{1}{2}\,(y + z)$

Appendix 2

This appendix contains the results of our tree-building using the methods described in the text. We used the unweighted matrix methods to test 12 trees using the data matrix of Goodman et al. for the haemoglobin α-chains of nine species.[26] The details are shown below. The trees include that generated by the unweighted pair group method (Fig. 10*a*) and the 'best-fit tree' of Goodman et al.[26] (Fig. 10*b*).sWe tested all the trees by the 'average per cent standard deviation' (APSD) and the 'total path length' (TPL) criteria, as shown on the diagrams (Figs. 10 and 11). If negative branches occurred, the tests were performed by summing distances by taking note of the sign (APSD-, TPL-) and by taking absolute values (APSD+, TPL+). The trees are drawn to a minimum mutation distance scale.

Man *Homo sapiens*	(Hs)	–							
Rhesus Monkey *Macaca mulatta*	(Mm)	5	–						
Greater Bush Baby *Galago crassicaudatus*	(Gc)	14	14	–					
Sifaka *Propithecus verreauxi*	(Pv)	23	24	25	–				
Brown Lemur *Lemur fulvus*	(Lf)	18	20	19	18	–			
Malayan Tree Shrew *Tupaia glis*	(Tg)	31	29	30	33	31	–		
House Mouse (Strain C57) *Mus musculus*	(Mus)	19	20	22	27	25	26	–	–
Horse (Hb fast 24F) *Equus caballus*	(Ec)	24	21	29	39	31	37	29	–
CHICKEN (Hb type AII) *Gallus domesticus*	(Gd)	45	46	49	57	54	60	51	53
		Hs	Mm	Gc	Pv	Lf	Tg	Mus	Ec

N.B. Some of these values do not agree with published sequences. The commonest cause of error is misassignment of acid-amides.

NOTES

1 Zuckerkandl, E. and Pauling, L. (1965), 'Evolutionary divergence and convergence in proteins', in Bryson, V. and Vogel, H.J. (eds.), *Evolving Genes and Proteins*, New York and London.

2 Eck, R.V. and Dayhoff, M.O. (1966), *Atlas of Protein Sequence and Structure*, Silver Spring: Nat. Biomed. Research Found.

3 *Taxon* is here used as a general term to cover those species, individuals, populations or other groupings under study. It is synonymous with Operational Taxonomic Unit as defined by Sokal and Sneath.[4]

4 Sokal, R. and Sneath, P.H.A. (1963), *Principles of Numerical Taxonomy*, San Francisco.

5 Sullivan, B. (1971), 'Structure, function and evolution of primate haemoglobins. II: A survey of the oxygen binding properties', *Comp. Biochem. Physiol.* 40, 359-80.

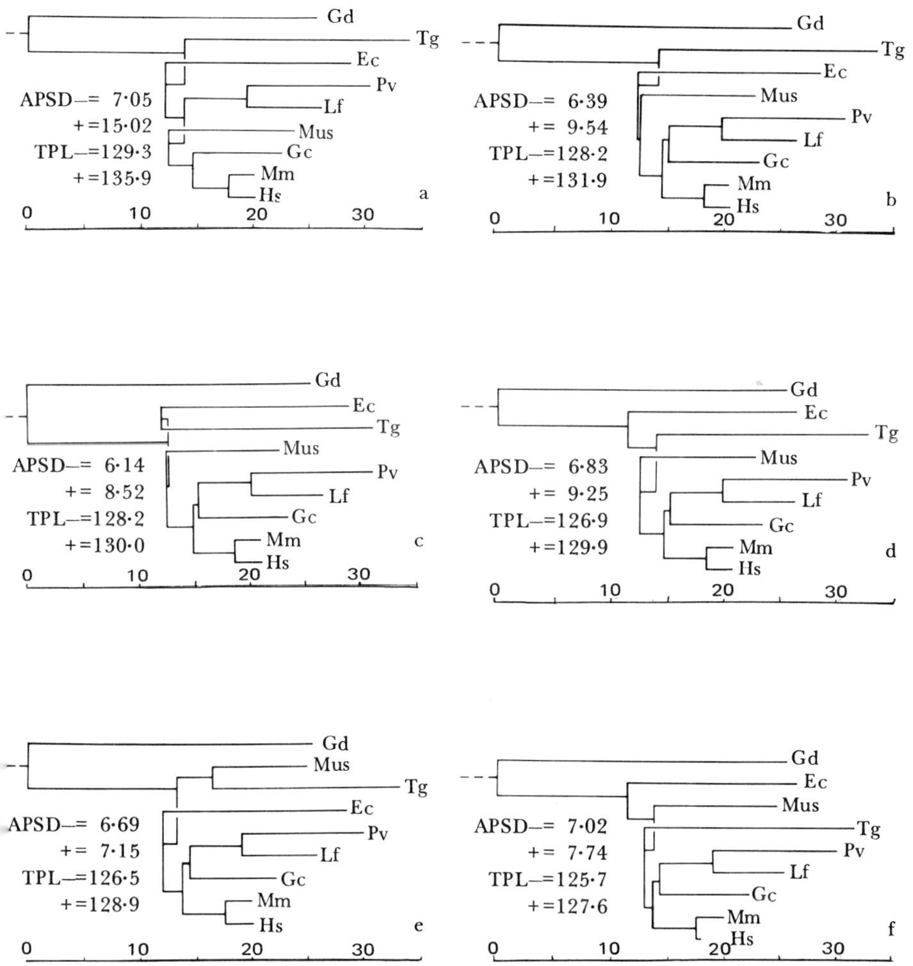

Figure 10. Examples of trees derived by unweighted matrix methods for haemoglobin α-chains.

6 Greer, J. and Perutz, M.F. (1971), 'Three dimensional structure of Haemoglobin Rainier', *Nat. New Biol.* 230, 261-4.

7 Kimura, M. (1968), 'Evolutionary rate at the molecular level', *Nature* 217, 624-6.

8 King, J.L. and Jukes, T.H. (1969), 'Non-Darwinian evolution', *Science*, 164, 788-98.

9 Napier, J.R. and Napier, P.H. (1967), *Handbook of Living Primates*, New York and London.

10 Fitch, W.M. and Margoliash, E. (1967), 'Construction of phylogenetic trees', *Science*, 155, 279-84.

11 Barnabas, J., Goodman, M. and Moore, G.W. (1971), 'Evolution of haemoglobin in primates and other therian mammals', *Comp. Biochem. Physiol.* 39, 455-82.

12 Rao, C.R. (1952), *Advanced Statistical Methods in Biometric Research*, New York.

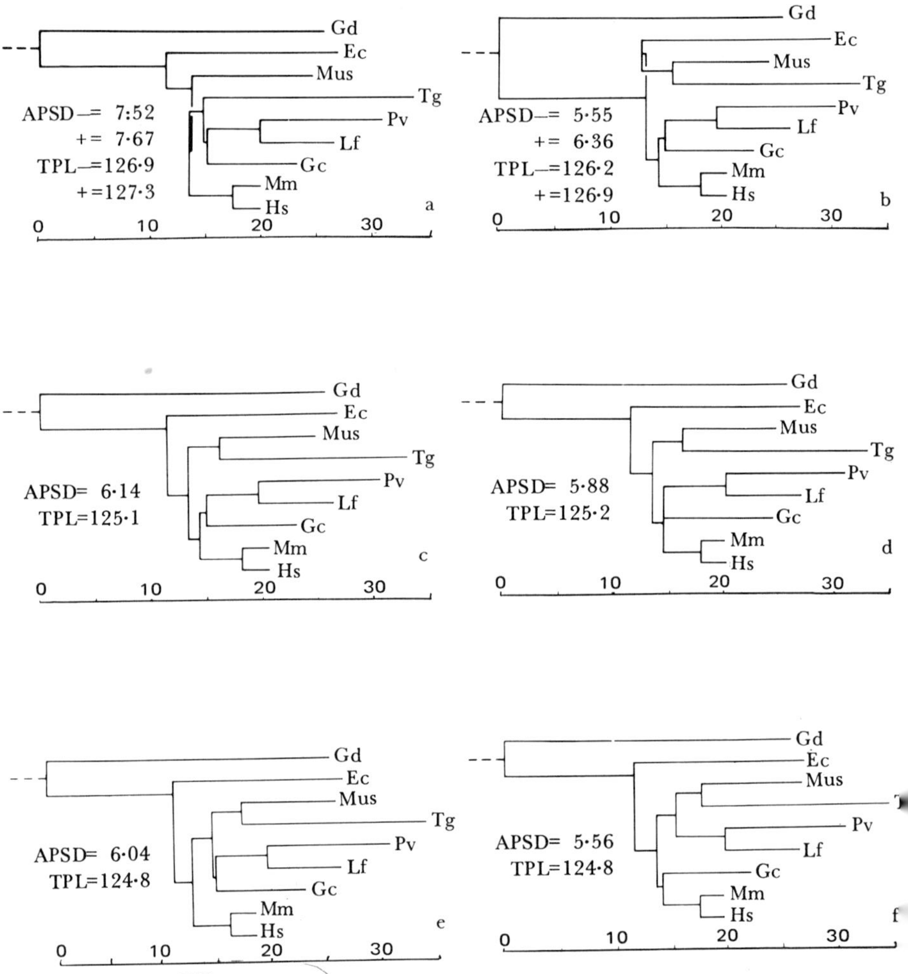

Figure 11. Further examples of trees derived by unweighted matrix method for haemoglobin α-chains.

13 Edwards, A.W.F. and Cavalli-Sforza, L.L. (1965), 'A method for cluster analysis', *Biometrics* 21, 362-75.

14 Caskey, C.T. (1970), 'The universal RNA genetic code', *Quart. Rev. Biophys.* 3, 295.

15 Cavalli-Sforza, L.L. and Edwards, A.W.F. (1967), 'Phylogenetic analysis, models and estimation procedures', *Am. J. Human Genet.* 19, 233-57.

16 Fitch, W.M., Margolish, E. and Gould, K.S. (1969), 'The construction of phylogenetic trees. II: How well do they reflect past history?', *Brookhaven Symp. Biol.* 21, 217-42.

17 Air, G.M., Thompson, E.O.P., Richardson, B.J. and Sharman, G.B. (1971), 'Amino acid sequences of kangaroo myoglobin and haemoglobin and the date of the Marsupial-Eutherian divergence', *Nature* 229, 391-4.

18 Dayhoff, M.O. (1969), *Atlas of Protein Sequence and Structure*, Silver Spring: Nat. Biomed. Research Found.

19 Dayhoff, M.O. (1969), 'Computer analysis of protein evolution', *Sci. Amer.* 221, 87.

20 Fitch, W.M. (1971), 'Towards defining the course of evolution: minimum change for a specific tree topology', *Syst. Zool.* 20, 406-16.

21 Boyer, S.H., Noyes, A.N., Vrablik, G.R., Donaldson, L.J., Shaefer, E.W., Gray, C.W. and Thurman, T.F. (1971), 'Silent haemoglobin alpha genes in apes: potential source of Thalassemia', *Science* 171, 182-5.

22 Boyer, S.H., Crosby, E.F., Noyes, A.N., Fuller, G.F., Leslie, S.E., Donaldson, L.J., Vrablik, G.R., Schaefer, E.W. and Thurman, T.F. (1971), 'Primate haemoglobins: some sequences and some proposals concerning evolution and mutation', *Biochem. Genet.* 5, 405-48.

23 Barnicot, N.A. and Wade, P.T. (1970), 'Protein structure and the systematics of Old World Monkeys', in Napier, J.R. and Napier, P.H. (eds.), *Old World Monkeys*, New York and London.

24 Buettner-Janusch, J. and Twichell, J.B. (1961), 'Alkali-resistant haemoglobins in prosimian primates', *Nature* 192, 669.

25 Buettner-Janusch, J. and Hill, R.L. (1965), 'Evolution of haemoglobin in primates', in Bryson, V. and Vogel, H.J. (eds.), *Evolving Genes and Proteins*, New York and London.

26 Goodman, M., Barnabas, J., Matsuda, G. and Moore, G.W. (1971), 'Molecular evolution in the descent of man', *Nature* 233, 604-13.

27 Buettner-Janusch, J. and Hill, R.L. (1965), 'Molecules and monkeys', *Science* 147, 836-42.

28 Hill, R.L. (1968), 'Unpublished sequences', in Sober, H. (ed.), *Handbook of Biochemistry*, Chemical Rubber Company. (N.B. There are errors in some sequences shown.)

29 Balani, A.S. and Barnabas, J. (1965), 'Polypeptide chains of buffalo haemoglobins', *Nature* 205, 1019-21.

30 Huisman, T.H.J., Adams, H.R., Dimock, M.O., Edwards, W.E. and Wilson, J.B. (1967), 'The structure of goat haemoglobins'. *J. Biol. Chem.* 242, 2534-41.

31 Hilse, K. and Popp, R.A. (1968), 'The structure of mouse haemoglobin', *Proc. Nat. Acad. Sci. U.S.* 61, 930-5.

32 Barnicot, N.A., Huehns, E.R. and Jolly, C.J. (1966), 'Biochemical studies on haemoglobin variants of the irus macaque', *Proc. Roy. Soc. Lond.* B. 165, 224-44.

33 Barnicot, N.A., Wade, P.T. and Cohen, P. (1970), 'Evidence for a second haemoglobin α-locus duplication in *Macaca irus*', *Nature* 228, 379-81.

34 Clegg, J.B. (1970), 'Horse haemoglobin polymorphism: evidence for two linked non-allelic α-chain genes', *Proc. Roy. Soc. Lond. B*, 176, 235-46.

35 Martin, R.D. (1968), 'Towards a new definition of primates', *Man* 3, 377-401.

36 Doolittle, R.F., this volume.

37 Doolittle, R.F., Wooding, G.L., Lin, Y. and Riley, M. (1971), 'Hominoid evolution as judged by fibrinopeptide structures', *J. Molec. Evolution* 1, 74-83.

38 Remane, A. (1956), 'Paläontologie und Evolution der Primaten', in Hofer, H., Schultz, A.H. and Stark, D. (eds.), *Primatologia*, Basel.

39 Jukes, T.H. (1971), 'Comparisons of the polypeptide chains of globins', *J. Molec. Evolution* 1, 46-62.

40 Since the seminar, a publication by Barnabas et al. has appeared (Barnabas, J., Goodman, M. and Moore, G.W. (1972), 'Descent of mammalian alpha globin chain sequences investigated by the maximum parsimony method', *J. Mol. Biol.* 69, 249-78). In this paper, many of the ideas suggested in our review are discussed. Topologies instead of trees are generated. A method for exchanging terminal branches has been devised such that, for a topology of N taxa 2(N-2) nearest neighbour interchange topologies exist (excluding the starting topology generated from the mutation distance matrix methods

described by Barnabas et al.).[11] A new 'statistical' test is used, named the Moore residual coefficient (MRC). This is essentially a 'sum of squared deviations' worked out for each of the interior branch points of the topology. Thus, the method generates a topology and estimates its MRC, after which the new ITERA programme generates the nearest neighbour topologies and their MRCs. The topology with lowest MRC is selected and the ITERA programme begun again, the cyclic procedure continuing until a 'final' topology is produced. When using *real* data, this is regarded as a 'valley' and the whole procedure begun again with a radically different initial topology, presumably selected by subjective criteria. They have found that several 'valleys' often exist. These 'best' topologies are subjected to an amplification of Fitch's 'maximum parsimony ancestral nucleotide reconstruction' method,[20] which itself owes much to Eck and Dayhoff's approach.[2,18,19] Ancestral reconstruction is probably near the limit of its operational efficiency. From the nucleotide ancestors produced at the nodes, they calculate the total mutation length of the topology. In doing so, they confirm some of our doubts by finding hidden mutations (e.g. Alanine (GAU or GAC) → Alanine (GAA or GAC) at α 53 (not α 54 as published) in mammalian haemoglobins).

Interestingly though, when constructing topologies from a wide range of haemoglobin α-chain sequences, three topologies, considered plausible on biological and paleontological grounds, are *not* produced by the iterative procedure (presumably having non-optimal MRC values or being undiscovered 'valleys'!). However they have total mutation lengths *lower* than the two best MRC topologies. This confirms our suspicions with regard to the choice of test of fit. We doubt whether such an oversimplified and incomplete data entry as the mutation distance matrix is worthy of so sophisticated a technique as the iterative procedure. The essential problem remains – how to select the 'best' topology, using only the sequences, and to be aware that you have it!

R. DOOLITTLE

Prosimian biology and protein evolution

Introduction

The biology of prosimians, like the biology of any group of organisms, is intrinsically interesting, if only because the creatures exist. Beyond that, however, prosimians are naturally of special interest because of their evolutionary relationship to our own species. In this article I am going to attempt to demonstrate that prosimians are also of special interest to students of protein evolution, and that the structure of their molecules, quite apart from shedding light on the molecular antiquity of man, may answer some fundamental questions about protein evolution in general.

Protein evolution

The elucidation of the structure of insulin in the early 1950s opened an entirely new aspect for comparative biology.[1] For the first time, it became possible to compare the amino acid sequences of proteins from different species. It was not surprising that, for any given protein, the more distant the relationship between the creatures compared, the more differences were found in their proteins. The unravelling of the genetic code firmly supported the general suppositions that single base substitutions in the genetic material (usually DNA) were directly reflected by single amino acid replacements in the proteins they prescribed. These were considered a major part of the grist upon which the mill of evolution turned.

It is generally assumed that the spontaneous occurrence of these single base substitutions is a random process which takes place at a more or less constant rate (although estimates of this rate vary from 10^{-6} to 10^{-9} in terms of the probability that a substitution will occur at any particular nucleotide location during a given replication).

Once a base substitution and amino acid replacement occur (the nature of the code is such that approximately 20% of base substitutions do not give rise to amino acid changes), the likelihood of persistence of that change must be considered. There are two schools of thought on this matter, which may be cast in their extreme forms as follows:

(*a*) most of the amino acid replacements which survive during the evolution of species are selected for in a Darwinian sense, the replacement being reproductively advantageous for that particular organism,

(*b*) most of those amino acid replacements which survive during the evolution of species are neutral changes, which are randomly fixed and have no bearing on reproductive success of the creatures in which they occur.[2,3]

It is important to know which of these arguments is nearer the truth. One question which can be posed to help settle this dispute is: Do primitive creatures have primitive proteins? This question can be better understood by reference to Fig. 1, in which the proteins of four related creatures (designated I, II, III and IV) are depicted upon a traditional phylogenetic tree. On the branch segments leading to these creatures are a series of ratios, m/M, n/N, o/O, etc., in which the denominators (capitals; M, N, etc.) represent the numbers of changes which *occur* in a particular protein, and the numerators (small letters; m, n, o, etc.) represent the fraction of these which *survive* during the evolution of the species (we would expect these to be very small fractions, regardless of the predominant process leading to survival). If the widely held assumption about the constancy of

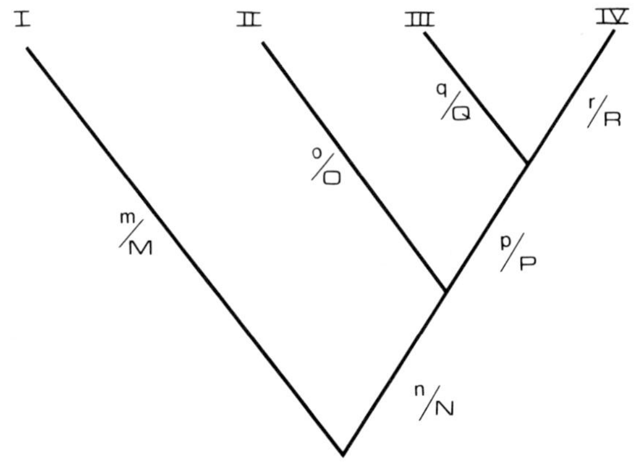

Figure 1. Cladogram showing lines of descent leading to a hypothetical protein in four different species (I, II, III and IV). Denominators (M, N, etc. represent changes which occur, whereas numerators (m, m, etc.) denote changes which actually survive evolution.

rates of occurrence is correct, then Σ M + N + O = Σ M + N + P + Q = Σ M + N + P + R, etc. The question we are trying to answer, however, has to do with the numerators of these ratios. We attempt to find if Σ m + n + o = Σ m + n + p + q = Σ m + n + p + r, etc., by comparing the amino acid sequences of proteins I, II, III and IV. If I is no more different from IV than it is from II and III, then it would appear that the number of surviving changes since II, III and IV had a common ancestor has been constant along their (summed) segments. On the other hand, let us assume that II is derived from a relatively primitive creature whose proteins have not significantly changed since it had a common ancestor with III and IV. In this case, the structures of the proteins I and II would be much more similar than I compared with III or IV. If 'old' organisms do indeed have old' proteins, then it is more difficult to accept the notion that the majority of amino acid replacements in proteins are randomly fixed, since the number of opportunities for fixation during historical time would necessarily have been equivalent to those for 'new' species.

Primitive primates

The argument can be cast in terms of primates and their traditional classification. It is generally agreed that primates can be divided into two major groups, the prosimians and the anthropoids, and that the latter can be subdivided into three subgroups, the New World monkeys, the Old World monkeys and the hominoids (Fig. 2). The question of 'primitiveness' is a delicate one, but, in simple terms, we would like to answer the same questions about the segmental lengths

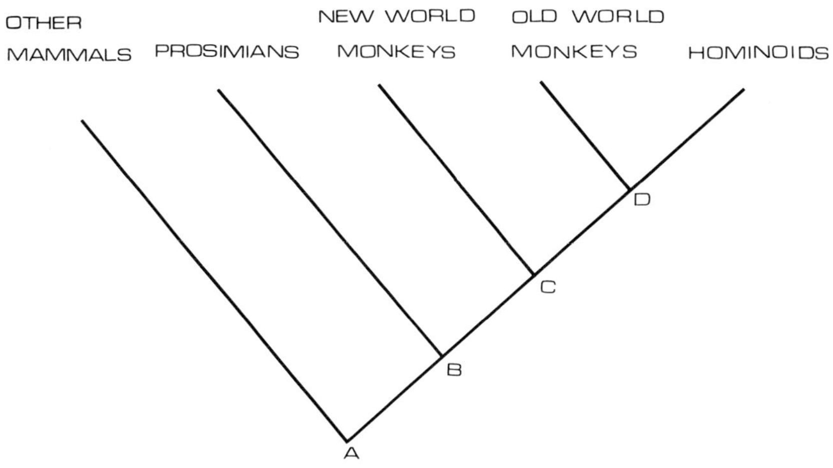

Figure 2. Simplified classification of major primate groups, A, B, C and D denote important common ancestors.

of the limbs of this classical phylogenetic tree as we posed for our protein comparisons. Do existing prosimians, for example, more closely resemble the common ancestor which existed at junction B than do the 'higher' primates? (Fig. 2). If the answer to this question is ever 'yes', then we would certainly like to examine that creature's proteins to find how divergent they are from the proteins of other primates on the one hand, and from those of non-primate mammals on the other.

Which proteins to examine?

Ideally, we would like to compare a number of different proteins for a large number of species from the groups we are studying. The practical limitations of the labour involved in amino acid sequence determinations usually demand that a choice be made restricting either the number of proteins examined or the number of species, or both. Different proteins evolve at different rates, and it is important to choose a protein for study consistent with the expected relationship of the groups being compared. For example, cytochrome c is a very conservative protein which changes very slowly during evolution (Fig. 3). On the other hand, it is a virtually ubiquitous material, and as such it is ideal for comparing distantly related creatures, and maps covering much of the animal and plant kingdoms have been prepared.[4] It is of little use for studying the relationships within a single order like the Primates, however, since there will probably be too few changes to arrive at any statistically significant conclusions. A faster-changing protein like haemoglobin (Fig. 3) would be much more suitable for the latter purpose.

At first glance, the fact that different proteins evolve at different rates (Fig. 3) might seem to argue against random fixation of amino acid replacements even before we begin our analysis. The point to be made, however, is that Darwinian *rejection* is certainly allowed by the random fixation hypothesis. In a slowly changing protein like cytochrome c, fewer of the spontaneously occurring amino acid replacements are neutral, and so fewer replacements are randomly fixed. This is quite different from thinking that a greater number of the replacements which occur in haemoglobins are selectively advantageous.

1. *Fibrinopeptides*

The fibrinopeptides play a unique role in the field of molecular evolution because of their extraordinary propensity for rapid change.[5] They are a pair of peptides which split off from vertebrate fibrinogen molecules during blood-clotting. Their relatively undemending rôle of preventing the spontaneous gelation of fibrinogen

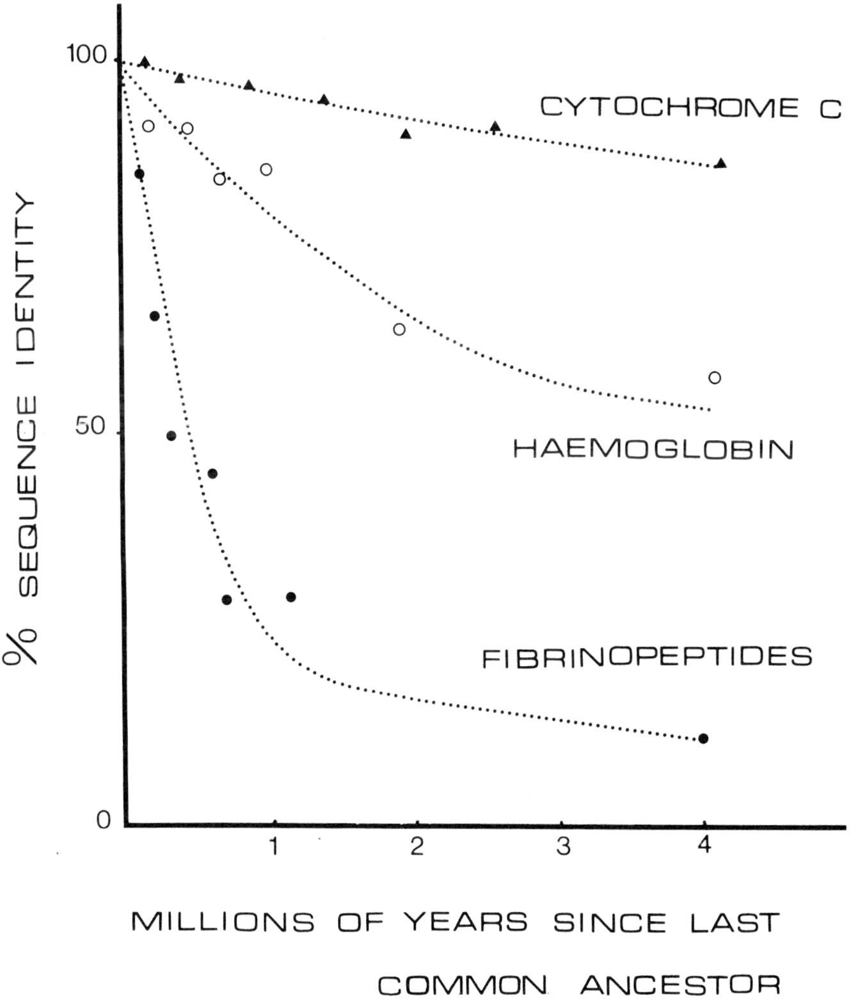

Figure 3. Differences in the rates of change of three different vertebrate proteins (only part of a protein in the case of fibrinopeptides).

molecules can evidently be accomplished by a wide variety of amino acid combinations, and amino acid replacements occur at a rate considerably faster than in the case of most other proteins or peptides (Fig. 3). To date, the fibrinopeptides of more than 45 mammalian species, representing 8 different orders, have been studied, either by Blombäck's group in Stockholm or in our laboratory in La Jolla,[6,7] where the emphasis has been on two particular orders: the artiodactyls (of which about 20 have now been studied) and the primates.[8,9] The advantage of studying fibrinopeptides for establishing the relationships between mammals stems from the fact that the changes occur frequently enough to allow step-by-step comparisons, whilst certain parts of the molecule are conservative enough to ensure proper alignment of the amino acid segments even in the face of considerable flux elsewhere.

The major drawback in the study of fibrinopeptides stems from the substantial amounts of blood necessary for their preparation. For example, in one of our early studies we required 30 litres of sheep's

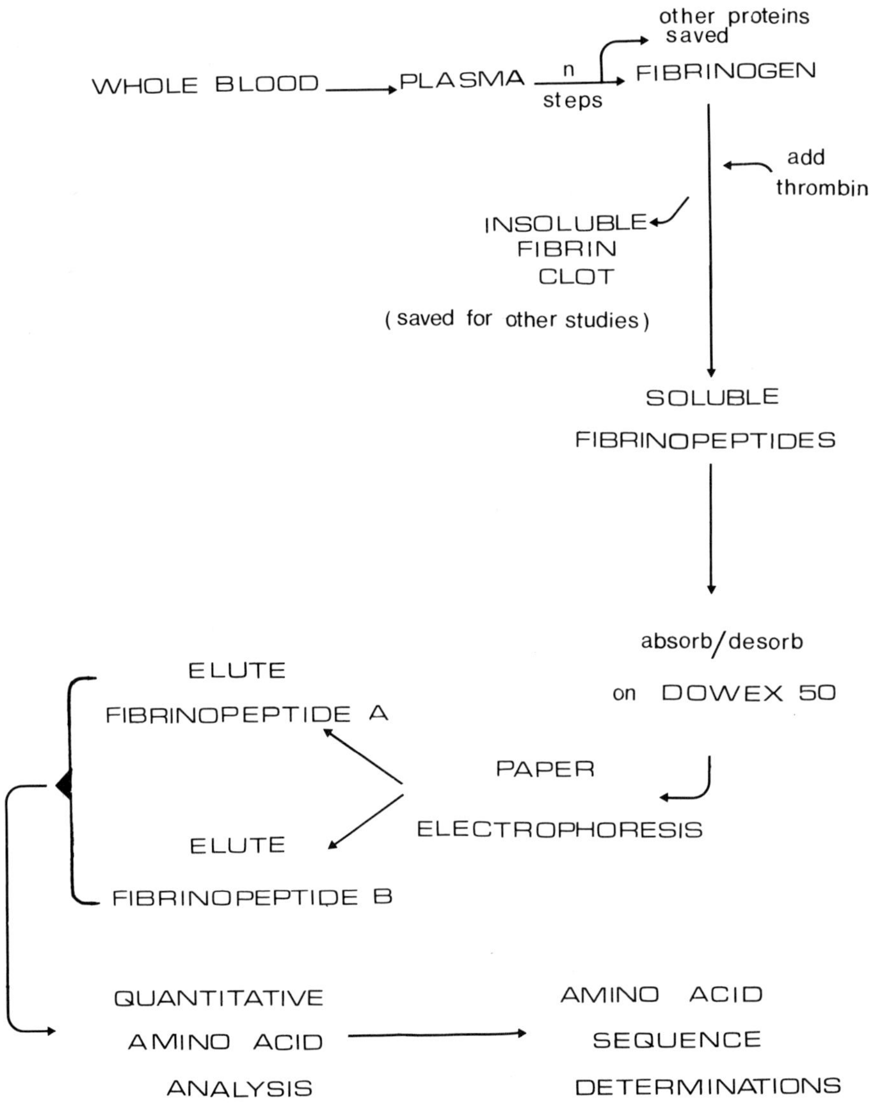

Figure 4. Diagram of steps involved in preparation of fibrinopeptides.

blood to prepare enough material to determine the sequences of the sheep fibrinopeptides A and B.[10] Techniques have improved over the years, however, and we are now able to establish the sequences of a new pair of fibrinopeptides with less than 100 ml. of blood plasma. However, this is still a considerable quantity when dealing with small, rare and expensive primates. Nevertheless, the blood plasma is quite stable when frozen, and pools from small animals can be accumulated over long periods of time. The general procedure used in our laboratory for the preparation of the fibrinopeptides is outlined in Fig. 4.

2. *Primate fibrinopeptides*

The structures of the twelve pairs of primate fibrinopeptides which have been studied so far are depicted cladistically in Figs. 5 and 6. The step-by-step amino acid replacements are readily localised, although the directions of the changes at the bases of the cladograms are necessarily equivocal. The relationships between the various species is completely consistent with other biochemical data, including immunologic criteria.[11,12,13] The advantage of these data lies in the dramatic visualisation of each of the genetic events leading to the present-day structures.

To date, only one prosimian has been studied in this survey.[9] The fibrinopeptides of the slow loris were prepared in our laboratory from pooled blood plasma stocks accumulated by Morris Goodman in Detroit. There are several very interesting features about the slow loris fibrinopeptides which have a bearing on our discussion of general protein evolution. In the first place, the two peptides, A and B, both differ considerably from the corresponding structures of the other 11 primates. Secondly, they contain certain structural features found in non-primate fibrinopeptides, including the rare amino acid tyrosine-O-sulfate – a residue found in most non-primate fibrinopeptide B, but never before found in a primate. The questions before us should now be clear: first, is the slow loris a primitive primate in a classical sense (i.e. does it resemble the prosimian-anthropoid common ancestor more than do existing anthropoid species), and, second, are its fibrinopeptides more primitive than those of the anthropoid fibrinopeptides which have been studied?

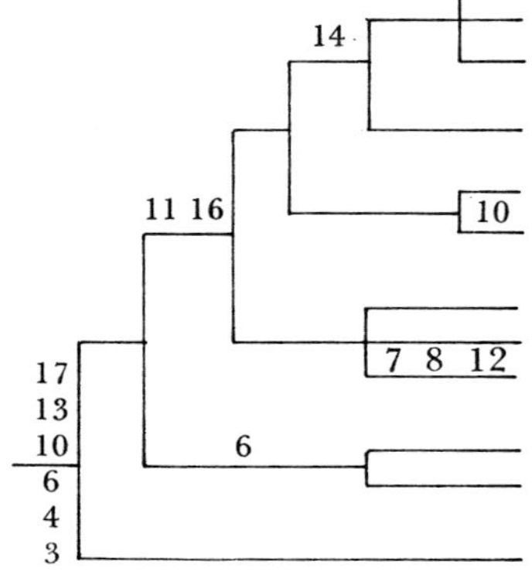

	18	17	16	15	14	13	12	11	10	9	8	7	6	5	4	3	2	1
Human			ALA	ASP	SER	GLY	GLU	GLY	ASP	PHE	LEU	ALA	GLU	GLY	GLY	GLY	VAL	ARG
Chimpanzee			ALA	ASP	SER	GLY	GLU	(GLY,	ASP)	PHE	(LEU,	ALA,	GLU,	GLY,	GLY,	GLY,	VAL)	ARG
Gorilla			ALA	ASP	SER	GLY	GLU	GLY	ASP	PHE	LEU	(ALA,	GLU,	GLY,	GLY,	GLY)	VAL	ARG
Orangutan			ALA	ASP	SER	GLY	GLU	GLY	ASP	PHE	LEU	(ALA	GLU	GLY	GLY	GLY	VAL	ARG
Siamang			ALA	ASP	THR	GLY	GLU	GLY	ASP	PHE	LEU	ALA	GLU	GLY	GLY	GLY	VAL	ARG
Gibbon			ALA	ASP	THR	GLY	GLU	(GLY,	GLU,	PHE	LEU	(ALA,	GLU,	GLY,	GLY,	GLY,	VAL)	ARG
Rhesus macaque			ALA	ASP	THR	GLY	GLU	GLY	ASP	PHE	LEU	ALA	GLU	GLY	GLY	GLY	VAL	ARG
Green monkey			ALA	ASP	THR	GLY	GLU	GLY	ASP	PHE	LEU	ALA	GLU	GLY	GLY	GLY	VAL	ARG
Drill			(ALA,	ASP,	THR,	GLY,	ASP,	GLY,	ASP,	PHE)	ILE	(THR,	GLU,	GLY,	GLY,	GLY)	VAL	ARG
Spider monkey			THR	ASP	THR	GLY	GLU	ASP	ASP	PHE	LEU	ALA	ALA	GLY	GLY	GLY	VAL	ARG
Cebus monkey			THR	(ASP,	THR,	GLY,	GLU,	ASP,	ASP,	PHE,	LEU,	ALA,	ALA,	GLY,	GLY,	GLY)	VAL	ARG
Slow loris	THR	ASP	THR	ASP	THR	ASP	GLU	ASP	GLY	PHE	LEU	ALA	LYS	GLY	ALA	ASP	VAL	ARG

Figure 5. Cladogram showing amino acid changes involved in the evolution of fibrinopeptides A of twelve primates. Numbers on left indicate position involved in change.

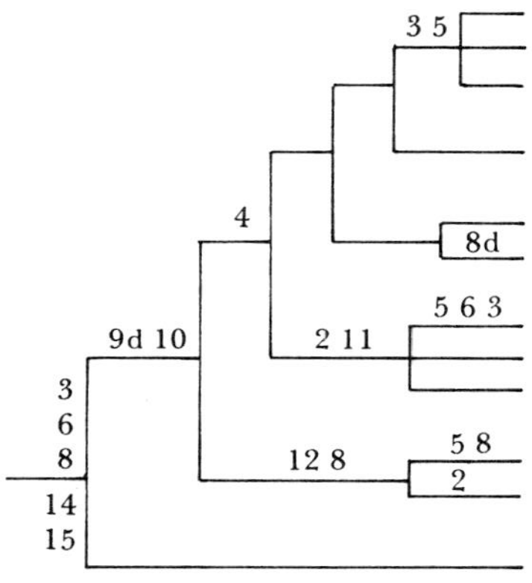

	15	14	13	12	11	10	9	8	7	6	5	4	3	2	1
Human		PCA	GLY	VAL	ASN	ASP	ASN	GLU	GLU	GLY	PHE	PHE	SER	ALA	ARG
Chimpanzee		PCA	(GLY,	VAL,	ASN,	ASP,	ASN,	GLU,	GLU,	GLY,	PHE)	PHE	SER	ALA	ARG
Gorilla		PCA	GLY	VAL	ASN	ASP	ASN	GLU	GLU	GLY	PHE	PHE	SER	ALA	ARG
Orangutan		PCA	GLY	VAL	ASN	ASP	ASN	GLU	GLU	GLY	LEU	PHE	GLY	ALA	ARG
Siamang		PCA	GLY	VAL	ASN	ASP	ASN	GLU	GLU	GLY	LEU	PHE	GLY	ALA	ARG
Gibbon		PCA	(GLY,	VAL,	ASX,	ASX,	ASX	–	GLX,	GLY,	LEU)	PHE	GLY*	ALA	ARG
Rhesus macaque		(———	———	? ———	———	———)	ASN	GLU	GLU	SER	PRO	PHE	SER	GLY	ARG
Green monkey		PCA	(GLY,	VAL,	ASX,	GLY)	ASN	GLU	GLU	GLY	LEU	PHE	GLY	GLY	ARG
Drill		PCA	(GLY,	VAL,	ASX,	GLY,	ASX,	GLX,	GLX,	GLY,	LEU)	PHE	GLY	GLY	ARG
Spider monkey		PCA	GLY	GLY	ASX	ASX	ASX	THR	GLX	GLY	ILE	LEU	GLY	ALA	ARG
Cebus monkey		PCA	(GLY,	GLY,	ASX,	ASX,	ASX)	LYS	GLU	GLY	LEU	LEU	GLY	VAL	ARG
Slow loris	ASP	GLX	VAL	ASX	ASX	*TYR	ASX	GLY	GLX	ARG	LEU	LEU	ASP	ALA	ARG

Figure 6. Cladogram showing amino acid changes involved in the evolution of fibrinopeptides B of twelve primates.

	18	17	16	15	14	13	12	11	10	9	8	7	6	5	4	3	2	1
HUMAN A			ALA	ASP	SER	GLY	GLU	GLY	ASP	PHE	LEU	ALA	GLU	GLY	GLY	GLY	VAL	ARG
SLOW LORIS A	THR	ASP	THR	ASP	THR	ASP	GLU	ASP	GLY	PHE	LEU	ALA	LYS	GLY	ALA	ASP	VAL	ARG
RABBIT A			VAL	ASP	PRO	GLY	GLU	SER	THR	PHE	ILE	ASP	GLU	GLY	ALA	THR	GLY	ARG
HUMAN B				PCA	GLY	VAL	ASN	ASP	ASN	–	GLU	GLU	GLY	PHE	PHE	SER	ALA	ARG
SLOW LORIS B				ASP	GLX	VAL	ASX	ASX	TYR*	ASX	GLY	GLX	ARG	LEU	LEU	ASP	ALA	ARG
RABBIT B						ALA	ASP	ASP	TYR*	ASP	ASP	GLU	VAL	LEU	PRO	ASP	ALA	ARG

Figure 7. Comparision of rabbit, slow loris and human fibrinopeptides.

Analysing the data

The analysis of comparative amino acid sequence data can take a variety of forms of greater or lesser sophistication. Certainly, the best of these utilise computers; but simple tabulations by hand are not particularly difficult. All of the approaches ultimately depend on the *Principle of Parsimony* for basic data-reduction in terms of positioning the amino acid replacements. Usually, a simple matrix is constructed of all the sequences being compared. In the case of the primate fibrinopeptides, a 12 x 12 matrix of amino acid replacements and the corresponding base substitutions can be readily prepared and the numbers of differences between all pairs of peptides tabulated. These data are easily transformed into phylogenetic trees of the sort discussed above, the limb-lengths of which are made proportional to the number of differences (Fig. 8). These 'mutational distances' necessarily involve certain arbitrary assumptions, especially with regard to the base of the tree, where the limbs are arbitrarily set to satisfy the sums expressed with regard to Fig. 1. This is just the issue we are trying to resolve, however, and it is obvious that we need to add another grouping (cf. Fig. 2) to answer our question vis-à-vis the slow loris. A non-judicious choice of non-primate group could be misleading, however, and ideally we would like a 'closely related' non-primate group for this comparison. (In theory, any non-primate should do; in practice, the numbers of neutral changes are limited. For example, human fibrinopeptides differ from ox fibrinopeptides by a minimum of 21 base changes; but

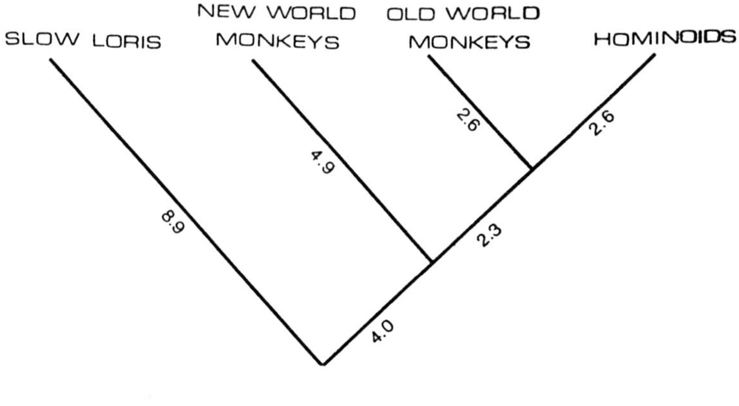

Figure 8. "Mutational distances" separating major primate groups based on a reconstruction of the amino acid sequences of 12 existing species.

pig fibrinopeptides differ from the ox peptides by 20 base changes, even though pigs and cattle are both in the same order.[7] It would be absurd to think that humans are only slightly more distantly related to the ox than is the pig, and a comparison of groups intermediate between primates and artiodactyls should confirm this.)

The slow loris fibrinopeptides are remarkably similar to rabbit fibrinopeptides (Fig. 7). This is especially true in the case of the B peptide, the slow loris have 9 of 13 amino acid residues in common with the rabbit, including the unusual tyrosine-O-sulfate residue. On the other hand, the loris B peptide has only 6 of 14 residues in common with the human B. The rabbit peptide B and human peptide B have only 4 of 13 residues in common. If the rabbit fibrinopeptides are used for the non-primate comparison on the phylogenetic tree, then, there is a suggestion that the slow loris does indeed have primitive fibrinopeptides. This would appear to be true, even though the comparison of the rabbit and slow loris fibrinopeptides is not entirely straightforward (Fig. 8).

Conclusion

There is an obvious need for a much greater body of data before one could conclude that prosimians are primitive primates with primitive proteins. The slow loris fibrinopeptide sequences are tantalising, however, because they fly in the face of trends of thought which are *au courant* among protein evolutionists. What one would like is to see if data from haemoglobin, for example, follow the same pattern. We would also like to examine the fibrinopeptides of more prosimians, including tarsiers, aye-ayes, other lemurs and tree shrews, as well as the classically interesting hedgehog. The most difficult aspect for us is obtaining the necessary blood samples, however. There is a certain amount of skill involved in the sequence determinations, of course, but this can be readily acquired.

In this article I hope to have made a convincing case for the usefulness of amino acid sequence studies on more prosimian fibrinopeptides. These conveniently small, fast-changing peptides offer a graphic demonstration of the step-by-step changes leading to the divergence of the various primate groups. They may also help to solve some of the basic problems plaguing the field of protein evolution.

Acknowledgments

This article was written while the author was a visiting scientist in the Department of Biochemistry, Oxford University, and was supported by a Career Development Award from the US Public Health Service.

NOTES

1 Brown, H., Sanger, F. and Kitai, R. (1955), 'Pig and sheep insulins', *Biochem. J.* 60, 556-65.

2 King, J.L. and Jukes, T.H. (1969), 'Non-Darwinian evolution', *Science* 164, 788-98.

3 Kimura, M. and Ohta, T. (1971), 'On the rate of molecular evolution', *J. Mol. Evol.* 1, 1-17.

4 Dayhoff, M.O. (1969), *Atlas of Protein Sequence and Structure*, vol. 4, Maryland.

5 Doolittle, R.F. and Blombäck, B. (1964), 'Amino acid sequence investigations of fibrinopeptides from various mammals: evolutionary implications', *Nature* 202, 147-52.

6 Blombäck, B. and Blombäck, M. (1968), 'Primary structure of animal proteins as a guide in taxonomic studies', in Hawkes, J.G. (ed.), *Chemotaxonomy and Serotaxonomy*, London.

7 Doolittle, R.F. (1970), 'Evolution of fibrinogen molecules', *Thromb. Diath. Haem.* Supp. 39, 25-42.

8 Doolittle, R.F., Wooding, G.L., Lin, Y. and Riley, M. (1971), 'Hominoid evolution as judged by fibrinopeptide structures', *J. Mol. Evol.* 1, 74-83.

9 Wooding, G.L. and Doolittle, R.F. (1972), 'Primate fibrinopeptides: evolutionary significance', *J. Hum. Evol.* 1, 553-63.

10 Blombäck, B. and Doolittle, R.F. (1963), 'Amino acid sequence studies on fibrinopeptides from several species', *Acta. Chem. Scand.* 17, 1819-22.

11 Goodman, M. (1963), 'Serological analysis of the systematics of recent hominoids', *Hum. Biol.* 35, 377-406.

12 Hafleigh, A.S. and Williams, C.A. Jr. (1966), 'Antigenic correspondence of serum albumins among the primates', *Science* 151, 1530-5.

13 Sarich, V. and Wilson, A.C. (1966), 'Quantitative immunochemistry and the evolution of primate albumins: micro-complement fixation', *Science* 154, 1563-6.

14 Mross, G.A. and Doolittle, R.F. (1967), 'Amino acid sequence studies on artiodactyl fibrinopeptides. II: Vicuna, elk, muntjak, pronghorn antelope, and water buffalo', *Arch. Biochem. Biophys.* 122, 674-84.

Index of authors

950, 957

General index